Function Notation (3-3, 3-5, 3-6)

$f(x)$ — Value of f at x
$(f \circ g)(x) = f[g(x)]$ — Composite function
$f^{-1}(x)$ — Value of inverse of f at x

Linear Equations and Functions (2-1, 3-2, 3-4)

$y = mx + b$ — Slope–intercept form
$(y - y_1) = m(x - x_1)$ — Point–slope form
$f(x) = mx + b$ — Linear function
$y = b$ — Horizontal line
$x = a$ — Vertical line

Polynomial and Rational Forms (1-2, 1-3, 1-4, 2-8, 3-4, 4-1–4-5)

$f(x) = ax^2 + bx + c$ — Quadratic function

$f(x) = a_n x^n + a_{n-1} x^{n-1} + \cdots + a_1 x + a_0,$
$a_n \neq 0, n$ a nonnegative integer — Polynomial function

$f(x) = \dfrac{p(x)}{q(x)},$ p and q polynomial functions, $q(x) \neq 0$ — Rational function

Exponential and Logarithmic Functions (5-1–5-5)

$f(x) = b^x, \quad b > 0, b \neq 1$ — Exponential function
$f(x) = \log_b x, \quad b > 0, b \neq 1$ — Logarithmic function
$y = \log_b x$ if and only if $x = b^y, \quad b > 0, b \neq 1$

Matrices and Determinants (9-1, 10-1–10-6)

$\begin{bmatrix} a & b & c \\ d & e & f \end{bmatrix}$ Matrix

$\begin{vmatrix} a & b & c \\ d & e & f \\ g & h & i \end{vmatrix}$ Determinant

Arithmetic Sequence (11-3)

$a_1, a_2, \ldots, a_n, \ldots$
$a_n - a_{n-1} = d$ — Common difference
$a_n = a_1 + (n - 1)d$ — nth-term formula
$S_n = a_1 + \cdots + a_n = \dfrac{n}{2}[2a_1 + (n - 1)d]$ — Sum of n terms

$S_n = \dfrac{n}{2}(a_1 + a_n)$

Geometric Sequence (11-3)

$a_1, a_2, \ldots, a_n, \ldots$
$\dfrac{a_n}{a_{n-1}} = r$ — Common ratio
$a_n = a_1 r^{n-1}$ — nth-term formula
$S_n = a_1 + \cdots + a_n = \dfrac{a_1 - a_1 r^n}{1 - r}, \quad r \neq 1$ — Sum of n terms
$S_n = \dfrac{a_1 - r a_n}{1 - r}, \quad r \neq 1$
$S_\infty = a_1 + a_2 + \cdots = \dfrac{a_1}{1 - r}, \quad |r| < 1$ — Sum of infinitely many terms

Factorial and Binomial Formulas (11-4)

$n! = n(n - 1) \cdots 2 \cdot 1, \quad n \in N$ — n factorial
$0! = 1$
$\dbinom{n}{r} = \dfrac{n!}{r!(n - r)!}, \quad 0 \leq r \leq n$
$(a + b)^n = \displaystyle\sum_{k=0}^{n} \binom{n}{k} a^{n-k} b^k, \quad n \geq 1$ — Binomial formula

Permutations and Combinations (11-5)

For $0 \leq r \leq n$:
$P_{n,r} = \dfrac{n!}{(n - r)!}$ — Permutation
$C_{n,r} = \dbinom{n}{r} = \dfrac{n!}{r!(n - r)!}$ — Combination

(Continued on back endpaper)

SEVENTH EDITION

COLLEGE ALGEBRA

Barnett, Ziegler & Byleen's Precalculus Series

College Algebra, Seventh Edition
This book is the same as *College Algebra with Trigonometry* without the three chapters on trigonometry.
ISBN 0-07-236868-3

College Algebra with Trigonometry, Seventh Edition
This book is the same as *College Algebra* with three chapters on trigonometry added. Comparing *College Algebra with Trigonometry* with *Precalculus*, *College Algebra with Trigonometry* has more intermediate algebra review and starts trigonometry with angles and right triangles.
ISBN 0-07-236869-1

Precalculus: Functions and Graphs, Fifth Edition
This book differs from *College Algebra with Trigonometry* in that *Precalculus* starts at a higher level, placing intermediate algebra review in the appendix; and it starts trigonometry with the unit circle and circular functions.
ISBN 0-07-236871-3

College Algebra: A Graphing Approach
This book is the same as *Precalculus: A Graphing Approach* without the three chapters on trigonometry. This text assumes the use of a graphing utility.
ISBN 0-07-005710-9

Precalculus: A Graphing Approach
This book is the same as *College Algebra: A Graphing Approach* with three additional chapters on trigonometry. This text assumes the use of a graphing utility.
ISBN 0-07-005717-6

SEVENTH EDITION

COLLEGE ALGEBRA

RAYMOND A. BARNETT
Merritt College

MICHAEL R. ZIEGLER
Marquette University

KARL E. BYLEEN
Marquette University

Boston   Burr Ridge, IL   Dubuque, IA   Madison, WI   New York   San Francisco   St. Louis
Bangkok   Bogotá   Caracas   Lisbon   London   Madrid
Mexico City   Milan   New Delhi   Seoul   Singapore   Sydney   Taipei   Toronto

McGraw-Hill Higher Education

A Division of The **McGraw-Hill** Companies

COLLEGE ALGEBRA, SEVENTH EDITION

Published by McGraw-Hill, an imprint of The McGraw-Hill Companies, Inc., 1221 Avenue of the Americas, New York, NY 10020. Copyright © 2001, 1999, 1993, 1989, 1984, 1979, 1974 by The McGraw-Hill Companies, Inc. All rights reserved. No part of this publication may be reproduced or distributed in any form or by any means, or stored in a database or retrieval system, without the prior written consent of The McGraw-Hill Companies, Inc., including, but not limited to, in any network or other electronic storage or transmission, or broadcast for distance learning.

Some ancillaries, including electronic and print components, may not be available to customers outside the United States.

This book is printed on acid-free paper.

1 2 3 4 5 6 7 8 9 0 VNH/VNH 0 9 8 7 6 5 4 3 2 1 0

ISBN 0–07–236868–3

Vice president and editor-in-chief: *Kevin T. Kane*
Publisher: *JP Lenney*
Senior sponsoring editor: *Maggie Rogers*
Developmental editor: *Michelle Munn*
Editorial assistant: *Allyndreth Cassidy*
Marketing manager: *Mary K. Kittell*
Project manager: *Sheila M. Frank*
Lead producer: *Steve Metz*
Senior production supervisor: *Sandra Hahn*
Designer: *K. Wayne Harms*
Cover/interior designer: *Rokusek Design*
Senior supplement coordinator: *David A. Welsh*
Compositor: *GTS Graphics, Inc.*
Typeface: *10/12 Times Roman*
Printer: *Von Hoffmann Press, Inc.*

Library of Congress Cataloging-in-Publication Data

Barnett, Raymond A.
    College algebra / Raymond A. Barnett, Michael R. Ziegler, Karl E. Byleen. — 7th ed.
        p   cm. — (Barnett, Ziegler & Byleen's precalculus series)
    Includes indexes.
    ISBN 0–07–236868–3 (acid-free paper)
    1. Algebra.   I. Ziegler, Michael R.   II. Byleen, Karl E.   III. Title.

QA154.2 .B35   2001
512.9—dc21   00–035143
                                                                         CIP

www.mhhe.com

ABOUT THE AUTHORS

Raymond A. Barnett, a native of and educated in California, received his B.A. in mathematical statistics from the University of California at Berkeley and his M.A. in mathematics from the University of Southern California. He has been a member of the Merritt College Mathematics Department and was chairman of the department for four years. Associated with four different publishers, Raymond Barnett has authored or co-authored eighteen textbooks in mathematics, most of which are still in use. In addition to international English editions, a number of books have been translated into Spanish. Co-authors include Michael Ziegler, Marquette University; Thomas Kearns, Northern Kentucky University; Charles Burke, City College of San Francisco; John Fujii, Merritt College; and Karl Byleen, Marquette University.

Michael R. Ziegler received his B.S. from Shippensburg State College and his M.S. and Ph.D. from the University of Delaware. After completing postdoctoral work at the University of Kentucky, he was appointed to the faculty of Marquette University where he currently holds the rank of Professor in the Department of Mathematics, Statistics, and Computer Science. Dr. Ziegler has published over a dozen research articles in complex analysis and has co-authored over a dozen undergraduate mathematics textbooks with Raymond Barnett and Karl Byleen.

Karl E. Byleen received his B.S., M.A., and Ph.D. degrees in mathematics from the University of Nebraska. He is currently an Associate Professor in the Department of Mathematics, Statistics, and Computer Science of Marquette University. He has published a dozen research articles on the algebraic theory of semigroups, and has coauthored several undergraduate mathematics textbooks with Raymond Barnett and Michael Ziegler.

CONTENTS

PREFACE

College Algebra is one of five books in the authors' precalculus series (see page ii for a brief comparison of all five books). The primary goal of this text is student comprehension. Great care has been taken to write a book that is mathematically correct and accessible to all students, regardless of their level of understanding. Emphasis is placed on problem solving, ideas, and computational skills rather than mathematical theory. Unless their inclusion adds significant insight into a particular concept, most derivations and proofs are omitted. The text presents concrete examples first, followed by a discussion of related general concepts and results.

Improvements in this edition evolved from the generous response of a large number of users of the last and previous editions, as well as survey results from instructors, mathematics departments, course outlines, and college catalogs. Fundamental to a book's growth and effectiveness is classroom use and feedback. Now in its seventh edition, *College Algebra* has had the benefit of having a substantial amount of both. The most significant change in this seventh edition is the extensive updating of end-of-section and end-of-chapter problem sets. Continuing in our goal of addressing all levels of students, these new problem sets are appropriately balanced across the routine, easy mechanical problems and the more difficult mechanics with some theory. (See the *Features* section within this preface for further explanation.) Additional Explore-Discuss boxes have also been included where appropriate, further emphasizing the emerging use of technology in mathematics.

Also new to this edition, a free Student tutorial CD-ROM accompanies every copy of this text. This text-specific SMART-CD includes an extended real-world application video to begin each chapter. At the section level, this tutorial CD-ROM uses video, descriptive discussion, interactive diagrams, and extended matched problems to reinforce a student's understanding and further exploration of key concepts in the text. Algorithmically generated diagnostic Pre-Tests, Practice, and Post-Tests are also included within each chapter for further reinforcement.

Due to extremely positive feedback from our users, we have continued our emphasis on learning tools that enable a student to easily understand the material being introduced. These tools include Examples with Matched Problems, Explore-Discuss boxes, important student aids such as Think and Caution boxes, and extensive Application problem sets for real-world understanding of concepts. Our Group Activities, Chapter Reviews, and Cumulative Review Exercises continue to reinforce the understanding of critical concepts throughout the text. (See the *Features* section within this preface for further discussion of these tools.)

Technology In the Text

The generic term "graphing utility" is used to refer to any of the various graphing calculators or computer software packages that might be available to a student using this book. TI-83 screen shots are used throughout the text to illustrate graphing principles, but the discussion and explanations lend themselves to any graphing utility being used. Although use of a graphing utility is optional, it is likely that many students and instructors will want to make use of one of these devices. To assist these students and instructors, optional graphing utility activities are included throughout the book beginning in Chapter 3. These activities include brief discussions in the text, examples or portions of examples solved on a graphing utility, and problems for the student to solve. All of the graphing utility material is optional and is clearly identified by the following symbol ⬚ and can be omitted without loss of continuity, if desired.

Technology Support

The Barnett precalculus series provides extensive technology-based support to assist in student comprehension. Each of the following supplements has been created specifically for this series, clarifying and exploring the concepts introduced within the text.

SMART-CD Student Tutorial

This CD-ROM tutorial is included with every copy of the text. In recognition of the visual, auditory, and interactive aspects of individual learning styles, this tutorial makes use of video, descriptive discussion, interactive diagrams, and extended matched problems. Each chapter on the CD-ROM begins with an extended real-world application video designed to convince even the most skeptical of students that mathematics is important. Algorithmically generated diagnostic Pre-Test, Practice, and Post-Test sections are included to direct students to particular sections where further practice is needed. Key concepts from each section have been identified in the tutorial portion of the CD-ROM, and interactive Matched Problems are included for concept mastery. This tutorial portion also includes forty-five Interactive Diagrams developed by Jere Confrey and Alan Maloney of Quest Math and Science Multimedia, Inc., designed specifically for use with the Barnett series. Each Interactive Diagram (ID) is a separate Java Applet that contains an illustration that can be manipulated by the user for further conceptual understanding of the topic presented. For each section of the text where an ID has been created, an icon ⬚ has been placed in the margin.

The SMART CD-ROM is available in both Windows and Macintosh platforms.

Website and Online Learning Center

In addition to the SMART-CD, an extensive text-specific website has been created to further enhance the technological support for this series. A visit to the website at *www.mhhe.com/barnett* will prove our commitment to supporting the ever-changing technology needs in mathematical instruction and learning. The following features are only a few of the many technological supplements available to students that can be accessed through this site:

- **NetTutor-Your Online Tutorial Service:** McGraw-Hill is proud to offer this web-based "homework hotline" via an Internet whiteboard. NetTutor provides live "office hours" with mathematics tutors every weekday, as well as a 24-hour response posting service for students accessing the service during off hours.

- **Graphing Calculator Workshops:** Keystroke guides have been created for each of the main calculators in use today. We will continue to update these guides as new calculators are introduced into the market.

- **Explore the Web Exercises, Additional Applications and Answers,** and **Practice Quizzes** are also available for further practice of the mathematical concepts being introduced.

The Barnett series website continues in its support of instructors using this text by providing PageOut for course-specific website creation, access to the course solutions consultants for the series, and extended instructor materials such as PowerPoint presentations and solutions to all even-numbered problems online.

FEATURES

The following pages serve as an introduction to the specific features included within this text.

Examples and Matched Problems

Integrated throughout the text, completely worked examples and practice problems are used to introduce concepts and demonstrate problem-solving techniques. Each Example is followed by a similar Matched Problem for the student to work through while reading the material. Answers to the matched problems are located at the end of each section, for easy reference. This active involvement in the learning process helps students develop a more thorough understanding of algebraic concepts and processes.

Technology

The generic term "graphing utility" is used to refer to any of the various graphing calculators or computer software packages that might be available to students using this book. The use of a graphing utility is optional within this text. To assist those that choose to use a graphing utility, optional activities are included throughout the book beginning in Chapter 3. These include brief discussions in the text, examples or portions of examples solved on a graphing utility, and problems for the student to solve. All optional graphing material is clearly identified by the following symbol and can

be omitted without loss of continuity, if desired.

---

176   3   Graphs and Functions

The **graph of an equation in two variables** is the graph of its solution set. In equation (1), we find that its solution set will have infinitely many elements and its graph will extend off any paper we might choose, no matter how large. Thus, **to sketch the graph of an equation,** we include enough points from its solution set so that the total graph is apparent. This process is called **point-by-point plotting.**

EXAMPLE 1   **Graphing an Equation Using Point-by-Point Plotting**

Sketch a graph of $y = x^2 - 4$.

Solution   We make up a table of solutions—ordered pairs of real numbers that satisfy the given equation.

x	-4	-3	-2	-1	0	1	2	3	4
y	12	5	0	-3	-4	-3	0	5	12

After plotting these solutions, if there are any portions of the graph that are unclear, we plot additional points until the shape of the graph is apparent. Then we join all these plotted points with a smooth curve, as shown in Figure 3. Arrowheads are used to indicate that the graph continues beyond the portion shown here with no significant changes in shape.

The resulting figure is called a *parabola*. Notice that if we fold the paper along the y axis, the right side will match the left side. We say that the graph is *symmetric with respect to the y axis* and call the y axis the *axis of the parabola*. More will be said about parabolas later in the text.

FIGURE 3

Matched Problem 1   Sketch a graph of $y^2 = x$.

We now use an electronic graphing device to check Example 1. We will refer to any electronic device capable of displaying graphs as a **graphing utility.** The two most common graphing utilities are hand-held graphing calculators and computers with appropriate software. This book contains a number of activities that use graphing utilities to emphasize the connection between graphic, numeric, and algebraic viewpoints. All these activities are clearly marked and easily omitted if no such device is available. Figure 4 shows the steps necessary to reproduce the graph in Figure 3 on a graphing utility.

FIGURE 4

```
Plot1 Plot2 Plot3
\Y1◼X²-4
\Y2=
\Y3=
\Y4=
\Y5=
\Y6=
\Y7=
```
Enter the equation.
(a)

```
WINDOW
Xmin=-5
Xmax=5
Xscl=1
Ymin=-5
Ymax=15
Yscl=1
Xres=1
```
Enter the window variables.
(b)

Graph the equation.
(c)

Exploration and Discussion

Interspersed at appropriate places in every section, Explore/Discuss boxes encourage students to think critically about mathematics and to explore key concepts in more detail. Verbalization of mathematical concepts, results, and processes is encouraged in these Explore/Discuss boxes, as well as in some matched problems, and in particular problems in almost every exercise set. Explore/Discuss material can be used in class or as an out-of-class activity.

EXPLORE-DISCUSS 1 To graph the equation $y = -x^3 + 2x$, we use point-by-point plotting to obtain the graph in Figure 5.

FIGURE 5

x	y
−1	−1
0	0
1	1

(A) Do you think this is the correct graph of the equation? If so, why? If not, why?

(B) Add points on the graph for $x = -2, -0.5, 0.5$, and 2.

(C) Now, what do you think the graph looks like? Sketch your version of the graph, adding more points as necessary.

(D) Write a short statement explaining any conclusions you might draw from parts A, B, and C.

Balanced Exercise Sets

College Algebra contains over 4300 problems. Each Exercise Set is designed so that an average or below-average student will experience success and a very capable student will be challenged. Exercise Sets are found at the end of each section in the text, and are divided into A (routine, easy mechanics), B (more difficult mechanics), and C (difficult mechanics and some theory) levels of difficulty so that students at all levels can be challenged. Problem numbers that appear in blue indicate exercises that require the students to apply their reasoning and writing skills to the solution of the problem.

188   3 Graphs and Functions

In Problems 29 and 30, use the graph to estimate to the nearest integer the missing coordinates of the indicated points. (Be sure you find all possible answers.)

29. (A) $(-3, ?)$
    (B) $(2, ?)$
    (C) $(?, 3)$
    (D) $(?, -1)$

30. (A) $(-4, ?)$
    (B) $(-1, ?)$
    (C) $(?, 1)$
    (D) $(?, 4)$

The figures in Problems 31 and 32 show a portion of a graph. Extend the given graph to one that exhibits the indicated type of symmetry.

31. (A) x axis only
    (B) y axis only
    (C) origin only
    (D) x and y axes

32. (A) x axis only
    (B) y axis only
    (C) origin only
    (D) x and y axes

B

In Problems 33–48, determine symmetry with respect to the x axis, y axis, or origin, if any exists, and graph.

* Check your graphs in Problems 33–48 by graphing on a graphing utility.

33. $y^2 = x + 2$ 34. $y^2 = x - 2$ 35. $y = x^2 + 1$
36. $y + 2 = x^2$ 37. $x^2 + 4y^2 = 4$ 38. $x^2 + 9y^2 = 9$

39. $4y^2 - x^2 = 1$ 40. $4x^2 - y^2 = 1$ 41. $y^3 = x$
42. $y = x^4$ 43. $y = 0.6x^2 - 4.5$
44. $x = 0.8y^2 - 3.5$ 45. $y = \sqrt{17 - x^2}$
46. $y = \sqrt{100 - 4x^2}$ 47. $y = x^{2/3}$
48. $y^{2/3} = x$

In Problems 49 and 50, use the Pythagorean theorem to show that the given points are the vertices of a right triangle. Find the area and the perimeter (to two decimal places) of the triangle. (Formulas related to triangles can be found in Appendix B.)

49. $(-3, 2), (1, -2), (8, 5)$
50. $(-4, -1), (0, 7), (6, -6)$
51. Find x such that $(x, 7)$ is 10 units from $(-4, 1)$.
52. Find x such that $(x, 2)$ is 4 units from $(3, -3)$.
53. Find y such that $(2, y)$ is 3 units from $(-1, 4)$.
54. Find y such that $(3, y)$ is 13 units from $(-9, 2)$.

In Problems 55–60, find the center and radius of the circle with the given equation. Graph the equation.

55. $(x + 4)^2 + (y - 2)^2 = 7$
56. $(x - 5)^2 + (y + 7)^2 = 15$
57. $x^2 + y^2 - 6x - 4y = 36$
58. $x^2 + y^2 - 2x - 10y = 55$
59. $x^2 + y^2 + 8x - 6y + 8 = 0$
60. $x^2 + y^2 + 4x + 10y + 15 = 0$

In Problems 61–64, graph the triangle with vertices A, B, and C and the triangle with vertices A′, B′, and C′ in the same coordinate system. Describe the relationship between the graphs of these triangles in terms of reflections.

61. $A(1, 1), B(7, 2), C(4, 6)$
    $A'(1, -1), B'(7, -2), C'(4, -6)$
62. $A(1, 1), B(7, 2), C(4, 6)$
    $A'(-1, 1), B'(-7, 2), C'(-4, 6)$
63. $A(1, 2), B(7, 2), C(4, 6)$
    $A'(-1, -1), B'(-7, -2), C'(-4, -6)$
64. $A(1, 2), B(1, 4), C(3, 4)$
    $A'(2, 1), B'(4, 1), C'(4, 3)$
    Hint: Add the graph of $y = x$ to your graph.

In Problems 65–68, solve for y, producing two equations, and then graph both of these equations in the same viewing window.

65. $x^2 + y^2 = 3$ 66. $x^2 + y^2 = 5$
67. $(x + 3)^2 + (y + 1)^2 = 2$ 68. $(x - 2)^2 + (y - 1)^2 = 3$

*Please note that use of a graphing utility is not required to complete these exercises. Checking with a g.u. is optional. If you do not have a g.u., you should still work these exercises.

6. $(x - 4)^2 + (y + 5)^2 = 16$; radius: 4, center: $(4, -5)$

EXERCISE 3-1

A

In Problems 1–10, give a verbal description of the indicated subset of the plane in terms of quadrants and axes.

1. $\{(x, y) \mid x = 0\}$ 2. $\{(x, y) \mid x > 0, y > 0\}$
3. $\{(x, y) \mid x < 0, y < 0\}$ 4. $\{(x, y) \mid y = 0\}$
5. $\{(x, y) \mid x > 0, y < 0\}$ 6. $\{(x, y) \mid y < 0, x \neq 0\}$
7. $\{(x, y) \mid xy < 0\}$ 8. $\{(x, y) \mid x < 0, y > 0\}$
9. $\{(x, y) \mid x > 0, y \neq 0\}$ 10. $\{(x, y) \mid xy > 0\}$

In Problems 11–18, determine symmetry with respect to the x axis, y axis, or origin, if any exists, and graph.

11. $y = 2x - 4$ 12. $y = \frac{1}{2}x + 1$
13. $y = \frac{1}{2}x$ 14. $y = 2x$
15. $|y| = x$ 16. $|y| = -x$
17. $|x| = |y|$ 18. $y = -x$

Find the distance be[tween ...]
19–22. Leave the a[...]

19. $(-5, -3), (4, 2)$
20. $(-6, 4), (2, -1)$
21. $(3, 5), (2, -4)$
22. $(2, -5), (-3, 1)$

In Problems 23–28, [...]
indicated center an[d ...]

23. $C(0, 0), r = 4$
24. $C(0, 0), r = 6$
25. $C(3, -2), r = 1$
26. $C(-4, 2), r = 5$
27. $C(2, 6), r = \sqrt{3}$
28. $C(-1, -3), r = $

Applications

One of the primary objectives of this book is to give the student substantial experience in modeling and solving real-world problems. Over 400 application exercises help convince even the most skeptical student that mathematics is relevant to everyday life. The most difficult application problems are marked with two stars (★★), the moderately difficult application problems with one star (★), and the easier application problems are not marked. An **Applications Index** is included in the back of this textbook to help locate particular applications.

Functional Use of Four Colors

The functional use of four colors improves the clarity of many illustrations, graphs, and developments, and guides students through certain critical steps.

---

222   3 Graphs and Functions

89. Physics—Rate. The distance in feet that an object falls in a vacuum is given by $s(t) = 16t^2$, where t is time in seconds. Find:
(A) $s(0), s(1), s(2), s(3)$
(B) $\dfrac{s(2 + h) - s(2)}{h}$
(C) What happens in part B when h tends to 0? Interpret physically.

90. Physics—Rate. An automobile starts from rest and travels along a straight and level road. The distance in feet traveled by the automobile is given by $s(t) = 10t^2$, where t is time in seconds. Find:
(A) $s(8), s(9), s(10), s(11)$
(B) $\dfrac{s(11 + h) - s(11)}{h}$
(C) What happens in part B as h tends to 0? Interpret physically.

91. Manufacturing. A candy box is to be made out of a piece of cardboard that measures 8 by 12 inches. Squares, x inches on a side, will be cut from each corner, and then the ends and sides will be folded down (see the figure). Find a formula for the volume of the box $V(x)$ in terms of x. From practical considerations, what is the domain of the function V?

92. Construction. A rancher has 20 miles of fencing to fence a rectangular piece of grazing land along a straight river. If no fence is required along the river and the sides perpendicular to the river are x miles long, find a formula for the area $A(x)$ of the rectangle in terms of x. From practical considerations, what is the domain of the function A?

★ **93. Construction.** The manager of an animal clinic wants to construct a kennel with four individual pens, as indicated in the figure. State law requires that each pen have a gate 3 feet wide and an area of 50 square feet. If x is the width of one pen, express the total amount of fencing $F(x)$ (excluding the gates) required for the construction of the kennel as a function of x. Complete the following table [round values of $F(x)$ to one decimal place]:

x	4	5	6	7
$F(x)$				

Figure for 93

★ **94. Architecture.** An architect wants to design a window with an area of 24 square feet in the shape of a rectangle surmounted by a semicircle, as indicated in the figure. If x is the width of the window, express the perimeter $P(x)$ of the window as a function of x. Complete the table below [round each value of $P(x)$ to one decimal place]:

x	4	5	6	7
$P(x)$				

★ **95. Construction.** A freshwater pipeline is to be run from a source on the edge of a lake to a small resort community on an island 8 miles offshore, as indicated in the figure. It costs $10,000 per mile to lay the pipe on land and $15,000 per mile to lay the pipe in the lake. Express the total cost $C(x)$ of constructing the pipeline as a function of x. From practical considerations, what is the domain of the function C?

★ **96. Weather.** An observation balloon is released at a point 10 miles from the station that receives its signal and rises vertically, as indicated in the figure. Express the distance $d(h)$ between the balloon and the receiving station as a function of the altitude h of the balloon.

Foundation for Calculus

As many students will use this book to prepare for a calculus course, examples and exercises that are especially pertinent to calculus are marked with an icon. $\displaystyle\int$

---

204   3 Graphs and Functions

B

In Problems 25–28, find the equation of the line passing through the given point with the given slope. Write the final answer in the slope-intercept form $y = mx + b$.

25. $(0, 3); m = -2$ **26.** $(4, 0); m = 3$
27. $(-5, 4); m = \frac{3}{2}$ **28.** $(2, -3); m = -\frac{4}{5}$

In Problems 29–34, find the equation of the line passing through the two given points. Write the final answer in the slope-intercept form $y = mx + b$ or in the form $x = c$.

29. $(2, 5); (4, -3)$ **30.** $(-1, 4); (3, 2)$
31. $(-3, 2); (-3, 5)$ **32.** $(0, 5); (2, 5)$
33. $(-4, 2); (0, 2)$ **34.** $(5, -4); (5, 6)$

In Problems 35–46, write an equation of the line that contains the indicated point and meets the indicated condition(s). Write the final answer in the standard form $Ax + By = C$, $A \geq 0$.

35. $(2, -1);$ parallel to $y = -3x + 7$
36. $(-3, 2);$ parallel to $y = 4x - 5$
37. $(0, -4);$ parallel to $2x + 3y = 9$

$\displaystyle\int$ *Problems 53–58 are calculus-related. Recall that a line tangent to a circle at a point is perpendicular to the radius drawn to that point (see the figure). Find the equation of the line tangent to the circle at the indicated point. Write the final answer in the standard form $Ax + By = C$, $A \geq 0$. Graph the circle and the tangent line on the same coordinate system.*

53. $x^2 + y^2 = 25, (3, 4)$
54. $x^2 + y^2 = 100, (-8, 6)$
55. $x^2 + y^2 = 50, (5, -5)$
56. $x^2 + y^2 = 80, (-4, -8)$
57. $(x - 3)^2 + (y + 4)^2 = 169, (8, -16)$
58. $(x + 5)^2 + (y - 9)^2 = 289, (-13, -6)$

Group Activities

A Group Activity is located at the end of each chapter and involves many of the concepts discussed in that chapter. These activities strongly encourage the verbalization of mathematical concepts, results, and processes. All of these special activities are highlighted to emphasize their importance.

CHAPTER 3 GROUP ACTIVITY   Mathematical Modeling in Business*

This group activity is concerned with analyzing a basic model for manufacturing and selling a product by using tables of data and linear regression to determine appropriate values for the constants a, b, m, and n in the following functions:

TABLE 1   Business Modeling Functions

Function	Definition	Interpretation
Price–demand	$p(x) = m - nx$	x is the number of items that can be sold at $\$p$ per item
Cost	$C(x) = a + bx$	Total cost of producing x items
Revenue	$R(x) = xp$	Total revenue from the sale of x items
	$= x(m - nx)$	
Profit	$P(x) = R(x) - C(x)$	Total profit from the sale of x items

A manufacturing company manufactures and sells mountain bikes. The management would like to have price–demand and cost functions for break-even and profit–loss analysis. Price–demand and cost functions may be established by collecting appropriate data at different levels of output, and then finding a model in the form of a basic elementary function (from our library of elementary functions) that "closely fits" the collected data. The financial department, using statistical techniques, arrived at the price–demand and cost data in Tables 2 and 3, where p is the wholesale price of a bike for a demand of x thousand bikes and C is the cost, in thousands of dollars, of producing and selling x thousand bikes.

TABLE 2   Price–Demand

x (thousand)	$p(\$)$
7	530

TABLE 3   Cost

x (thousand)	C (thousand $)
5	2,100
12	2,940

Graphs and Illustrations

All graphs in this text are computer generated to ensure mathematical accuracy. Graphing utility screens displayed in the text are actual output from a graphing calculator.

      ints are equi... ...om the x a...

3. The origin if $(-a, -b)$ is on the graph whenever (a, b) is on the graph—the two points are equidistant from the origin on a line through the origin.

Figure 8 illustrates these three types of symmetry.

FIGURE 8 Symmetry.

EXPLORE-DISCUSS 2   If a graph possesses two of the three types of symmetry in Definition 1, must it also possess the third? Explain.

Boldface Type

Boldface type is used to introduce new terms and highlight important comments.

Think Boxes

Think boxes are dashed boxes used to enclose steps that are usually performed mentally.

Caution Boxes

Caution boxes appear throughout the text to indicate where student errors often occur.

Annotation

Annotation of examples and developments, in small colored type, is found throughout the text to help students through critical stages.

3-3 Functions **217**

In addition to evaluating functions at specific numbers, it is important to be able to evaluate functions at expressions that involve one or more variables. For example, the **difference quotient**

$$\frac{f(x + h) - f(x)}{h} \qquad x \text{ and } x + h \text{ in the domain of } f, h \neq 0$$

is studied extensively in a calculus course.

 EXAMPLE 6 **Evaluating and Simplifying a Difference Quotient**

For $f(x) = x^2 + 4x + 5$, find and simplify:

(A) $f(x + h)$ (B) $\dfrac{f(x + h) - f(x)}{h}, h \neq 0$

Solution (A) To find $f(x + h)$, we replace x with $x + h$ everywhere it appears in the equation that defines f and simplify:

$$f(x + h) = (x + h)^2 + 4(x + h) + 5$$
$$= x^2 + 2xh + h^2 + 4x + 4h + 5$$

(B) Using the result of part A, we get

$$\frac{f(x + h) - f(x)}{h} = \frac{x^2 + 2xh + h^2 + 4x + 4h + 5 - (x^2 + 4x + 5)}{h}$$

$$= \frac{x^2 + 2xh + h^2 + 4x + 4h + 5 - x^2 - 4x - 5}{h}$$

$$= \frac{2xh + h^2 + 4h}{h} = \boxed{\frac{h(2x + h + 4)}{h}} = 2x + h + 4$$

Matched Problem 6 Repeat Example 6 for $f(x) = x^2 + 3x + 7$.

CAUTION 1. If f is a function, then the symbol $f(x + h)$ represents the value of f at the number $x + h$ and must be evaluated by replacing the independent variable in the equation that defines f with the expression $x + h$, as we did in Example 6. Do not confuse this notation with the familiar algebraic notation for multiplication:

$$f(x + h) \neq fx + fh \qquad f(x + h) \text{ is function notation.}$$
$$4(x + h) = 4x + 4h \qquad 4(x + h) \text{ is algebraic multiplication notation.}$$

Screened Boxes

Screened boxes are used to highlight important definitions, theorems, results, and step-by-step processes.

Properties of a Quadratic Function and Its Graph

Given a quadratic function and the form obtained by completing the square

$$f(x) = ax^2 + bx + c = a(x - h)^2 + k \qquad a \neq 0$$

we summarize general properties as follows:

1. The graph of f is a parabola:

$a > 0$
Opens upward

$a < 0$
Opens downward

2. Vertex: (h, k) (Parabola rises on one side of the vertex and falls on the other.)
3. Axis (of symmetry): $x = h$ (Parallel to y axis)

Chapter Review

Chapter Review sections are provided at the end of each chapter and include a thorough review of all the important terms and symbols. This recap is followed by a comprehensive set of review exercises.

---

5. Find a and b for $r = 0.1\overline{9}$ and then find a terminating decimal expansion for r.

Chapter 1 Review

1-1   ALGEBRA AND REAL NUMBERS

A **set** is a collection of objects called **elements** or **members** of the set. Sets are usually described by **listing** the elements or by stating a **rule** that determines the elements. A set may be **finite** or **infinite**. A set with no elements is called the **empty set** or the

null set and is denoted $\varnothing$. A **variable** is a symbol that represents unspecified elements from a **replacement set**. A **constant** is a symbol for a single object. If each element of set A is also in set B, we say A is a **subset** of B and write $A \subset B$.

Real numbers:

$N \subset Z \subset Q \subset R$

Real numbers (R)

Rational numbers (Q) Irrational numbers (I)

Integers (Z) Noninteger ratios of integers

Natural numbers (N) Zero   Negatives of natural numbers

Real number line:

Origin

Negative real numbers   Positive real numbers

Coordinate   Positive direction

Cumulative Review Exercise

A Cumulative Review Exercise is provided after every second or third chapter, for additional reinforcement.

...gh. A ...her use... to form...
pattern of isosceles triangles along the wall of a building

Cumulative Review Exercise Chapters 1 and 2

Work through all the problems in this cumulative review and check answers in the back of the book. Answers to all review problems are there, and following each answer is a number in italics indicating the section in which that type of problem is discussed. Where weaknesses show up, review appropriate sections in the text.

A

1. Solve for x: $\dfrac{7x}{5} - \dfrac{3 + 2x}{2} = \dfrac{x - 10}{3} + 2$

2. Replace each question mark with an appropriate expression that will illustrate the use of the indicated real number property:
(A) Distributive: $c(a + b) = ?$
(B) Associative $(+)$: $(a + b) + c = ?$
(C) Commutative $(\cdot)$: $(a + b)c = ?$

3. Solve for x and y: $2x - 3y = 8$
                          $4x + y = 2$

In Problems 4–8, perform the indicated operation and simplify.

4. $4x(x - 3) + 2(3x + 5)$ **5.** $3x(x - 7) - (5x - 9)$

6. $(2x - 5y)(3x + 4y)$ **7.** $(2a - 3b)(2a + 3b)$

8. $(5m + 2n)^2$

Solve and graph Problems 9–11.

9. $2(3 - y) + 4 \le 5 - y$ **10.** $|x - 2| < 7$

11. $x^2 + 3x \ge 10$

In Problems 12 and 13, write each polynomial in a completely factored form relative to the integers. If the polynomial is prime relative to the integers, say so.

12. $x^2 - 3x + 10$ **13.** $6t^2 + 7t - 5$

In Problems 14 and 15, perform the indicated operations and reduce to lowest terms. Represent all compound fractions as simple fractions reduced to lowest terms.

14. $\dfrac{6}{x^2 - 3x} - \dfrac{4}{x^2 - 2x}$ **15.** $\dfrac{\dfrac{4}{y} - y}{\dfrac{2}{y^2} - \dfrac{1}{2}}$

16. Perform the indicated operations and write the answer in standard form:
(A) $(2 - 3i) - (-5 + 7i)$
(B) $(1 + 4i)(3 - 5i)$
(C) $\dfrac{5 + i}{2 + 3i}$

Simplify Problems 17–19 and write answers using positive exponents only. All variables represent positive real numbers.

17. $3x^2(xy^2)^5$ **18.** $\dfrac{(2xy^2)^3}{4x^4y^4}$ **19.** $(a^4b^{-6})^{3/2}$

Solve Problems 20–23.

20. $3x^2 = -12x$ **21.** $4x^2 - 20 = 0$

22. $x^2 - 6x + 2 = 0$ **23.** $x - \sqrt{12 - x} = 0$

24. Change to radical form: $5a^{3/4}$

25. Change to rational exponent form: $2\sqrt[5]{x^4y^3}$

SUPPLEMENTS

A comprehensive set of ancillary materials for both the student and the instructor is available for use with this text.

Student's Solutions Manual

This supplement is available for sale to the student, and includes detailed solutions to all odd-numbered problems and most review exercises.

Instructor's Solutions Manual

This manual provides solutions to even-numbered problems and answers to all problems in the text.

Instructor's Resource Manual

This supplement provides transparency masters and sample tests for each chapter in the text.

Print and Computerized Testbanks

A Computerized Testbank is available that provides a variety of formats to enable the instructor to create tests using both algorithmically generated test questions and those from a static testbank. This testing system enables the instructor to choose questions either manually or randomly by section, question types, difficulty level, and other criteria. This testing software supports online testing and is available for PC and Macintosh computers. A softcover print version of the testbank includes the static questions found in the computerized version.

Barnett/Ziegler/Byleen Video Series

Course videotapes, created specifically for this series, provide students with additional reinforcement of the topics presented in the book. These videos are keyed to the text and feature an effective combination of learning techniques, including personal instruction, state-of-the-art graphics, and real-world applications.

Course Solutions

Fully integrated multimedia, a full-scale Online Learning Center, and a Course Integration Guide are designed specifically to help you with your individual college algebra course needs. Assembled by an expert in your field of study, this printed manual fully integrates the numerous products available to accompany *College Algebra*. The Course Integration Guide will also contain detailed solutions for the Group Activities found in the text, a description of each Interactive Diagram, and detailed solutions to the exploratory questions that accompany each Interactive Diagram.

SMART-CD Student Tutorial

This CD-ROM tutorial is included with every copy of the text. In recognition of the visual, auditory, and interactive aspects of individual learning styles, this tutorial makes use of video, descriptive discussion, interactive diagrams, and extended matched problems. Each chapter on the CD-ROM begins with an extended real-world application video designed to convince even the most skeptical of students that mathematics is important. Algorithmically generated diagnostic Pre-Test, Practice, and Post-Test sections are included to direct students to particular sections where further practice is needed. Key concepts from each section have been identified in the tutorial portion of the CD-ROM, and interactive Matched Problems are included for concept mastery. This tutorial portion also includes forty-five Interactive Diagrams developed by Jere Confrey and Alan Maloney of Quest Math and Science Multimedia, Inc., designed specifically for use with the Barnett series. Each Interactive Diagram (ID) is a separate Java Applet that contains an illustration that can be manipulated by the user for further conceptual understanding of the topic presented. For each section of the text where an ID has been created, an icon ⊙ has been placed in the margin. The SMART-CD is available in both Windows and Macintosh platforms.

Website and Online Learning Center

An extensive text-specific website has been created to further enhance the technological support for this series. Visit the site at **www.mhhe.com/barnett** to explore the numerous features available to both students and instructors. Among the many tools available to students is NetTutor-Your Online Tutorial Service, a "homework hotline" via an Internet whiteboard. NetTutor provides live "office hours" with mathematics tutors every weekday, as well as a 24-hour response posting service for students accessing the service during off hours. The Graphing Calculator Workshop features keystroke guides for each of the main calculators in use today. Instructors have access to PageOut for course-specific website creation, extended instructor materials such as PowerPoint presentations, and solutions to all even-numbered problems online, and access to the course solutions consultants for the series.

For further information about these or any supplements, please contact your local McGraw-Hill sales representative, or visit our website at **www.mhhe.com/barnett**.

Accuracy

Because of the careful checking and proofing by a number of mathematics instructors (acting independently), the authors and publisher believe this book to be substantially error free. For any errors remaining, the authors would be grateful if they were sent to: Karl E. Byleen, 9322 W. Garden Court, Hales Corners, WI 53130; or by email, to *byleen@execpc.com.*

Acknowledgments

In addition to the authors, many others are involved in the successful publication of a book. We wish to thank personally:

Hossein Hamedani and Caroline Woods for providing a careful and thorough check of all the mathematical calculations in the book, the student solutions manual, and the answer manual (a tedious but extremely important job); all of the supplements authors for developing the supplemental manuals that are so important to the success of a text; Jeanne Wallace for accurately and efficiently producing many of the manuals that supplement the text; George Morris and his staff at Scientific Illustrators for their effective illustrations and accurate graphs; Maggie Rogers, Sheila Frank, Michelle Munn, and all the other people at McGraw-Hill who contributed their efforts to the production of this book. Producing this new edition with the help of all these extremely competent people has been a most satisfying experience.

We also wish to thank the following manuscript reviewers: Yungchen Cheng, S.W. Missouri State University; Allan Cochran, University of Arkansas; Wayne Ehler, Anne Arundel Community College; Abdulla Elusta, Broward Community College; Betsy Farber, Bucks County Community College; Jeffrey Graham, Western Carolina University; Selwyn Hollis, Armstrong Atlantic State University; Randal Hoppens, Blinn College; Linda Horner, Broward Community College–N. Campus; Beverly Reed, Kent State University; Mike Rosenthal, Florida International University; Robert Woodside, East Carolina State University; Mary Wright, Southern Illinois University. Chapter reviewers: Diane Abbott, Bowling Green State University; Victor Akatsa, Chicago State University; Judith Beauford, University of the Incarnate Word; George Bradley, Duquesne University; Ronald Brent, University of Massachusetts-Lowell; Jeff Brown, University of North Carolina at Wilmington; Kimberly Brown, Tarrant County Junior College; Ed Buman, Creighton University; Roxanne Byrne, University of Colorado at Denver; Magdalena Caproiu, Antelope Valley College; Harold Carda, South Dakota School of Mines and Technology; Allan Cochran, University of Arkansas; Barbara Cortzen, DePaul University; Patrick Crowe, Johnson City Community College; Elizabeth Chu, Sussex County Community College; Marilyn Danchanko, Camria County Area Community College; Karin Deck, University of North Carolina at Wilmington; Joe Di Costanzo, Johnson County Community College; Michelle Diehl, University of New Mexico; Jeff Dinitz, University of Vermont; Charles Douglas, South Georgia College; Mary Ehlers, Seattle University; Laura Fernelius, University of Wisconsin at Oshkosh; Larry Friesen, Butler County Community College; Doris Fuller, Virginia State University; Dan Gardner, Elgin Community College; Rebecca Gehrke-Griswold, Jefferson Community College-SW; Sheryl Griffith, Iowa Central Community College; Vernon Gwaltney, John Tyler Community College; Rutger Hangelbroek, Western Illinois University; Keith Howell, Southwestern Michigan College; Brian Jackson, Connor's State College; Juan Carlos Jimenez, Springfield Technical Community College; Nancy Johnson, Manatee Community College; Klaus Kaiser, University of Houston; Rahim Karimpour, Southern Illinois University at Edwardsville; Karla Karstens, University of Vermont; Thomas Keller, Southwest Texas State University; Warren Koepp, Texas A&M at Commerce; Kent Kromarek, University of Wisconsin-Waukesha; Sonja Kung, University of Wisconsin at Stevens Point; Mark Lesperance, Kansas State University; Carol Lucas, University of Kansas; Richard Mason, Indian Hills Community College; Marilyn McCollum, North Carolina State University; Gael Mericle, Minnesota State University-Mankato; Michael Montano, Riverside Community College; Patricia Moreland, Cowley County Community College; Robert Moyer, Fort Valley State University; Arumudam Muhundan, Manatee Community College; Milt Myers, Delaware County Community College; Richard Nadel, Florida International University; Roxie Novak, Radford University; Vicki Partin, Lexington Community College; Gloria Phoenix, North Carolina Agricultural & Technical State University; Donna Russo, Quincy College; Mohammed Saleem, San Jose State University; Donna Salinardi, University of Connecticut; Jean Sanders, University of Wisconsin at Platteville; Radha Sankaran, Passaic County Community College; Ellen Scheiber, Drexel University; Patricia Schmidt, University of Pittsburgh at Greensburg; Samkar Sethuraman, Augusta State University; James Shockley, Virginia Polytechnic Institute; Debbye Stapleton, Georgia Southern University; Lora Stewart, Cosumnes River College; Joseph Sukta, Moraine Valley Community College; Ralph Svetic, Winona State University; Hussain Talibi, Tuskegee University; Stuart Thomas, University of Oregon; Richard Vinson, University of South Alabama; James Ward, Pensacola Junior College; Lyndon Weberg, University of Wisconsin; Chad Wheatley, Delaware Technical & Community College; Edward White, Frostburg State University; Clifton Whyburn, University of Houston; Tom Williams, Rowan-Cabarrus Community College; Terry Wilson, San Jacinto College; Zhanbo Yang, University of the Incarnate Word; George Zazi, Chicago State University; Cathleen Zucco-Teveloff, Rutgers University.

R. A. Barnett
M. R. Ziegler
K. E. Byleen

CHAPTER DEPENDENCIES

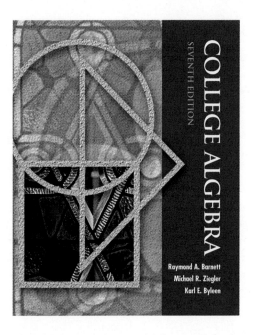

McGraw-Hill is proud to offer an exciting new suite of multimedia products and services called Course Solutions.

Designed specifically to help you with your individual course needs, **Course Solutions** will assist you in integrating your syllabus with our premier titles and state-of-the-art new media tools that support them.

At the heart of Course Solutions you'll find:

- Fully integrated multimedia
- A full-scale Online Learning Center
- A Course Integration Guide

As well as these unparalleled services:

- McGraw-Hill Learning Architecture
- McGraw-Hill Course Consultant Service
- Visual Resource Library (VRL)
- Image Licensing
- McGraw-Hill Student Tutorial Service
- McGraw-Hill Instructor Syllabus Service
- PageOut Lite
- PageOut: The Course Website Development Center
- Other Delivery Options

Course Solutions truly has the solutions to your every teaching need. Read on to learn how we can specifically help you with your classroom challenges.

SPECIAL ATTENTION

to your specific needs.

These "perks" are all part of the extra service delivered through McGraw-Hill's Course Solutions:

McGraw-Hill Learning Architecture

Each McGraw-Hill *Online Learning Center* is ready to be ported into our *McGraw-Hill Learning Architecture*—a full course management software system for *Local Area Networks* and *Distance Learning Classes*. Developed in conjunction with Top Class software, *McGraw-Hill Learning Architecture* is a powerful course management system available upon special request.

McGraw-Hill Course Consultant Service

In addition to the *Course Integration Guide*, instructors using **Course Solutions** textbooks can access a special curriculum-based *Course Consultant Service,* via a web-based threaded discussion list within each *Online Learning Center.* A **McGraw-Hill Course Solutions Consultant** will personally help you—as a text adopter—integrate this text and media into your course to fit your specific needs. This content-based service is offered in addition to our usual software support services.

McGraw-Hill Instructor Syllabus Service

For *new* adopters of **Course Solutions** textbooks, McGraw-Hill will help correlate all text, supplement, and appropriate materials and services to your course syllabus. Simply call your McGraw-Hill sales representative for assistance.

PageOut Lite

Free to **Course Solutions** textbook adopters, *PageOut Lite* is perfect for instructors who want to create their own website. In just a few minutes, even novices can turn their syllabus into a website using *PageOut Lite.*

PageOut: The Course Website Development Center

For those who want the benefits of *PageOut Lite's* no-hassle approach to site development, but with even more features, we offer *PageOut: The Course Website Development Center.*

    PageOut shares many of *PageOut Lite's* features, but its syllabus page lets you create links that will take your students to your original material, other website addresses, and to *McGraw-Hill Online Learning Center* content. This means you can assign *Online Learning Center* content within your syllabus-based website. *PageOut's* grade book function will tell you when each student has taken a quiz or worked through an exercise, automatically recording his or her scores for you. *PageOut* also features a discussion board list where you and your students can exchange questions and post announcements, as well as an area for students to build personal web pages. Contact your McGraw-Hill sales representative for assistance.

Other Delivery Options

Online Learning Centers are also compatible with a number of full-service online course delivery systems or outside educational service providers. For a current list of compatible delivery systems, contact your McGraw-Hill sales representative.

And for your students . . .

McGraw-Hill Student Tutorial Service

Within each *Online Learning Center* resides a **FREE** *Student Tutorial Service.* This web-based "homework hotline"—available via a threaded discussion list—features guaranteed, 24-hour response time on weekdays. Students can also receive immediate tutorial assistance via an Internet whiteboard during regularly scheduled NetTutor office hours.

www.mhhe.com/barnett

TO THE STUDENT

The following suggestions are made to help you get the most out of this book and your efforts.

As you study the text we suggest a five step-process. For each section:

1. Read a mathematical development.   ⎫
2. Work through the illustrative examples. ⎬   Repeat the 1-2-3 cycle until the section is finished.
3. Work the matched problem.   ⎭
4. Review the main ideas in the section.
5. Work the assigned exercises at the end of the section.

All of this should be done with plenty of paper, pencils, a calculator, and a wastebasket at hand. In fact, no mathematics text should be read without pencil and paper in hand; mathematics is not a spectator sport. Just as you cannot learn to swim by watching someone else swim, you cannot learn mathematics by simply reading worked examples—you must work problems, lots of them.

If you have a graphing calculator or access to a computer with mathematical software, such as Maple or Mathematica, you should pay particular attention to the remarks, explore-discuss boxes, and exercises marked with ▨ . This is optional material that we have included to help you learn to make effective use of technology as part of the problem-solving process. If you do not have one of these devices, please omit material so marked, since it may involve calculations that cannot be done by hand.

If you have difficulty with the course, then, in addition to doing the regular assignments, spend more time on the examples and matched problems and work more A exercises, even if they are not assigned. If you find the course too easy, then work more C exercises and applied problems, even if they are not assigned.

Raymond A. Barnett
Michael R. Ziegler
Karl E. Byleen

Basic Algebraic Operations

CHAPTER 1

Algebra is often referred to as "generalized arithmetic." In arithmetic we deal with the basic arithmetic operations of addition, subtraction, multiplication, and division performed on specific numbers. In algebra we continue to use all that we know in arithmetic, but, in addition, we reason and work with symbols that represent one or more numbers. In this chapter we review some important basic algebraic operations usually studied in earlier courses. The material may be studied systematically before commencing with the rest of the book or reviewed as needed.

SECTION 1-1   Algebra and Real Numbers

- Sets
- The Set of Real Numbers
- The Real Number Line
- Basic Real Number Properties
- Further Properties
- Fraction Properties

The rules for manipulating and reasoning with symbols in algebra depend, in large measure, on properties of the real numbers. In this section we look at some of the important properties of this number system. To make our discussions here and elsewhere in the text clearer and more precise, we first introduce a few useful notions about sets.

- **Sets** Georg Cantor (1845–1918) developed a theory of sets as an outgrowth of his studies on infinity. His work has become a milestone in the development of mathematics.

Our use of the word "set" will not differ appreciably from the way it is used in everyday language. Words such as "set," "collection," "bunch," and "flock" all convey the same idea. Thus, we think of a **set** as a collection of objects with the important property that we can tell whether any given object is or is not in the set.

Each object in a set is called an **element,** or **member,** of the set. Symbolically,

$a \in A$	means	"a is an element of set A"	$3 \in \{1, 3, 5\}$
$a \notin A$	means	"a is not an element of set A"	$2 \notin \{1, 3, 5\}$

Capital letters are often used to represent sets and lowercase letters to represent elements of a set.

A set is **finite** if the number of elements in the set can be counted and **infinite** if there is no end in counting its elements. A set is **empty** if it contains no elements. The empty set is also called the **null** set and is denoted by $\varnothing$. It is important to observe that the empty set is *not* written as $\{\varnothing\}$.

A set is usually described in one of two ways—by **listing** the elements between braces, { }, or by enclosing within braces a **rule** that determines its elements. For example, if D is the set of all numbers x such that $x^2 = 4$, then using the listing method we write

$$D = \{-2, 2\} \quad \text{Listing method}$$

or, using the rule method we write

$$D = \{x \mid x^2 = 4\} \quad \text{Rule method}$$

Note that in the rule method, the vertical bar $\mid$ represents "such that," and the entire symbolic form $\{x \mid x^2 = 4\}$ is read, "The set of all x such that $x^2 = 4$."

The letter x introduced in the rule method is a *variable.* In general, a **variable** is a symbol that is used as a placeholder for the elements of a set with two or more elements. This set is called the **replacement set** for the variable. A **constant,** on the other hand, is a symbol that names exactly one object. The symbol "8" is a constant, since it always names the number eight.

If each element of set A is also an element of set B, we say that A is a **subset** of set B, and we write

$$A \subset B \quad \{1, 5\} \subset \{1, 3, 5\}$$

Note that the definition of a subset allows a set to be a subset of itself.

Since the empty set $\varnothing$ has no elements, every element of $\varnothing$ is also an element of any given set. Thus, the empty set is a subset of every set. For example,

$$\varnothing \subset \{1, 3, 5\} \quad \text{and} \quad \varnothing \subset \{2, 4, 6\}$$

If two sets A and B have exactly the same elements, the sets are said to be **equal,** and we write

$$A = B \quad \{4, 2, 6\} = \{6, 4, 2\}$$

Notice that the order of listing elements in a set does not matter.

We can now begin our discussion of the real number system. Additional set concepts will be introduced as needed.

• The Set of Real Numbers

The real number system is the number system you have used most of your life. Informally, a **real number** is any number that has a decimal representation. Table 1 on the next page describes the set of real numbers and some of its important subsets. Figure 1 illustrates how these sets of numbers are related to each other.

Figure 1 Real numbers and important subsets.

TABLE 1   The Set of Real Numbers

Symbol	Name	Description	Examples
N	Natural numbers	Counting numbers (also called positive integers)	1, 2, 3, . . .
Z	Integers	Natural numbers, their negatives, and 0	. . . , $-2, -1, 0, 1, 2, . . .$
Q	Rational numbers	Numbers that can be represented as a/b, where a and b are integers and $b \neq 0$; decimal representations are repeating or terminating	$-4, 0, 1, 25, \frac{-3}{5}, \frac{2}{3}, 3.67, -0.33\overline{3},*$ 5.272727
I	Irrational numbers	Numbers that can be represented as nonrepeating and nonterminating decimal numbers	$\sqrt{2}, \pi, \sqrt[3]{7}, 1.414213 . . . ,$ 2.71828182 . . .
R	Real numbers	Rational numbers and irrational numbers	

*The overbar indicates that the number (or block of numbers) repeats indefinitely.

● The Real Number Line

A one-to-one correspondence exists between the set of real numbers and the set of points on a line. That is, each real number corresponds to exactly one point, and each point to exactly one real number. A line with a real number associated with each point, and vice versa, as in Figure 2, is called a **real number line,** or simply a **real line.** Each number associated with a point is called the **coordinate** of the point. The point with coordinate 0 is called the **origin.** The arrow on the right end of the line indicates a positive direction. The coordinates of all points to the right of the origin are called **positive real numbers,** and those to the left of the origin are called **negative real numbers.** The real number 0 is neither positive nor negative.

Figure 2 A real number line.

● Basic Real Number Properties

We now take a look at some of the basic properties of real numbers. (See the box on the next page.)

You are already familiar with the **commutative properties** for addition and multiplication. They indicate that the order in which the addition or multiplication of two numbers is performed doesn't matter. For example,

$$4 + 5 = 5 + 4 \qquad \text{and} \qquad 4 \cdot 5 = 5 \cdot 4$$

Is there a commutative property relative to subtraction or division? That is, does $x - y = y - x$ or does $x \div y = y \div x$ for all real numbers x and y (division by 0 excluded)? The answer is no, since, for example,

$$7 - 5 \neq 5 - 7 \qquad \text{and} \qquad 6 \div 3 \neq 3 \div 6$$

Basic Properties of the Set of Real Numbers

Let R be the set of real numbers, and let x, y, and z be arbitrary elements of R.

Addition Properties

Closure: $x + y$ is a unique element in R.

Associative: $(x + y) + z = x + (y + z)$

Commutative: $x + y = y + x$

Identity: 0 is the additive identity; that is, $0 + x = x + 0 = x$ for all x in R, and 0 is the only element in R with this property.

Inverse: For each x in R, $-x$ is its unique additive inverse; that is, $x + (-x) = (-x) + x = 0$, and $-x$ is the only element in R relative to x with this property.

Multiplication Properties

Closure: xy is a unique element in R.

Associative: $(xy)z = x(yz)$

Commutative: $xy = yx$

Identity: 1 is the multiplicative identity; that is, for x in R, $(1)x = x(1) = x$, and 1 is the only element in R with this property.

Inverse: For each x in R, $x \neq 0$, $1/x$ is its unique multiplicative inverse; that is, $x(1/x) = (1/x)x = 1$, and $1/x$ is the only element in R relative to x with this property.

Combined Property

Distributive: $x(y + z) = xy + xz$ $(x + y)z = xz + yz$

When computing

$$2 + 5 + 3 \quad \text{or} \quad 2 \cdot 5 \cdot 3$$

why don't we need parentheses to indicate which two numbers are to be added or multiplied first? The answer is to be found in the **associative properties.** These properties allow us to write

$$(2 + 5) + 3 = 2 + (5 + 3) \quad \text{and} \quad (2 \cdot 5) \cdot 3 = 2 \cdot (5 \cdot 3)$$

so it doesn't matter how we group numbers relative to either operation. Is there an associative property for subtraction or division? The answer is no, since, for example,

$$(8 - 4) - 2 \neq 8 - (4 - 2) \qquad \text{and} \qquad (8 \div 4) \div 2 \neq 8 \div (4 \div 2)$$

Evaluate both sides of these equations to see why.

Conclusion

Relative to addition, **commutativity** and **associativity** permit us to change the order of addition at will and insert or remove parentheses as we please. The same is true for multiplication, but not for subtraction and division.

What number added to a given number will give that number back again? What number times a given number will give that number back again? The answers are 0 and 1, respectively. Because of this, 0 and 1 are called the **identity elements** for the real numbers. Hence, for any real numbers x and y,

$$7 + 0 = 7 \qquad 0 + (x + y) = x + y \qquad \text{0 is the additive identity.}$$

$$1 \cdot 6 = 6 \qquad 1(x + y) = x + y \qquad \text{1 is the multiplicative identity.}$$

We now consider **inverses.** For each real number x, there is a unique real number $-x$ such that $x + (-x) = 0$. The number $-x$ is called the **additive inverse** of x, or the **negative** of x. For example, the additive inverse of 4 is -4, since $4 + (-4) = 0$. The additive inverse of -4 is $-(-4) = 4$, since $-4 + [-(-4)] = 0$. It is important to remember:

$-x$ is not necessarily a negative number; it is positive if x is negative and negative if x is positive.

For each nonzero real number x there is a unique real number $1/x$ such that $x(1/x) = 1$. The number $1/x$ is called the **multiplicative inverse** of x, or the **reciprocal** of x. For example, the multiplicative inverse of 7 is $\frac{1}{7}$, since $7(\frac{1}{7}) = 1$. Also note that 7 is the multiplicative inverse of $\frac{1}{7}$. The number 0 has no multiplicative inverse.

We now turn to a real number property that involves both multiplication and addition. Consider the two computations:

$$3(4 + 2) = 3(6) = 18$$

$$3(4) + 3(2) = 12 + 6 = 18$$

Thus,

$$3(4 + 2) = 3(4) + 3(2)$$

and we say that multiplication by 3 *distributes* over the sum $(4 + 2)$. In general, multiplication **distributes** over addition in the real number system. Two more illustrations are given below:

$$2(x + y) = 2x + 2y \qquad (3 + 5)x = 3x + 5x$$

EXAMPLE 1   **Using Real Number Properties**

Which real number property justifies the indicated statement?

Statement	**Property Illustrated**
(A) $(7x)y = 7(xy)$	Associative $(\cdot)$
(B) $a(b + c) = (b + c)a$	Commutative $(\cdot)$
(C) $(2x + 3y) + 5y = 2x + (3y + 5y)$	Associative $(+)$
(D) $(x + y)(a + b) = (x + y)a + (x + y)b$	Distributive
(E) If $a + b = 0$, then $b = -a$.	Inverse $(+)$

Matched Problem 1*   Which real number property justifies the indicated statement?

(A) $4 + (2 + x) = (4 + 2) + x$ (B) $(a + b) + c = c + (a + b)$
(C) $3x + 7x = (3 + 7)x$ (D) $(2x + 3y) + 0 = 2x + 3y$
(E) If $ab = 1$, then $b = 1/a$.

● **Further Properties**   Subtraction and division can be defined in terms of addition and multiplication, respectively:

DEFINITION 1   **Subtraction and Division**

For all real numbers a and b:

Subtraction: $a - b = a + (-b)$ $(-5) - (-3) = (-5) + (3) = -2$

Division: $b\overline{)a} = a \div b = \dfrac{a}{b} = a\left(\dfrac{1}{b}\right)$ $b \neq 0$ $3 \div 2 = 3\left(\dfrac{1}{2}\right)$

Thus, to subtract b from a, add the negative of b to a. To divide a by b, multiply a by the reciprocal of b. Note that division by 0 is not defined, since 0 does not have a reciprocal. It is important to remember:

Division by 0 is never allowed.

The following properties of negatives can be proved using the preceding properties and definitions.

*Answers to matched problems in a given section are found near the end of the section, before the exercise set.

Theorem 1   **Properties of Negatives**

For all real numbers a and b:

1. $-(-a) = a$

2. $(-a)b = -(ab) = a(-b) = -ab$

3. $(-a)(-b) = ab$

4. $(-1)a = -a$

5. $\dfrac{-a}{b} = -\dfrac{a}{b} = \dfrac{a}{-b}$ $b \neq 0$

6. $\dfrac{-a}{-b} = -\dfrac{-a}{b} = -\dfrac{a}{-b} = \dfrac{a}{b}$ $b \neq 0$

We now state an important theorem involving 0.

Theorem 2   **Zero Properties**

For all real numbers a and b:

1. $a \cdot 0 = 0$

2. $ab = 0$ if and only if $a = 0$ or $b = 0$ or both

EXAMPLE 2   **Using Negative and Zero Properties**

Which real number property or definition justifies each statement?

Statement	Property or Definition Illustrated
(A) $3 - (-2) = 3 + [-(-2)] = 5$	Subtraction (Definition 1 and Theorem 1, part 1)
(B) $-(-2) = 2$	Negatives (Theorem 1, part 1)
(C) $-\dfrac{-3}{2} = \dfrac{3}{2}$	Negatives (Theorem 1, part 6)
(D) $\dfrac{5}{-2} = -\dfrac{5}{2}$	Negatives (Theorem 1, part 5)
(E) If $(x-3)(x+5) = 0$, then either $x - 3 = 0$ or $x + 5 = 0$.	Zero (Theorem 2, part 2)

Matched Problem 2   Which real number property or definition justifies each statement?

(A) $\dfrac{3}{5} = 3\left(\dfrac{1}{5}\right)$ (B) $(-5)(2) = -(5 \cdot 2)$ (C) $(-1)3 = -3$

(D) $\dfrac{-7}{9} = -\dfrac{7}{9}$ (E) If $x + 5 = 0$, then $(x - 3)(x + 5) = 0$.

EXPLORE-DISCUSS 1   In general, a set of numbers is closed under an operation if performing the operation on numbers in the set always produces another number in the set. For example, the real numbers are closed under addition, multiplication, subtraction, and division, excluding division by 0. Replace each ? in the following tables with T (true) or F (false), and illustrate each false statement with an example. (See Table 1 for the definitions of the sets N, Z, I, Q, and R.)

	Closed under Addition	Closed under Multiplication
N	?	?
Z	?	?
Q	?	?
I	?	?
R	T	T

	Closed under Subtraction	Closed under Division*
N	?	?
Z	?	?
Q	?	?
I	?	?
R	T	T

*Excluding division by 0.

● **Fraction Properties**   Recall that the quotient $a \div b$, $b \neq 0$, written in the form a/b is called a **fraction.** The quantity a is called the **numerator** and the quantity b is the **denominator.**

Theorem 3	Fraction Properties

For all real numbers a, b, c, d, and k (division by 0 excluded):

1. $\dfrac{a}{b} = \dfrac{c}{d}$ if and only if $ad = bc$

$\dfrac{4}{6} = \dfrac{6}{9}$ since $4 \cdot 9 = 6 \cdot 6$

2. $\dfrac{ka}{kb} = \dfrac{a}{b}$ **3.** $\dfrac{a}{b} \cdot \dfrac{c}{d} = \dfrac{ac}{bd}$ **4.** $\dfrac{a}{b} \div \dfrac{c}{d} = \dfrac{a}{b} \cdot \dfrac{d}{c}$

$\dfrac{7 \cdot 3}{7 \cdot 5} = \dfrac{3}{5}$ $\dfrac{3}{5} \cdot \dfrac{7}{8} = \dfrac{3 \cdot 7}{5 \cdot 8}$ $\dfrac{2}{3} \div \dfrac{5}{7} = \dfrac{2}{3} \cdot \dfrac{7}{5}$

5. $\dfrac{a}{b} + \dfrac{c}{b} = \dfrac{a + c}{b}$ **6.** $\dfrac{a}{b} - \dfrac{c}{b} = \dfrac{a - c}{b}$ **7.** $\dfrac{a}{b} + \dfrac{c}{d} = \dfrac{ad + bc}{bd}$

$\dfrac{3}{6} + \dfrac{5}{6} = \dfrac{3 + 5}{6}$ $\dfrac{7}{8} - \dfrac{3}{8} = \dfrac{7 - 3}{8}$ $\dfrac{2}{3} + \dfrac{3}{5} = \dfrac{2 \cdot 5 + 3 \cdot 3}{3 \cdot 5}$

Answers to Matched Problems

1. (A) Associative ($+$)   (B) Commutative ($+$)   (C) Distributive   (D) Identity ($+$)
     (E) Inverse ($\cdot$)

2. (A) Division (Definition 1)   (B) Negatives (Theorem 1, part 2)
     (C) Negatives (Theorem 1, part 4)   (D) Negatives (Theorem 1, part 5)
     (E) Zero (Theorem 2, part 1)

EXERCISE 1-1

All variables represent real numbers.

A

In Problems 1–8, indicate true (T) or false (F).

1. $4 \in \{3, 4, 5\}$ **2.** $6 \in \{2, 4, 6\}$

3. $3 \notin \{3, 4, 5\}$ **4.** $7 \notin \{2, 4, 6\}$

5. $\{1, 2\} \subset \{1, 3, 5\}$ **6.** $\{2, 6\} \subset \{2, 4, 6\}$

7. $\{7, 3, 5\} \subset \{3, 5, 7\}$ **8.** $\{7, 3, 5\} = \{3, 5, 7\}$

In Problems 9–14, replace each question mark with an appropriate expression that will illustrate the use of the indicated real number property.

9. Commutative property ($+$): $x + 7 = ?$

10. Commutative property ($\cdot$): $uv = ?$

11. Associative property ($\cdot$): $x(yz) = ?$

12. Associative property ($+$): $3 + (7 + y) = ?$

13. Identity property ($+$): $0 + 9m = ?$

14. Identity property ($\cdot$): $1(u + v) = ?$

In Problems 15–26, each statement illustrates the use of one of the following properties or definitions. Indicate which one.

Commutative ($+$, $\cdot$)	Subtraction
Associative ($+$, $\cdot$)	Division
Distributive	Negatives (Theorem 1)
Identity ($+$, $\cdot$)	Zero (Theorem 2)
Inverse ($+$, $\cdot$)	

15. $x + ym = x + my$ **16.** $7(3m) = (7 \cdot 3)m$

17. $7u + 9u = (7 + 9)u$ **18.** $-\dfrac{u}{-v} = \dfrac{u}{v}$

19. $(-2)(\frac{1}{-2}) = 1$ **20.** $8 - 12 = 8 + (-12)$

21. $w + (-w) = 0$ **22.** $5 \div (-6) = 5(\frac{1}{-6})$

23. $3(xy + z) + 0 = 3(xy + z)$

24. $ab(c + d) = abc + abd$

25. $\dfrac{-x}{-y} = \dfrac{x}{y}$

26. $(x + y) \cdot 0 = 0$

B

Write each set in Problems 27–32 using the listing method; that is, list the elements between braces. If the set is empty, write $\varnothing$.

27. $\{x \mid x$ is an even integer between -3 and $5\}$

28. $\{x \mid x$ is an odd integer between -4 and $6\}$

29. $\{x \mid x$ is a letter in "status"$\}$

30. $\{x \mid x$ is a letter in "consensus"$\}$

31. $\{x \mid x$ is a month starting with B$\}$

32. $\{x \mid x$ is a month with 32 days$\}$

33. The set $S_1 = \{a\}$ has only two subsets, S_1 and $\varnothing$. How many subsets does each of the following sets have?
(A) $S_2 = \{a, b\}$
(B) $S_3 = \{a, b, c\}$
(C) $S_4 = \{a, b, c, d\}$

34. Based on the results in Problem 33, how many subsets do you think a set with n elements will have?

In Problems 35–42, each statement illustrates the use of one of the following properties or definitions. Indicate which one.

Commutative $(+, \cdot)$ Subtraction
Associative $(+, \cdot)$ Division
Distributive   Negatives (Theorem 1)
Identity $(+, \cdot)$ Zero (Theorem 2)
Inverse $(+, \cdot)$

35. $-(-y) + 2x = y + 2x$

36. $(ab)(ba) = (ab)(ab)$

37. $(wz)(zw) = w[z(zw)]$

38. $s + (t + 2) = (s + t) + 2$

39. $(n + 2)(m + 3) = n(m + 3) + 2(m + 3)$

40. $p(r - 1) + q(r - 1) = (p + q)(r - 1)$

41. $(2x - 3)(3x + 5) = 0$ if and only if $2x - 3 = 0$ or $3x + 5 = 0$

42. $\dfrac{-y}{-(1 - y)} = \dfrac{y}{1 - y}$

43. If $ab = 0$, does either a or b have to be 0?

44. If $ab = 1$, does either a or b have to be 1?

45. Indicate which of the following are true:
(A) All natural numbers are integers.
(B) All real numbers are irrational.
(C) All rational numbers are real numbers.

46. Indicate which of the following are true:
(A) All integers are natural numbers.
(B) All rational numbers are real numbers.
(C) All natural numbers are rational numbers.

47. Give an example of a rational number that is not an integer.

48. Give an example of a real number that is not a rational number.

In Problems 49 and 50, list the subset of S consisting of (A) natural numbers, (B) integers, and (C) rational numbers.

49. $S = \{-3, -\frac{2}{3}, 0, 1, \sqrt{3}, \frac{9}{5}, 12\}$

50. $S = \{-\sqrt{5}, -1, -\frac{1}{2}, 2, \sqrt{7}, 6, \frac{25}{3}\}$

In Problems 51 and 52, use a calculator* to express each number as a decimal fraction. Classify each decimal number as terminating, repeating, or nonrepeating and nonterminating. Identify the pattern of repeated digits in any repeating decimal numbers.

51. (A) $\frac{8}{9}$ (B) $\frac{3}{11}$ (C) $\sqrt{5}$ (D) $\frac{11}{8}$

52. (A) $\frac{13}{6}$ (B) $\sqrt{21}$ (C) $\frac{7}{16}$ (D) $\frac{29}{111}$

53. Indicate true (T) or false (F), and for each false statement find real number replacements for a and b that will provide a counterexample. For all real numbers a and b:
(A) $a + b = b + a$
(B) $a - b = b - a$
(C) $ab = ba$
(D) $a \div b = b \div a$

54. Indicate true (T) or false (F), and for each false statement find real number replacements for a, b, and c that will provide a counterexample. For all real numbers a, b, and c:
(A) $(a + b) + c = a + (b + c)$
(B) $(a - b) - c = a - (b - c)$
(C) $a(bc) = (ab)c$
(D) $(a \div b) \div c = a \div (b \div c)$

C

55. If $A = \{1, 2, 3, 4\}$ and $B = \{2, 4, 6\}$, find:
(A) $\{x \mid x \in A$ or $x \in B\}$
(B) $\{x \mid x \in A$ and $x \in B\}$

56. If $F = \{-2, 0, 2\}$ and $G = \{-1, 0, 1, 2\}$, find:
(A) $\{x \mid x \in F$ or $x \in G\}$
(B) $\{x \mid x \in F$ and $x \in G\}$

*Later in the book you will encounter optional exercises that require a graphing calculator. If you have such a calculator, you can certainly use it here. Otherwise, any scientific calculator will be sufficient for the problems in this chapter.

57. If $c = 0.151515\ldots$, then $100c = 15.1515\ldots$ and

$$100c - c = 15.1515\ldots - 0.151515\ldots$$

$$99c = 15$$

$$c = \tfrac{15}{99} = \tfrac{5}{33}$$

Proceeding similarly, convert the repeating decimal $0.090909\ldots$ into a fraction. (All repeating decimals are rational numbers, and all rational numbers have repeating decimal representations.)

58. Repeat Problem 57 for $0.181818\ldots$.

59. To see how the distributive property is behind the mechanics of long multiplication, compute each of the following and compare:

Long Multiplication	Use of the Distributive Property
23	$23 \cdot 12$
$\times\ 12$	$= 23(2 + 10)$
	$= 23 \cdot 2 + 23 \cdot 10 =$

60. For a and b real numbers, justify each step using a property in this section.

Statement		Reason
1. $(a + b) + (-a) = (-a) + (a + b)$		**1.**
2. $= [(-a) + a] + b$		**2.**
3. $= 0 + b$		**3.**
4. $= b$		**4.**

SECTION **1-2** **Polynomials: Basic Operations**

- Natural Number Exponents
- Polynomials
- Combining Like Terms
- Addition and Subtraction
- Multiplication
- Combined Operations
- Application

In this section we review the basic operations on *polynomials,* a mathematical form encountered frequently throughout mathematics. We start the discussion with a brief review of natural number exponents. Integer and rational exponents and their properties will be discussed in detail in subsequent sections.

- **Natural Number Exponents**

The definition of a **natural number exponent** is given below:

DEFINITION 1 **Natural Number Exponent**

For n a natural number and a any real number:

$$a^n = \underbrace{a \cdot a \cdot \cdots \cdot a}_{n \text{ factors of } a} \qquad 2^4 = \underbrace{2 \cdot 2 \cdot 2 \cdot 2}_{4 \text{ factors of } 2}$$

Also, the **first property of exponents** is stated as follows:

Theorem 1	**First Property of Exponents**

For any natural numbers m and n, and any real number a:

$$a^m a^n = a^{m+n} \quad (3x^5)(2x^7) \boxed{= (3 \cdot 2)x^{5+7}}^* = 6x^{12}$$

• **Polynomials**   **Algebraic expressions** are formed by using constants and variables and the algebraic operations of addition, subtraction, multiplication, division, raising to powers, and taking roots. Some examples are

$$\sqrt[3]{x^3 + 5} \qquad 5x^4 + 2x^2 - 7$$

$$x + y - 7 \qquad (2x - y)^2$$

$$\frac{x - 5}{x^2 + 2x - 5} \qquad 1 + \cfrac{1}{1 + \cfrac{1}{x}}$$

An algebraic expression involving only the operations of addition, subtraction, multiplication, and raising to natural number powers on variables and constants is called a **polynomial.** Some examples are

$$2x - 3 \qquad 4x^2 - 3x + 7$$

$$x - 2y \qquad 5x^3 - 2x^2 - 7x + 9$$

$$5 \qquad x^2 - 3xy + 4y^2$$

$$0 \qquad x^3 - 3x^2y + xy^2 + 2y^7$$

In a polynomial, a variable cannot appear in a denominator, as an exponent, or within a radical. Accordingly, a **polynomial in one variable** x is constructed by adding or subtracting constants and terms of the form ax^n, where a is a real number and n is a natural number. A **polynomial in two variables** x and y is constructed by adding and subtracting constants and terms of the form $ax^m y^n$, where a is a real number and m and n are natural numbers. Polynomials in three or more variables are defined in a similar manner.

Polynomial forms can be classified according to their *degree*. If a term in a polynomial has only one variable as a factor, then the **degree of that term** is the power of the variable. If two or more variables are present in a term as factors, then the **degree of the term** is the sum of the powers of the variables. The **degree of a polynomial** is the degree of the nonzero term with the highest degree in the polynomial. Any nonzero constant is defined to be a **polynomial of degree 0.** The number 0 is also a polynomial but is not assigned a degree.

*Throughout the book, dashed boxes—called **think boxes**—are used to represent steps that are usually performed mentally.

EXAMPLE 1   **Polynomials and Nonpolynomials**

(A) Polynomials in one variable:

$$x^2 - 3x + 2 \qquad 6x^3 - \sqrt{2}x - \tfrac{1}{3}$$

(B) Polynomials in several variables:

$$3x^2 - 2xy + y^2 \qquad 4x^3y^2 - \sqrt{3}xy^2z^5$$

(C) Nonpolynomials:

$$\sqrt{2x} - \frac{3}{x} + 5 \qquad \frac{x^2 - 3x + 2}{x - 3} \qquad \sqrt{x^2 - 3x + 1}$$

(D) The degree of the first term in $6x^3 - \sqrt{2}x - \tfrac{1}{3}$ is 3, the degree of the second term is 1, the degree of the third term is 0, and the degree of the whole polynomial is 3.

(E) The degree of the first term in $4x^3y^2 - \sqrt{3}xy^2$ is 5, the degree of the second term is 3, and the degree of the whole polynomial is 5.

Matched Problem 1   (A) Which of the following are polynomials?

$$3x^2 - 2x + 1 \qquad \sqrt{x - 3} \qquad x^2 - 2xy + y^2 \qquad \frac{x - 1}{x^2 + 2}$$

(B) Given the polynomial $3x^5 - 6x^3 + 5$, what is the degree of the first term? The second term? The whole polynomial?

(C) Given the polynomial $6x^4y^2 - 3xy^3$, what is the degree of the first term? The second term? The whole polynomial?

In addition to classifying polynomials by degree, we also call a single-term polynomial a **monomial,** a two-term polynomial a **binomial,** and a three-term polynomial a **trinomial.**

$$\tfrac{5}{2}x^2y^3 \qquad\qquad \text{Monomial}$$
$$x^3 + 4.7 \qquad\qquad \text{Binomial}$$
$$x^4 - \sqrt{2}x^2 + 9 \qquad \text{Trinomial}$$

● **Combining Like Terms**   We start with a word about *coefficients.* A constant in a term of a polynomial, including the sign that precedes it, is called the **numerical coefficient,** or simply, the **coefficient,** of the term. If a constant doesn't appear, or only a + sign appears, thick, write an algebraic expression in terms of x that represents the volume of the plastic used to construct the container. Simplify the expression. [*Recall:* The volume V of a sphere of radius r is given by $V = \frac{4}{3}\pi r^3$.]

68. Packaging. A cubical container for shipping computer components is formed by coating a metal mold with polystyrene. If the metal mold is a cube with sides x centimeters long and the polystyrene coating is 2 centimeters thick, write an algebraic expression in terms of x that represents the volume of the polystyrene used to construct the container. Simplify the expression. [*Recall:* The volume V of a cube with sides of length t is given by $V = t^3$.]

0.3 cm

x cm

Figure for 67

SECTION 1-3 Polynomials: Factoring

- Factoring—What Does It Mean?
- Common Factors and Factoring by Grouping
- Factoring Second-Degree Polynomials
- More Factoring

• Factoring—What Does It Mean?

A **factor of a number** is one of two or more numbers whose product is the given number. Similarly, a **factor of an algebraic expression** is one of two or more algebraic expressions whose product is the given algebraic expression. For example,

$$30 = 2 \cdot 3 \cdot 5 \qquad \text{2, 3, and 5 are each factors of 30.}$$

$$x^2 - 4 = (x - 2)(x + 2) \qquad (x - 2) \text{ and } (x + 2) \text{ are each factors of } x^2 - 4.$$

The process of writing a number or algebraic expression as the product of other numbers or algebraic expressions is called **factoring.** We start our discussion of factoring with the positive integers.

An integer such as 30 can be represented in a factored form in many ways. The products

$$6 \cdot 5 \qquad (\tfrac{1}{2})(10)(6) \qquad 15 \cdot 2 \qquad 2 \cdot 3 \cdot 5$$

all yield 30. A particularly useful way of factoring positive integers greater than 1 is in terms of *prime* numbers.

DEFINITION 1 **Prime and Composite Numbers**

An integer greater than 1 is **prime** if its only positive integer factors are itself and 1. An integer greater than 1 that is not prime is called a **composite number.** The integer 1 is neither prime nor composite.

Examples of prime numbers: 2, 3, 5, 7, 11, 13

Examples of composite numbers: 4, 6, 8, 9, 10, 12

In the array below, cross out all multiples of 2, except 2 itself. Then cross out all multiples of 3, except 3 itself. Repeat this for each integer in the array that has not yet been crossed out. Describe the set of numbers that remains when this process is completed.

$$
\begin{array}{cccccccccc}
1 & 2 & 3 & 4 & 5 & 6 & 7 & 8 & 9 & 10 & 11 & 12 & 13 & 14 & 15 & 16 & 17 & 18 & 19 & 20 \\
21 & 22 & 23 & 24 & 25 & 26 & 27 & 28 & 29 & 30 & 31 & 32 & 33 & 34 & 35 & 36 & 37 & 38 & 39 & 40 \\
41 & 42 & 43 & 44 & 45 & 46 & 47 & 48 & 49 & 50 & 51 & 52 & 53 & 54 & 55 & 56 & 57 & 58 & 59 & 60 \\
61 & 62 & 63 & 64 & 65 & 66 & 67 & 68 & 69 & 70 & 71 & 72 & 73 & 74 & 75 & 76 & 77 & 78 & 79 & 80 \\
81 & 82 & 83 & 84 & 85 & 86 & 87 & 88 & 89 & 90 & 91 & 92 & 93 & 94 & 95 & 96 & 97 & 98 & 99 & 100
\end{array}
$$

This process is referred to as the **sieve of Eratosthenes.** (Eratosthenes was a Greek mathematician and astronomer who was a contemporary of Archimedes, circa 200 B.C.)

A composite number is said to be **factored completely** if it is represented as a product of prime factors. The only factoring of 30 given above that meets this condition is $30 = 2 \cdot 3 \cdot 5$.

EXAMPLE 1   **Factoring a Composite Number**

Write 60 in completely factored form.

Solution
$$60 = 6 \cdot 10 = 2 \cdot 3 \cdot 2 \cdot 5 = 2^2 \cdot 3 \cdot 5$$

or

$$60 = 5 \cdot 12 = 5 \cdot 4 \cdot 3 = 2^2 \cdot 3 \cdot 5$$

or

$$60 = 2 \cdot 30 = 2 \cdot 2 \cdot 15 = 2^2 \cdot 3 \cdot 5$$

Matched Problem 1   Write 180 in completely factored form.

Notice in Example 1 that we end up with the same prime factors for 60 irrespective of how we progress through the factoring process. This illustrates an important property of integers:

Theorem 1   **The Fundamental Theorem of Arithmetic**

Each integer greater than 1 is either prime or can be expressed uniquely, except for the order of factors, as a product of prime factors.

We can also write polynomials in completely factored form. A polynomial such as $2x^2 - x - 6$ can be written in factored form in many ways. The products

$$(2x + 3)(x - 2) \qquad 2(x^2 - \tfrac{1}{2}x - 3) \qquad 2(x + \tfrac{3}{2})(x - 2)$$

all yield $2x^2 - x - 6$. A particularly useful way of factoring polynomials is in terms of prime polynomials.

DEFINITION 2   **Prime Polynomials**

A polynomial of degree greater than 0 is said to be **prime** relative to a given set of numbers if: (*1*) all of its coefficients are from that set of numbers; and (*2*) it cannot be written as a product of two polynomials of positive degree having coefficients from that set of numbers.

> Relative to the set of integers:
>
> $x^2 - 2$ is prime
> $x^2 - 9$ is not prime, since $x^2 - 9 = (x - 3)(x + 3)$

[*Note:* The set of numbers most frequently used in factoring polynomials is the set of integers.]

A nonprime polynomial is said to be **factored completely relative to a given set of numbers** if it is written as a product of prime polynomials relative to that set of numbers.

Our objective in this section is to review some of the standard factoring techniques for polynomials with integer coefficients. In Chapter 4 we will treat in detail the topic of factoring polynomials of higher degree with arbitrary coefficients.

● **Common Factors and Factoring by Grouping**

The next example illustrates the use of the distributive properties in factoring.

EXAMPLE 2   **Factoring Out Common Factors**

Factor out, relative to the integers, all factors common to all terms:

(A) $2x^3y - 8x^2y^2 - 6xy^3$ (B) $2x(3x - 2) - 7(3x - 2)$

Solutions   (A) $2x^3y - 8x^2y^2 - 6xy^3 \quad \boxed{= (2xy)x^2 - (2xy)4xy - (2xy)3y^2}$

$$= 2xy(x^2 - 4xy - 3y^2)$$

(B) $2x(3x - 2) - 7(3x - 2) \quad \boxed{= 2x(3x - 2) - 7(3x - 2)}$

$$= (2x - 7)(3x - 2)$$

Matched Problem 2   Factor out, relative to the integers, all factors common to all terms:

(A) $3x^3y - 6x^2y^2 - 3xy^3$ (B) $3y(2y + 5) + 2(2y + 5)$

EXAMPLE 3   Factoring Out Common Factors

Factor completely relative to the integers:

$$4(2x + 7)(x - 3)^2 + 2(2x + 7)^2(x - 3)$$

Solution

$$4(2x + 7)(x - 3)^2 + 2(2x + 7)^2(x - 3)$$
$$= 2(2x + 7)(x - 3)[2(x - 3) + (2x + 7)]$$
$$= 2(2x + 7)(x - 3)(2x - 6 + 2x + 7)$$
$$= 2(2x + 7)(x - 3)(4x + 1)$$

Matched Problem 3   Factor completely relative to the integers:

$$4(2x + 5)(3x + 1)^2 + 6(2x + 5)^2(3x + 1)$$

Some polynomials can be factored by first grouping terms in such a way that we obtain an algebraic expression that looks something like Example 2B. We can then complete the factoring by the method used in that example.

EXAMPLE 4   Factoring by Grouping

Factor completely, relative to the integers, by grouping:

(A) $3x^2 - 6x + 4x - 8$ (B) $wy + wz - 2xy - 2xz$
(C) $3ac + bd - 3ad - bc$

Solutions   (A) $3x^2 - 6x + 4x - 8$

$= (3x^2 - 6x) + (4x - 8)$ Group the first two and last two terms.
$= 3x(x - 2) + 4(x - 2)$ Remove common factors from each group.
$= (3x + 4)(x - 2)$ Factor out the common factor $(x - 2)$.

(B) $wy + wz - 2xy - 2xz$

$= (wy + wz) - (2xy + 2xz)$ Group the first two and last two terms—be careful of signs.
$= w(y + z) - 2x(y + z)$ Remove common factors from each group.
$= (w - 2x)(y + z)$ Factor out the common factor $(y + z)$.

(C) $3ac + bd - 3ad - bc$

In parts (A) and (B) the polynomials are arranged in such a way that grouping the first two terms and the last two terms leads to common factors. In this problem neither the first two terms nor the last two terms have a common factor. Sometimes rearranging terms will lead to a factoring by grouping. In this case, we interchange the second and fourth terms to obtain a problem comparable to part (B), which can be factored as follows:

$$3ac - bc - 3ad + bd = (3ac - bc) - (3ad - bd)$$
$$= c(3a - b) - d(3a - b)$$
$$= (c - d)(3a - b)$$

Matched Problem 4   Factor completely, relative to the integers, by grouping:

(A) $2x^2 + 6x + 5x + 15$ (B) $2pr + ps - 6qr - 3qs$
(C) $6wy - xz - 2xy + 3wz$

• Factoring Second-Degree Polynomials

We now turn our attention to factoring second-degree polynomials of the form

$$2x^2 - 5x - 3   \quad \text{and} \quad   2x^2 + 3xy - 2y^2$$

into the product of two first-degree polynomials with integer coefficients. The following example will illustrate an approach to the problem.

EXAMPLE 5   **Factoring Second-Degree Polynomials**

Factor each polynomial, if possible, using integer coefficients:

(A) $2x^2 + 3xy - 2y^2$ (B) $x^2 - 3x + 4$ (C) $6x^2 + 5xy - 4y^2$

Solutions   (A) $2x^2 + 3xy - 2y^2 = (2x + \ y)(x - \ y)$ Put in what we know. Signs must be opposite. (We can reverse this choice if we get $-3xy$ instead of $+3xy$ for the middle term.)

$$\begin{array}{c} \uparrow \qquad \uparrow \\ ? \qquad ? \end{array}$$

Now, what are the factors of 2 (the coefficient of y^2)?

$$\frac{2}{\begin{array}{l} 1 \cdot 2 \\ 2 \cdot 1 \end{array}}$$

$(2x + y)(x - 2y) = 2x^2 - 3xy - 2y^2$
$(2x + 2y)(x - y) = 2x^2 - 2y^2$

The first choice gives us $-3xy$ for the middle term—close, but not there—so we reverse our choice of signs to obtain

$$2x^2 + 3xy - 2y^2 = (2x - y)(x + 2y)$$

(B) $x^2 - 3x + 4 = (x - \quad)(x - \quad)$ Signs must be the same because the third term is positive and must be negative because the middle term is negative.

$$\frac{4}{2 \cdot 2}$$ $(x - 2)(x - 2) = x^2 - 4x + 4$

$1 \cdot 4$ $(x - 1)(x - 4) = x^2 - 5x + 4$

$4 \cdot 1$ $(x - 4)(x - 1) = x^2 - 5x + 4$

No choice produces the middle term; hence $x^2 - 3x + 4$ is not factorable using integer coefficients.

(C) $6x^2 + 5xy - 4y^2 = (\quad x + \quad y)(\quad x - \quad y)$

$$\uparrow \quad \uparrow \quad \uparrow \quad \uparrow$$
$$? \quad ? \quad ? \quad ?$$

The signs must be opposite in the factors, because the third term is negative. We can reverse our choice of signs later if necessary. We now write all factors of 6 and of 4:

$$\begin{array}{cc} \dfrac{6}{2 \cdot 3} & \dfrac{4}{2 \cdot 2} \\ 3 \cdot 2 & 1 \cdot 4 \\ 1 \cdot 6 & 4 \cdot 1 \\ 6 \cdot 1 & \end{array}$$

and try each choice on the left with each on the right—a total of 12 combinations that give us the first and last terms in the polynomial $6x^2 + 5xy - 4y^2$. The question is: Does any combination also give us the middle term, $5xy$? After trial and error and, perhaps, some educated guessing among the choices, we find that $3 \cdot 2$ matched with $4 \cdot 1$ gives us the correct middle term. Thus,

$$6x^2 + 5xy - 4y^2 = (3x + 4y)(2x - y)$$

If none of the 24 combinations (including reversing our sign choice) had produced the middle term, then we would conclude that the polynomial is not factorable using integer coefficients.

Matched Problem 5   Factor each polynomial, if possible, using integer coefficients:

(A) $x^2 - 8x + 12$ (B) $x^2 + 2x + 5$
(C) $2x^2 + 7xy - 4y^2$ (D) $4x^2 - 15xy - 4y^2$

• **More Factoring**   The factoring formulas listed below will enable us to factor certain polynomial forms that occur frequently.

Special Factoring Formulas

1. $u^2 + 2uv + v^2 = (u + v)^2$ **Perfect Square**

2. $u^2 - 2uv + v^2 = (u - v)^2$ **Perfect Square**

3. $u^2 - v^2 = (u - v)(u + v)$ **Difference of Squares**

4. $u^3 - v^3 = (u - v)(u^2 + uv + v^2)$ **Difference of Cubes**

5. $u^3 + v^3 = (u + v)(u^2 - uv + v^2)$ **Sum of Cubes**

The formulas in the box can be established by multiplying the factors on the right.

CAUTION   Note that we did not list a special factoring formula for the sum of two squares. In general,

$$u^2 + v^2 \neq (au + bv)(cu + dv)$$

for any choice of real number coefficients a, b, c, and d. In the next chapter we will see that $u^2 + v^2$ can be factored using complex numbers.

EXAMPLE 6   **Using Special Factoring Formulas**

Factor completely relative to the integers:

(A) $x^2 + 6xy + 9y^2$ (B) $9x^2 - 4y^2$ (C) $8m^3 - 1$ (D) $x^3 + y^3z^3$

Solutions   (A) $x^2 + 6xy + 9y^2\ \boxed{= x^2 + 2(x)(3y) + (3y)^2}\ = (x + 3y)^2$

(B) $9x^2 - 4y^2\ \boxed{= (3x)^2 - (2y)^2}\ = (3x - 2y)(3x + 2y)$

(C) $8m^3 - 1\ \boxed{\begin{aligned} &= (2m)^3 - 1^3 \\ &= (2m - 1)[(2m)^2 + (2m)(1) + 1^2] \end{aligned}}$

$= (2m - 1)(4m^2 + 2m + 1)$

(D) $x^3 + y^3z^3\ \boxed{= x^3 + (yz)^3}$

$= (x + yz)(x^2 - xyz + y^2z^2)$

Matched Problem 6   Factor completely relative to the integers:

(A) $4m^2 - 12mn + 9n^2$ (B) $x^2 - 16y^2$ (C) $z^3 - 1$ (D) $m^3 + n^3$

EXPLORE-DISCUSS 2   (A) Verify the following factor formulas for $u^4 - v^4$:

$$u^4 - v^4 = (u - v)(u + v)(u^2 + v^2)$$
$$= (u - v)(u^3 + u^2v + uv^2 + v^3)$$

(B) Discuss the pattern in the following formulas:

$$u^2 - v^2 = (u - v)(u + v)$$
$$u^3 - v^3 = (u - v)(u^2 + uv + v^2)$$
$$u^4 - v^4 = (u - v)(u^3 + u^2v + uv^2 + v^3)$$

(C) Use the pattern you discovered in part (B) to write similar formulas for $u^5 - v^5$ and $u^6 - v^6$. Verify your formulas by multiplication.

We complete this section by considering factoring that involves combinations of the preceding techniques as well as a few additional ones. Generally speaking:

When asked to factor a polynomial, we first take out all factors common to all terms, if they are present, and then proceed as above until all factors are prime.

EXAMPLE 7   **Combining Factoring Techniques**

Factor completely relative to the integers:

(A) $18x^3 - 8x$ (B) $x^2 - 6x + 9 - y^2$ (C) $4m^3n - 2m^2n^2 + 2mn^3$
(D) $2t^4 - 16t$ (E) $2y^4 - 5y^2 - 12$

Solutions   (A) $18x^3 - 8x = 2x(9x^2 - 4)$
$$= 2x(3x - 2)(3x + 2)$$

(B) $x^2 - 6x + 9 - y^2$

$= (x^2 - 6x + 9) - y^2$ Group the first three terms.

$= (x - 3)^2 - y^2$ Factor $x^2 - 6x + 9$.

$= [(x - 3) - y][(x - 3) + y]$ Difference of squares

$= (x - 3 - y)(x - 3 + y)$

(C) $4m^3n - 2m^2n^2 + 2mn^3 = 2mn(2m^2 - mn + n^2)$

(D) $2t^4 - 16t = 2t(t^3 - 8)$
$$= 2t(t - 2)(t^2 + 2t + 4)$$

(E) $2y^4 - 5y^2 - 12 = (2y^2 + 3)(y^2 - 4)$
$$= (2y^2 + 3)(y - 2)(y + 2)$$

Matched Problem 7   Factor completely relative to the integers:

(A) $3x^3 - 48x$ (B) $x^2 - y^2 - 4y - 4$
(C) $3u^4 - 3u^3v - 9u^2v^2$ (D) $3m^4 - 24mn^3$
(E) $3x^4 - 5x^2 + 2$

Answers to Matched Problems

1. $2^2 \cdot 3^2 \cdot 5$ 2. (A) $3xy(x^2 - 2xy - y^2)$ (B) $(3y + 2)(2y + 5)$
3. $2(2x + 5)(3x + 1)(12x + 17)$
4. (A) $(2x + 5)(x + 3)$ (B) $(p - 3q)(2r + s)$ (C) $(3w - x)(2y + z)$
5. (A) $(x - 2)(x - 6)$ (B) Not factorable using integers   (C) $(2x - y)(x + 4y)$
     (D) $(4x + y)(x - 4y)$
6. (A) $(2m - 3n)^2$ (B) $(x - 4y)(x + 4y)$ (C) $(z - 1)(z^2 + z + 1)$
     (D) $(m + n)(m^2 - mn + n^2)$
7. (A) $3x(x - 4)(x + 4)$ (B) $(x - y - 2)(x + y + 2)$ (C) $3u^2(u^2 - uv - 3v^2)$
     (D) $3m(m - 2n)(m^2 + 2mn + 4n^2)$ (E) $(3x^2 - 2)(x - 1)(x + 1)$

EXERCISE **1-3**

A
_____

In Problems 1–8, factor out, relative to the integers, all factors common to all terms.

1. $6x^4 - 8x^3 - 2x^2$
2. $6m^4 - 9m^3 - 3m^2$
3. $10x^3y + 20x^2y^2 - 15xy^3$
4. $8u^3v - 6u^2v^2 + 4uv^3$
5. $5x(x + 1) - 3(x + 1)$
6. $7m(2m - 3) + 5(2m - 3)$
7. $2w(y - 2z) - x(y - 2z)$
8. $a(3c + d) - 4b(3c + d)$

In Problems 9–16, factor completely relative to integers.

9. $x^2 - 2x + 3x - 6$
10. $2y^2 - 6y + 5y - 15$
11. $6m^2 + 10m - 3m - 5$
12. $5x^2 - 40x - x + 8$
13. $2x^2 - 4xy - 3xy + 6y^2$
14. $3a^2 - 12ab - 2ab + 8b^2$
15. $8ac + 3bd - 6bc - 4ad$
16. $3pr - 2qs - qr + 6ps$

In Problems 17–28, factor completely relative to the integers. If a polynomial is prime relative to the integers, say so.

17. $2x^2 + x - 3$
18. $3y^2 - 8y - 3$
19. $x^2 + 3x - 8$
20. $u^2 + 4uv - 12v^2$
21. $m^2 + m - 6$
22. $x^2 + 3xy - 10y^2$
23. $4a^2 - 9b^2$
24. $x^2 + 4y^2$
25. $4x^2 - 20x + 25$
26. $a^2b^2 - c^2$
27. $a^2b^2 + c^2$
28. $9x^2 - 4$

B
_____

In Problems 29–42, factor completely relative to the integers. If a polynomial is prime relative to the integers, say so.

29. $6x^2 + 48x + 72$
30. $3z^2 - 28z + 48$

31. $2y^3 - 22y^2 + 48y$

32. $2x^4 - 24x^3 + 40x^2$

33. $16x^2y - 8xy + y$

34. $4xy^2 - 12xy + 9x$

35. $6s^2 + 7st + 3t^2$

36. $6m^2 - mn - 12n^2$

37. $x^3y - 9xy^3$

38. $4u^3v - uv^3$

39. $3m^3 - 6m^2 + 15m$

40. $2x^3 - 2x^2 + 8x$

41. $m^3 + n^3$

42. $r^3 - t^3$

 Problems 43–50 are calculus-related. Factor completely relative to the integers.

43. $2x(x + 1)^4 + 4x^2(x + 1)^3$

44. $(x - 1)^3 + 3x(x - 1)^2$

45. $6(3x - 5)(2x - 3)^2 + 4(3x - 5)^2(2x - 3)$

46. $2(x - 3)(4x + 7)^2 + 8(x - 3)^2(4x + 7)$

47. $5x^4(9 - x)^4 - 4x^5(9 - x)^3$

48. $3x^4(x - 7)^2 + 4x^3(x - 7)^3$

49. $2(x + 1)(x^2 - 5)^2 + 4x(x + 1)^2(x^2 - 5)$

50. $4(x - 3)^3(x^2 + 2)^3 + 6x(x - 3)^4(x^2 + 2)^2$

In Problems 51–56, factor completely relative to the integers. In polynomials involving more than three terms, try grouping the terms in various combinations as a first step. If a polynomial is prime relative to the integers, say so.

51. $(a - b)^2 - 4(c - d)^2$ **52.** $(x + 2)^2 + 9$

53. $2am - 3an + 2bm - 3bn$

54. $15ac - 20ad + 3bc - 4bd$

55. $3x^2 - 2xy - 4y^2$ **56.** $5u^2 + 4uv - v^2$

C

In Problems 57–72, factor completely relative to the integers. In polynomials involving more than three terms, try grouping the terms in various combinations as a first step. If a polynomial is prime relative to the integers, say so.

57. $x^3 - 3x^2 - 9x + 27$

58. $x^3 - x^2 - x + 1$

59. $a^3 - 2a^2 - a + 2$

60. $t^3 - 2t^2 + t - 2$

61. $4(A + B)^2 - 5(A + B) - 5$

62. $6(x - y)^2 + 23(x - y) - 4$

63. $m^4 - n^4$ **64.** $y^4 - 3y^2 - 4$

65. $y^4 - 3y^2 - 5$ **66.** $27a^2 + a^5b^3$

67. $m^2 + 2mn + n^2 - m - n$

68. $y^2 - 2xy + x^2 - y + x$

69. $18a^3 - 8a(x^2 + 8x + 16)$

70. $25(4x^2 - 12xy + 9y^2) - 9a^2b^2$

71. $x^4 + 2x^2 + 1 - x^2$

72. $a^4 + 2a^2b^2 + b^4 - a^2b^2$

$\int$ **APPLICATIONS**

73. Construction. A rectangular open-topped box is to be constructed out of 20-inch-square sheets of thin cardboard by cutting x-inch squares out of each corner and bending the sides up as indicated in the figure. Express each of the following quantities as a polynomial in both factored and expanded form.

(A) The area of cardboard after the corners have been removed.

(B) The volume of the box.

74. Construction. A rectangular open-topped box is to be constructed out of 9- by 16-inch sheets of thin cardboard by cutting x-inch squares out of each corner and bending the sides up. Express each of the following quantities as a polynomial in both factored and expanded form.

(A) The area of cardboard after the corners have been removed.

(B) The volume of the box.

SECTION 1-4 Rational Expressions: Basic Operations

- Reducing to Lowest Terms
- Multiplication and Division
- Addition and Subtraction
- Compound Fractions

We now turn our attention to fractional forms. A quotient of two algebraic expressions, division by 0 excluded, is called a **fractional expression.** If both the numerator and denominator of a fractional expression are polynomials, the fractional expression is called a **rational expression.** Some examples of rational expressions are the following (recall, a nonzero constant is a polynomial of degree 0):

$$\frac{x - 2}{2x^2 - 3x + 5} \qquad \frac{1}{x^4 - 1} \qquad \frac{3}{x} \qquad \frac{x^2 + 3x - 5}{1}$$

In this section we discuss basic operations on rational expressions, including multiplication, division, addition, and subtraction.

Since variables represent real numbers in the rational expressions we are going to consider, the properties of real number fractions summarized in Section 1-1 play a central role in much of the work that we will do.

Even though not always explicitly stated, we always assume that variables are restricted so that division by 0 is excluded.

● **Reducing to Lowest Terms**

We start this discussion by restating the **fundamental property of fractions** (from Theorem 3 in Section 1-1):

Fundamental Property of Fractions

If a, b, and k are real numbers with $b, k \neq 0$, then

$$\frac{ka}{kb} = \frac{a}{b} \qquad \frac{2 \cdot 3}{2 \cdot 4} = \frac{3}{4} \qquad \frac{(x - 3)2}{(x - 3)x} = \frac{2}{x}$$
$$x \neq 0, x \neq 3$$

Using this property from left to right to eliminate all common factors from the numerator and the denominator of a given fraction is referred to as **reducing a fraction to lowest terms.** We are actually dividing the numerator and denominator by the same nonzero common factor.

Using the property from right to left—that is, multiplying the numerator and the denominator by the same nonzero factor—is referred to as **raising a fraction to higher terms.** We will use the property in both directions in the material that follows.

We say that a rational expression is **reduced to lowest terms** if the numerator and denominator do not have any factors in common. Unless stated to the contrary, factors will be relative to the integers.

EXAMPLE 1   **Reducing Rational Expressions**

Reduce each rational expression to lowest terms.

(A) $\dfrac{x^2 - 6x + 9}{x^2 - 9} = \dfrac{(x-3)^2}{(x-3)(x+3)}$ Factor numerator and denominator completely. Divide numerator and denominator by $(x - 3)$; this is a valid operation as long as $x \neq 3$ and $x \neq -3$.

$$= \dfrac{x-3}{x+3}$$

(B) $\dfrac{x^3 - 1}{x^2 - 1} = \dfrac{\overset{1}{\cancel{(x-1)}}(x^2 + x + 1)}{\underset{1}{\cancel{(x-1)}}(x+1)}$ Dividing numerator and denominator by $(x - 1)$ can be indicated by drawing lines through both $(x - 1)$'s and writing the resulting quotients, 1's.

$$= \dfrac{x^2 + x + 1}{x + 1} \qquad x \neq -1 \text{ and } x \neq 1$$

Matched Problem 1   Reduce each rational expression to lowest terms.

(A) $\dfrac{6x^2 + x - 2}{2x^2 + x - 1}$ (B) $\dfrac{x^4 - 8x}{3x^3 - 2x^2 - 8x}$

EXAMPLE 2   **Reducing a Rational Expression**

∫ Reduce the following rational expression to lowest terms.

$$\frac{6x^5(x^2 + 2)^2 - 4x^3(x^2 + 2)^3}{x^8} = \frac{2x^3(x^2 + 2)^2[3x^2 - 2(x^2 + 2)]}{x^8}$$

$$= \frac{\overset{1}{2x^3}(x^2 + 2)^2(x^2 - 4)}{\underset{x^5}{x^8}}$$

$$= \frac{2(x^2 + 2)^2(x - 2)(x + 2)}{x^5}$$

Matched Problem 2   Reduce the following rational expression to lowest terms.

$$\frac{6x^4(x^2 + 1)^2 - 3x^2(x^2 + 1)^3}{x^6}$$

CAUTION   Remember to always factor the numerator and denominator first, then divide out any *common factors*. Do not indiscriminately eliminate *terms* that appear in both the numerator and the denominator. For example,

$$\frac{2x^3 + y^2}{y^2} \neq \frac{\overset{1}{\cancel{2x^3 + y^2}}}{\underset{1}{\cancel{y^2}}} = 2x^3 + 1$$

Since the term y^2 is not a factor of the numerator, it cannot be eliminated. In fact, $(2x^3 + y^2)/y^2$ is already reduced to lowest terms.

• Multiplication and Division

Since we are restricting variable replacements to real numbers, multiplication and division of rational expressions follow the rules for multiplying and dividing real number fractions (Theorem 3 in Section 1-1).

Multiplication and Division

If a, b, c, and d are real numbers with b, $d \neq 0$, then:

1. $\dfrac{a}{b} \cdot \dfrac{c}{d} = \dfrac{ac}{bd}$ $\dfrac{2}{3} \cdot \dfrac{x}{x-1} = \dfrac{2x}{3(x-1)}$

2. $\dfrac{a}{b} \div \dfrac{c}{d} = \dfrac{a}{b} \cdot \dfrac{d}{c}$ $c \neq 0$ $\dfrac{2}{3} \div \dfrac{x}{x-1} = \dfrac{2}{3} \cdot \dfrac{x-1}{x}$

EXPLORE-DISCUSS 1   Write a verbal description of the process of multiplying two fractions. Do the same for the quotient of two fractions.

EXAMPLE 3   **Multiplying and Dividing Rational Expressions**

Perform the indicated operations and reduce to lowest terms.

(A) $\dfrac{10x^3y}{3xy + 9y} \cdot \dfrac{x^2 - 9}{4x^2 - 12x} = \dfrac{\overset{5x^2}{\cancel{10x^3y}}}{\underset{3 \cdot 1}{\cancel{3y}(x + 3)}} \cdot \dfrac{\overset{1 \cdot 1}{(x-3)(x+3)}}{\underset{2 \cdot 1}{\cancel{4x}(x-3)}}$ Factor numerators and denominators; then divide any numerator and any denominator with a like common factor.

$$= \frac{5x^2}{6}$$

(B) $\dfrac{4 - 2x}{4} \div (x - 2) = \dfrac{\overset{1}{\cancel{2}(2 - x)}}{\underset{2}{\cancel{4}}} \cdot \dfrac{1}{x - 2}$ $x - 2$ is the same as $\dfrac{x-2}{1}$.

$$= \frac{2 - x}{2(x - 2)} = \frac{\overset{-1}{-(\cancel{x - 2})}}{2(\underset{1}{\cancel{x - 2}})} \qquad \begin{array}{l} b - a = -(a - b), \\ \text{a useful change in} \\ \text{some problems.} \end{array}$$

$$= -\frac{1}{2}$$

(C) $\dfrac{2x^3 - 2x^2y + 2xy^2}{x^3y - xy^3} \div \dfrac{x^3 + y^3}{x^2 + 2xy + y^2}$

$$= \frac{\overset{2}{\cancel{2x}}(\overset{1}{\cancel{x^2 - xy + y^2}})}{\underset{y}{\cancel{xy}}(\cancel{x + y})(x - y)} \cdot \frac{\overset{1}{(\cancel{x + y})^2}}{(\underset{1}{\cancel{x + y}})(\underset{1}{\cancel{x^2 - xy + y^2}})}$$

$$= \frac{2}{y(x - y)}$$

Matched Problem 3   Perform the indicated operations and reduce to lowest terms.

(A) $\dfrac{12x^2y^3}{2xy^2 + 6xy} \cdot \dfrac{y^2 + 6y + 9}{3y^3 + 9y^2}$ (B) $(4 - x) \div \dfrac{x^2 - 16}{5}$

(C) $\dfrac{m^3 + n^3}{2m^2 + mn - n^2} \div \dfrac{m^3n - m^2n^2 + mn^3}{2m^3n^2 - m^2n^3}$

● Addition and Subtraction

Again, because we are restricting variable replacements to real numbers, addition and subtraction of rational expressions follow the rules for adding and subtracting real number fractions (Theorem 3 in Section 1-1).

> ### Addition and Subtraction
>
> For a, b, and c real numbers with $b \neq 0$:
>
> **1.** $\dfrac{a}{b} + \dfrac{c}{b} = \dfrac{a + c}{b}$ $\dfrac{x}{x - 3} + \dfrac{2}{x - 3} = \dfrac{x + 2}{x - 3}$
>
> **2.** $\dfrac{a}{b} - \dfrac{c}{b} = \dfrac{a - c}{b}$ $\dfrac{x}{2xy^2} - \dfrac{x - 4}{2xy^2} = \dfrac{x - (x - 4)}{2xy^2}$

Thus, we add rational expressions with the same denominators by adding or subtracting their numerators and placing the result over the common denominator. If the denominators are not the same, we raise the fractions to higher terms, using the fundamental property of fractions to obtain common denominators, and then proceed as described.

Even though any common denominator will do, our work will be simplified if the least common denominator (LCD) is used. Often, the LCD is obvious, but if it is not, the steps in the box describe how to find it.

The Least Common Denominator (LCD)

The LCD of two or more rational expressions is found as follows:

1. Factor each denominator completely.

2. Identify each different prime factor from all the denominators.

3. Form a product using each different factor to the highest power that occurs in any one denominator. This product is the LCD.

EXAMPLE 4 **Adding and Subtracting Rational Expressions**

Combine into a single fraction and reduce to lowest terms.

(A) $\dfrac{3}{10} + \dfrac{5}{6} - \dfrac{11}{45}$ (B) $\dfrac{4}{9x} - \dfrac{5x}{6y^2} + 1$

(C) $\dfrac{x+3}{x^2 - 6x + 9} - \dfrac{x+2}{x^2 - 9} - \dfrac{5}{3 - x}$

Solutions   (A) To find the LCD, factor each denominator completely:

$$\left.\begin{array}{l} 10 = 2 \cdot 5 \\ 6 = 2 \cdot 3 \\ 45 = 3^2 \cdot 5 \end{array}\right\} \text{LCD} = 2 \cdot 3^2 \cdot 5 = 90$$

Now use the fundamental property of fractions to make each denominator 90:

$$\frac{3}{10} + \frac{5}{6} - \frac{11}{45} = \frac{9 \cdot 3}{9 \cdot 10} + \frac{15 \cdot 5}{15 \cdot 6} - \frac{2 \cdot 11}{2 \cdot 45}$$

$$= \frac{27}{90} + \frac{75}{90} - \frac{22}{90}$$

$$= \frac{27 + 75 - 22}{90} = \frac{80}{90} = \frac{8}{9}$$

(B) $\left.\begin{array}{l} 9x = 3^2 x \\ 6y^2 = 2 \cdot 3y^2 \end{array}\right\} \text{LCD} = 2 \cdot 3^2 xy^2 = 18xy^2$

$$\frac{4}{9x} - \frac{5x}{6y^2} + 1 = \frac{2y^2 \cdot 4}{2y^2 \cdot 9x} - \frac{3x \cdot 5x}{3x \cdot 6y^2} + \frac{18xy^2}{18xy^2}$$

$$= \frac{8y^2 - 15x^2 + 18xy^2}{18xy^2}$$

(C) $\dfrac{x+3}{x^2-6x+9} - \dfrac{x+2}{x^2-9} - \dfrac{5}{3-x} = \dfrac{x+3}{(x-3)^2} - \dfrac{x+2}{(x-3)(x+3)} + \dfrac{5}{x-3}$

> *Note:* $-\dfrac{5}{3-x} = -\dfrac{5}{-(x-3)} = \dfrac{5}{x-3}$ We have again used the fact that $a - b = -(b - a)$.

The LCD $= (x-3)^2(x+3)$. Thus,

$$\dfrac{(x+3)^2}{(x-3)^2(x+3)} - \dfrac{(x-3)(x+2)}{(x-3)^2(x+3)} + \dfrac{5(x-3)(x+3)}{(x-3)^2(x+3)}$$

$$= \dfrac{(x^2+6x+9) - (x^2-x-6) + 5(x^2-9)}{(x-3)^2(x+3)} \qquad \text{Be careful of sign errors here.}$$

$$= \dfrac{x^2+6x+9 - x^2+x+6 + 5x^2-45}{(x-3)^2(x+3)}$$

$$= \dfrac{5x^2+7x-30}{(x-3)^2(x+3)}$$

Matched Problem 4   Combine into a single fraction and reduce to lowest terms.

(A) $\dfrac{5}{28} - \dfrac{1}{10} + \dfrac{6}{35}$

(B) $\dfrac{1}{4x^2} - \dfrac{2x+1}{3x^3} + \dfrac{3}{12x}$

(C) $\dfrac{y-3}{y^2-4} - \dfrac{y+2}{y^2-4y+4} - \dfrac{2}{2-y}$

EXPLORE-DISCUSS 2   What is the value of $\dfrac{\frac{16}{4}}{2}$?

What is the result of entering $16 \div 4 \div 2$ on a calculator?

What is the difference between $16 \div (4 \div 2)$ and $(16 \div 4) \div 2$?

How could you use fraction bars to distinguish between these two cases when writing $\dfrac{\frac{16}{4}}{2}$?

● **Compound Fractions**   A fractional expression with fractions in its numerator, denominator, or both is called a **compound fraction**. It is often necessary to represent a compound fraction as a simple fraction—that is (in all cases we will consider), as the quotient of two poly-nomials. The process does not involve any new concepts. It is a matter of applying old concepts and processes in the right sequence. We will illustrate two approaches to the problem, each with its own merits, depending on the particular problem under consideration.

EXAMPLE 5   **Simplifying Compound Fractions**

Express as a simple fraction reduced to lowest terms:

$$\frac{\dfrac{2}{x} - 1}{\dfrac{4}{x^2} - 1}$$

Solution   *Method 1.* Multiply the numerator and denominator by the LCD of all fractions in the numerator and denominator—in this case, x^2. (We are multiplying by $1 = x^2/x^2$.)

$$\frac{x^2\left(\dfrac{2}{x} - 1\right)}{x^2\left(\dfrac{4}{x^2} - 1\right)} = \frac{x^2\dfrac{2}{x} - x^2}{x^2\dfrac{4}{x^2} - x^2} = \frac{2x - x^2}{4 - x^2} = \frac{x(2 - x)}{(2 + x)(2 - x)}$$

$$= \frac{x}{2 + x}$$

Method 2. Write the numerator and denominator as single fractions. Then treat as a quotient.

$$\frac{\dfrac{2}{x} - 1}{\dfrac{4}{x^2} - 1} = \frac{\dfrac{2 - x}{x}}{\dfrac{4 - x^2}{x^2}} = \frac{2 - x}{x} \div \frac{4 - x^2}{x^2} = \frac{2 - x}{x} \cdot \frac{x^2}{(2 - x)(2 + x)}$$

$$= \frac{x}{2 + x}$$

Matched Problem 5   Express as a simple fraction reduced to lowest terms. Use the two methods described in Example 5.

$$\frac{1 + \dfrac{1}{x}}{x - \dfrac{1}{x}}$$

EXAMPLE 6   **Simplifying Compound Fractions**

Express as a simple fraction reduced to lowest terms:

$$\frac{\dfrac{y}{x^2} - \dfrac{x}{y^2}}{\dfrac{y}{x} - \dfrac{x}{y}}$$

Solution   Using the first method described in Example 5, we have

$$\frac{x^2y^2\left(\dfrac{y}{x^2} - \dfrac{x}{y^2}\right)}{x^2y^2\left(\dfrac{y}{x} - \dfrac{x}{y}\right)} = \frac{x^2y^2\dfrac{y}{x^2} - x^2y^2\dfrac{x}{y^2}}{x^2y^2\dfrac{y}{x} - x^2y^2\dfrac{x}{y}} = \frac{y^3 - x^3}{xy^3 - x^3y} = \frac{\overset{1}{(y - x)}(y^2 + xy + x^2)}{xy\underset{1}{(y - x)}(y + x)}$$

$$= \frac{y^2 + xy + x^2}{xy(y + x)}$$

Matched Problem 6   Express as a simple fraction reduced to lowest terms. Use the first method described in Example 5.

$$\frac{\dfrac{a}{b} - \dfrac{b}{a}}{\dfrac{a}{b} + 2 + \dfrac{b}{a}}$$

Answers to Matched Problems

1. (A) $\dfrac{3x + 2}{x + 1}$ (B) $\dfrac{x^2 + 2x + 4}{3x + 4}$ 2. $\dfrac{3(x^2 + 1)^2(x + 1)(x - 1)}{x^4}$

3. (A) $2x$ (B) $\dfrac{-5}{x + 4}$ (C) mn

4. (A) $\dfrac{1}{4}$ (B) $\dfrac{3x^2 - 5x - 4}{12x^3}$ (C) $\dfrac{2y^2 - 9y - 6}{(y - 2)^2(y + 2)}$

5. $\dfrac{1}{x - 1}$ 6. $\dfrac{a - b}{a + b}$

EXERCISE **1-4**

A _____

In Problems 1–20, perform the indicated operations and reduce answers to lowest terms. Represent any compound fractions as simple fractions reduced to lowest terms.

1. $\left(\dfrac{d^5}{3a} \div \dfrac{d^2}{6a^2}\right) \cdot \dfrac{a}{4d^3}$ 2. $\dfrac{d^5}{3a} \div \left(\dfrac{d^2}{6a^2} \cdot \dfrac{a}{4d^3}\right)$

3. $\dfrac{2y}{18} - \dfrac{-1}{28} - \dfrac{y}{42}$ 4. $\dfrac{x^2}{12} + \dfrac{x}{18} - \dfrac{1}{30}$

5. $\dfrac{3x + 8}{4x^2} - \dfrac{2x - 1}{x^3} - \dfrac{5}{8x}$

6. $\dfrac{4m - 3}{18m^3} + \dfrac{3}{4m} - \dfrac{2m - 1}{6m^2}$

7. $\dfrac{2x^2 - 3x - 2}{x^2 - 4} \div (2x + 1)$

8. $\dfrac{3x^2 + x - 2}{3x^2 - 2x} \div (3x^2 + 5x + 2)$

9. $\dfrac{2m + 3n}{2m^2 - 5mn - 12n^2} \div \dfrac{2m^2 - 7mn - 4n^2}{m^2 - 8mn + 16n^2}$

10. $\dfrac{4y^2 - 4y + 1}{2y^2 + 5y - 3} \div \dfrac{2y^2 - 3y - 2}{2y^2 + 7y + 3}$

11. $\dfrac{2a - b}{a^2 - b^2} - \dfrac{2a + 3b}{a^2 + 2ab + b^2}$

12. $\dfrac{x + 2}{x^2 - 1} - \dfrac{x - 2}{(x - 1)^2}$

13. $m + 2 - \dfrac{m - 2}{m - 1}$ **14.** $\dfrac{x + 1}{x - 1} + x$

15. $\dfrac{3}{x - 2} - \dfrac{2}{2 - x}$ **16.** $\dfrac{1}{a - 3} - \dfrac{2}{3 - a}$

17. $\dfrac{3}{y + 2} + \dfrac{2}{y - 2} - \dfrac{4y}{y^2 - 4}$

18. $\dfrac{4x}{x^2 - y^2} + \dfrac{3}{x + y} - \dfrac{2}{x - y}$

19. $\dfrac{\dfrac{x^2}{y^2} - 1}{\dfrac{x}{y} + 1}$ **20.** $\dfrac{\dfrac{4}{x} - x}{\dfrac{2}{x} - 1}$

B

$\int$ *Problems 21–26 are calculus-related. Reduce each fraction to lowest terms.*

21. $\dfrac{6x^3(x^2 + 2)^2 - 2x(x^2 + 2)^3}{x^4}$

22. $\dfrac{4x^4(x^2 + 3) - 3x^2(x^2 + 3)^2}{x^6}$

23. $\dfrac{2x(1 - 3x)^3 + 9x^2(1 - 3x)^2}{(1 - 3x)^6}$

24. $\dfrac{2x(2x + 3)^4 - 8x^2(2x + 3)^3}{(2x + 3)^8}$

25. $\dfrac{-2x(x + 4)^3 - 3(3 - x^2)(x + 4)^2}{(x + 4)^6}$

26. $\dfrac{3x^2(x + 1)^3 - 3(x^3 + 4)(x + 1)^2}{(x + 1)^6}$

In Problems 27–40, perform the indicated operations and reduce answers to lowest terms. Represent any compound fractions as simple fractions reduced to lowest terms.

27. $\dfrac{y}{y^2 - 2y - 8} - \dfrac{2}{y^2 - 5y + 4} + \dfrac{1}{y^2 + y - 2}$

28. $\dfrac{x}{x^2 - 9x + 18} + \dfrac{x - 8}{x - 6} + \dfrac{x + 4}{x - 3}$

29. $\dfrac{16 - m^2}{m^2 + 3m - 4} \cdot \dfrac{m - 1}{m - 4}$

30. $\dfrac{x + 1}{x(1 - x)} \cdot \dfrac{x^2 - 2x + 1}{x^2 - 1}$

31. $\dfrac{x + 7}{ax - bx} + \dfrac{y + 9}{by - ay}$

32. $\dfrac{c + 2}{5c - 5} - \dfrac{c - 2}{3c - 3} + \dfrac{c}{1 - c}$

33. $\dfrac{x^2 - 16}{2x^2 + 10x + 8} \div \dfrac{x^2 - 13x + 36}{x^3 + 1}$

34. $\left(\dfrac{x^3 - y^3}{y^3} \cdot \dfrac{y}{x - y}\right) \div \dfrac{x^2 + xy + y^2}{y^2}$

35. $\dfrac{x^2 - xy}{xy + y^2} \div \left(\dfrac{x^2 - y^2}{x^2 + 2xy + y^2} \div \dfrac{x^2 - 2xy + y^2}{x^2y + xy^2}\right)$

36. $\left(\dfrac{x^2 - xy}{xy + y^2} \div \dfrac{x^2 - y^2}{x^2 + 2xy + y^2}\right) \div \dfrac{x^2 - 2xy + y^2}{x^2y + xy^2}$

37. $\left(\dfrac{x}{x^2 - 16} - \dfrac{1}{x + 4}\right) \div \dfrac{4}{x + 4}$

38. $\left(\dfrac{3}{x - 2} - \dfrac{1}{x + 1}\right) \div \dfrac{x + 4}{x - 2}$

39. $\dfrac{1 + \dfrac{2}{x} - \dfrac{15}{x^2}}{1 + \dfrac{4}{x} - \dfrac{5}{x^2}}$ **40.** $\dfrac{\dfrac{x}{y} - 2 + \dfrac{y}{x}}{\dfrac{x}{y} - \dfrac{y}{x}}$

$\int$ *Problems 41–44 are calculus-related. Perform the indicated operations and reduce answers to lowest terms. Represent any compound fractions as simple fractions reduced to lowest terms.*

41. $\dfrac{\dfrac{1}{x + h} - \dfrac{1}{x}}{h}$ **42.** $\dfrac{\dfrac{1}{(x + h)^2} - \dfrac{1}{x^2}}{h}$

43. $\dfrac{\dfrac{(x + h)^2}{x + h + 2} - \dfrac{x^2}{x + 2}}{h}$ **44.** $\dfrac{\dfrac{2x + 2h + 3}{x + h} - \dfrac{2x + 3}{x}}{h}$

In Problems 45–52, imagine that the indicated "solutions" were given to you by a student whom you were tutoring in this class.

(A) *Is the solution correct? If the solution is incorrect, explain what is wrong and how it can be corrected.*

(B) *Show a correct solution for each incorrect solution.*

45. $\dfrac{x^2 + 5x + 4}{x + 4} = \dfrac{x^2 + 5x}{x} = x + 5$

46. $\dfrac{x^2 - 2x - 3}{x - 3} = \dfrac{x^2 - 2x}{x} = x - 2$

∫ **47.** $\dfrac{(x + h)^2 - x^2}{h} = (x + 1)^2 - x^2 = 2x + 1$

∫ **48.** $\dfrac{(x + h)^3 - x^3}{h} = (x + 1)^3 - x^3 = 3x^2 + 3x + 1$

49. $\dfrac{x^2 - 2x}{x^2 - x - 2} + x - 2 = \dfrac{x^2 - 2x + x - 2}{x^2 - x - 2} = 1$

50. $\dfrac{2}{x - 1} - \dfrac{x + 3}{x^2 - 1} = \dfrac{2x + 2 - x - 3}{x^2 - 1} = \dfrac{1}{x + 1}$

51. $\dfrac{2x^2}{x^2 - 4} - \dfrac{x}{x - 2} = \dfrac{2x^2 - x^2 - 2x}{x^2 - 4} = \dfrac{x}{x + 2}$

52. $x + \dfrac{x - 2}{x^2 - 3x + 2} = \dfrac{x + x - 2}{x^2 - 3x + 2} = \dfrac{2}{x - 2}$

C _____

In Problems 53–56, perform the indicated operations and reduce answers to lowest terms. Represent any compound fractions as simple fractions reduced to lowest terms.

53. $\dfrac{y - \dfrac{y^2}{y - x}}{1 + \dfrac{x^2}{y^2 - x^2}}$

54. $\dfrac{\dfrac{s^2}{s - t} - s}{\dfrac{t^2}{s - t} + t}$

55. $2 - \dfrac{1}{1 - \dfrac{2}{a + 2}}$

56. $1 - \dfrac{1}{1 - \dfrac{1}{1 - \dfrac{1}{x}}}$

In Problems 57 and 58, a, b, c, and d represent real numbers.

57. (A) Prove that d/c is the multiplicative inverse of c/d ($c, d \neq 0$).

   (B) Use part (A) to prove that

$$\frac{a}{b} \div \frac{c}{d} = \frac{a}{b} \cdot \frac{d}{c} \qquad b, c, d \neq 0$$

58. Prove that

$$\frac{a}{b} + \frac{c}{b} = \frac{a + c}{b} \qquad b \neq 0$$

SECTION 1-5 Integer Exponents

- Integer Exponents
- Scientific Notation

The French philosopher/mathematician René Descartes (1596–1650) is generally credited with the introduction of the very useful exponent notation "x^n." This notation as well as other improvements in algebra may be found in his *Geometry*, published in 1637.

   In Section 1-2 we introduced the natural number exponent as a short way of writing a product involving the same factors. In this section we will expand the meaning of exponent to include all integers so that exponential forms of the following types will all have meaning:

$$7^5 \qquad 5^{-4} \qquad 3.14^0$$

• Integer Exponents   Definition 1 generalizes exponent notation to include 0 and negative integer exponents.

DEFINITION 1 a^n, *n* an integer and *a* a real number

    1. For *n* a positive integer:

$$a^n = a \cdot a \cdot \cdots \cdot a \qquad 3^5 = 3 \cdot 3 \cdot 3 \cdot 3 \cdot 3$$
$$\underset{n \text{ factors of } a}{}$$

    2. For *n* = 0:

$$a^0 = 1 \qquad a \neq 0 \qquad 132^0 = 1$$
$$0^0 \text{ is not defined}$$

    3. For *n* a negative integer:

$$a^n = \frac{1}{a^{-n}} \qquad a \neq 0 \qquad 7^{-3} \; \boxed{= \frac{1}{7^{-(-3)}}} = \frac{1}{7^3}$$

    Note: In general, it can be shown that for *all* integers *n*

$$a^{-n} = \frac{1}{a^n} \qquad\qquad a^{-5} = \frac{1}{a^5} \qquad a^{-(-3)} = \frac{1}{a^{-3}}$$

EXAMPLE 1   **Using the Definition of Integer Exponents**

Write each part as a decimal fraction or using positive exponents.

(A) $(u^3v^2)^0 = 1 \quad u \neq 0, \quad v \neq 0$ (B) $10^{-3} = \dfrac{1}{10^3} = \dfrac{1}{1,000} = 0.001$

(C) $x^{-8} = \dfrac{1}{x^8}$ (D) $\dfrac{x^{-3}}{y^{-5}} \; \boxed{= \dfrac{x^{-3}}{1} \cdot \dfrac{1}{y^{-5}} = \dfrac{1}{x^3} \cdot \dfrac{y^5}{1}} = \dfrac{y^5}{x^3}$

Matched Problem 1   Write parts (A)–(D) as decimal fractions and parts (E) and (F) with positive exponents.

(A) 636^0 (B) $(x^2)^0 \quad x \neq 0$ (C) 10^{-5}

(D) $\dfrac{1}{10^{-3}}$ (E) $\dfrac{1}{x^{-4}}$ (F) $\dfrac{u^{-7}}{v^{-3}}$

    The basic properties of integer exponents are summarized in Theorem 1. The proof of this theorem involves *mathematical induction,* which is discussed in Chapter 8.

Theorem 1 **Properties of Integer Exponents**

For n and m integers and a and b real numbers:

1. $a^m a^n = a^{m+n}$ $\qquad\qquad$ $a^5 a^{-7} \;\boxed{= a^{5+(-7)}}\; = a^{-2}$

2. $(a^n)^m = a^{mn}$ $\qquad\qquad$ $(a^3)^{-2} \;\boxed{= a^{(-2)3}}\; = a^{-6}$

3. $(ab)^m = a^m b^m$ $\qquad\qquad$ $(ab)^3 = a^3 b^3$

4. $\left(\dfrac{a}{b}\right)^m = \dfrac{a^m}{b^m}$ $\quad b \neq 0$ $\qquad$ $\left(\dfrac{a}{b}\right)^4 = \dfrac{a^4}{b^4}$

5. $\dfrac{a^m}{a^n} = \begin{cases} a^{m-n} \\ \dfrac{1}{a^{n-m}} \end{cases}$ $\quad a \neq 0$ $\qquad$ $\dfrac{a^3}{a^{-2}} = a^{3-(-2)} = a^5$

$\qquad\qquad\qquad\qquad\qquad\qquad\qquad\qquad$ $\dfrac{a^3}{a^{-2}} = \dfrac{1}{a^{-2-3}} = \dfrac{1}{a^{-5}}$

EXPLORE-DISCUSS 1 Property 1 in Theorem 1 can be expressed verbally as follows:

To find the product of two exponential forms with the same base, add the exponents and use the same base.

Express the other properties in Theorem 1 verbally. Decide which you find easier to remember, a formula or a verbal description.

EXAMPLE 2 **Using Exponent Properties**

Simplify using exponent properties, and express answers using positive exponents only.*

(A) $(3a^5)(2a^{-3}) \;\boxed{= (3 \cdot 2)(a^5 a^{-3})}\; = 6a^2$

(B) $\dfrac{6x^{-2}}{8x^{-5}} \;\boxed{= \dfrac{3x^{-2-(-5)}}{4}}\; = \dfrac{3x^3}{4}$

(C) $-4y^3 - (-4y)^3 \;\boxed{= -4y^3 - (-4)^3 y^3}$

$\qquad\qquad\qquad = -4y^3 + 64y^3 = 60y^3$

*By "simplify" we mean eliminate common factors from numerators and denominators and reduce to a minimum the number of times a given constant or variable appears in an expression. We ask that answers be expressed using positive exponents only in order to have a definite form for an answer. Later (in this section and elsewhere) we will encounter situations where we will want negative exponents in a final answer.

Matched Problem 2   Simplify using exponent properties, and express answers using positive exponents only.

(A) $(5x^{-3})(3x^4)$ (B) $\dfrac{9y^{-7}}{6y^{-4}}$ (C) $2x^4 - (-2x)^4$

CAUTION   Be careful when using the relationship $a^{-n} = \dfrac{1}{a^n}$:

$$ab^{-1} \neq \frac{1}{ab} \qquad\qquad ab^{-1} = \frac{a}{b} \quad \text{and} \quad (ab)^{-1} = \frac{1}{ab}$$

$$\frac{1}{a+b} \neq a^{-1} + b^{-1} \qquad \frac{1}{a+b} = (a+b)^{-1} \quad \text{and} \quad \frac{1}{a} + \frac{1}{b} = a^{-1} + b^{-1}$$

Do not confuse properties 1 and 2 in Theorem 1:

$$a^3 a^4 \neq a^{3 \cdot 4} \qquad\qquad a^3 a^4 = a^{3+4} = a^7 \qquad \text{property 1, Theorem 1}$$

$$(a^3)^4 \neq a^{3+4} \qquad\qquad (a^3)^4 = a^{3 \cdot 4} = a^{12} \qquad \text{property 2, Theorem 1}$$

From the definition of negative exponents and the five properties of exponents, we can easily establish the following properties, which are used very frequently when dealing with exponent forms.

Theorem 2   **Further Exponent Properties**

For a and b any real numbers and m, n, and p any integers (division by 0 excluded):

1. $(a^m b^n)^p = a^{pm} b^{pn}$ **2.** $\left(\dfrac{a^m}{b^n}\right)^p = \dfrac{a^{pm}}{b^{pn}}$

3. $\dfrac{a^{-n}}{b^{-m}} = \dfrac{b^m}{a^n}$ **4.** $\left(\dfrac{a}{b}\right)^{-n} = \left(\dfrac{b}{a}\right)^n$

Proof   We prove properties 1 and 4 in Theorem 2 and leave the proofs of 2 and 3 to you.

1. $(a^m b^n)^p = (a^m)^p (b^n)^p \qquad$ property 3, Theorem 1

$ = a^{pm} b^{pn} \qquad\qquad$ property 2, Theorem 1

4. $\left(\dfrac{a}{b}\right)^{-n} = \dfrac{a^{-n}}{b^{-n}}$ property 4, Theorem 1

$\phantom{\left(\dfrac{a}{b}\right)^{-n}} = \dfrac{b^n}{a^n}$ property 3, Theorem 2

$\phantom{\left(\dfrac{a}{b}\right)^{-n}} = \left(\dfrac{b}{a}\right)^n$ property 4, Theorem 1

EXAMPLE 3   **Using Exponent Properties**

Simplify using exponent properties, and express answers using positive exponents only.

(A) $(2a^{-3}b^2)^{-2} = 2^{-2}a^6b^{-4} = \dfrac{a^6}{4b^4}$

(B) $\left(\dfrac{a^3}{b^5}\right)^{-2} = \dfrac{a^{-6}}{b^{-10}} = \dfrac{b^{10}}{a^6}$ or $\left(\dfrac{a^3}{b^5}\right)^{-2} = \left(\dfrac{b^5}{a^3}\right)^2 = \dfrac{b^{10}}{a^6}$

(C) $\dfrac{4x^{-3}y^{-5}}{6x^{-4}y^3} = \dfrac{2x^{-3-(-4)}}{3y^{3-(-5)}} = \dfrac{2x}{3y^8}$

(D) $\left(\dfrac{m^{-3}m^3}{n^{-2}}\right)^{-2} \boxed{= \left(\dfrac{m^{-3+3}}{n^{-2}}\right)^{-2} = \left(\dfrac{m^0}{n^{-2}}\right)^{-2}} = \left(\dfrac{1}{n^{-2}}\right)^{-2} = \dfrac{1}{n^4}$

(E) $(x + y)^{-3} = \dfrac{1}{(x + y)^3}$

Matched Problem 3   Simplify using exponent properties, and express answers using positive exponents only.

(A) $(3x^4y^{-3})^{-2}$ (B) $\left(\dfrac{x^2}{y^4}\right)^{-3}$ (C) $\dfrac{6m^{-2}n^3}{15m^{-1}n^{-2}}$

(D) $\left(\dfrac{x^{-3}}{y^4y^{-4}}\right)^{-3}$ (E) $\dfrac{1}{(a - b)^{-2}}$

In simplifying exponent forms there is often more than one sequence of steps that will lead to the same result (see Example 3B). Use whichever sequence of steps makes sense to you.

EXAMPLE 4   **Simplifying a Compound Fraction**

Express as a simple fraction reduced to lowest terms:

APPLICATIONS

71. **Earth Science.** If the mass of the earth is approximately 6.1×10^{27} grams and each gram is 2.2×10^{-3} pound, what is the mass of the earth in pounds?

72. **Biology.** In 1929 Vernadsky, a biologist, estimated that all the free oxygen of the earth weighs 1.5×10^{21} grams and that it is produced by life alone. If 1 gram is approximately 2.2×10^{-3} pound, what is the weight of the free oxygen in pounds?

73. **Computer Science.** If a computer can perform a single operation in 10^{-10} second, how many operations can it perform in 1 second? In 1 minute? Compute answers to three significant digits.

★ 74. **Computer Science.** If electricity travels in a computer circuit at the speed of light (1.86×10^5 miles per second), how far will electricity travel in the superconducting computer

(see Problem 73) in the time it takes it to perform one operation? (Size of circuits is a critical problem in computer design.) Give the answer in miles, feet, and inches (1 mile = 5,280 feet). Compute answers to three significant digits.

75. **Economics.** If in the United States in 1999 the national debt was about \$5,680,000,000,000 and the population was about 274,000,000, estimate to three significant digits each individual's share of the national debt. Write your answer in scientific notation and in standard decimal form.

76. **Economics.** If in the United States in 1999 the gross national product (GNP) was about \$8,870,000,000,000 and the population was about 274,000,000, estimate to three significant digits the GNP per person. Write your answer in scientific notation and in standard decimal form.

SECTION **1-6** **Rational Exponents**

- Roots of Real Numbers
- Rational Exponents

We now know what symbols such as 3^5, 2^{-3}, and 7^0 mean; that is, we have defined a^n, where n is any integer and a is a real number. But what do symbols such as $4^{1/2}$ and $7^{2/3}$ mean? In this section we will extend the definition of exponent to the rational numbers. Before we can do this, however, we need a precise knowledge of what is meant by "a root of a number."

• Roots of Real Numbers

Perhaps you recall that a **square root** of a number b is a number c such that $c^2 = b$, and a **cube root** of a number b is a number d such that $d^3 = b$.

What are the square roots of 9?

$$3 \text{ is a square root of } 9, \text{ since } 3^2 = 9.$$

$$-3 \text{ is a square root of } 9, \text{ since } (-3)^2 = 9.$$

Thus, 9 has two real square roots, one the negative of the other.

What are the cube roots of 8?

$$2 \text{ is a cube root of } 8, \text{ since } 2^3 = 8.$$

And 2 is the only real number with this property. In general:

DEFINITION 1   **Definition of an *n*th Root**

For a natural number n and a and b real numbers:

a is an nth root of b if $a^n = b$ 3 is a fourth root of 81, since $3^4 = 81$

EXPLORE-DISCUSS 1   Is -4 a cube root of -64?

Is either 8 or -8 a square root of -64?

Can you find any real number b with the property that $b^2 = -64$? [*Hint:* Consider the sign of b^2 for $b > 0$ and $b < 0$.]

How many real square roots of 4 exist? Of 5? Of -9? How many real fourth roots of 5 exist? Of -5? How many real cube roots of 27 are there? Of -27? The following important theorem (which we state without proof) answers these questions.

Theorem 1   **Number of Real *n*th Roots of a Real Number *b****

	n even	*n* odd
b positive	Two real nth roots -3 and 3 are both fourth roots of 81	One real nth root 2 is the only real cube root of 8
b negative	No real nth root -9 has no real square roots	One real nth root -2 is the only real cube root of -8

Thus, 4 and 5 have two real square roots each, and -9 has none. There are two real fourth roots of 5 and none for -5. And 27 and -27 have one real cube root each. What symbols do we use to represent these roots? We turn to this question now.

● **Rational Exponents**   If all exponent properties are to continue to hold even if some of the exponents are rational numbers, then

$$(5^{1/3})^3 = 5^{3/3} = 5 \qquad \text{and} \qquad (7^{1/2})^2 = 7^{2/2} = 7$$

Since Theorem 1 states that the number 5 has one real cube root, it seems reasonable to use the symbol $5^{1/3}$ to represent this root. On the other hand, Theorem 1 states that 7 has two real square roots. Which real square root of 7 does $7^{1/2}$ represent? We answer this question in the following definition.

---

*In this section we limit our discussion to real roots of real numbers. After the real numbers are extended to the complex numbers (see Section 2-5), additional roots may be considered. For example, it turns out that 1 has three cube roots: in addition to the real number 1, there are two other cube roots of 1 in the complex number system.

DEFINITION 2 $b^{1/n}$, **Principal nth Root**

For n a natural number and b a real number,

$$b^{1/n} \text{ is the } \textbf{principal } n\textbf{th root of } b$$

defined as follows:

1. If n is even and b is positive, then $b^{1/n}$ represents the positive nth root of b.

$$16^{1/2} = 4 \qquad \text{not } -4 \text{ and } 4.$$
$$-16^{1/2} = -4 \qquad -16^{1/2} \text{ and } (-16)^{1/2} \text{ are not the same.}$$

2. If n is even and b is negative, then $b^{1/n}$ does not represent a real number. (More will be said about this case later.)

$$(-16)^{1/2} \text{ is not real.}$$

3. If n is odd, then $b^{1/n}$ represents the real nth root of b (there is only one).

$$32^{1/5} = 2 \qquad (-32)^{1/5} = -2$$

4. $0^{1/n} = 0 \qquad 0^{1/9} = 0 \qquad 0^{1/6} = 0$

EXAMPLE 1   **Principal nth Roots**

(A) $9^{1/2} = 3$

(B) $-9^{1/2} = -3$ Compare parts (B) and (C).

(C) $(-9)^{1/2}$ is not a real number.   (D) $27^{1/3} = 3$

(E) $(-27)^{1/3} = -3$ (F) $0^{1/7} = 0$

Matched Problem 1   Find each of the following:

(A) $4^{1/2}$ (B) $-4^{1/2}$ (C) $(-4)^{1/2}$

(D) $8^{1/3}$ (E) $(-8)^{1/3}$ (F) $0^{1/8}$

How should a symbol such as $7^{2/3}$ be defined? If the properties of exponents are to hold for rational exponents, then $7^{2/3} = (7^{1/3})^2$; that is, $7^{2/3}$ must represent the square of the cube root of 7. This leads to the following general definition:

DEFINITION 3 $b^{m/n}$ and $b^{-m/n}$, **Rational Number Exponent**

For m and n natural numbers and b any real number (except b cannot be negative when n is even):

$$b^{m/n} = (b^{1/n})^m \qquad \text{and} \qquad b^{-m/n} = \frac{1}{b^{m/n}}$$

$$4^{3/2} = (4^{1/2})^3 = 2^3 = 8 \qquad 4^{-3/2} = \frac{1}{4^{3/2}} = \frac{1}{8} \qquad (-4)^{3/2} \text{ is not real}$$

$$(-32)^{3/5} = [(-32)^{1/5}]^3 = (-2)^3 = -8$$

We have now discussed $b^{m/n}$ for all rational numbers m/n and real numbers b. It can be shown, though we will not do so, that all five properties of exponents listed in Theorem 1 in Section 1-5 continue to hold for rational exponents as long as we avoid even roots of negative numbers. With the latter restriction in effect, the following useful relationship is an immediate consequence of the exponent properties:

Theorem 2   **Rational Exponent Property**

For m and n natural numbers and b any real number (except b cannot be negative when n is even):

$$b^{m/n} = \begin{cases} (b^{1/n})^m \\ (b^m)^{1/n} \end{cases} \qquad 8^{2/3} = \begin{cases} (8^{1/3})^2 \\ (8^2)^{1/3} \end{cases}$$

EXPLORE-DISCUSS 2   Find the contradiction in the following chain of equations:

$$-1 = (-1)^{2/2} = [(-1)^2]^{1/2} = 1^{1/2} = 1 \qquad\qquad (1)$$

Where did we try to use Theorem 2? Why was this not correct?

The three exponential forms in Theorem 2 are equal as long as only real numbers are involved. But if b is negative and n is even, then $b^{1/n}$ is not a real number and Theorem 2 does not necessarily hold, as illustrated in Explore-Discuss 2. One way to avoid this difficulty is to assume that m and n have no common factors.

EXAMPLE 2   **Using Rational Exponents**

Simplify, and express answers using positive exponents only. All letters represent positive real numbers.

(A) $8^{2/3} = (8^{1/3})^2 = 2^2 = 4$ or $8^{2/3} = (8^2)^{1/3} = 64^{1/3} = 4$

(B) $(-8)^{5/3} = [(-8)^{1/3}]^5 = (-2)^5 = -32$

(C) $(3x^{1/3})(2x^{1/2}) = 6x^{1/3+1/2} = 6x^{5/6}$

(D) $\left(\dfrac{4x^{1/3}}{x^{1/2}}\right)^{1/2} = \dfrac{4^{1/2}x^{1/6}}{x^{1/4}} = \dfrac{2}{x^{1/4-1/6}} = \dfrac{2}{x^{1/12}}$

(E) $(u^{1/2} - 2v^{1/2})(3u^{1/2} + v^{1/2}) = 3u - 5u^{1/2}v^{1/2} - 2v$

Matched Problem 2   Simplify, and express answers using positive exponents only. All letters represent positive real numbers.

(A) $9^{3/2}$ (B) $(-27)^{4/3}$ (C) $(5y^{3/4})(2y^{1/3})$ (D) $(2x^{-3/4}y^{1/4})^4$

(E) $\left(\dfrac{8x^{1/2}}{x^{2/3}}\right)^{1/3}$ (F) $(2x^{1/2} + y^{1/2})(x^{1/2} - 3y^{1/2})$

EXAMPLE 3   **Evaluating Rational Exponential Forms with a Calculator**

Evaluate to four significant digits using a calculator. (Refer to the instruction book for your particular calculator to see how exponential forms are evaluated.)

(A) $11^{3/4}$ (B) $3.1046^{-2/3}$ (C) $(0.000\,000\,008\,437)^{3/11}$

Solutions   (A) First change $\frac{3}{4}$ to the standard decimal form 0.75; then evaluate $11^{0.75}$ using a calculator.

$$11^{3/4} = 6.040$$

(B) $3.1046^{-2/3} = 0.4699$

(C) $(0.000\,000\,008\,437)^{3/11} = (8.437 \times 10^{-9})^{3/11}$

$$= 0.006\,281$$

Matched Problem 3   Evaluate to four significant digits using a calculator.

(A) $2^{3/8}$ (B) $57.28^{-5/6}$ (C) $(83,240,000,000)^{5/3}$

EXAMPLE 4   **Simplifying Fractions Involving Rational Exponents**

$\int$ Write the following expression as a simple fraction reduced to lowest terms and without negative exponents:

$$\frac{(1 + x^2)^{1/2}(2x) - x^2(\frac{1}{2})(1 + x^2)^{-1/2}(2x)}{1 + x^2}$$

Solution   The negative exponent indicates the presence of a fraction in the numerator. Multiply numerator and denominator by $(1 + x^2)^{1/2}$ to eliminate the negative exponent and simplify.

$$\frac{(1 + x^2)^{1/2}(2x) - x^2(\frac{1}{2})(1 + x^2)^{-1/2}(2x)}{1 + x^2} \cdot \frac{(1 + x^2)^{1/2}}{(1 + x^2)^{1/2}}$$

$$= \frac{2x(1 + x^2) - x^3}{(1 + x^2)^{3/2}} = \frac{2x + 2x^3 - x^3}{(1 + x^2)^{3/2}} = \frac{2x + x^3}{(1 + x^2)^{3/2}}$$

$$= \frac{x(2 + x^2)}{(1 + x^2)^{3/2}}$$

Matched Problem 4   Write the following expression as a simple fraction reduced to lowest terms and without negative exponents:

$$\frac{x^2(\frac{1}{2})(1 + x^2)^{-1/2}(2x) - (1 + x^2)^{1/2}(2x)}{x^4}$$

Answers to Matched Problems

1. (A) 2   (B) −2   (C) Not real   (D) 2   (E) −2   (F) 0
2. (A) 27   (B) 81   (C) $10y^{13/12}$ (D) $16y/x^3$ (E) $2/x^{1/18}$ (F) $2x - 5x^{1/2}y^{1/2} - 3y$
3. (A) 1.297   (B) 0.034 28   (C) 1.587×10^{18}
4. $-(2 + x^2)/[x^3(1 + x^2)^{1/2}]$

EXERCISE 1-6

All variables represent positive real numbers unless otherwise stated.

A

In Problems 1–12, evaluate each expression that results in a rational number.

1. $25^{1/2}$ **2.** $27^{1/3}$ **3.** $9^{3/2}$

4. $8^{2/3}$ **5.** $-16^{1/2}$ **6.** $64^{2/3}$

7. $(-16)^{1/2}$ **8.** $(-64)^{2/3}$ **9.** $(\frac{27}{125})^{2/3}$

10. $(\frac{25}{36})^{3/2}$ **11.** $4^{-5/2}$ **12.** $9^{-5/2}$

Simplify Problems 13–20, and express answers using positive exponents only.

13. $a^{1/3}a^{4/3}$ **14.** $b^{2/5}b^{4/5}$ **15.** $c^{3/5}c^{-1/5}$

16. $d^{1/5}d^{-3/5}$ **17.** $(u^{-5})^{1/10}$ **18.** $(v^{-3/4})^8$

19. $(16x^8y^{-4})^{1/4}$ **20.** $(27x^{-6}y^9)^{1/3}$

B

Simplify Problems 21–30, and express answers using positive exponents only.

21. $\left(\frac{a^{-3}}{b^4}\right)^{1/12}$ **22.** $\left(\frac{m^{-2/3}}{n^{-1/2}}\right)^{-6}$ **23.** $\left(\frac{4x^{-2}}{y^4}\right)^{-1/2}$

24. $\left(\frac{w^4}{9x^{-2}}\right)^{-1/2}$ **25.** $\left(\frac{8a^{-4}b^3}{27a^2b^{-3}}\right)^{1/3}$ **26.** $\left(\frac{25x^5y^{-1}}{16x^{-3}y^{-5}}\right)^{1/2}$

27. $\frac{8x^{-1/3}}{12x^{1/4}}$ **28.** $\frac{6a^{3/4}}{15a^{-1/3}}$ **29.** $\left(\frac{a^{2/3}b^{-1/2}}{a^{1/2}b^{1/2}}\right)^2$

30. $\left(\frac{x^{-1/3}y^{1/2}}{x^{-1/4}y^{1/3}}\right)^6$

In Problems 31–38, multiply, and express answers using positive exponents only.

31. $2m^{1/3}(3m^{2/3} - m^6)$ **32.** $3x^{3/4}(4x^{1/4} - 2x^8)$

33. $(a^{1/2} + 2b^{1/2})(a^{1/2} - 3b^{1/2})$

34. $(3u^{1/2} - v^{1/2})(u^{1/2} - 4v^{1/2})$

35. $(2x^{1/2} - 3y^{1/2})(2x^{1/2} + 3y^{1/2})$

36. $(5m^{1/2} + n^{1/2})(5m^{1/2} - n^{1/2})$

37. $(x^{1/2} + 2y^{1/2})^2$ **38.** $(3x^{1/2} - y^{1/2})^2$

In Problems 39–46, evaluate to four significant digits using a calculator. (Refer to the instruction book for your calculator to see how exponential forms are evaluated.)

39. $15^{5/4}$ **40.** $22^{3/2}$ **41.** $103^{-3/4}$

42. $827^{-3/8}$ **43.** $2.876^{8/5}$ **44.** $37.09^{7/3}$

45. $(0.000\,000\,077\,35)^{-2/7}$ **46.** $(491,300,000,000)^{7/4}$

Problems 47–50 illustrate common errors involving rational exponents. In each case, find numerical examples that show that the left side is not always equal to the right side.

47. $(x + y)^{1/2} \neq x^{1/2} + y^{1/2}$ 48. $(x^3 + y^3)^{1/3} \neq x + y$

49. $(x + y)^{1/3} \neq \dfrac{1}{(x + y)^3}$ 50. $(x + y)^{-1/2} \neq \dfrac{1}{(x + y)^2}$

∫ *Problems 51–56 are calculus-related. Write each problem in the form $ax^p + bx^q$, where a and b are real numbers and p and q are rational numbers. For example,*

$$\frac{2x^{1/3} + 4}{4x} \quad \boxed{= \frac{2x^{1/3}}{4x} + \frac{4}{4x} = \frac{1}{2}x^{1/3-1} + x^{-1}}$$

$$= \frac{1}{2}x^{-2/3} + x^{-1}$$

51. $\dfrac{12x^{1/2} - 3}{4x^{1/2}}$ 52. $\dfrac{x^{2/3} + 2}{2x^{1/3}}$ 53. $\dfrac{3x^{2/3} + x^{1/2}}{5x}$

54. $\dfrac{2x^{3/4} + 3x^{1/3}}{3x}$ 55. $\dfrac{x^2 - 4x^{1/2}}{2x^{1/3}}$ 56. $\dfrac{2x^{1/3} - x^{1/2}}{4x^{1/2}}$

C

In Problems 57–60, m and n represent positive integers. Simplify and express answers using positive exponents.

57. $(a^{3/n}b^{3/m})^{1/3}$ 58. $(a^{n/2}b^{n/3})^{1/n}$

59. $(x^{m/4}y^{n/3})^{-12}$ 60. $(a^{m/3}b^{n/2})^{-6}$

61. If possible, find a real value of x such that:
(A) $(x^2)^{1/2} \neq x$ (B) $(x^2)^{1/2} = x$ (C) $(x^3)^{1/3} \neq x$

62. If possible, find a real value of x such that:
(A) $(x^2)^{1/2} \neq -x$ (B) $(x^2)^{1/2} = -x$ (C) $(x^3)^{1/3} = -x$

63. If n is even and b is negative, then $b^{1/n}$ is not real. If m is odd, n is even, and b is negative, is $(b^m)^{1/n}$ real?

64. If we assume that m is odd and n is even, is it possible that one of $(b^{1/n})^m$ and $(b^m)^{1/n}$ is real and the other is not?

∫ *Problems 65–68 are calculus-related. Simplify by writing each expression as a simple fraction reduced to lowest terms and without negative exponents.*

65. $\dfrac{(2x - 1)^{1/2} - (x + 2)(\frac{1}{2})(2x - 1)^{-1/2}(2)}{2x - 1}$

66. $\dfrac{(x - 1)^{1/2} - x(\frac{1}{2})(x - 1)^{-1/2}}{x - 1}$

67. $\dfrac{2(3x - 1)^{1/3} - (2x + 1)(\frac{1}{3})(3x - 1)^{-2/3}(3)}{(3x - 1)^{2/3}}$

68. $\dfrac{(x + 2)^{2/3} - x(\frac{2}{3})(x + 2)^{-1/3}}{(x + 2)^{4/3}}$

APPLICATIONS

69. **Economics.** The number of units N of a finished product produced from the use of x units of labor and y units of capital for a particular Third World country is approximated by

$$N = 10x^{3/4}y^{1/4} \qquad \text{Cobb-Douglas equation}$$

Estimate how many units of a finished product will be produced using 256 units of labor and 81 units of capital.

70. **Economics.** The number of units N of a finished product produced by a particular automobile company where x units of labor and y units of capital are used is approximated by

$$N = 50x^{1/2}y^{1/2} \qquad \text{Cobb-Douglas equation}$$

Estimate how many units will be produced using 256 units of labor and 144 units of capital.

71. **Braking Distance.** R. A. Moyer of Iowa State College found, in comprehensive tests carried out on 41 wet pavements, that the braking distance d (in feet) for a particular automobile traveling at v miles per hour was given approximately by

$$d = 0.0212v^{7/3}$$

Approximate the braking distance to the nearest foot for the car traveling on wet pavement at 70 miles per hour.

72. **Braking Distance.** Approximately how many feet would it take the car in Problem 71 to stop on wet pavement if it were traveling at 50 miles per hour? (Compute answer to the nearest foot.)

SECTION 1-7 Radicals

- From Rational Exponents to Radicals, and Vice Versa
- Properties of Radicals
- Simplifying Radicals
- Sums and Differences
- Products
- Rationalizing Operations

What do the following algebraic expressions have in common?

$$2^{1/2} \qquad 2x^{2/3} \qquad \frac{1}{x^{1/2} + y^{1/2}}$$

$$\sqrt{2} \qquad 2\sqrt[3]{x^2} \qquad \frac{1}{\sqrt{x} + \sqrt{y}}$$

Each vertical pair represents the same quantity, one in rational exponent form and the other in *radical form.* There are occasions when it is more convenient to work with radicals than with rational exponents, or vice versa. In this section we see how the two forms are related and investigate some basic operations on radicals.

● From Rational Exponents to Radicals, and Vice Versa

We start this discussion by defining an **nth-root radical:**

DEFINITION 1 $\sqrt[n]{b}$, *nth-Root Radical*

For n a natural number greater than 1 and b a real number, we define $\sqrt[n]{b}$ to be the **principal nth root of b** (see Definition 2 in Section 1-6); that is,

$$\sqrt[n]{b} = b^{1/n}$$

If $n = 2$, we write $\sqrt{b}$ in place of $\sqrt[2]{b}$

$$\sqrt{25} \boxed{= 25^{1/2}} = 5 \qquad \sqrt[5]{32} \boxed{= 32^{1/5}} = 2$$

$$-\sqrt{25} \boxed{= -25^{1/2}} = -5 \qquad \sqrt[5]{-32} \boxed{= (-32)^{1/5}} = -2$$

$$\sqrt{-25} \text{ is not real} \qquad \sqrt[4]{0} = 0^{1/4} = 0$$

The symbol $\sqrt{}$ is called a **radical,** n is called the **index,** and b is called the **radicand.**

As stated above, it is often an advantage to be able to shift back and forth between rational exponent forms and radical forms. The following relationships, which are direct consequences of Definition 1 and Theorem 2 in Section 1-6, are useful in this regard:

Rational Exponent/Radical Conversions

For m and n positive integers $(n > 1)$, and b not negative when n is even,

$$b^{m/n} = \begin{cases} \boxed{(b^m)^{1/n}} = \sqrt[n]{b^m} \\ \boxed{(b^{1/n})^m} = (\sqrt[n]{b})^m \end{cases} \qquad 2^{2/3} = \begin{cases} \sqrt[3]{2^2} \\ (\sqrt[3]{2})^2 \end{cases}$$

Note: Unless stated to the contrary, all variables in the rest of the discussion are restricted so that all quantities involved are real numbers.

EXPLORE-DISCUSS 1   In each of the following, evaluate both radical forms.

$$16^{3/2} = \sqrt{16^3} = (\sqrt{16})^3$$
$$27^{2/3} = \sqrt[3]{27^2} = (\sqrt[3]{27})^2$$

Which radical conversion form is easier to use if you are performing the calculations by hand?

EXAMPLE 1   **Rational Exponents/Radical Conversions**

Change from rational exponent form to radical form.

(A) $x^{1/7} = \sqrt[7]{x}$

(B) $(3u^2v^3)^{3/5} = \sqrt[5]{(3u^2v^3)^3}$ or $(\sqrt[5]{3u^2v^3})^3$ The first is usually preferred.

(C) $y^{-2/3} = \dfrac{1}{y^{2/3}} = \dfrac{1}{\sqrt[3]{y^2}}$ or $\sqrt[3]{y^{-2}}$ or $\sqrt[3]{\dfrac{1}{y^2}}$

Change from radical form to rational exponent form.

(D) $\sqrt[5]{6} = 6^{1/5}$ (E) $-\sqrt[3]{x^2} = -x^{2/3}$ (F) $\sqrt{x^2 + y^2} = (x^2 + y^2)^{1/2}$

Matched Problem 1   Change from rational exponent form to radical form.

(A) $u^{1/5}$ (B) $(6x^2y^5)^{2/9}$ (C) $(3xy)^{-3/5}$

Change from radical form to rational exponent form.

(D) $\sqrt[4]{9u}$ (E) $-\sqrt[7]{(2x)^4}$ (F) $\sqrt[3]{x^3 + y^3}$

● **Properties of**   The process of changing and simplifying radical expressions is aided by the intro-
Radicals   duction of several properties of radicals that follow directly from exponent properties considered earlier.

Theorem 1   **Properties of Radicals**

For n a natural number greater than 1, and x and y positive real numbers:

1. $\sqrt[n]{x^n} = x$ $\sqrt[3]{x^3} = x$

2. $\sqrt[n]{xy} = \sqrt[n]{x}\sqrt[n]{y}$ $\sqrt[5]{xy} = \sqrt[5]{x}\sqrt[5]{y}$

3. $\sqrt[n]{\dfrac{x}{y}} = \dfrac{\sqrt[n]{x}}{\sqrt[n]{y}}$ $\sqrt[4]{\dfrac{x}{y}} = \dfrac{\sqrt[4]{x}}{\sqrt[4]{y}}$

EXAMPLE 2   **Simplifying Radicals**

Simplify:

(A) $\sqrt[5]{(3x^2y)^5} = 3x^2y$

(B) $\sqrt{10}\sqrt{5} = \sqrt{50} = \sqrt{25 \cdot 2} = \sqrt{25}\sqrt{2} = 5\sqrt{2}$

(C) $\sqrt[3]{\dfrac{x}{27}} = \dfrac{\sqrt[3]{x}}{\sqrt[3]{27}} = \dfrac{\sqrt[3]{x}}{3}$ or $\dfrac{1}{3}\sqrt[3]{x}$

Matched Problem 2   Simplify:

(A) $\sqrt[7]{(u^2 + v^2)^7}$ (B) $\sqrt{6}\sqrt{2}$ (C) $\sqrt[3]{\dfrac{x^2}{8}}$

CAUTION   In general, properties of radicals can be used to simplify terms raised to powers, not sums of terms raised to powers. Thus, for x and y positive real numbers,

$$\sqrt{x^2 + y^2} \neq \sqrt{x^2} + \sqrt{y^2} = x + y$$

but

$$\sqrt{x^2 + 2xy + y^2} = \sqrt{(x + y)^2} = x + y$$

● **Simplifying Radicals**   The properties of radicals provide us with the means of changing algebraic expressions containing radicals to a variety of equivalent forms. One form that is often useful is a *simplified form*. An algebraic expression that contains radicals is said to be in **simplified form** if all four of the conditions listed in the following definition are satisfied.

DEFINITION 2   **Simplified (Radical) Form**

1. No radicand (the expression within the radical sign) contains a factor to a power greater than or equal to the index of the radical.

   For example, $\sqrt{x^5}$ violates this condition.

2. No power of the radicand and the index of the radical have a common factor other than 1.

   For example, $\sqrt[6]{x^4}$ violates this condition.

3. No radical appears in a denominator.

   For example, $y/\sqrt{x}$ violates this condition.

4. No fraction appears within a radical.

   For example, $\sqrt{\frac{3}{5}}$ violates this condition.

EXAMPLE 3   **Finding Simplified Form**

Express radicals in simplified form.

(A) $\sqrt{12x^3y^5z^2} = \sqrt{(4x^2y^4z^2)(3xy)}$ Condition 1 is not met.

$\qquad\quad = \sqrt{(2xy^2z)^2(3xy)}$ $x^{pm}y^{pn} = (x^my^n)^p$

$\qquad\quad = \sqrt{(2xy^2z)^2}\sqrt{3xy}$ $\sqrt[n]{xy} = \sqrt[n]{x}\sqrt[n]{y}$

$\qquad\quad = 2xy^2z\sqrt{3xy}$ $\sqrt[n]{x^n} = x$

(B) $\sqrt[3]{6x^2y}\sqrt[3]{4x^5y^2}\;\;= \sqrt[3]{(6x^2y)(4x^5y^2)}$ $\sqrt[n]{x}\sqrt[n]{y} = \sqrt[n]{xy}$

$\qquad\qquad\quad = \sqrt[3]{24x^7y^3}$

$\qquad\qquad\quad = \sqrt[3]{(8x^6y^3)(3x)}$ Condition 1 is not met.

$\qquad\qquad\quad = \sqrt[3]{(2x^2y)^3(3x)}$ $x^{pm}y^{pn} = (x^my^n)^p$

$\qquad\qquad\quad = \sqrt[3]{(2x^2y)^3}\sqrt[3]{3x}$ $\sqrt[n]{xy} = \sqrt[n]{x}\sqrt[n]{y}$

$\qquad\qquad\quad = 2x^2y\sqrt[3]{3x}$ $\sqrt[n]{x^n} = x$

(C) $\sqrt[6]{16x^4y^2} = [(4x^2y)^2]^{1/6}$ Condition 2 is not met.

$\qquad\qquad = (4x^2y)^{2/6}$ Note the convenience of using rational exponents.

$\qquad\qquad = (4x^2y)^{1/3}$

$\qquad\qquad = \sqrt[3]{4x^2y}$

(D) $\sqrt[3]{\sqrt{27}} = [(3^3)^{1/2}]^{1/3}$

$\qquad\quad = (3^3)^{1/6} = 3^{3/6}\;\; = 3^{1/2} = \sqrt{3}$

Matched Problem 3   Express radicals in simplified form.

(A) $\sqrt{18x^5y^2z^3}$ (B) $\sqrt[4]{27a^3b^3}\sqrt[4]{3a^5b^3}$ (C) $\sqrt[9]{8x^6y^3}$ (D) $\sqrt{\sqrt[3]{4}}$

• Sums and Differences

Algebraic expressions involving radicals often can be simplified by adding and subtracting terms that contain exactly the same radical expressions. We proceed in essentially the same way as we do when we combine like terms in polynomials. The distributive property of real numbers plays a central role in this process.

EXAMPLE 4   **Combining Like Terms**

Combine as many terms as possible:

(A) $5\sqrt{3} + 4\sqrt{3} \;\boxed{= (5+4)\sqrt{3}}\; = 9\sqrt{3}$

(B) $2\sqrt[3]{xy^2} - 7\sqrt[3]{xy^2} \;\boxed{= (2-7)\sqrt[3]{xy^2}}\; = -5\sqrt[3]{xy^2}$

(C) $3\sqrt{xy} - 2\sqrt[3]{xy} + 4\sqrt{xy} - 7\sqrt[3]{xy} \;\boxed{= 3\sqrt{xy} + 4\sqrt{xy} - 2\sqrt[3]{xy} - 7\sqrt[3]{xy}}$
$$= 7\sqrt{xy} - 9\sqrt[3]{xy}$$

Matched Problem 4   Combine as many terms as possible:

(A) $6\sqrt{2} + 2\sqrt{2}$ (B) $3\sqrt[5]{2x^2y^3} - 8\sqrt[5]{2x^2y^3}$

(C) $5\sqrt[3]{mn^2} - 3\sqrt{mn} - 2\sqrt[3]{mn^2} + 7\sqrt{mn}$

• Products

We will now consider several types of special products that involve radicals. The distributive property of real numbers plays a central role in our approach to these problems.

EXAMPLE 5   **Multiplication with Radical Forms**

Multiply and simplify:

(A) $\sqrt{2}(\sqrt{10} - 3) = \sqrt{2}\sqrt{10} - \sqrt{2}\cdot 3 = \sqrt{20} - 3\sqrt{2} = 2\sqrt{5} - 3\sqrt{2}$

(B) $(\sqrt{2} - 3)(\sqrt{2} + 5) = \sqrt{2}\sqrt{2} - 3\sqrt{2} + 5\sqrt{2} - 15$
$$= 2 + 2\sqrt{2} - 15$$
$$= 2\sqrt{2} - 13$$

(C) $(\sqrt{x} - 3)(\sqrt{x} + 5) = \sqrt{x}\sqrt{x} - 3\sqrt{x} + 5\sqrt{x} - 15$
$$= x + 2\sqrt{x} - 15$$

EQUATIONS AND INEQUALITIES

CHAPTER 2

One of the important uses of algebra is the solving of equations and inequalities. In this chapter we look at techniques for solving linear and nonlinear equations and inequalities. In addition, we consider a number of applications that can be solved using these techniques. Additional techniques for solving polynomial equations will be discussed in Chapter 4.

SECTION 2-1   Linear Equations and Applications

- Equations
- Solving Linear Equations
- A Strategy for Solving Word Problems
- Number and Geometric Problems
- Rate–Time Problems
- Mixture Problems
- Some Final Observations on Linear Equations

• Equations

An **algebraic equation** is a mathematical statement that relates two algebraic expressions involving at least one variable. Some examples of equations with x as the variable are

$$3x - 2 = 7 \qquad \frac{1}{1 + x} = \frac{x}{x - 2}$$

$$2x^2 - 3x + 5 = 0 \qquad \sqrt{x + 4} = x - 1$$

The **replacement set,** or **domain,** for a variable is defined to be the set of numbers that are permitted to replace the variable.

Assumption   On Domains of Variables

Unless stated to the contrary, we assume that the domain for a variable is the set of those real numbers for which the algebraic expressions involving the variable are real numbers.

For example, the domain for the variable x in the expression

$$2x - 4$$

is R, the set of all real numbers, since $2x - 4$ represents a real number for all replacements of x by real numbers. The domain of x in the equation

$$\frac{1}{x} = \frac{2}{x - 3}$$

is the set of all real numbers except 0 and 3. These values are excluded because the left member is not defined for $x = 0$ and the right member is not defined for $x = 3$.

The left and right members represent real numbers for all other replacements of x by real numbers.

The **solution set** for an equation is defined to be the set of elements in the domain of the variable that makes the equation true. Each element of the solution set is called a **solution,** or **root,** of the equation. To **solve an equation** is to find the solution set for the equation.

An equation is called an **identity** if the equation is true for all elements from the domain of the variable. An equation is called a **conditional equation** if it is true for certain domain values and false for others. For example,

$$2x - 4 = 2(x - 2) \quad \text{and} \quad \frac{5}{x^2 - 3x} = \frac{5}{x(x - 3)}$$

are identities, since both equations are true for all elements from the respective domains of their variables. On the other hand, the equations

$$3x - 2 = 5 \quad \text{and} \quad \frac{2}{x - 1} = \frac{1}{x}$$

are conditional equations, since, for example, neither equation is true for the domain value 2.

Knowing what we mean by the solution set of an equation is one thing; finding it is another. To this end we introduce the idea of equivalent equations. Two equations are said to be **equivalent** if they both have the same solution set for a given replacement set. A basic technique for solving equations is to perform operations on equations that produce simpler equivalent equations, and to continue the process until an equation is reached whose solution is obvious.

Application of any of the properties of equality given in Theorem 1 will produce equivalent equations.

Theorem 1	**Properties of Equality**

For a, b, and c any real numbers:

1. If $a = b$, then $a + c = b + c$.	**Addition Property**
2. If $a = b$, then $a - c = b - c$.	**Subtraction Property**
3. If $a = b$, then $ca = cb$, $c \neq 0$.	**Multiplication Property**
4. If $a = b$, then $\dfrac{a}{c} = \dfrac{b}{c}$, $c \neq 0$.	**Division Property**
5. If $a = b$, then either may replace the other in any statement without changing the truth or falsity of the statement.	**Substitution Property**

• Solving Linear Equations

We now turn our attention to methods of solving *first-degree,* or *linear, equations* in one variable.

DEFINITION 1   **Linear Equation in One Variable**

Any equation that can be written in the form

$$ax + b = 0 \qquad a \neq 0 \qquad \textbf{Standard Form}$$

where a and b are real constants and x is a variable, is called a **linear**, or **first-degree, equation** in one variable.

$5x - 1 = 2(x + 3)$ is a linear equation, since it can be written in the standard form $3x - 7 = 0$.

EXAMPLE 1   **Solving a Linear Equation**

Solve $5x - 9 = 3x + 7$ and check.

Solution   We use the properties of equality to transform the given equation into an equivalent equation whose solution is obvious.

$$5x - 9 = 3x + 7 \qquad \text{Original equation}$$
$$5x - 9 + 9 = 3x + 7 + 9 \qquad \text{Add 9 to both sides.}$$
$$5x = 3x + 16 \qquad \text{Combine like terms.}$$
$$5x - 3x = 3x + 16 - 3x \qquad \text{Subtract } 3x \text{ from both sides.}$$
$$2x = 16 \qquad \text{Combine like terms.}$$
$$\frac{2x}{2} = \frac{16}{2} \qquad \text{Divide both sides by 2.}$$
$$x = 8 \qquad \text{Simplify.}$$

The solution set for this last equation is obvious:

$$\text{Solution set: } \{8\}$$

And since the equation $x = 8$ is equivalent to all the preceding equations in our solution, $\{8\}$ is also the solution set for all these equations, including the original equation. [*Note:* If an equation has only one element in its solution set, we generally use the last equation (in this case, $x = 8$) rather than set notation to represent the solution.]

Check

$$5x - 9 = 3x + 7 \qquad \text{Original equation}$$
$$5(8) - 9 \overset{?}{=} 3(8) + 7 \qquad \text{Substitute } x = 8.$$
$$40 - 9 \overset{?}{=} 24 + 7 \qquad \text{Simplify each side.}$$
$$31 \overset{\checkmark}{=} 31 \qquad \text{A true statement}$$

Matched Problem 1   Solve and check: $7x - 10 = 4x + 5$

A = lw

w

l

FIGURE 1 Area of a rectangle.

We frequently encounter equations involving more than one variable. For example, if l and w are the length and width of a rectangle, respectively, the area of the rectangle is given by (see Fig. 1).

$$A = lw$$

Depending on the situation, we may want to solve this equation for l or w. To solve for w, we simply consider A and l to be constants and w to be a variable. Then the equation $A = lw$ becomes a linear equation in w which can be solved easily by dividing both sides by l:

$$w = \frac{A}{l} \qquad l \neq 0$$

EXAMPLE 2   **Solving an Equation with More Than One Variable**

Solve for P in terms of the other variables: $A = P + Prt$

Solution

$$A = P + Prt \qquad \text{Think of } A, r, \text{ and } t \text{ as constants.}$$

$$A = P(1 + rt) \qquad \text{Factor to isolate } P.$$

$$\frac{A}{1 + rt} = P \qquad \text{Divide both sides by } 1 + rt.$$

$$P = \frac{A}{1 + rt} \qquad \text{Restriction: } 1 + rt \neq 0$$

Matched Problem 2   Solve for F in terms of C: $C = \frac{5}{9}(F - 32)$

● **A Strategy for Solving Word Problems**

A great many practical problems can be solved using algebraic techniques—so many, in fact, that there is no one method of attack that will work for all. However, we can formulate a strategy that will help you organize your approach.

> **Strategy for Solving Word Problems**
>
> 1. Read the problem carefully—several times if necessary—that is, until you understand the problem, know what is to be found, and know what is given.

> **2.** Let one of the unknown quantities be represented by a variable, say x, and try to represent all other unknown quantities in terms of x. This is an important step and must be done carefully.
>
> **3.** If appropriate, draw figures or diagrams and label known and unknown parts.
>
> **4.** Look for formulas connecting the known quantities to the unknown quantities.
>
> **5.** Form an equation relating the unknown quantities to the known quantities.
>
> **6.** Solve the equation and write answers to *all* questions asked in the problem.
>
> **7.** Check and interpret all solutions in terms of the original problem—not just the equation found in step 5—since a mistake may have been made in setting up the equation in step 5.

The remaining examples in this section contain solutions to a variety of word problems illustrating both the process of setting up word problems and the techniques used to solve the resulting equations. It is suggested that you cover up a solution, try solving the problem yourself, and uncover just enough of a solution to get you going again in case you get stuck. After successfully completing an example, try the matched problem. After completing the section in this way, you will be ready to attempt a fairly large variety of applications.

● **Number and Geometric Problems**

The first examples introduce the process of setting up and solving word problems in a simple mathematical context. Following these, the examples are of a more substantive nature.

EXAMPLE 3 **Setting Up and Solving a Word Problem**

Find four consecutive even integers such that the sum of the first three exceeds the fourth by 8.

Solution   Let $x =$ the first even integer, then

$$x \qquad x + 2 \qquad x + 4 \qquad \text{and} \qquad x + 6$$

represent four consecutive even integers starting with the even integer x. (Remember, even integers increase by 2.) The phrase "the sum of the first three exceeds the fourth by 8" translates into an equation:

$$\text{Sum of the first three} = \text{Fourth} + \text{Excess}$$
$$x + (x + 2) + (x + 4) = (x + 6) + 8$$
$$3x + 6 = x + 14$$
$$2x = 8$$
$$x = 4$$

The four consecutive integers are 4, 6, 8, and 10.

Check

$$4 + 6 + 8 = \underline{\begin{array}{r} 18 \\ -8 \\ \hline 10 \end{array}}$$

  Sum of first three
  Excess
  Fourth

Matched Problem 3   Find three consecutive odd integers such that 3 times their sum is 5 more than 8 times the middle one.

EXPLORE-DISCUSS 1   According to property 1 of Theorem 1, multiplying both sides of an equation by a nonzero number always produces an equivalent equation. By what number would you choose to multiply both sides of the following equation to eliminate all the fractions?

$$\frac{x + 1}{3} - \frac{x}{4} = \frac{1}{2}$$

If you did not choose 12, the LCD of all the fractions in this equation, you could still solve the resulting equation, but with more effort. (For a discussion of LCDs and how to find them, see Section 1-4.)

EXAMPLE 4   **Using a Diagram in the Solution of a Word Problem**

If one side of a triangle is one-third the perimeter, the second side is one-fifth the perimeter, and the third side is 7 meters, what is the perimeter of the triangle?

Solution   Let p = the perimeter. Draw a triangle and label the sides, as shown in Figure 2. Then

$$p = a + b + c$$

$$p = \frac{p}{3} + \frac{p}{5} + 7$$

$p = a + b + c$

7 meters

FIGURE 2

$$15 \cdot p = 15\left(\frac{p}{3} + \frac{p}{5} + 7\right) \qquad \text{Multiply both sides by 15, the LCD. This and the next step usually can be done mentally.}$$

$$15p = 15 \cdot \frac{p}{3} + 15 \cdot \frac{p}{5} + 15 \cdot 7$$

$$15p = 5p + 3p + 105$$

$$7p = 105$$

$$p = 15$$

The perimeter is 15 meters.

Check

$$\frac{p}{3} = \frac{15}{3} = 5 \qquad \text{Side 1}$$

$$\frac{p}{5} = \frac{15}{5} = 3 \qquad \text{Side 2}$$

$$\underline{\hspace{3cm} 7 \qquad \text{Side 3}}$$

$$15 \text{ meters} \qquad \text{Perimeter}$$

Matched Problem 4   If one side of a triangle is one-fourth the perimeter, the second side is 7 centimeters, and the third side is two-fifths the perimeter, what is the perimeter?

CAUTION   A very common error occurs about now—students tend to confuse *algebraic expressions* involving fractions with *algebraic equations* involving fractions. Consider these two problems:

(A) Solve: $\dfrac{x}{2} + \dfrac{x}{3} = 10$ (B) Add: $\dfrac{x}{2} + \dfrac{x}{3} + 10$

The problems look very much alike but are actually very different. To solve the equation in (A) we multiply both sides by 6 (the LCD) to clear the fractions. This works so well for equations that students want to do the same thing for problems like (B). The only catch is that (B) is not an equation, and the multiplication property of equality does not apply. If we multiply (B) by 6, we simply obtain an expression 6 times as large as the original! Compare the following:

(A) $\dfrac{x}{2} + \dfrac{x}{3} = 10$ (B) $\dfrac{x}{2} + \dfrac{x}{3} + 10$

$$\boxed{6 \cdot \frac{x}{2} + 6 \cdot \frac{x}{3} = 6 \cdot 10} \qquad \boxed{= \frac{3 \cdot x}{3 \cdot 2} + \frac{2 \cdot x}{2 \cdot 3} + \frac{6 \cdot 10}{6 \cdot 1}}$$

$$3x + 2x = 60 \qquad\qquad = \frac{3x}{6} + \frac{2x}{6} + \frac{60}{6}$$

$$5x = 60 \qquad\qquad\qquad$$

$$x = 12 \qquad\qquad\qquad = \frac{5x + 60}{6}$$

● Rate–Time Problems   There are many types of quantity–rate–time problems and distance–rate–time problems. In general, if Q is the quantity of something produced (kilometers, words, parts, and so on) in T units of time (hours, years, minutes, seconds, and so on), then the formulas given in the box are relevant.

Quantity–Rate–Time Formulas

$$R = \frac{Q}{T} \qquad \text{Rate} = \frac{\text{Quantity}}{\text{Time}}$$

$$Q = RT \qquad \text{Quantity} = (\text{Rate})(\text{Time})$$

$$T = \frac{Q}{R} \qquad \text{Time} = \frac{\text{Quantity}}{\text{Rate}}$$

If Q is distance D, then

$$R = \frac{D}{T} \qquad D = RT \qquad T = \frac{D}{R}$$

[*Note: R* is an average or uniform rate.]

EXAMPLE 5   **A Distance–Rate–Time Problem**

The distance along a shipping route between San Francisco and Honolulu is 2,100 nautical miles. If one ship leaves San Francisco at the same time another leaves Honolulu, and if the former travels at 15 knots* and the latter at 20 knots, how long will it take the two ships to rendezvous? How far will they be from Honolulu and San Francisco at that time?

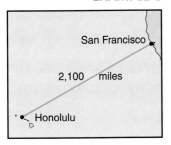

Solution   Let T = number of hours until both ships meet. Draw a diagram and label known and unknown parts. Both ships will have traveled the same amount of time when they meet.

$$D_1 = 20T \qquad D_2 = 15T$$

20 knots   15 knots

H   Meeting   SF

$$\begin{pmatrix} \text{Distance ship 1} \\ \text{from Honolulu} \\ \text{travels to} \\ \text{meeting point} \end{pmatrix} + \begin{pmatrix} \text{Distance ship 2} \\ \text{from San Francisco} \\ \text{travels to} \\ \text{meeting point} \end{pmatrix} = \begin{pmatrix} \text{Total distance} \\ \text{from Honolulu} \\ \text{to San Francisco} \end{pmatrix}$$

$$
\begin{array}{ccccc}
D_1 & + & D_2 & = & 2{,}100 \\
20T & + & 15T & = & 2{,}100 \\
    &   & 35T & = & 2{,}100 \\
    &   & T   & = & 60
\end{array}
$$

Therefore, it takes 60 hours, or 2.5 days, for the ships to meet.

*15 knots means 15 nautical miles per hour. There are 6,076.1 feet in 1 nautical mile.

$$\text{Distance from Honolulu} = 20 \cdot 60 = 1{,}200 \text{ nautical miles}$$

$$\text{Distance from San Francisco} = 15 \cdot 60 = 900 \text{ nautical miles}$$

Check

$$1{,}200 + 900 = 2{,}100 \text{ nautical miles}$$

Matched Problem 5
An old piece of equipment can print, stuff, and label 38 mailing pieces per minute. A newer model can handle 82 per minute. How long will it take for both pieces of equipment to prepare a mailing of 6,000 pieces? [*Note:* The mathematical form is the same as in Example 5.]

Some equations involving variables in a denominator can be transformed into linear equations. We may proceed in essentially the same way as in the preceding example; however, we must exclude any value of the variable that will make a denominator 0. With these values excluded, we may multiply through by the LCD even though it contains a variable, and, according to Theorem 1, the new equation will be equivalent to the old.

EXAMPLE 6   **A Distance–Rate–Time Problem**

An excursion boat takes 1.5 times as long to go 360 miles up a river than to return. If the boat cruises at 15 miles per hour in still water, what is the rate of the current?

Solution   Let

360 miles

$$x = \text{Rate of current (in miles per hour)}$$

$$15 - x = \text{Rate of boat upstream}$$

$$15 + x = \text{Rate of boat downstream}$$

$$\text{Time upstream} = (1.5)(\text{Time downstream})$$

$$\frac{\text{Distance upstream}}{\text{Rate upstream}} = (1.5)\frac{\text{Distance downstream}}{\text{Rate downstream}} \qquad \text{Recall: } T = \frac{D}{R}$$

$$\frac{360}{15 - x} = (1.5)\frac{360}{15 + x} \qquad x \neq 15,\ x \neq -15$$

$$\frac{360}{15 - x} = \frac{540}{15 + x}$$

$$360(15 + x) = 540(15 - x) \qquad \begin{array}{l}\text{Multiply both sides by}\\ (15 - x)(15 + x).\end{array}$$

$$5{,}400 + 360x = 8{,}100 - 540x$$

$$900x = 2{,}700$$

$$x = 3$$

Therefore, the rate of the current is 3 miles per hour. The check is left to the reader.

Matched Problem 6 A jetliner takes 1.2 times as long to fly from Paris to New York (3,600 miles) as to return. If the jet cruises at 550 miles per hour in still air, what is the average rate of the wind blowing in the direction of Paris from New York?

EXPLORE-DISCUSS 2 Consider the following solution:

$$\frac{x}{x-2} + 2 = \frac{2x-2}{x-2}$$

$$x + 2x - 4 = 2x - 2$$

$$x = 2$$

Is $x = 2$ a root of the original equation? If not, why? Discuss the importance of excluding values that make a denominator 0 when solving equations.

EXAMPLE 7 **A Quantity–Rate–Time Problem**

An advertising firm has an old computer that can prepare a whole mailing in 6 hours. With the help of a newer model the job is complete in 2 hours. How long would it take the newer model to do the job alone?

Solution Let x = time (in hours) for the newer model to do the whole job alone.

$$\left(\begin{array}{c}\text{Part of job completed}\\\text{in a given length of time}\end{array}\right) = (\text{Rate})(\text{Time})$$

$$\text{Rate of old model} = \frac{1}{6}\text{ job per hour}$$

$$\text{Rate of new model} = \frac{1}{x}\text{ job per hour}$$

$$\left(\begin{array}{c}\text{Part of job completed}\\\text{by old model}\\\text{in 2 hours}\end{array}\right) + \left(\begin{array}{c}\text{Part of job completed}\\\text{by new model}\\\text{in 2 hours}\end{array}\right) = 1\text{ whole job}$$

$$\left(\begin{array}{c}\text{Rate of}\\\text{old model}\end{array}\right)\left(\begin{array}{c}\text{Time of}\\\text{old model}\end{array}\right) + \left(\begin{array}{c}\text{Rate of}\\\text{new model}\end{array}\right)\left(\begin{array}{c}\text{Time of}\\\text{new model}\end{array}\right) = 1 \quad \text{Recall: } Q = RT$$

$$\frac{1}{6}(2) \qquad + \qquad \frac{1}{x}(2) \qquad = 1 \quad x \neq 0$$

$$\frac{1}{3} + \frac{2}{x} = 1$$

$$x + 6 = 3x$$

$$-2x = -6$$

$$x = 3$$

Therefore, the new computer could do the job alone in 3 hours.

Check

Part of job completed by old model in 2 hours $= 2(\frac{1}{6}) = \frac{1}{3}$
Part of job completed by new model in 2 hours $= 2(\frac{1}{3}) = \frac{2}{3}$
Part of job completed by both models in 2 hours $= \overline{1}$

Matched Problem 7   Two pumps are used to fill a water storage tank at a resort. One pump can fill the tank by itself in 9 hours, and the other can fill it in 6 hours. How long will it take both pumps operating together to fill the tank?

● **Mixture Problems**   A variety of applications can be classified as mixture problems. Even though the problems come from different areas, their mathematical treatment is essentially the same.

EXAMPLE 8   **A Mixture Problem**

How many liters of a mixture containing 80% alcohol should be added to 5 liters of a 20% solution to yield a 30% solution?

Solution   Let x = amount of 80% solution used.

$$\begin{pmatrix} \text{Amount of} \\ \text{alcohol in} \\ \text{first solution} \end{pmatrix} + \begin{pmatrix} \text{Amount of} \\ \text{alcohol in} \\ \text{second solution} \end{pmatrix} = \begin{pmatrix} \text{Amount of} \\ \text{alcohol in} \\ \text{mixture} \end{pmatrix}$$

$$0.8x \quad + \quad 0.2(5) \quad = \quad 0.3(x + 5)$$

$$0.8x + 1 = 0.3x + 1.5$$
$$0.5x = 0.5$$
$$x = 1$$

Add 1 liter of the 80% solution.

Check

	Liters of solution	Liters of alcohol	Percent alcohol
First solution	1	$0.8(1) = 0.8$	80
Second solution	5	$0.2(5) = 1$	20
Mixture	6	1.8	1.8/6 = 0.3, or 30%

Matched Problem 8

A chemical storeroom has a 90% acid solution and a 40% acid solution. How many centiliters of the 90% solution should be added to 50 centiliters of the 40% solution to yield a 50% solution?

• Some Final Observations on Linear Equations

It can be shown that any equation that can be written in the form

$$ax + b = 0 \qquad a \neq 0 \tag{1}$$

with no restrictions on x, has exactly one solution, and the solution can be found as follows:

$$ax + b = 0$$
$$ax = -b \qquad \text{Subtraction property of equality}$$
$$x = \frac{-b}{a} \qquad \text{Division property of equality}$$

Requiring $a \neq 0$ in equation (1) is an important restriction, because without it we are able to write equations with first-degree members that have no solutions or infinitely many solutions. For example,

$$2x - 3 = 2x + 5$$

has no solution, and

$$3x - 4 = 5 + 3(x - 3)$$

has infinitely many solutions. Try to solve each equation to see what happens.

Answers to Matched Problems

1. $x = 5$ **2.** $F = \frac{9}{5}C + 32$ **3.** 3, 5, 7   **4.** 20 centimeters
5. 50 minutes   **6.** 50 miles per hour   **7.** 3.6 hours   **8.** 12.5 centiliters

EXERCISE 2-1

A

In Problems 1–16, solve each equation.

1. $4(x + 5) = 6(x - 2)$ **2.** $3(y - 4) + 2y = 18$

3. $5 + 4(w - 1) = 2w + 2(w + 4)$

4. $4 - 3(t + 2) + t = 5(t - 1) - 7t$

5. $5 - \dfrac{3a - 4}{5} = \dfrac{7 - 2a}{2}$ **6.** $\dfrac{3b}{7} + \dfrac{2b - 5}{2} = -4$

7. $\dfrac{x}{2} + \dfrac{2x - 1}{3} = \dfrac{3x + 4}{4}$ **8.** $\dfrac{x}{5} + \dfrac{3x - 1}{2} = \dfrac{6x + 5}{4}$

9. $0.1(t + 0.5) + 0.2t = 0.3(t - 0.4)$

10. $0.1(w + 0.5) + 0.2w = 0.2(w - 0.4)$

11. $0.35(s + 0.34) + 0.15s = 0.2s - 1.66$

12. $0.35(u + 0.34) - 0.15u = 0.2u - 1.66$

13. $\dfrac{2}{y} + \dfrac{5}{2} = 4 - \dfrac{2}{3y}$ **14.** $\dfrac{3 + w}{6w} = \dfrac{1}{2w} + \dfrac{4}{3}$

15. $\dfrac{z}{z - 1} = \dfrac{1}{z - 1} + 2$ **16.** $\dfrac{t}{t - 1} = \dfrac{2}{t - 1} + 2$

B

In Problems 17–24, solve each equation.

17. $\dfrac{2m}{5} + \dfrac{m - 4}{6} = \dfrac{4m + 1}{4} - 2$

18. $\dfrac{3(n - 2)}{5} + \dfrac{2n + 3}{6} = \dfrac{4n + 1}{9} + 2$

19. $1 - \dfrac{x - 3}{x - 2} = \dfrac{2x - 3}{x - 2}$ **20.** $\dfrac{2x - 3}{x + 1} = 2 - \dfrac{3x - 1}{x + 1}$

21. $\dfrac{6}{y + 4} + 1 = \dfrac{5}{2y + 8}$ **22.** $\dfrac{4y}{y - 3} + 5 = \dfrac{12}{y - 3}$

23. $\dfrac{3a - 1}{a^2 + 4a + 4} - \dfrac{3}{a^2 + 2a} = \dfrac{3}{a}$

24. $\dfrac{1}{b - 5} - \dfrac{10}{b^2 - 5b + 25} = \dfrac{1}{b + 5}$

In Problems 25–28, use a calculator to solve each equation to 3 significant digits.

25. $3.142x - 0.4835(x - 4) = 6.795$

26. $0.0512x + 0.125(x - 2) = 0.725x$

27. $\dfrac{2.32x}{x - 2} - \dfrac{3.76}{x} = 2.32$ **28.** $\dfrac{6.08}{x} + 4.49 = \dfrac{4.49x}{x + 3}$

In Problems 29–36, solve for the indicated variable in terms of the other variables.

29. $a_n = a_1 + (n - 1)d$ for d (arithmetic progressions)

30. $F = \frac{9}{5}C + 32$ for C (temperature scale)

31. $\dfrac{1}{f} = \dfrac{1}{d_1} + \dfrac{1}{d_2}$ for f (simple lens formula)

32. $\dfrac{1}{R} = \dfrac{1}{R_1} + \dfrac{1}{R_2}$ for R_1 (electric circuit)

33. $A = 2ab + 2ac + 2bc$ for a (surface area of a rectangular solid)

34. $A = 2ab + 2ac + 2bc$ for c

35. $y = \dfrac{2x - 3}{3x + 5}$ for x **36.** $x = \dfrac{3y + 2}{y - 3}$ for y

In Problems 37 and 38, imagine that the indicated "solutions" were given to you by a student whom you were tutoring in this class. Is the solution right or wrong? If the solution is wrong, explain what is wrong and show a correct solution.

37.
$$\dfrac{x}{x - 3} + 4 = \dfrac{2x - 3}{x - 3}$$
$$x + 4x - 12 = 2x - 3$$
$$x = 3$$

38.
$$\dfrac{x^2 + 1}{x - 1} = \dfrac{x^2 + 4x - 3}{x - 1}$$
$$x^2 + 1 = x^2 + 4x - 3$$
$$x = 1$$

C

In Problems 39–41, solve for x.

39. $\dfrac{x - \dfrac{1}{x}}{1 + \dfrac{1}{x}} = 3$ **40.** $\dfrac{x - \dfrac{1}{x}}{x + 1 - \dfrac{2}{x}} = 1$

41. $\dfrac{x + 1 - \dfrac{2}{x}}{1 - \dfrac{1}{x}} = x + 2$

42. Solve for y in terms of x: $\dfrac{y}{1-y} = \left(\dfrac{x}{1-x}\right)^3$

43. Solve for x in terms of y: $y = \dfrac{a}{1 + \dfrac{b}{x+c}}$

44. Let m and n be real numbers with m larger than n. Then there exists a positive real number p such that $m = n + p$. Find the fallacy in the following argument:

$$m = n + p$$

$$(m-n)m = (m-n)(n+p)$$

$$m^2 - mn = mn + mp - n^2 - np$$

$$m^2 - mn - mp = mn - n^2 - np$$

$$m(m-n-p) = n(m-n-p)$$

$$m = n$$

APPLICATIONS

These problems are grouped according to subject area. As before, the most difficult problems are marked with two stars (★★), the moderately difficult problems are marked with one star (★), and the easier problems are not marked.

Numbers

45. Find a number such that 10 less than two-thirds the number is one-fourth the number.

46. Find a number such that 6 more than one-half the number is two-thirds the number.

47. Find four consecutive even integers so that the sum of the first three is 2 more than twice the fourth.

48. Find three consecutive even integers so that the first plus twice the second is twice the third.

Geometry

49. Find the dimensions of a rectangle with a perimeter of 54 meters, if its length is 3 meters less than twice its width.

50. A rectangle 24 meters long has the same area as a square that is 12 meters on a side. What are the dimensions of the rectangle?

51. Find the perimeter of a triangle if one side is 16 feet, another side is two-sevenths the perimeter, and the third side is one-third the perimeter.

52. Find the perimeter of a triangle if one side is 11 centimeters, another side is one-fifth the perimeter, and the third side is one-fourth the perimeter.

Business and Economics

53. The sale price on a camera after a 20% discount is $72. What was the price before the discount?

54. A stereo store marks up each item it sells 60% above wholesale price. What is the wholesale price on a cassette player that retails at $144?

55. One employee of a computer store is paid a base salary of $2,150 a month plus an 8% commission on all sales over $7,000 during the month. How much must the employee sell in 1 month to earn a total of $3,170 for the month?

56. A second employee of the computer store in Problem 55 is paid a base salary of $1,175 a month plus a 5% commission on all sales during the month.
  (A) How much must this employee sell in 1 month to earn a total of $3,170 for the month?
  (B) Determine the sales level where both employees receive the same monthly income. If employees can select either of these payment methods, how would you advise an employee to make this selection?

Earth Science

★ **57.** In 1984, the Soviets led the world in drilling the deepest hole in the Earth's crust—more than 12 kilometers deep. They found that below 3 kilometers the temperature T increased 2.5°C for each additional 100 meters of depth.
  (A) If the temperature at 3 kilometers is 30°C and x is the depth of the hole in kilometers, write an equation using x that will give the temperature T in the hole at any depth beyond 3 kilometers.
  (B) What would the temperature be at 15 kilometers? (The temperature limit for their drilling equipment was about 300°C.)
  (C) At what depth (in kilometers) would they reach a temperature of 280°C?

★ **58.** Because air is not as dense at high altitudes, planes require a higher ground speed to become airborne. A rule of thumb is 3% more ground speed per 1,000 feet of elevation, assuming no wind and no change in air temperature. (Compute numerical answers to 3 significant digits.)
  (A) Let

$$V_S = \text{Takeoff ground speed at sea level for a particular plane (in miles per hour)}$$

$$A = \text{Altitude above sea level (in thousands of feet)}$$

$$V = \text{Takeoff ground speed at altitude } A \text{ for the same plane (in miles per hour)}$$

   Write a formula relating these three quantities.
  (B) What takeoff ground speed would be required at Lake Tahoe airport (6,400 feet), if takeoff ground speed at San Francisco airport (sea level) is 120 miles per hour?

(C) If a landing strip at a Colorado Rockies hunting lodge (8,500 feet) requires a takeoff ground speed of 125 miles per hour, what would be the takeoff ground speed in Los Angeles (sea level)?

(D) If the takeoff ground speed at sea level is 135 miles per hour and the takeoff ground speed at a mountain resort is 155 miles per hour, what is the altitude of the mountain resort in thousands of feet?

★★ **59.** An earthquake emits a primary wave and a secondary wave. Near the surface of the Earth the primary wave travels at about 5 miles per second, and the secondary wave travels at about 3 miles per second. From the time lag between the two waves arriving at a given seismic station, it is possible to estimate the distance to the quake. Suppose a station measures a time difference of 12 seconds between the arrival of the two waves. How far is the earthquake from the station? (The *epicenter* can be located by obtaining distance bearings at three or more stations.)

★★ **60.** A ship using sound-sensing devices above and below water recorded a surface explosion 39 seconds sooner on its underwater device than on its above-water device. If sound travels in air at about 1,100 feet per second and in water at about 5,000 feet per second, how far away was the explosion?

Life Science

61. A naturalist for a fish and game department estimated the total number of trout in a certain lake using the popular capture–mark–recapture technique. She netted, marked, and released 200 trout. A week later, allowing for thorough mixing, she again netted 200 trout and found 8 marked ones among them. Assuming that the ratio of marked trout to the total number in the second sample is the same as the ratio of all marked fish in the first sample to the total trout population in the lake, estimate the total number of fish in the lake.

62. Repeat Problem 61 with a first (marked) sample of 300 and a second sample of 180 with only 6 marked trout.

Chemistry

★ **63.** How many gallons of distilled water must be mixed with 50 gallons of 30% alcohol solution to obtain a 25% solution?

★ **64.** How many gallons of hydrochloric acid must be added to 12 gallons of a 30% solution to obtain a 40% solution?

★ **65.** A chemist mixes distilled water with a 90% solution of sulfuric acid to produce a 50% solution. If 5 liters of distilled water is used, how much 50% solution is produced?

★ **66.** A fuel oil distributor has 120,000 gallons of fuel with 0.9% sulfur content, which exceeds pollution control standards of 0.8% sulfur content. How many gallons of fuel oil with a 0.3% sulfur content must be added to the 120,000 gallons to

obtain fuel oil that will comply with the pollution control standards?

Rate–Time

★ **67.** An old computer can do the weekly payroll in 5 hours. A newer computer can do the same payroll in 3 hours. The old computer starts on the payroll, and after 1 hour the newer computer is brought on-line to work with the older computer until the job is finished. How long will it take both computers working together to finish the job? (Assume the computers operate independently.)

★ **68.** One pump can fill a gasoline storage tank in 8 hours. With a second pump working simultaneously, the tank can be filled in 3 hours. How long would it take the second pump to fill the tank operating alone?

★★ **69.** The cruising speed of an airplane is 150 miles per hour (relative to the ground). You wish to hire the plane for a 3-hour sightseeing trip. You instruct the pilot to fly north as far as he can and still return to the airport at the end of the allotted time.

(A) How far north should the pilot fly if the wind is blowing from the north at 30 miles per hour?

(B) How far north should the pilot fly if there is no wind?

★ **70.** Suppose you are at a river resort and rent a motor boat for 5 hours starting at 7 A.M. You are told that the boat will travel at 8 miles per hour upstream and 12 miles per hour returning. You decide that you would like to go as far up the river as you can and still be back at noon. At what time should you turn back, and how far from the resort will you be at that time?

Music

★ **71.** A major chord in music is composed of notes whose frequencies are in the ratio 4:5:6. If the first note of a chord has a frequency of 264 hertz (middle C on the piano), find the frequencies of the other two notes. [*Hint:* Set up two proportions using 4:5 and 4:6.]

★ **72.** A minor chord is composed of notes whose frequencies are in the ratio 10:12:15. If the first note of a minor chord is A, with a frequency of 220 hertz, what are the frequencies of the other two notes?

Psychology

73. In an experiment on motivation, Professor Brown trained a group of rats to run down a narrow passage in a cage to receive food in a goal box. He then put a harness on each rat and connected it to an overhead wire attached to a scale. In this way he could place the rat different distances from the food and measure the pull (in grams) of the rat toward the food. He found that the relationship between motivation (pull) and position was given approximately by the equation

$$p = -\tfrac{1}{5}d + 70 \qquad 30 \le d \le 170$$

where pull p is measured in grams and distance d in centimeters. When the pull registered was 40 grams, how far was the rat from the goal box?

74. Professor Brown performed the same kind of experiment as described in Problem 73, except that he replaced the food in the goal box with a mild electric shock. With the same kind of apparatus, he was able to measure the avoidance strength relative to the distance from the object to be avoided. He found that the avoidance strength a (measured in grams) was related to the distance d that the rat was from the shock (measured in centimeters) approximately by the equation

$$a = -\tfrac{4}{3}d + 230 \qquad 30 \le d \le 170$$

If the same rat were trained as described in this problem and in Problem 73, at what distance (to one decimal place) from the goal box would the approach and avoidance strengths be the same? (What do you think the rat would do at this point?)

Puzzle

75. An oil-drilling rig in the Gulf of Mexico stands so that one-fifth of it is in sand, 20 feet of it is in water, and two-thirds of it is in the air. What is the total height of the rig?

76. During a camping trip in the North Woods in Canada, a couple went one-third of the way by boat, 10 miles by foot, and one-sixth of the way by horse. How long was the trip?

★★ **77.** After exactly 12 o'clock noon, what time will the hands of a clock be together again?

SECTION 2-2 Systems of Linear Equations and Applications

- Systems of Equations
- Substitution
- Applications

- **Systems of Equations**

In the preceding section, we solved word problems by introducing a single variable representing one of the unknown quantities and then tried to represent all other unknown quantities in terms of this variable. In certain word problems, it is more convenient to introduce several variables, find equations relating these variables, and then solve the resulting system of equations. For example, if a 12-foot board is cut into two pieces so that one piece is 2 feet longer than the other, then letting

$$x = \text{Length of the longer piece}$$
$$y = \text{Length of the shorter piece}$$

we see that x and y must satisfy both the following equations:

$$x + y = 12$$
$$x - y = 2$$

We now have a system of two linear equations in two variables. Thus, we can solve this problem by finding all pairs of numbers x and y that satisfy both equations.

In general, we are interested in solving linear systems of the type:

$$ax + by = h \qquad \text{System of two linear equations in two variables}$$
$$cx + dy = k$$

where x and y are variables and a, b, c, d, h, and k are real constants. A pair of numbers $x = x_0$ and $y = y_0$ is a **solution** of this system if each equation is satisfied by the pair. The set of all such pairs of numbers is called the **solution set** for the system. To **solve** a system is to find its solution set. In this section, we will restrict our discussion to simple solution techniques for systems of two linear equations in two variables. Larger systems and more sophisticated solution methods will be discussed in Chapter 6.

● **Substitution**   To solve a system by **substitution,** we first choose one of the two equations in a system and solve for one variable in terms of the other. (We make a choice that avoids fractions, if possible.) Then we substitute the result in the other equation and solve the resulting linear equation in one variable. Finally, we substitute this result back into the expression obtained in the first step to find the second variable. We illustrate this process by returning to the board problem stated at the beginning of the section.

EXAMPLE 1   **Solving a System by Substitution**

Solve the board problem by solving the system

$$x + y = 12$$
$$x - y = 2$$

Solution   Solve either equation for one variable and substitute into the remaining equation. We choose to solve the first equation for y in terms of x:

$$x + y = 12 \qquad \text{Solve the first equation for } y \text{ in terms of } x.$$
$$y = \underbrace{12 - x} \qquad \text{Substitute into the second equation.}$$

$$x - y = 2$$
$$x - (12 - x) = 2$$
$$x - 12 + x = 2$$
$$2x = 14$$
$$x = 7$$

Now, replace x with 7 in $y = 12 - x$:

$$y = 12 - x$$
$$y = 12 - 7$$
$$y = 5$$

Thus, the longer board measures 7 feet and the shorter board measures 5 feet.

Check
$$x + y = 12 \qquad x - y = 2$$
$$7 + 5 \overset{?}{=} 12 \qquad 7 - 5 \overset{?}{=} 2$$
$$12 \overset{\checkmark}{=} 12 \qquad 2 \overset{\checkmark}{=} 2$$

Matched Problem 1 Solve by substitution and check: $x - y = 3$
$$x + 2y = -3$$

EXAMPLE 2 **Solving a System by Substitution**

Solve by substitution and check: $2x - 3y = 7$
$$3x - y = 7$$

Solution To avoid fractions, we choose to solve the second equation for y:

$$3x - y = 7 \qquad \text{Solve for } y \text{ in terms of } x.$$
$$-y = -3x + 7$$
$$y = 3x - 7 \qquad \text{Substitute into first equation.}$$
$$2x - 3y = 7 \qquad \text{First equation}$$
$$2x - 3(3x - 7) = 7 \qquad \text{Solve for } x.$$
$$2x - 9x + 21 = 7$$
$$-7x = -14$$
$$x = 2 \qquad \text{Substitute } x = 2 \text{ in } y = 3x - 7.$$
$$y = 3x - 7$$
$$y = 3(2) - 7$$
$$y = -1$$

Thus, the solution is $x = 2$ and $y = -1$.

Check
$$2x - 3y = 7 \qquad\qquad 3x - y = 7$$
$$2(2) - 3(-1) \overset{?}{=} 7 \qquad 3(2) - (-1) \overset{?}{=} 7$$
$$7 \overset{\checkmark}{=} 7 \qquad\qquad 7 \overset{\checkmark}{=} 7$$

Matched Problem 2 Solve by substitution and check: $3x - 4y = 18$
$$2x + y = 1$$

EXPLORE-DISCUSS 1   Use substitution to solve each of the following systems. Discuss the nature of the solution sets you obtain.

$$x + 3y = 4 \qquad x + 3y = 4$$
$$2x + 6y = 7 \qquad 2x + 6y = 8$$

● **Applications**   The following examples illustrate the advantages of using systems of equations and substitution in solving word problems.

EXAMPLE 3   **Diet**

An individual wants to use milk and orange juice to increase the amount of calcium and vitamin A in her daily diet. An ounce of milk contains 38 milligrams of calcium and 56 micrograms* of vitamin A. An ounce of orange juice contains 5 milligrams of calcium and 60 micrograms of vitamin A. How many ounces of milk and orange juice should she drink each day to provide exactly 550 milligrams of calcium and 1,200 micrograms of vitamin A?

Solution   First we define the relevant variables:

$$x = \text{Number of ounces of milk}$$
$$y = \text{Number of ounces of orange juice}$$

Next we summarize the given information in a table. It is convenient to organize the tables so that the quantities represented by the variables correspond to columns in the table (rather than to rows), as shown.

	Milk	Orange Juice	Total Needed
Calcium	38	5	550
Vitamin A	56	60	1,200

Now we use the information in the table to form equations involving x and y:

$$\begin{pmatrix}\text{Calcium in } x \text{ oz} \\ \text{of milk}\end{pmatrix} \qquad \begin{pmatrix}\text{Calcium in } y \text{ oz} \\ \text{of orange juice}\end{pmatrix} \qquad \begin{pmatrix}\text{Total calcium} \\ \text{needed (mg)}\end{pmatrix}$$
$$38x \qquad + \qquad 5y \qquad = \qquad 550$$
$$\begin{pmatrix}\text{Vitamin A in } x \text{ oz} \\ \text{of milk}\end{pmatrix} \qquad \begin{pmatrix}\text{Vitamin A in } y \text{ oz} \\ \text{of orange juice}\end{pmatrix} \qquad \begin{pmatrix}\text{Total vitamin A} \\ \text{needed } (\mu g)\end{pmatrix}$$
$$56x \qquad + \qquad 60y \qquad = \qquad 1{,}200$$

*A microgram (μg) is one-millionth (10^{-6}) of a gram.

$$5y = 550 - 38x \qquad \text{Solve first equation for } y.$$

$$y = 110 - 7.6x \qquad\qquad\qquad\qquad (1)$$

$$56x + 60(110 - 7.6x) = 1,200 \qquad \text{Substitute for } y \text{ in second equation.}$$

$$56x + 6,600 - 456x = 1,200$$

$$-400x = -5,400$$

$$x = 13.5 \qquad \text{Substitute in (1).}$$

$$y = 110 - 7.6(13.5)$$

$$y = 7.4$$

Drinking 13.5 ounces of milk and 7.4 ounces of orange juice each day will provide the required amounts of calcium and vitamin A.

Check

$$38x + 5y = 550 \qquad\qquad 56x + 60y = 1,000$$

$$38(13.5) + 5(7.4) \overset{?}{=} 500 \qquad 56(13.5) + 60(7.4) \overset{?}{=} 1,200$$

$$500 \overset{\checkmark}{=} 500 \qquad\qquad\qquad 1,200 \overset{\checkmark}{=} 1,200$$

Matched Problem 3

An individual wants to use cottage cheese and yogurt to increase the amount of protein and calcium in his daily diet. An ounce of cottage cheese contains 3 grams of protein and 12 milligrams of calcium. An ounce of yogurt contains 1 gram of protein and 44 milligrams of calcium. How many ounces of cottage cheese and yogurt should he eat each day to provide exactly 57 grams of protein and 840 milligrams of calcium?

EXAMPLE 4   **Airspeed**

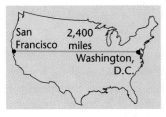

An airplane makes the 2,400-mile trip from Washington, D.C., to San Francisco in 7.5 hours and makes the return trip in 6 hours. Assuming that the plane travels at a constant airspeed and that the wind blows at a constant rate from west to east, find the plane's airspeed and the wind rate.

Solution

Let x represent the airspeed of the plane and let y represent the rate at which the wind is blowing (both in miles per hour). The ground speed of the plane is determined by combining these two rates; that is,

$$x - y = \text{Ground speed flying east to west (headwind)}$$

$$x + y = \text{Ground speed flying west to east (tailwind)}$$

Applying the familiar formula $D = RT$ to each leg of the trip leads to the following system of equations:

$$2,400 = 7.5(x - y) \qquad \text{From Washington to San Francisco}$$

$$2,400 = 6(x + y) \qquad \text{From San Francisco to Washington}$$

After simplification, we have

$$x - y = 320$$
$$x + y = 400$$

Solve using substitution:

$$x = y + 320 \qquad \text{Solve first equation for } x.$$
$$y + 320 + y = 400 \qquad \text{Substitute in second equation.}$$
$$2y = 80$$
$$y = 40 \text{ mph} \qquad \text{Wind rate}$$
$$x = 40 + 320$$
$$x = 360 \text{ mph} \qquad \text{Airspeed}$$

Check

$$2{,}400 = 7.5(x - y) \qquad\qquad 2{,}400 = 6(x + y)$$
$$2{,}400 \overset{?}{=} 7.5(360 - 40) \qquad 2{,}400 \overset{?}{=} 6(360 + 40)$$
$$2{,}400 \overset{\checkmark}{=} 2{,}400 \qquad\qquad 2{,}400 \overset{\checkmark}{=} 2{,}400$$

Matched Problem 4   A boat takes 8 hours to travel 80 miles upstream and 5 hours to return to its starting point. Find the speed of the boat in still water and the speed of the current.

EXAMPLE 5   **Supply and Demand**

The quantity of a product that people are willing to buy during some period of time depends on its price. Generally, the higher the price, the less the demand; the lower the price, the greater the demand. Similarly, the quantity of a product that a supplier is willing to sell during some period of time also depends on the price. Generally, a supplier will be willing to supply more of a product at higher prices and less of a product at lower prices. The simplest supply and demand model is a linear model.

Suppose we are interested in analyzing the sale of cherries each day in a particular city. Using special analytic techniques (regression analysis) and collected data, an analyst arrives at the following price–demand and price–supply equations:

$$p = -0.3q + 5 \qquad \text{Demand equation (consumer)}$$
$$p = 0.06q + 0.68 \qquad \text{Supply equation (supplier)}$$

where q represents the quantity in thousands of pounds and p represents the price in dollars. For example, we see that consumers will purchase 11 thousand pounds ($q = 11$) when the price is $p = -0.3(11) + 5 = \$1.70$ per pound. On the other hand, suppliers will be willing to supply 17 thousand pounds of cherries at \$1.70 per pound (solve $1.7 = 0.06q + 0.68$ for q). Thus, at \$1.70 per pound the suppliers are willing to supply more cherries than the consumers are willing to purchase. The supply exceeds the demand at that price, and the price will come down. At what price will cherries stabilize for the day? That is, at what price will supply equal demand? This price, if it exists, is called the **equilibrium price,** and the quantity sold at that price is called the **equilibrium quantity.** How do we find these quantities? We solve the linear system

$$p = -0.3q + 5 \qquad \text{Demand equation}$$

$$p = 0.06q + 0.68 \qquad \text{Supply equation}$$

using substitution (substituting $p = -0.3q + 5$ into the second equation).

$$-0.3q + 5 = 0.06q + 0.68$$

$$-0.36q = -4.32$$

$$q = 12 \text{ thousand pounds} \qquad \text{(equilibrium quantity)}$$

Now substitute $q = 12$ back into either of the original equations in the system and solve for p (we choose the second equation):

$$p = 0.06(12) + 0.68$$

$$p = \$1.40 \text{ per pound} \qquad \text{(equilibrium price)}$$

Matched Problem 5   The price–demand and price–supply equations for strawberries in a certain city are

$$p = -0.2q + 4 \qquad \text{Demand equation}$$

$$p = 0.04q + 1.84 \qquad \text{Supply equation}$$

where q represents the quantity in thousands of pounds and p represents the price in dollars.

Find the equilibrium quantity and the equilibrium price.

EXAMPLE 6   **Cost and Revenue**

A publisher is planning to produce a new textbook. The **fixed costs** (reviewing, editing, typesetting, and so on) are $320,000, and the **variable costs** (printing, sales commissions, and so on) are $31.25 per book. The wholesale price (the amount received by the publisher) will be $43.75 per book. How many books must the publisher sell to **break even;** that is, so that costs will equal revenues?

Solution   If x represents the number of books printed and sold, then the cost and revenue equations for the publisher are

$$y = 320,000 + 31.25x \qquad \text{Cost equation}$$

$$y = 43.75x \qquad \text{Revenue equation}$$

The publisher breaks even when costs equal revenues. We can find when this occurs by solving this system. Using the second equation to substitute for y in the first equation, we have

$$43.75x = 320,000 + 31.25x$$

$$12.5x = 320,000$$

$$x = 25,600$$

Thus, the publisher will break even when 25,600 books are printed and sold.

Matched Problem 6   A computer software company is planning to market a new word processor. The fixed costs (design, programming, and so on) are $720,000, and the variable costs (disk duplication, manual production, and so on) are $25.40 per copy. The wholesale price of the word processor will be $44.60 per copy. How many copies of the word processor must the company manufacture and sell to break even?

Answers to Matched Problems

1. $x = 1, y = -2$ **2.** $x = 2, y = -3$
3. 13.9 oz cottage cheese, 15.3 oz of yogurt
4. Boat: 13 mph; current: 3 mph
5. Equilibrium quantity = 9 thousand pounds; Equilibrium price = $2.20 per pound
6. 37,500 copies

EXERCISE **2-2**

A

Solve Problems 1–6 by substitution.

1. $y = 3x + 5$ **2.** $y = 5x - 8$ **3.** $4x + 5y = -8$
$\quad y = 4x + 7$ $\quad y = 2x + 4$ $\quad 3x + \ y = 5$

4. $2x + 7y = 4$ **5.** $6x + 11y = 16$ **6.** $\quad 9x + 7y = 8$
$\quad x + 3y = 1$ $\quad 2x - \ 3y = -8$ $\quad -4x + 3y = 27$

B

Solve Problems 7–16 by substitution.

7. $3s - \ 5t = -30$ **8.** $2s - \ 9t = 22$
$\quad 7s + 11t = 32$ $\quad 8s - 11t = 48$

9. $13m + 3n = 10$ **10.** $12m - 11n = 2$
$\quad 7m - 3n = -10$ $\quad 18m + \ 7n = 3$

11. $y = 5.46x$ **12.** $y = 7.15x$
$\quad y = 6,300 + 4.86x$ $\quad y = 13,860 + 3.85x$

13. $0.7u - 0.2v = 0.5$ **14.** $0.4u - 0.9v = 0.71$
$\quad 0.3u - 0.5v = 0.09$ $\quad 0.8u - 0.2v = 0.3$

15. $\frac{6}{5}a + \frac{3}{2}b = -4$ **16.** $\frac{3}{8}a - \frac{1}{2}b = 2$
$\quad \frac{1}{3}a - \frac{5}{4}b = 1$ $\quad \frac{9}{10}a + \frac{2}{5}b = 4$

17. In the process of solving a system by substitution, suppose you encounter a contradiction, such as $0 = 1$. How would you describe the solutions to such a system? Illustrate your ideas with the system

$$x - 2y = -3$$

$$-2x + 4y = 7$$

18. In the process of solving a system by substitution, suppose you encounter an identity, such as $0 = 0$. How would you describe the solutions to such a system? Illustrate your ideas with the system

$$x - 2y = -3$$
$$-2x + 4y = 6$$

C

In Problems 19 and 20, solve each system for p and q in terms of x and y. Explain how you could check your solution and then perform the check.

19. $x = 2 + p - 2q$
    $y = 3 - p + 3q$

20. $x = -1 + 2p - q$
    $y = \quad 4 - \quad p + q$

Problems 21 and 22 refer to the system

$$ax + by = h$$
$$cx + dy = k$$

where x and y are variables and a, b, c, d, h, and k are real constants.

21. Solve the system for x and y in terms of the constants a, b, c, d, h, and k. Clearly state any assumptions you must make about the constants during the solution process.

22. Discuss the nature of solutions to systems that do not satisfy the assumptions you made in Problem 21.

APPLICATIONS

23. **Airspeed.** It takes a private airplane 8.75 hours to make the 2,100-mile flight from Atlanta to Los Angeles and 5 hours to make the return trip. Assuming that the wind blows at a constant rate from Los Angeles to Atlanta, find the airspeed of the plane and the wind rate.

24. **Airspeed.** A plane carries enough fuel for 20 hours of flight at an airspeed of 150 miles per hour. How far can it fly into a 30 mph headwind and still have enough fuel to return to its starting point? (This distance is called the *point of no return.*)

25. **Rate–Time.** A crew of eight can row 20 kilometers per hour in still water. The crew rows upstream and then returns to its starting point in 15 minutes. If the river is flowing at 2 km/h, how far upstream did the crew row?

26. **Rate–Time.** It takes a boat 2 hours to travel 20 miles down a river and 3 hours to return upstream to its starting point. What is the rate of the current in the river?

27. **Chemistry.** A chemist has two solutions of hydrochloric acid in stock: a 50% solution and an 80% solution. How much of each should be used to obtain 100 milliliters of a 68% solution?

28. **Business.** A jeweler has two bars of gold alloy in stock, one of 12 carats and the other of 18 carats (24-carat gold is pure gold, 12-carat is $\frac{12}{24}$ pure, 18-carat gold is $\frac{18}{24}$ pure, and so on). How many grams of each alloy must be mixed to obtain 10 grams of 14-carat gold?

29. **Break-Even Analysis.** It costs a small recording company $17,680 to prepare a compact disc. This is a one-time fixed cost that covers recording, package design, and so on. Variable costs, including such things as manufacturing, marketing, and royalties, are $4.60 per CD. If the CD is sold to music shops for $8 each, how many must be sold for the company to break even?

30. **Break-Even Analysis.** A videocassette manufacturer has determined that its weekly cost equation is $C = 3,000 + 10x$, where x is the number of cassettes produced and sold each week. If cassettes are sold for $15 each to distributors, how many must be sold each week for the manufacturer to break even? (Refer to Problem 29.)

31. **Finance.** Suppose you have $12,000 to invest. If part is invested at 10% and the rest at 15%, how much should be invested at each rate to yield 12% on the total amount invested?

32. **Finance.** An investor has $20,000 to invest. If part is invested at 8% and the rest at 12%, how much should be invested at each rate to yield 11% on the total amount invested?

33. **Production.** A supplier for the electronics industry manufactures keyboards and screens for graphing calculators at plants in Mexico and Taiwan. The hourly production rates at each plant are given in the table. How many hours should each plant be operated to exactly fill an order for 4,000 keyboards and screens?

Plant	Keyboards	Screens
Mexico	40	32
Taiwan	20	32

34. **Production.** A company produces Italian sausages and bratwursts at plants in Green Bay and Sheboygan. The hourly production rates at each plant are given in the table. How many hours should each plant be operated to exactly fill an order for 62,250 Italian sausages and 76,500 bratwursts?

Plant	Italian sausage	Bratwurst
Green Bay	800	800
Sheboygan	500	1,000

35. Nutrition. Animals in an experiment are to be kept on a strict diet. Each animal is to receive, among other things, 20 grams of protein and 6 grams of fat. The laboratory technician is able to purchase two food mixes of the following compositions: Mix A has 10% protein and 6% fat; mix B has 20% protein and 2% fat. How many grams of each mix should be used to obtain the right diet for a single animal?

36. Nutrition. A fruit grower can use two types of fertilizer in an orange grove, brand A and brand B. Each bag of brand A contains 8 pounds of nitrogen and 4 pounds of phosphoric acid. Each bag of brand B contains 7 pounds of nitrogen and 7 pounds of phosphoric acid. Tests indicate that the grove needs 720 pounds of nitrogen and 500 pounds of phosphoric acid. How many bags of each brand should be used to provide the required amounts of nitrogen and phosphoric acid?

★ **37. Supply and Demand.** At $0.60 per bushel, the daily supply for wheat is 450 bushels and the daily demand is 645 bushels. When the price is raised to $0.90 per bushel, the daily supply increases to 750 bushels and the daily demand decreases to 495 bushels. Assume that the supply and demand equations are linear.
(A) Find the supply equation. [*Hint:* Write the supply equation in the form $p = aq + b$ and solve for a and b.]
(B) Find the demand equation.
(C) Find the equilibrium price and quantity.

★ **38. Supply and Demand.** At $1.40 per bushel, the daily supply for soybeans is 1,075 bushels and the daily demand is 580 bushels. When the price falls to $1.20 per bushel, the daily supply decreases to 575 bushels and the daily demand increases to 980 bushels. Assume that the supply and demand equations are linear.
(A) Find the supply equation. [See the hint in Problem 37.]
(B) Find the demand equation.
(C) Find the equilibrium price and quantity.

★ **39. Physics.** An object dropped off the top of a tall building falls vertically with constant acceleration. If s is the distance of the object above the ground (in feet) t seconds after its release, then s and t are related by an equation of the form

$$s = a + bt^2$$

where a and b are constants. Suppose the object is 180 feet above the ground 1 second after its release and 132 feet above the ground 2 seconds after its release.
(A) Find the constants a and b.
(B) How high is the building?
(C) How long does the object fall?

★ **40. Physics.** Repeat Problem 39 if the object is 240 feet above the ground after 1 second and 192 feet above the ground after 2 seconds.

★ **41. Earth Science.** An earthquake emits a primary wave and a secondary wave. Near the surface of the Earth the primary wave travels at about 5 miles per second and the secondary wave at about 3 miles per second. From the time lag between the two waves arriving at a given receiving station, it is possible to estimate the distance to the quake. (The *epicenter* can be located by obtaining distance bearings at three or more stations.) Suppose a station measured a time difference of 16 seconds between the arrival of the two waves. How long did each wave travel, and how far was the earthquake from the station?

★ **42. Earth Science.** A ship using sound-sensing devices above and below water recorded a surface explosion 6 seconds sooner by its underwater device than its above-water device. Sound travels in air at about 1,100 feet per second and in seawater at about 5,000 feet per second.
(A) How long did it take each sound wave to reach the ship?
(B) How far was the explosion from the ship?

SECTION 2-3 **Linear Inequalities**

- Inequality Relations and Interval Notation
- Solving Linear Inequalities
- Applications

We now turn to the problem of solving linear inequalities in one variable, such as

$$3(x - 5) \geq 5(x + 7) - 10 \qquad \text{and} \qquad -4 \leq 3 - 2x < 7$$

• **Inequality Relations and Interval Notation**

The above mathematical forms involve the **inequality,** or **order, relation**—that is, "less than" and "greater than" relations. Just as we use = to replace the words "is equal to," we use the **inequality symbols** < and > to represent "is less than" and "is greater than," respectively.

While it probably seems obvious to you that

$$2 < 4 \qquad 5 > 0 \qquad 25{,}000 > 1$$

are true, it may not seem as obvious that

$$-4 < -2 \qquad 0 > -5 \qquad -25{,}000 < -1$$

To make the inequality relation precise so that we can interpret it relative to all real numbers, we need a precise definition of the concept.

DEFINITION 1

$a < b$ **and** $b > a$

For a and b real numbers, we say that a **is less than** b or b **is greater than** a and write

$$a < b \qquad \text{or} \qquad b > a$$

if there exists a positive real number p such that $a + p = b$ (or equivalently, $b - a = p$).

We certainly expect that if a positive number is added to *any* real number, the sum is larger than the original. That is essentially what the definition states.

When we write

$$a \le b$$

we mean $a < b$ or $a = b$ and say a **is less than or equal to b.** When we write

$$a \ge b$$

we mean $a > b$ or $a = b$ and say a **is greater than or equal to b.**

The inequality symbols < and > have a very clear geometric interpretation on the real number line. If $a < b$, then a is to the left of b; if $c > d$, then c is to the right of d (Fig.1).

It is an interesting and useful fact that for any two real numbers a and b, either $a < b$, or $a > b$, or $a = b$. This is called the **trichotomy property** of real numbers.

The double inequality $a < x \le b$ means that $x > a$ *and* $x \le b$; that is, x is between a and b, including b but not including a. The set of all real numbers x satisfying the inequality $a < x \le b$ is called an **interval** and is represented by $(a, b]$. Thus,

$$(a, b] = \{x \mid a < x \le b\}*$$

FIGURE 1 $a < b, c > d.$

---

*In general, $\{x \mid P(x)\}$ represents the set of all x such that statement $P(x)$ is true. To express this set verbally, just read the vertical bar as "such that."

The number a is called the **left endpoint** of the interval, and the symbol "(" indicates that a is not included in the interval. The number b is called the **right endpoint** of the interval, and the symbol "]" indicates that b is included in the interval. Other types of intervals of real numbers are shown in Table 1.

TABLE 1 Interval Notation

Interval notation	Inequality notation	Line graph	Type
$[a, b]$	$a \leq x \leq b$		Closed
$[a, b)$	$a \leq x < b$		Half-open
$(a, b]$	$a < x \leq b$		Half-open
(a, b)	$a < x < b$		Open
$[b, \infty)$	$x \geq b$		Closed
(b, ∞)	$x > b$		Open
$(-\infty, a]$	$x \leq a$		Closed
$(-\infty, a)$	$x < a$		Open

Note that the symbol "∞," read "infinity," used in Table 1 is not a numeral. When we write $[b, \infty)$, we are simply referring to the interval starting at b and continuing indefinitely to the right. We would never write $[b, \infty]$ or $b \leq x \leq \infty$, since ∞ cannot be used as an endpoint of an interval. The interval $(-\infty, \infty)$ represents the set of real numbers R, since its graph is the entire real number line.

CAUTION   It is important to note that

$$5 > x \geq -3 \quad \text{is equivalent to } [-3, 5) \text{ and not to } (5, -3]$$

In interval notation, the smaller number is always written to the left. Thus, it may be useful to rewrite the inequality as $-3 \leq x < 5$ before rewriting it in interval notation.

EXAMPLE 1   **Graphing Intervals and Inequalities**

Write each of the following in inequality notation and graph on a real number line:

(A) $[-2, 3)$ (B) $(-4, 2)$ (C) $[-2, \infty)$ (D) $(-\infty, 3)$

Solutions   (A) $-2 \leq x < 3$

(B) $-4 < x < 2$

(C) $x \geq -2$

(D) $x < 3$

Matched Problem 1   Write each of the following in interval notation and graph on a real number line:

(A) $-3 < x \leq 3$ (B) $2 \geq x \geq -1$ (C) $x > 1$ (D) $x \leq 2$

EXPLORE-DISCUSS 1   Example 1C shows the graph of the inequality $x \geq -2$. What is the graph of $x < -2$? What is the corresponding interval? Describe the relationship between these sets.

Since intervals are sets of real numbers, the set operations of *union* and *intersection* are often useful when working with intervals. The **union** of sets A and B, denoted by $A \cup B$, is the set formed by combining all the elements of A and all the elements of B. The **intersection** of sets A and B, denoted by $A \cap B$, is the set of elements of A that are also in B. Symbolically:

DEFINITION 2   **Union and Intersection**

Union: $A \cup B = \{x \mid x \text{ is in } A \text{ or } x \text{ is in } B\}$
$\{1, 2, 3\} \cup \{2, 3, 4, 5\} = \{1, 2, 3, 4, 5\}$

Intersection: $A \cap B = \{x \mid x \text{ is in } A \text{ and } x \text{ is in } B\}$
$\{1, 2, 3\} \cap \{2, 3, 4, 5\} = \{2, 3\}$

EXAMPLE 2   **Graphing Unions and Intersections of Intervals**

If $A = [-2, 3]$, $B = (1, 6)$, and $C = (4, \infty)$, graph the indicated sets and write as a single interval, if possible:

(A) $A \cup B$ and $A \cap B$ (B) $A \cup C$ and $A \cap C$

Solution   (A)

$A = [-2, 3]$

$B = (1, 6)$

$A \cup B = [-2, 6)$

$A \cap B = (1, 3]$

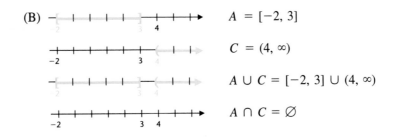

(B)

$A = [-2, 3]$

$C = (4, \infty)$

$A \cup C = [-2, 3] \cup (4, \infty)$

$A \cap C = \varnothing$

Matched Problem 2   If $D = [-4, 1)$, $E = (-1, 3]$, and $F = [2, \infty)$, graph the indicated sets and write as a single interval, if possible:

(A) $D \cup E$ (B) $D \cap E$ (C) $E \cup F$ (D) $E \cap F$

EXPLORE-DISCUSS 2   Replace ? with $<$ or $>$ in each of the following.

(A) $-1 \; ? \; 3$ and $2(-1) \; ? \; 2(3)$

(B) $-1 \; ? \; 3$ and $-2(-1) \; ? \; -2(3)$

(C) $12 \; ? \; -8$ and $\dfrac{12}{4} \; ? \; \dfrac{-8}{4}$

(D) $12 \; ? \; -8$ and $\dfrac{12}{-4} \; ? \; \dfrac{-8}{-4}$

Based on these examples, describe verbally the effect of multiplying both sides of an inequality by a number.

• **Solving Linear Inequalities**

We now turn to the problem of solving linear inequalities in one variable, such as

$$2(2x + 3) < 6(x - 2) + 10 \qquad \text{and} \qquad -3 < 2x + 3 \leq 9$$

The **solution set** for an inequality is the set of all values of the variable that make the inequality a true statement. Each element of the solution set is called a **solution** of the inequality. To **solve an inequality** is to find its solution set. Two inequalities are **equivalent** if they have the same solution set for a given replacement set. Just as with equations, we perform operations on inequalities that produce simpler equivalent inequalities, and continue the process until an inequality is reached whose solution is obvious. The properties of inequalities given in Theorem 1 can be used to produce equivalent inequalities.

Theorem 1   **Inequality Properties**

For a, b, and c any real numbers:

1. If $a < b$ and $b < c$, then $a < c$.   **Transitive Property**
2. If $a < b$, then $a + c < b + c$.   **Addition Property**
   $$-2 < 4 \qquad -2 + 3 < 4 + 3$$
3. If $a < b$, then $a - c < b - c$.   **Subtraction Property**
   $$-2 < 4 \qquad -2 - 3 < 4 - 3$$
4. If $a < b$ and c is positive, then $ca < cb$.   **Multiplication Property**
   $$-2 < 4 \qquad\qquad 3(-2) < 3(4)$$ **(Note difference between**
5. If $a < b$ and c is negative, then $ca > cb$.   **4 and 5.)**
   $$-2 < 4 \qquad\qquad (-3)(-2) > (-3)(4)$$

6. If $a < b$ and c is positive, then $\dfrac{a}{c} < \dfrac{b}{c}$.
   $$-2 < 4 \qquad\qquad \frac{-2}{2} < \frac{4}{2}$$ **Division Property**
                                                              (Note difference between
7. If $a < b$ and c is negative, then $\dfrac{a}{c} > \dfrac{b}{c}$.   **6 and 7.)**
   $$-2 < 4 \qquad\qquad \frac{-2}{-2} > \frac{4}{-2}$$

Similar properties hold if each inequality sign is reversed, or if $<$ is replaced with $\leq$ and $>$ is replaced with $\geq$. Thus, we find that we can perform essentially the same operations on inequalities that we perform on equations. When working with inequalities, however, we have to be particularly careful of the use of the multiplication and division properties.

The order of the inequality reverses if we multiply or divide both sides of an inequality statement by a negative number.

EXPLORE-DISCUSS 3   Properties of equality are easily summarized: We can add, subtract, multiply, or divide both sides of an equation by any nonzero real number to produce an equivalent equation. Write a similar summary for the properties of inequalities.

Now let's see how the inequality properties are used to solve linear inequalities. Several examples will illustrate the process.

EXAMPLE 3   **Solving a Linear Inequality**

Solve and graph: $2(2x + 3) - 10 < 6(x - 2)$

Solution

$$2(2x + 3) - 10 < 6(x - 2)$$

$$4x + 6 - 10 < 6x - 12 \qquad \text{Simplify left and right sides.}$$

$$4x - 4 < 6x - 12$$

$$\boxed{4x - 4 + 4 < 6x - 12 + 4} \qquad \text{Addition property}$$

$$4x < 6x - 8$$

$$\boxed{4x - 6x < 6x - 8 - 6x} \qquad \text{Subtraction property}$$

$$-2x < -8$$

$$\boxed{\frac{-2x}{-2} > \frac{-8}{-2}} \qquad \begin{array}{l}\text{Division property—note}\\\text{that order reverses}\\\text{since } -2 \text{ is negative}\end{array}$$

$$x > 4 \qquad \text{or} \qquad (4, \infty) \qquad \text{Solution set}$$

Graph of solution set

Matched Problem 3   Solve and graph: $3(x - 1) \geq 5(x + 2) - 5$

EXAMPLE 4   Solving a Linear Inequality Involving Fractions

Solve and graph: $\dfrac{2x - 3}{4} + 6 \geq 2 + \dfrac{4x}{3}$

Solution

$$\frac{2x - 3}{4} + 6 \geq 2 + \frac{4x}{3}$$

$$\boxed{12 \cdot \frac{2x - 3}{4} + 12 \cdot 6 \geq 12 \cdot 2 + 12 \cdot \frac{4x}{3}} \qquad \begin{array}{l}\text{Multiply both sides by 12,}\\\text{the LCD.}\end{array}$$

$$3(2x - 3) + 72 \geq 24 + 4(4x)$$

$$6x - 9 + 72 \geq 24 + 16x$$

$$6x + 63 \geq 24 + 16x$$

$$-10x \geq -39$$

$$x \leq 3.9 \qquad \text{or} \qquad (-\infty, 3.9] \qquad \begin{array}{l}\text{Order reverses since both}\\\text{sides are divided by } -10, \text{ a}\\\text{negative number.}\end{array}$$

SECTION 2-4 Absolute Value in Equations and Inequalities

- Absolute Value and Distance
- Absolute Value in Equations and Inequalities
- Absolute Value and Radicals

This section discusses solving absolute value equations and inequalities.

Absolute Value and Distance

We start with a geometric definition of absolute value. If a is the coordinate of a point on a real number line, then the distance from the origin to a is represented by $|a|$ and is referred to as the **absolute value** of a. Thus, $|5| = 5$, since the point with coordinate 5 is five units from the origin, and $|-6| = 6$, since the point with coordinate -6 is six units from the origin (Fig.1).

FIGURE 1 Absolute value.

Symbolically, and more formally, we define absolute value as follows:

DEFINITION 1   **Absolute Value**

$$|x| = \begin{cases} x & \text{if } x \geq 0 \\ -x & \text{if } x < 0 \end{cases} \qquad \begin{matrix} |4| = 4 \\ |-3| = -(-3) = 3 \end{matrix}$$

[*Note:* $-x$ is positive if x is negative.]

Both the geometric and nongeometric definitions of absolute value are useful, as will be seen in the material that follows. Remember:

The absolute value of a number is never negative.

EXAMPLE 1   **Absolute Value of a Real Number**

(A) $|\pi - 3| = \pi - 3$ Since $\pi \approx 3.14$, $\pi - 3$ is positive.

(B) $|3 - \pi| = -(3 - \pi) = \pi - 3$ Since $3 - \pi$ is negative

---

Matched Problem 1   Write without the absolute value sign:

(A) $|8|$ (B) $|\sqrt[3]{9} - 2|$ (C) $|-\sqrt{2}|$ (D) $|2 - \sqrt[3]{9}|$

Following the same reasoning used in Example 1, the next theorem can be proved (see Problem 79 in Exercise 2-4).

Theorem 1

For all real numbers a and b,

$$|b - a| = |a - b|$$

We use this result in defining the distance between two points on a real number line.

DEFINITION 2

Distance between Points A and B

Let A and B be two points on a real number line with coordinates a and b, respectively. The **distance between A and B** is given by

$$d(A, B) = |b - a|$$

This distance is also called the **length of the line segment** joining A and B.

EXAMPLE 2   **Distance between Points on a Number Line**

Find the distance between points A and B with coordinates a and b, respectively, as given.

(A) $a = 4,\ b = 9$ (B) $a = 9,\ b = 4$ (C) $a = 0,\ b = 6$

Solutions

(A)

$$d(A, B) = |9 - 4| = |5| = 5$$

(B)

$$d(A, B) = |4 - 9| = |-5| = 5$$

(C)

$$d(A, B) = |6 - 0| = |6| = 6$$

It should be clear, since $|b - a| = |a - b|$, that

$$d(A, B) = d(B, A)$$

Thus, for x any real number,

$$\sqrt{x^2} = \begin{cases} x & \text{if } x \geq 0 \\ -x & \text{if } x < 0 \end{cases}$$

But this is exactly how we defined $|x|$ at the beginning of this section (see Definition 1). Thus, for x any real number,

$$\sqrt{x^2} = |x|$$

Answers to Matched Problems

1. (A) 8   (B) $\sqrt[3]{9} - 2$ (C) $\sqrt{2}$ (D) $\sqrt[3]{9} - 2$
2. (A) 4   (B) 4   (C) 6   (D) 11   (E) 8   (F) 15
3. (A) x is a number whose distance from -2 is 6.
      $x = -8, 4$ or $\{-8, 4\}$
   (B) x is a number whose distance from -2 is less than 6.
      $-8 < x < 4$ or $(-8, 4)$
   (C) x is a number whose distance from -2 is less than 6, but x cannot equal -2.
      $-8 < x < 4$, $x \neq -2$, or $(-8, -2) \cup (-2, 4)$
   (D) x is a number whose distance from -2 is greater than 6.
      $x < -8$ or $x > 4$, or $(-\infty, -8) \cup (4, \infty)$
4. (A) $x = -\frac{7}{2}, \frac{9}{2}$ or $\{-\frac{7}{2}, \frac{9}{2}\}$ (B) $-7 \leq x \leq 7$ or $[-7, 7]$ (C) $-4 \leq x \leq 2$ or $[-4, 2]$
   (D) $-2 < x < 7$ or $(-2, 7)$
5. (A) $x \leq -5$ or $x \geq 5$, or $(-\infty, -5] \cup [5, \infty)$ (B) $x < -\frac{1}{2}$ or $x > 2$, or $(-\infty, -\frac{1}{2}) \cup (2, \infty)$
   (C) $x < -2$ or $x > \frac{22}{5}$, or $(-\infty, -2) \cup (\frac{22}{5}, \infty)$
6. $x = -\frac{1}{4}, \frac{9}{2}$ or $\{-\frac{1}{4}, \frac{9}{2}\}$

EXERCISE 2-4

A

In Problems 1–8, simplify, and write without absolute value signs. Do not replace radicals with decimal approximations.

1. $|\sqrt{5}|$

2. $\left|-\frac{3}{4}\right|$

3. $|(-6) - (-2)|$

4. $|(-2) - (-6)|$

5. $|5 - \sqrt{5}|$

6. $|\sqrt{7} - 2|$

7. $|\sqrt{5} - 5|$

8. $|2 - \sqrt{7}|$

In Problems 9–12, find the distance between points A and B with coordinates a and b respectively, as given.

9. $a = -7$, $b = 5$

10. $a = 3$, $b = 12$

11. $a = 5$, $b = -7$

12. $a = -9$, $b = -17$

In Problems 13–18, use the number line below to find the indicated distances.

13. $d(B, O)$

14. $d(A, B)$

15. $d(O, B)$

16. $d(B, A)$

17. $d(B, C)$

18. $d(D, C)$

Write each of the statements in Problems 19–28 as an absolute value equation or inequality.

19. x is 4 units from 3.

20. y is 3 units from 1.

21. m is 5 units from -2.

22. n is 7 units from -5.

23. x is less than 5 units from 3.

24. z is less than 8 units from -2.

25. p is more than 6 units from -2.

26. c is no greater than 7 units from -3.

27. q is no less than 2 units from 1.

28. d is no more than 4 units from 5.

B

In Problems 29–44, solve, interpret geometrically, and graph. When applicable, write answers using both inequality notation and interval notation.

29. $|x| \leq 7$ 30. $|t| \leq 5$ 31. $|x| \geq 7$

32. $|x| \geq 5$ 33. $|y - 5| = 3$ 34. $|t - 3| = 4$

35. $|y - 5| < 3$ 36. $|t - 3| < 4$ 37. $|y - 5| > 3$

38. $|t - 3| > 4$ 39. $|u + 8| = 3$ 40. $|x + 1| = 5$

41. $|u + 8| \leq 3$ 42. $|x + 1| \leq 5$ 43. $|u + 8| \geq 3$

44. $|x + 1| \geq 5$

In Problems 45–62, solve each equation or inequality. When applicable, write answers using both inequality notation and interval notation.

45. $|3x - 7| \leq 4$ 46. $|5y + 2| \geq 8$

47. $|4 - 2t| > 6$ 48. $|10 + 4s| < 6$

49. $|7m + 11| = 3$ 50. $|4 - 5n| \leq 8$

51. $|\frac{1}{2}w - \frac{3}{4}| < 2$ 52. $|\frac{1}{3}z + \frac{5}{6}| = 1$

53. $|0.2u + 1.7| \geq 0.5$ 54. $|0.5v - 2.5| > 1.6$

55. $|\frac{9}{5}C + 32| < 31$ 56. $|\frac{5}{9}(F - 32)| < 40$

57. $\sqrt{x^2} < 2$ 58. $\sqrt{m^2} > 3$

59. $\sqrt{(1 - 3t)^2} \leq 2$ 60. $\sqrt{(3 - 2x)^2} < 5$

61. $\sqrt{(2t - 3)^2} > 3$ 62. $\sqrt{(3m + 5)^2} \geq 4$

C

∫ Problems 63–66 are calculus-related. Solve and graph. Write each solution using interval notation.

63. $0 < |x - 3| < 0.1$ 64. $0 < |x - 5| < 0.01$

65. $0 < |x - c| < d$ 66. $0 < |x - 4| < d$

In Problems 67–76, for what values of x does each hold?

67. $|x - 2| = x - 2$ 68. $|x + 4| = -(x + 4)$

69. $|2x - 3| = 3 - 2x$ 70. $|3x - 9| = 3x - 9$

71. $|3x + 5| = 2x + 6$ 72. $|7 - 2x| = 5 - x$

73. $|x| + |x + 3| = 3$ 74. $|x| - |x - 5| = 5$

75. $|2x + 7| - |6 - 3x| = 8$ 76. $|3x + 1| + |3 - 2x| = 11$

∫ 77. What are the possible values of $\dfrac{x}{|x|}$?

∫ 78. What are the possible values of $\dfrac{|x - 1|}{x - 1}$?

79. Prove that $|b - a| = |a - b|$ for all real numbers a and b.

80. Prove that $|x|^2 = x^2$ for all real numbers x.

81. Prove that the average of two numbers is between the two numbers; that is, if $m < n$, then

$$m < \frac{m + n}{2} < n$$

82. Prove that for $m < n$,

$$d\left(m, \frac{m + n}{2}\right) = d\left(\frac{m + n}{2}, n\right)$$

83. Prove that $|-m| = |m|$.

84. Prove that $|m| = |n|$ if and only if $m = n$ or $m = -n$.

85. Prove that for $n \neq 0$,

$$\left|\frac{m}{n}\right| = \frac{|m|}{|n|}$$

86. Prove that $|mn| = |m||n|$.

87. Prove that $-|m| \leq m \leq |m|$.

88. Prove the **triangle inequality:**

$$|m + n| \leq |m| + |n|$$

Hint: Use Problem 87 to show that

$$-|m| - |n| \leq m + n \leq |m| + |n|$$

89. If a and b are real numbers, prove that the maximum of a and b is given by

$$\max(a, b) = \tfrac{1}{2}[a + b + |a - b|]$$

90. Prove that the minimum of a and b is given by

$$\min(a, b) = \tfrac{1}{2}[a + b - |a - b|]$$

APPLICATIONS

91. **Statistics.** Inequalities of the form

$$\left|\frac{x - m}{s}\right| < n$$

occur frequently in statistics. If $m = 45.4$, $s = 3.2$, and $n = 1$, solve for x.

92. **Statistics.** Repeat Problem 91 for $m = 28.6$, $s = 6.5$, and $n = 2$.

★ **93. Business.** The daily production P in an automobile assembly plant is within 20 units of 500 units. Express the daily production as an absolute value inequality.

★ **94. Chemistry.** In a chemical process, the temperature T is to be kept within 10°C of 200°C. Express this restriction as an absolute value inequality.

95. Approximation. The area A of a region is approximately equal to 12.436. The error in this approximation is less than 0.001. Describe the possible values of this area both with an absolute value inequality and with interval notation.

96. Approximation. The volume V of a solid is approximately

equal to 6.94. The error in this approximation is less than 0.02. Describe the possible values of this volume both with an absolute value inequality and with interval notation.

★ **97. Significant Digits.** If $N = 2.37$ represents a measurement, then we assume an accuracy of 2.37 ± 0.005. Express the accuracy assumption using an absolute value inequality.

★ **98. Significant Digits.** If $N = 3.65 \times 10^{-3}$ is a number from a measurement, then we assume an accuracy of $3.65 \times 10^{-3} \pm 5 \times 10^{-6}$. Express the accuracy assumption using an absolute value inequality.

SECTION 2-5 Complex Numbers

- Introductory Remarks
- The Complex Number System
- Complex Numbers and Radicals

● **Introductory Remarks**

The Pythagoreans (500–275 B.C.) found that the simple equation

$$x^2 = 2 \tag{1}$$

had no rational number solutions. If equation (1) were to have a solution, then a new kind of number had to be invented—an irrational number. The irrational numbers $\sqrt{2}$ and $-\sqrt{2}$ are both solutions to equation (1). Irrational numbers were not put on a firm mathematical foundation until the nineteenth century. The rational and irrational numbers together constitute the real number system.

Is there any need to consider another number system? Yes, if we want the simple equation

$$x^2 = -1$$

to have a solution. If x is any real number, then $x^2 \geq 0$. Thus, $x^2 = -1$ cannot have any real number solutions. Once again a new type of number must be invented, a number whose square can be negative. These new numbers are among the numbers called *complex numbers*. The complex numbers evolved over a long period of time, but, like the real numbers, it was not until the nineteenth century that they were placed on a firm mathematical foundation.

● **The Complex Number System**

We start the development of the complex number system by defining a complex number and several special types of complex numbers. We then define equality, addition, and multiplication in this system, and from these definitions the important special properties and operational rules for addition, subtraction, multiplication, and division will follow.

TABLE 1	Brief History of Complex Numbers	

Approximate date	Person	Event
50	Heron of Alexandria	First recorded encounter of a square root of a negative number
850	Mahavira of India	Said that a negative has no square root, since it is not a square
1545	Cardano of Italy	Solutions to cubic equations involved square roots of negative numbers.
1637	Descartes of France	Introduced the terms *real* and *imaginary*
1748	Euler of Switzerland	Used i for $\sqrt{-1}$
1832	Gauss of Germany	Introduced the term *complex number*

DEFINITION 1   **Complex Number**

A **complex number** is a number of the form

$$a + bi \qquad \textbf{Standard Form}$$

where a and b are real numbers and i is called the **imaginary unit.**

The imaginary unit i introduced in Definition 1 is not a real number. It is a special symbol used in the representation of the elements in this new complex number system.

Some examples of complex numbers are

$$3 - 2i \qquad \tfrac{1}{2} + 5i \qquad 2 - \tfrac{1}{3}i$$
$$0 + 3i \qquad 5 + 0i \qquad 0 + 0i$$

Particular kinds of complex numbers are given special names as follows:

DEFINITION 2   **Names for Particular Kinds of Complex Numbers**

Imaginary Unit: $\qquad\qquad i$

Complex Number: $\qquad\quad a + bi \qquad a$ and b real numbers

Imaginary Number: $\qquad a + bi \qquad b \neq 0$

Pure Imaginary Number: $\quad 0 + bi = bi \qquad b \neq 0$

Real Number: $\qquad\qquad a + 0i = a$

Zero: $\qquad\qquad\qquad 0 + 0i = 0$

Conjugate of $a + bi$: $\qquad a - bi$

EXAMPLE 1   **Special Types of Complex Numbers**

Given the list of complex numbers:

$$3 - 2i \qquad \tfrac{1}{2} + 5i \qquad 2 - \tfrac{1}{3}i$$

$$0 + 3i = 3i \qquad 5 + 0i = 5 \qquad 0 + 0i = 0$$

(A) List all the imaginary numbers, pure imaginary numbers, real numbers, and zero.
(B) Write the conjugate of each.

Solutions   (A) Imaginary numbers: $3 - 2i, \tfrac{1}{2} + 5i, 2 - \tfrac{1}{3}i, 3i$
Pure imaginary numbers: $0 + 3i = 3i$
Real numbers: $5 + 0i = 5, 0 + 0i = 0$
Zero: $0 + 0i = 0$
(B) $3 + 2i$ $\qquad\qquad \tfrac{1}{2} - 5i \qquad\qquad 2 + \tfrac{1}{3}i$
$0 - 3i = -3i \qquad 5 - 0i = 5 \qquad 0 - 0i = 0$

Matched Problem 1   Given the list of complex numbers:

$$6 + 7i \qquad\qquad \sqrt{2} - \tfrac{1}{3}i \qquad\qquad 0 - i = -i$$

$$0 + \tfrac{2}{3}i = \tfrac{2}{3}i \qquad -\sqrt{3} + 0i = -\sqrt{3} \qquad 0 - 0i = 0$$

(A) List all the imaginary numbers, pure imaginary numbers, real numbers, and zero.
(B) Write the conjugate of each.

In Definition 2, notice that we identify a complex number of the form $a + 0i$ with the real number a, a complex number of the form $0 + bi, b \neq 0$, with the **pure imaginary number** bi, and the complex number $0 + 0i$ with the real number 0. Thus, a real number is also a complex number, just as a rational number is also a real number. Any complex number that is not a real number is called an **imaginary number.** If we combine the set of all real numbers with the set of all imaginary numbers, we obtain C, **the set of complex numbers.** The relationship of the complex number system to the other number systems we have studied is shown in Figure 1.

FIGURE 1 Complex numbers and important subsets

To use complex numbers, we must know how to add, subtract, multiply, and divide them. We start by defining equality, addition, and multiplication.

DEFINITION 3   **Equality and Basic Operations**

1. **Equality:** $a + bi = c + di$ if and only if $a = c$ and $b = d$
2. **Addition:** $(a + bi) + (c + di) = (a + c) + (b + d)i$
3. **Multiplication:** $(a + bi)(c + di) = (ac - bd) + (ad + bc)i$

In Section 1-1 we listed the basic properties of the real number system. Using Definition 3, it can be shown that the complex number system possesses the same properties. That is:

1. Addition and multiplication of complex numbers are commutative and associative operations.

2. There is an additive identity and a multiplicative identity for complex numbers.

3. Every complex number has an additive inverse or negative.

4. Every nonzero complex number has a multiplicative inverse or reciprocal.

5. Multiplication distributes over addition.

As a consequence of these properties, we can manipulate complex number symbols of the form $a + bi$ just like we manipulate binomials of the form $a + bx$, as long as we remember that i is a special symbol for the imaginary unit, not for a real number. Thus, you will not have to memorize the definitions of addition and multiplication of complex numbers. We now discuss these operations and some of their properties. Others will be considered in Exercise 2-5.

EXAMPLE 2   **Addition of Complex Numbers**

Carry out each operation and express the answer in standard form:

(A) $(2 - 3i) + (6 + 2i)$ (B) $(-5 + 4i) + (0 + 0i)$

Solution   (A) We could apply the definition of addition directly, but it is easier to use complex number properties.

$(2 - 3i) + (6 + 2i) = 2 - 3i + 6 + 2i$ Remove parentheses.

$= (2 + 6) + (-3 + 2)i$ Combine like terms.

$= 8 - i$

1. $a + bi = c + di$ if and only if $a = c$ and $b = d$

2. $(a + bi) + (c + di) = (a + c) + (b + d)i$

3. $(a + bi)(c + di) = (ac - bd) + (ad + bc)i$

Since complex numbers obey the same commutative, associative, and distributive properties as real numbers, most operations with complex numbers are performed by using these properties and the fact that $i^2 = -1$. The **property of conjugates,**

$$(a + bi)(a - bi) = a^2 + b^2$$

can be used to find **reciprocals** and **quotients.** If $a > 0$, then the **principal square root of the negative real number** $-a$ is $\sqrt{-a} = i\sqrt{a}$.

2-6   QUADRATIC EQUATIONS AND APPLICATIONS

A **quadratic equation** in **standard form** is an equation that can be written in the form

$$ax^2 + bx + c = 0 \qquad a \neq 0$$

where x is a variable and a, b, and c are constants. Methods of solution include:

1. Factoring and using the **zero property:**

$m \cdot n = 0$ if and only if $m = 0$ or $n = 0$ (or both)

2. Using the **square root property:**

If $A^2 = C$, then $A = \pm\sqrt{C}$

3. Completing the square:

$$x^2 + bx + \left(\frac{b}{2}\right)^2 = \left(x + \frac{b}{2}\right)^2$$

4. Using the **quadratic formula:**

$$x = \frac{-b \pm \sqrt{b^2 - 4ac}}{2a}$$

If the **discriminant** $b^2 - 4ac$ is positive, the equation has two distinct **real roots;** if the discriminant is 0, the equation has one real **double root;** and if the discriminant is negative, the equation has two **imaginary roots,** each the conjugate of the other.

2-7   EQUATIONS REDUCIBLE TO QUADRATIC FORM

A **square root radical** can be eliminated from an equation by isolating the radical on one side of the equation and squaring both sides of the equation. The new equation formed by squaring both sides may have **extraneous solutions.** Consequently, **every solution of the new equation must be checked in the original equation to eliminate extraneous solutions.** If an equation contains more than one radical, then the process of isolating a radical and squaring both sides can be repeated until all radicals are eliminated. If a substitution transforms an equation into the form $au^2 + bu + c = 0$, where u is an expression in some other variable, then the equation is an **equation of quadratic type** that can be solved by quadratic methods.

2-8   POLYNOMIAL AND RATIONAL INEQUALITIES

An inequality is in **standard form** if the right side is 0. If the left side is a **polynomial,** then the **real zeros** of this polynomial divide the real number line into intervals with the property that the polynomial has constant sign over each interval. Selecting a **test number** in each interval and constructing a **sign chart** produces the solution to the inequality. If the left side of an inequality is a **rational expression** of the form P/Q, where P and Q are polynomials, then the real zeros of both polynomials are used to divide the real number line into intervals over which P/Q has constant sign. Since **division by zero is never allowed,** the real zeros of Q must always be excluded from the solution set.

Chapter 2 Review Exercise

Work through all the problems in this chapter review and check answers in the back of the book. Answers to all review problems are there, and following each answer is a number in italics indicating the section in which that type of problem is discussed. Where weaknesses show up, review appropriate sections in the text.

A

Solve Problems 1–3.

1. $0.05x + 0.25(30 - x) = 3.3$

2. $\dfrac{5x}{3} - \dfrac{4 + x}{2} = \dfrac{x - 2}{4} + 1$

3. $y = 4x - 9$
$ y = -x + 6$

Solve and graph Problems 4–8.

4. $3(2 - x) - 2 \leq 2x - 1$ **5.** $|y + 9| < 5$

6. $|3 - 2x| \leq 5$ **7.** $x^2 + x < 20$

8. $x^2 \geq 4x + 21$

9. Perform the indicated operations and write the answers in standard form:
   (A) $(-3 + 2i) + (6 - 8i)$
   (B) $(3 - 3i)(2 + 3i)$
   (C) $\dfrac{13 - i}{5 - 3i}$

Solve Problems 10–16.

10. $2x^2 - 7 = 0$

11. $2x^2 = 4x$

12. $2x^2 = 7x - 3$

13. $m^2 + m + 1 = 0$

14. $y^2 = \frac{3}{2}(y + 1)$

15. $\sqrt{5x - 6} - x = 0$

16. $3x + 2y = 5$
    $4x - y = 14$

17. For what values of x does $\sqrt{3 - 5x}$ represent a real number?

B

Solve Problems 18–20.

18. $\dfrac{7}{2 - x} = \dfrac{10 - 4x}{x^2 + 3x - 10}$

19. $\dfrac{u - 3}{2u - 2} = \dfrac{1}{6} - \dfrac{1 - u}{3u - 3}$

20. $5m + 6n = 2$
    $4m - 9n = 20$

Solve and graph Problems 21–25.

21. $\dfrac{x + 3}{8} \leq 5 - \dfrac{2 - x}{3}$

22. $|3x - 8| > 2$

23. $\dfrac{1}{x} < 2$

24. $\dfrac{3}{x - 4} \leq \dfrac{2}{x - 3}$

25. $\sqrt{(1 - 2m)^2} \leq 3$

26. For what real values of x does the expression below represent a real number?
$$\sqrt{\dfrac{x + 4}{2 - x}}$$

27. If the coordinates of A and B on a real number line are -8 and -2, respectively, find:
    (A) $d(A, B)$ (B) $d(B, A)$

28. Perform the indicated operations and write the final answers in standard form:
    (A) $(3 + i)^2 - 2(3 + i) + 3$ (B) i^{27}

29. Convert to $a + bi$ forms, perform the indicated operations, and write the final answers in standard form:
    (A) $(2 - \sqrt{-4}) - (3 - \sqrt{-9})$
    (B) $\dfrac{2 - \sqrt{-1}}{3 + \sqrt{-4}}$ (C) $\dfrac{4 + \sqrt{-25}}{\sqrt{-4}}$

Solve Problems 30–35.

30. $\left(u + \dfrac{5}{2}\right)^2 = \dfrac{5}{4}$

31. $1 + \dfrac{3}{u^2} = \dfrac{2}{u}$

32. $\dfrac{x}{x^2 - x - 6} - \dfrac{2}{x - 3} = 3$

33. $2x^{2/3} - 5x^{1/3} - 12 = 0$ 34. $m^4 + 5m^2 - 36 = 0$

35. $\sqrt{y - 2} - \sqrt{5y + 1} = -3$

Use a calculator to solve Problems 36–40, and compute to two decimal places.

36. $2.15x - 3.73(x - 0.93) = 6.11x$

37. $-1.52 \leq 0.77 - 2.04x \leq 5.33$

38. $\dfrac{3.77 - 8.47i}{6.82 - 7.06i}$

39. $6.09x^2 + 4.57x - 8.86 = 0$

40. $15.2x + 5.6y = 20$
    $2.5x + 7.5y = 10$

Solve Problems 41–43 for the indicated variable in terms of the other variables.

41. $P = M - Mdt$ for M (mathematics of finance)

42. $P = EI - RI^2$ for I (electrical engineering)

43. $x = \dfrac{4y + 5}{2y + 1}$ for y

44. Find the error in the following "solution" and then find the correct solution.
$$\dfrac{4}{x^2 - 4x + 3} = \dfrac{3}{x^2 - 3x + 2}$$
$$4x^2 - 12x + 8 = 3x^2 - 12x + 9$$
$$x^2 = 1$$
$$x = -1 \quad \text{or} \quad x = 1$$

45. Consider the quadratic equation
$$x^2 - 6x + c = 0$$
where c is a real number. Discuss the relationship between the values of c and the three types of roots listed in Table 1 in Section 2-6.

C

46. For what values of a and b is the inequality $a + b < b - a$ true?

47. If a and b are negative numbers and $a > b$, then is a/b greater than 1 or less than 1?

48. Solve for x in terms of y: $y = \dfrac{1}{1 - \dfrac{1}{1 - x}}$

49. Solve and graph: $0 < |x - 6| < d$

Solve Problems 50 and 51.

50. $2x^2 = \sqrt{3}x - \frac{1}{2}$

51. $4 = 8x^{-2} - x^{-4}$

52. Evaluate: $(a + bi)\left(\dfrac{a}{a^2 + b^2} - \dfrac{b}{a^2 + b^2}i\right)$, $a, b \neq 0$

Solve Problems 53–55.

53. $2x > \dfrac{x^2}{5} + 5$ **54.** $\dfrac{x^2}{4} + 4 \geq 2x$

55. $\left|x - \dfrac{8}{x}\right| \geq 2$

56. Solve for u and v in terms of x and y and check.

$$x = 2 + 3u + 7v$$
$$y = -3 + 2u + 5v$$

57. Discuss the nature of the solution sets for each of the following systems:

(A) $\begin{aligned} 2x - y &= -5 \\ -6x + 3y &= 15 \end{aligned}$ (B) $\begin{aligned} 2x - y &= -5 \\ -6x + 3y &= 10 \end{aligned}$

APPLICATIONS

58. Numbers. Find a number such that subtracting its reciprocal from the number gives $\frac{16}{15}$.

59. Sports Medicine. The following quotation was found in a sports medicine handout: "The idea is to raise and sustain your heart rate to 70% of its maximum safe rate for your age. One way to determine this is to subtract your age from 220 and multiply by 0.7."
(A) If H is the maximum safe sustained heart rate (in beats per minute) for a person of age A (in years), write a formula relating H and A.
(B) What is the maximum safe sustained heart rate for a 20-year-old?
(C) If the maximum safe sustained heart rate for a person is 126 beats per minute, how old is the person?

★ **60. Chemistry.** A chemical storeroom has an 80% alcohol solution and a 30% alcohol solution. How many milliliters of each should be used to obtain 50 milliliters of a 60% solution?

★ **61. Rate–Time.** An excursion boat takes 2 hours longer to go 45 miles up a river than to return. If the boat's speed in still water is 12 miles per hour, what is the rate of the current?

★ **62. Rate–Time.** A crew of four practices by rowing up a river for a fixed distance and then returning to their starting point. The river has a current of 3 km/h.
(A) Currently the crew can row 15 km/h in still water. If it takes them 25 minutes to make the round-trip, how far upstream did they row?
(B) After some additional practice the crew cuts the round-trip time to 23 minutes. What is their still-water speed now? Round answers to one decimal place.
(C) If the crew wants to increase their still-water speed to 18 km/h, how fast must they make the round-trip? Express answer in minutes rounded to one decimal place.

63. Nutrition. A fruit grower can use two types of fertilizer in an orange grove, brand A and brand B. Each bag of brand A contains 8 pounds of nitrogen and 9 pounds of phosphoric acid. Each bag of brand B contains 4 pounds of nitrogen and 7 pounds of phosphoric acid. Tests indicate that the grove needs 860 pounds of nitrogen and 1,080 pounds of phosphoric acid. How many bags of each brand should be used to provide the required amounts of nitrogen and phosphoric acid?

64. Cost Analysis. Cost equations for manufacturing companies are often quadratic in nature. If the cost equation for manufacturing inexpensive calculators is $C = x^2 - 10x + 31$, where C is the cost of manufacturing x units per week (both in thousands), find:
(A) The output for a $15 thousand weekly cost
(B) The output for a $6 thousand weekly cost

65. Break-Even Analysis. The manufacturing company in Problem 64 sells its calculators to wholesalers for $3 each. Thus, its revenue equation is $R = 3x$, where R is revenue and x is the number of units sold per week (both in thousands). Find the break-even point(s) for the company—that is, the output at which revenue equals cost.

★ **66. Profit Analysis.** Referring to Problems 64 and 65, find all output levels for which a profit will result—that is, for which $R > C$.

★ **67. Chemistry.** If the temperature T of a solution must be kept within 5°C of 110°C, express this restriction as an absolute value inequality.

★ **68. Design.** The pages of a textbook have uniform margins of 2 centimeters on all four sides (see the figure on the next page). If the area of the entire page is 480 square centimeters and the area of the printed portion is 320 square centimeters, find the dimensions of the page.

Figure for 68

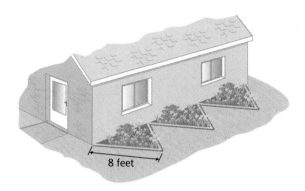

(see the figure). If the area of each triangle is 24 square feet, find the base correct to two decimal places.

8 feet

★ **69. Design.** A landscape designer uses 8-foot timbers to form a pattern of isosceles triangles along the wall of a building

---

Cumulative Review Exercise Chapters 1 and 2

Work through all the problems in this cumulative review and check answers in the back of the book. Answers to all review problems are there, and following each answer is a number in italics indicating the section in which that type of problem is discussed. Where weaknesses show up, review appropriate sections in the text.

A

1. Solve for x: $\dfrac{7x}{5} - \dfrac{3 + 2x}{2} = \dfrac{x - 10}{3} + 2$

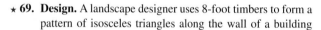

2. Replace each question mark with an appropriate expression that will illustrate the use of the indicated real number property:
(A) Distributive: $c(a + b) = ?$
(B) Associative $(+)$: $(a + b) + c = ?$
(C) Commutative $(\cdot)$: $(a + b)c = ?$

3. Solve for x and y: $2x - 3y = 8$
$\qquad\qquad\qquad\quad 4x + y = 2$

In Problems 4–8, perform the indicated operation and simplify.

4. $4x(x - 3) + 2(3x + 5)$ **5.** $3x(x - 7) - (5x - 9)$

6. $(2x - 5y)(3x + 4y)$ **7.** $(2a - 3b)(2a + 3b)$

8. $(5m + 2n)^2$

Solve and graph Problems 9–11.

9. $2(3 - y) + 4 \le 5 - y$ **10.** $|x - 2| < 7$

11. $x^2 + 3x \ge 10$

In Problems 12 and 13, write each polynomial in a completely factored form relative to the integers. If the polynomial is prime relative to the integers, say so.

12. $x^2 - 3x + 10$ **13.** $6t^2 + 7t - 5$

In Problems 14 and 15, perform the indicated operations and reduce to lowest terms. Represent all compound fractions as simple fractions reduced to lowest terms.

14. $\dfrac{6}{x^2 - 3x} - \dfrac{4}{x^2 - 2x}$ **15.** $\dfrac{\dfrac{4}{y} - y}{\dfrac{2}{y^2} - \dfrac{1}{2}}$

16. Perform the indicated operations and write the answer in standard form:
(A) $(2 - 3i) - (-5 + 7i)$
(B) $(1 + 4i)(3 - 5i)$
(C) $\dfrac{5 + i}{2 + 3i}$

Simplify Problems 17–19 and write answers using positive exponents only. All variables represent positive real numbers.

17. $3x^2(xy^2)^5$ **18.** $\dfrac{(2xy^2)^3}{4x^4y^4}$ **19.** $(a^4b^{-6})^{3/2}$

Solve Problems 20–23.

20. $3x^2 = -12x$ **21.** $4x^2 - 20 = 0$

22. $x^2 - 6x + 2 = 0$ **23.** $x - \sqrt{12 - x} = 0$

24. Change to radical form: $5a^{3/4}$

25. Change to rational exponent form: $2\sqrt[5]{x^2y^3}$

Simplify Problems 26–28.

26. $xy^2 \sqrt[3]{x^4 y^8}$ **27.** $\sqrt{2xy^3} \sqrt{6x^2 y}$ **28.** $\dfrac{3 + \sqrt{2}}{2 + 3\sqrt{2}}$

29. For what values of x does $\sqrt{2 + 3x}$ represent a real number?

B

30. Write using the listing method:

$$\{x \mid x \text{ is a prime factor of } 60\}$$

31. Indicate true (T) or false (F):
  (A) If $ab = 1$, then either $a = 1$ or $b = 1$ or both.
  (B) If $ab = 0$, then either $a = 0$ or $b = 0$ or both.
  (C) A real number with a repeating decimal expansion is an irrational number.

Solve Problems 32–35.

32. $\dfrac{x + 3}{2x + 2} + \dfrac{5x + 2}{3x + 3} = \dfrac{5}{6}$ **33.** $\dfrac{3}{x} = \dfrac{6}{x + 1} - \dfrac{1}{x - 1}$

34. $2x + 1 = 3\sqrt{2x - 1}$

35. $2x - 3y = 9$
  $4x + 2y = 23$

36. Indicate which of the following algebraic expressions are polynomials. State the degree of each polynomial.
  (A) $5x^2 - \sqrt{3}x + 7$
  (B) $4y^3 + \sqrt{5}y - 11$
  (C) $3x^3 + 2x^2 y^2 + 5y^3$
  (D) $2x^{-2} + 3x^{-1} - 4$

In Problems 37–39, perform the indicated operations and simplify.

37. $(a + 2b)(2a + b) - (a - 2b)(2a - b)$

∫ **38.** $3(x + h)^2 - 4(x + h) - (3x^2 - 4x)$

39. $(4m + 3n)^3$

In Problems 40–42, write in a completely factored form relative to the integers.

40. $(2y + 4)^2 - y^2$ **41.** $a^3 + 3a^2 - 4a - 12$

∫ **42.** $3x^4(x + 1)^2 + 4x^3(x + 1)^3$

In Problems 43–45, perform the indicated operations and reduce to lowest terms. Represent all compound fractions as simple fractions reduced to lowest terms.

∫ **43.** $\dfrac{-3x^4(x + 1)^2 + 4x^3(x + 1)^3}{(x + 1)^6}$

44. $\dfrac{2a + 4b}{(a^2 - b^2)} + \dfrac{3a}{a^2 - 3ab + 2b^2} - \dfrac{3b}{a^2 - ab - 2b^2}$

45. $\dfrac{2 - \dfrac{4}{2 - \dfrac{x}{y}}}{2 - \dfrac{4}{2 + \dfrac{x}{y}}}$

Check the solutions in Problems 46 and 47. If a solution is wrong, explain what is wrong and how it can be corrected, and then show a correct solution.

46. $\dfrac{x^2 - x - 2}{x - 2} = \dfrac{x^2 - x}{x}$

$$= x - 1$$

47. $\dfrac{5}{x^2 + 5x + 6} = \dfrac{3}{x^2 + 3x + 2}$

$$5x^2 + 15x + 10 = 3x^2 + 15x + 18$$

$$x^2 = 4$$

$$x = -2 \quad \text{or} \quad x = 2$$

Solve and graph Problems 48–50.

48. $|4x - 9| > 3$ **49.** $\sqrt{(3m - 4)^2} \le 2$

50. $\dfrac{2}{x + 1} \ge \dfrac{1}{x - 2}$

51. For what real values of x does the expression below represent a real number?

$$\dfrac{\sqrt{x - 2}}{x - 4}$$

52. Perform the indicated operations and write the final answers in standard form:
  (A) $(2 - 3i)^2 - (4 - 5i)(2 - 3i) - (2 + 10i)$
  (B) $\frac{3}{5} + \frac{4}{5}i + \dfrac{1}{\frac{3}{5} + \frac{4}{5}i}$ (C) i^{35}

53. Convert to $a + bi$ forms, perform the indicated operations, and write the final answers in standard form:
  (A) $(5 + 2\sqrt{-9}) - (2 - 3\sqrt{-16})$
  (B) $\dfrac{2 + 7\sqrt{-25}}{3 - \sqrt{-1}}$ (C) $\dfrac{12 - \sqrt{-64}}{\sqrt{-4}}$

54. Evaluate to four significant digits using a calculator:
  (A) 0.9274^{23}
  (B) $\sqrt[7]{12.47}$
  (C) $\sqrt[5]{\sqrt[3]{9} + 4}$
  (D) $\dfrac{(52,180,000,000)(0.000\ 000\ 002\ 973)}{0.000\ 000\ 000\ 000\ 271}$

In Problems 55–58, perform the indicated operations and express answers in simplified form. All radicands represent positive real numbers.

55. $3xy^2\sqrt[3]{16x^5y^7}$

56. $\sqrt[5]{\dfrac{7a^3}{16b^4}}$

57. $\sqrt[3]{\sqrt{8y^9}}$

58. $\dfrac{t}{\sqrt{t+9}-3}$

59. Evaluate each expression on a calculator and determine which pairs have the same value. Verify these results algebraically.

  (A) $\sqrt{7+4\sqrt{3}}$ (B) $2+\sqrt{3}$ (C) $\sqrt[3]{26+15\sqrt{3}}$

60. Write in the form $ax^p + bx^q$, where a and b are real numbers and p and q are rational numbers:

  (A) $\dfrac{2x^3+10}{5x^2}$ (B) $\dfrac{2\sqrt{x}-5}{4\sqrt{x}}$

Solve Problems 61–64.

61. $1+\dfrac{14}{y^2}=\dfrac{6}{y}$

62. $4x^{2/3}-4x^{1/3}-3=0$

63. $u^4+u^2-12=0$

64. $\sqrt{8t-2}-2\sqrt{t}=1$

Use a calculator to solve Problems 65–67. Compute answers to two decimal places.

65. $-3.45 < 1.86 - 0.33x \le 7.92$

66. $2.35x^2 + 10.44x - 16.47 = 0$

67. $12.5x + 2.5y = 20$
    $3.5x + 8.7y = 10$

68. Solve for y in terms of x:

$$\dfrac{x-2}{x+1}=\dfrac{2y+1}{y-2}$$

69. Solve for s and t in terms of x and y and check:

$$x = -1 + 5s + 2t$$
$$y = 2 + 2s + t$$

C

70. Evaluate $x^2 - x - 1$ for $x = \dfrac{1}{2} - \dfrac{\sqrt{5}}{2}$.

71. Evaluate $x^2 - x + 2$ for $x = \dfrac{1}{2} - \dfrac{i}{2}\sqrt{7}$.

72. Simplify $(x+1)^3 - (x-1)^3$.

73. For what values of a and b is the inequality $a - b < b - a$ true?

74. Factor completely relative to the integers:

$$9b^4 - 16b^2(a^2 - 2a + 1)$$

75. Simplify:

$$\dfrac{1 - \dfrac{1}{1+\dfrac{1}{1+m}}}{\dfrac{m}{2-m}+\dfrac{m}{2+m}}$$

76. Solve for y in terms of x:

$$\dfrac{x+y}{y-\dfrac{x+y}{x-y}}=1$$

77. Find all roots: $3x^2 = 2\sqrt{2}x - 1$.

78. Consider the quadratic equation

$$x^2 + bx + 1 = 0$$

where b is a real number. Discuss the relationship between the values of b and the three types of roots listed in Table 1 in Section 2-6.

79. Find all roots: $\sqrt{3-2x}-\sqrt{x+7}=\sqrt{x+4}$

80. Simplify and express answers using positive exponents only:
  (A) $(a^{1/4}-b^{1/4})(a^{1/4}+b^{1/4})(a^{1/2}+b^{1/2})$
  (B) $\left(\dfrac{x^{-1}y+xy^{-1}}{xy^{-1}-x^{-1}y}\right)^{-1}$

81. Rationalize the numerator: $\dfrac{\sqrt[3]{8+h}-2}{h}$.

82. Write in standard form: $\dfrac{a+bi}{a-bi}$, $a, b \neq 0$.

83. Solve: $\left|\dfrac{x+4}{x}\right| < 3$.

84. Write in simplified form: $\sqrt[n]{\left(\dfrac{x^{2n^2}}{x^{4n}}\right)^{1/(n-2)}}$, $n > 2$.

APPLICATIONS

85. Numbers. Find a number such that the number exceeds its reciprocal by $\frac{3}{2}$.

86. Rate–Time. An older conveyor belt working alone can fill a railroad car with ore in 14 minutes. The older belt and a newer belt operating together can fill the car in 6 minutes. How long will it take the newer belt working alone to fill the car with ore?

★ **87. Rate–Time.** A boat travels upstream for 35 miles and then returns to its starting point. If the round-trip took 4.8 hours and the boat's speed in still water is 15 miles per hour, what is the speed of the current?

★ **88. Chemistry.** How many gallons of distilled water must be mixed with 24 gallons of a 90% sulfuric acid solution to obtain a 60% solution?

89. Break-Even Analysis. The publisher's fixed costs for the production of a new cookbook are $41,800. Variable costs are $4.90 per book. If the book is sold to bookstores for $9.65, how many must be sold for the publisher to break even?

90. Finance. An investor instructs a broker to purchase a certain stock whenever the price per share p of the stock is within $10 of $200. Express this instruction as an absolute value inequality.

91. Supply and Demand. Suppose the supply and demand equations for Styrofoam "cheesehead" hats in Green Bay, Wisconsin, for a particular week are

$$p = 5.5 + 0.002q \quad \text{Supply equation}$$
$$p = 22 - 0.001q \quad \text{Demand equation}$$

where p is the price in dollars and q is the number of cheeseheads. Find the equilibrium price and quantity.

92. Construction. A square sandbox with no bottom is constructed out of concrete (see the figure). The walls of the sandbox are 6 inches thick and 1 foot high. If the outer dimension of the sandbox is x feet, write an algebraic expression in terms of x that represents the volume of the concrete used to construct the sandbox. Simplify the expression.

6 inches

1 foot

x feet

x feet

93. Profit and Loss Analysis. At a price of $$p$ per unit, the marketing department in a company estimates that the weekly cost C and the weekly revenue R, in thousands of dollars, will be given by the equations

$$C = 88 - 12p \quad \text{Cost equation}$$
$$R = 15p - 2p^2 \quad \text{Revenue equation}$$

Find the prices for which the company has:
(A) A profit   (B) A loss

★ **94. Shipping.** A ship leaves port A, sails east to port B, and then north to port C, a total distance of 115 miles. The next day the ship sails directly from port C back to port A, a distance of 85 miles. Find the distance between ports A and B and between ports B and C.

★ **95. Physics.** The distance s above the ground (in feet) of an object dropped from a hot-air balloon t seconds after it is released is given by

$$s = a + bt^2$$

where a and b are constants. Suppose the object is 2,100 feet above the ground 5 seconds after its release and 900 feet above the ground 10 seconds after its release.
(A) Find the constants a and b.
(B) How high is the balloon?
(C) How long does the object fall?

GRAPHS AND FUNCTIONS

CHAPTER 3

The function concept is one of the most important ideas in mathematics. The study of mathematics beyond the most elementary level requires a firm understanding of a basic list of elementary functions, their properties, and their graphs. In the first two sections of this chapter we consider some basic geometric concepts, including the graphs of circles and straight lines. In the remaining sections we introduce the important concept of a function, discuss basic properties, and consider operations that can be performed with functions. As we progress through this and subsequent chapters of the book, we will encounter a number of different types of elementary functions. A thorough understanding of the definitions, graphs, and properties of these elementary functions will provide you with a set of tools that should become a part of your mathematical toolbox for use in this and most future courses or activities that involve mathematics.

SECTION 3-1 Basic Tools; Circles

- Cartesian Coordinate System
- Graphing: Point by Point
- Symmetry
- Distance between Two Points
- Circles

In this section we develop some of the basic tools used in analytic geometry and apply these tools to the graphing of equations and to the derivation of the equation of a circle.

• Cartesian Coordinate System

Just as a real number line is formed by establishing a one-to-one correspondence between the points on a line and the elements in the set of real numbers, we can form a **real plane** by establishing a one-to-one correspondence between the points in a plane and elements in the set of all ordered pairs of real numbers. This can be done by means of a Cartesian coordinate system.

Recall that to form a **Cartesian** or **rectangular coordinate system,** we select two real number lines, one horizontal and one vertical, and let them cross through their origins as indicated in Figure 1. Up and to the right are the usual choices for the positive directions. These two number lines are called the **horizontal axis** and the **vertical axis,** or, together, the **coordinate axes.** The horizontal axis is usually referred to as the x **axis** and the vertical axis as the y **axis,** and each is labeled accordingly. Other labels may be used in certain situations. The coordinate axes divide the plane into four parts called **quadrants,** which are numbered counterclockwise from I to IV (see Fig. 1).

Now we want to assign *coordinates* to each point in the plane. Given an arbitrary point P in the plane, pass horizontal and vertical lines through the point (Fig. 2). The vertical line will intersect the horizontal axis at a point with coordinate a, and the horizontal line will intersect the vertical axis at a point with coordinate b. These two

FIGURE 1 Cartesian coordinate system.

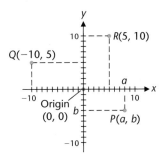

FIGURE 2 Coordinates in a plane.

numbers written as the ordered pair* (a, b) form the **coordinates** of the point P. The first coordinate a is called the **abscissa** of P; the second coordinate b is called the **ordinate** of P. The abscissa of Q in Figure 2 is -10, and the ordinate of Q is 5. The coordinates of a point can also be referenced in terms of the axis labels. The **x coordinate** of R in Figure 2 is 5, and the **y coordinate** of R is 10. The point with coordinates $(0, 0)$ is called the **origin.**

The procedure we have just described assigns to each point P in the plane a unique pair of real numbers (a, b). Conversely, if we are given an ordered pair of real numbers (a, b), then, reversing this procedure, we can determine a unique point P in the plane. Thus:

> There is a one-to-one correspondence between the points in a plane and the elements in the set of all ordered pairs of real numbers.

This is often referred to as the **fundamental theorem of analytic geometry.**

• Graphing: Point by Point

The fundamental theorem of analytic geometry allows us to look at algebraic forms geometrically and to look at geometric forms algebraically. We begin by considering an algebraic form, an equation in two variables:

$$y = x^2 - 4 \qquad (1)$$

A **solution** to equation (1) is an ordered pair of real numbers (a, b) such that

$$b = a^2 - 4$$

The **solution set** of equation (1) is the set of all its solutions. More formally,

Solution set of equation (1): $\{(x, y) \mid y = x^2 - 4\}$

To find a solution to equation (1) we simply replace one of the variables with a number and solve for the other variable. For example, if $x = 2$, then $y = 2^2 - 4 = 0$, and the ordered pair $(2, 0)$ is a solution. Similarly, if $y = 5$, then $5 = x^2 - 4$, $x^2 = 9$, $x = \pm 3$, and the ordered pairs $(3, 5)$ and $(-3, 5)$ are solutions.

Sometimes replacing one variable with a number and solving for the other variable will introduce imaginary numbers. For example, if $y = -5$ in equation (1), then

$$-5 = x^2 - 4$$
$$x^2 = -1$$
$$x = \pm\sqrt{-1} = \pm i$$

Thus, $(-i, 5)$ and $(i, 5)$ are solutions to $y = x^2 - 4$. However, the coordinates of a point in a rectangular coordinate system must be real numbers. Therefore, **when graphing an equation, we only consider those values of the variables that produce real solutions to the equation.**

---

*An **ordered pair** of real numbers is a pair of numbers in which the order is specified. We now use (a, b) as the coordinates of a point in a plane. In Chapter 2 we used (a, b) to represent an interval on a real number line. These concepts are not the same. You must always interpret the symbol (a, b) in terms of the context in which it is used.

The **graph of an equation in two variables** is the graph of its solution set. In equation (1), we find that its solution set will have infinitely many elements and its graph will extend off any paper we might choose, no matter how large. Thus, **to sketch the graph of an equation,** we include enough points from its solution set so that the total graph is apparent. This process is called **point-by-point plotting.**

EXAMPLE 1   **Graphing an Equation Using Point-by-Point Plotting**

Sketch a graph of $y = x^2 - 4$.

Solution   We make up a table of solutions—ordered pairs of real numbers that satisfy the given equation.

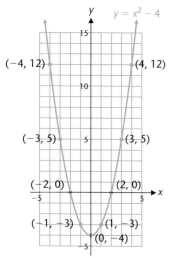

x	-4	-3	-2	-1	0	1	2	3	4
y	12	5	0	-3	-4	-3	0	5	12

After plotting these solutions, if there are any portions of the graph that are unclear, we plot additional points until the shape of the graph is apparent. Then we join all these plotted points with a smooth curve, as shown in Figure 3. Arrowheads are used to indicate that the graph continues beyond the portion shown here with no significant changes in shape.

The resulting figure is called a *parabola*. Notice that if we fold the paper along the y axis, the right side will match the left side. We say that the graph is *symmetric with respect to the y axis* and call the y axis the *axis of the parabola*. More will be said about parabolas later in the text.

FIGURE 3

Matched Problem 1   Sketch a graph of $y^2 = x$.

 We now use an electronic graphing device to check Example 1. We will refer to any electronic device capable of displaying graphs as a **graphing utility.** The two most common graphing utilities are hand-held graphing calculators and computers with appropriate software. This book contains a number of activities that use graphing utilities to emphasize the connection between graphic, numeric, and algebraic viewpoints. All these activities are clearly marked and easily omitted if no such device is available. Figure 4 shows the steps necessary to reproduce the graph in Figure 3 on a graphing utility.

FIGURE 4

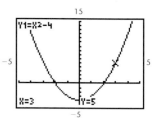

Enter the equation.
(a)

Enter the window variables.
(b)

Graph the equation.
(c)

EXPLORE-DISCUSS 1 To graph the equation $y = -x^3 + 2x$, we use point-by-point plotting to obtain the graph in Figure 5.

FIGURE 5

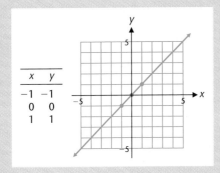

(A) Do you think this is the correct graph of the equation? If so, why? If not, why?

(B) Add points on the graph for $x = -2, -0.5, 0.5,$ and 2.

(C) Now, what do you think the graph looks like? Sketch your version of the graph, adding more points as necessary.

(D) Write a short statement explaining any conclusions you might draw from parts A, B, and C.

 (E) Compare your version of the graph with one produced on a graphing utility.

The use of graphs to illustrate relationships between quantities is commonplace. Estimating the coordinates of points on a graph provides specific examples of this relationship, even if no equation for the graph is available. The next example illustrates this process.

EXAMPLE 2 **Ozone Levels**

The ozone level is measured in parts per billion (ppb). The ozone level during a 12-hour period in a suburb of Milwaukee, Wisconsin, on a particular summer day is given in Figure 6 on page 178 (*source:* Wisconsin Department of Natural Resources). Use this graph to estimate the following ozone levels to the nearest integer and times to the nearest quarter hour.

(A) The ozone level at 6 P.M.
(B) The highest ozone level and the time when it occurs.
(C) The time(s) when the ozone level is 90 ppb.

FIGURE 6 Ozone level.

Solution (A) The ordinate of the point on the graph with abscissa 6 is 97 ppb (see Fig. 7).
(B) The highest ozone level is 109 ppb at 3 P.M.
(C) The ozone level is 90 ppb at about 12:30 P.M. and again at 10 P.M.

FIGURE 7

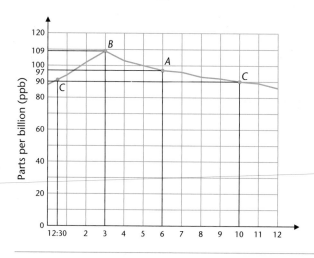

Matched Problem 2   Use Figure 6 to estimate the following ozone levels to the nearest integer and times to the nearest quarter hour.

(A) The ozone level at 7 P.M.
(B) The time(s) when the ozone level is 100 ppb.

An important aspect of this course, and later in calculus, is the development of tools that can be used to analyze graphs, whether using point-by-point plotting or a graphing utility. A particularly useful tool is *symmetry*, which we now discuss.

• **Symmetry**   We noticed that the graph of $y = x^2 - 4$ in Example 1 is *symmetric with respect to the y axis;* that is, the two parts of the graph coincide if the paper is folded along the *y* axis. Similarly, we say that a graph is *symmetric with respect to the x axis* if the parts above and below the *x* axis coincide when the paper is folded along the *x* axis. In general, we define symmetry with respect to the *y* axis, *x* axis, and origin as follows:

DEFINITION 1   **Symmetry**

A graph is **symmetric with respect to:**

1. **The y axis** if $(-a, b)$ is on the graph whenever (a, b) is on the graph—the two points are equidistant from the *y* axis.

2. **The x axis** if $(a, -b)$ is on the graph whenever (a, b) is on the graph—the two points are equidistant from the *x* axis.

3. **The origin** if $(-a, -b)$ is on the graph whenever (a, b) is on the graph— the two points are equidistant from the origin on a line through the origin.

Figure 8 illustrates these three types of symmetry.

FIGURE 8 Symmetry.

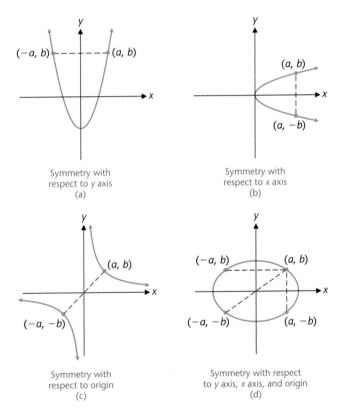

Symmetry with
respect to *y* axis
(a)

Symmetry with
respect to *x* axis
(b)

Symmetry with
respect to origin
(c)

Symmetry with respect
to *y* axis, *x* axis, and origin
(d)

EXPLORE-DISCUSS 2   If a graph possesses two of the three types of symmetry in Definition 1, must it also possess the third? Explain.

Given an equation, if we can determine the symmetry properties of its graph ahead of time, we can save a lot of time and energy in sketching the graph. For example, we know that the graph of $y = x^2 - 4$ in Example 1 is symmetric with respect to the y axis, so we can carefully sketch only the right side of the graph; then reflect the result across the y axis to obtain the whole sketch—the point-by-point plotting is cut in half!

The tests for symmetry are given in Theorem 1. These tests are easily applied and are very helpful aids to graphing. Recall, two equations are equivalent if they have the same solution set.

Theorem 1   Tests for Symmetry

Symmetry with respect to the:	Equation is equivalent when:
y axis	x is replaced with $-x$
x axis	y is replaced with $-y$
Origin	x and y are replaced with $-x$ and $-y$

EXAMPLE 3   Using Symmetry as an Aid to Graphing

Test for symmetry and graph:

(A) $y = x^3$ (B) $y = |x|$ (C) $x^2 + 4y^2 = 36$

Solution   (A) **Symmetry tests for $y = x^3$.**

Test y Axis Replace x with $-x$:	**Test x Axis** Replace y with $-y$:	**Test Origin** Replace x with $-x$ and y with $-y$:
$y = (-x)^3$	$-y = x^3$	$-y = (-x)^3$
$y = -x^3$	$y = -x^3$	$-y = -x^3$
		$y = x^3$

The only test that produces an equivalent equation is replacing x with $-x$ and y with $-y$. Thus, the only symmetry property for the graph of $y = x^3$ is symmetry with respect to the origin.

Graph. Note that positive values of x produce positive values for y and negative values of x produce negative values for y. Therefore, the graph occurs in the first and third quadrants. We make a careful sketch in the first quadrant; then reflect these points through the origin to obtain the complete sketch shown in Figure 9.

Solutions   (A) $(h, k) = (-3, 6)$ and $r = 4$:

$$(x - h)^2 + (y - k)^2 = r^2$$
$$[x - (-3)]^2 + (y - 6)^2 = 4^2$$
$$(x + 3)^2 + (y - 6)^2 = 16$$

(B) $(h, k) = (0, 0)$ and $r = 4$:

$$x^2 + y^2 = r^2$$
$$x^2 + y^2 = 4^2$$
$$x^2 + y^2 = 16$$

To graph the equation, locate the center $C(-3, 6)$ and draw a circle of radius 4 (Fig. 15).

To graph the equation, locate the center at the origin and draw a circle of radius 4 (Fig. 16).

$(x + 3)^2 + (y - 6)^2 = 16$

FIGURE 15

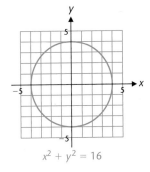

$x^2 + y^2 = 16$

FIGURE 16

Matched Problem 5   Find the equation of a circle with radius 3 and center at:

(A) $(3, -2)$ (B) $(0, 0)$

Graph each equation.

To graph the circle in Figure 16 on a graphing utility, we must graph two equations: $y_1 = \sqrt{16 - x^2}$ and $y_2 = -\sqrt{16 - x^2}$ [Fig. 17(a)]. Notice that the graph does not look circular, because the units on the x axis are physically longer than the units on the y axis in the rectangular viewing window. To rectify this, we must choose the window variables so that a length of one unit on the x axis equals a length of one unit on the y axis. The resulting window, called a **squared window,** displays circles that look circular [Fig. 17(b)]. Most graphing utilities have a routine, usually denoted by Zoom Square or something similar, that will adjust the window variables automatically to produce a squared window. Consult your manual for details.

FIGURE 17

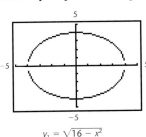

$y_1 = \sqrt{16 - x^2}$
$y_2 = -\sqrt{16 - x^2}$
(a)

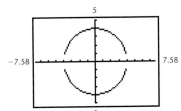

A squared window
(b)

Also notice that the graphs in Figure 17 have gaps near $x = -4$ and $x = 4$ due to the discrepancy between the actual coordinates of a point and the screen coordinates of the pixel containing that point. (Try tracing along the graph of $y_1 = \sqrt{16 - x^2}$ and observe what happens when x gets close to -4 or 4.) It is important to remember that graphing utilities are low-resolution devices that produce only rough approximations to graphs. It is up to you to visualize the correct appearance of a graph and fill in any missing gaps.

EXAMPLE 6 **Finding the Center and Radius of a Circle**

Find the center and radius of the circle with equation $x^2 + y^2 + 6x - 4y = 23$.

Solution We transform the equation into the form $(x - h)^2 + (y - k)^2 = r^2$ by completing the square relative to x and relative to y (see Section 2-6). From this standard form we can determine the center and radius.

$$x^2 + y^2 + 6x - 4y = 23$$
$$(x^2 + 6x \quad) + (y^2 - 4y \quad) = 23$$
$$(x^2 + 6x + 9) + (y^2 - 4y + 4) = 23 + 9 + 4 \quad \text{Complete the squares.}$$
$$(x + 3)^2 + (y - 2)^2 = 36$$
$$[x - (-3)]^2 + (y - 2)^2 = 6^2$$

Center: $C(h, k) = C(-3, 2)$

Radius: $r = \sqrt{36} = 6$

Matched Problem 6 Find the center and radius of the circle with equation $x^2 + y^2 - 8x + 10y = -25$.

Answers to Matched Problems

1.

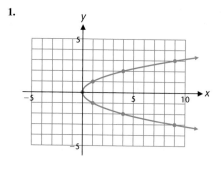

2. (A) 96 ppb (B) 1:45 P.M. and 5 P.M.

★ **85. Oceanography.** After about 9 hours of a steady wind, the height of waves in the ocean is approximately linearly related to the duration of time the wind has been blowing. During a storm with 50-knot winds, the wave height after 9 hours was found to be 23 feet, and after 24 hours it was 40 feet.

(A) If t is time after the 50-knot wind started to blow and h is the wave height in feet, write a linear equation that expresses height h in terms of time t.

(B) How long will the wind have been blowing for the waves to be 50 feet high?

Express all calculated quantities to three significant digits.

86. Oceanography. As a diver descends into the ocean, pressure increases linearly with depth. The pressure is 15 pounds per square inch on the surface and 30 pounds per square inch 33 feet below the surface.

(A) If p is the pressure in pounds per square inch and d is the depth below the surface in feet, write an equation that expresses p in terms of d.

(B) How deep can a scuba diver go if the safe pressure for his equipment and experience is 40 pounds per square inch?

★ **87. Medicine.** Cardiovascular research has shown that above the 210 cholesterol level, each 1% increase in cholesterol level increases coronary risk 2%. For a particular age group, the coronary risk at a 210 cholesterol level is found to be 0.160 and at a level of 231 the risk is found to be 0.192.

(A) Find a linear equation that expresses risk R in terms of cholesterol level C.

(B) What is the risk for a cholesterol level of 260?

(C) What is the slope of the graph of the equation found in part A? Interpret verbally.

Express all calculated quantities to three significant digits.

★ **88. Demographics.** The average number of persons per household in the United States has been shrinking steadily for as long as statistics have been kept and is approximately linear with respect to time. In 1900, there were about 4.76 persons per household and in 1990, about 2.5.

(A) If N represents the average number of persons per household and t represents the number of years since 1900, write a linear equation that expresses N in terms of t.

(B) What is the predicted household size in the year 2000?

Express all calculated quantities to three significant digits.

SECTION **3-3** **Functions**

- Definition of a Function
- Functions Defined by Equations
- Function Notation
- Application
- A Brief History of the Function Concept

The idea of correspondence plays a central role in the formulation of the function concept. You have already had experiences with correspondences in everyday life. For example:

To each person there corresponds an age.

To each item in a store there corresponds a price.

To each automobile there corresponds a license number.

To each circle there corresponds an area.

To each number there corresponds its cube.

One of the most important aspects of any science (managerial, life, social, physical, computer, etc.) is the establishment of correspondences among various types of phenomena. Once a correspondence is known, predictions can be made. A chemist can use a gas law to predict the pressure of an enclosed gas, given its temperature. An engineer can use a formula to predict the deflections of a beam subject to different loads. A computer scientist can use formulas to compare the efficiency of algorithms for sorting data stored in a computer. An economist would like to be able to predict interest rates, given the rate of change of the money supply. And so on.

● **Definition of a Function**

What do all the preceding examples have in common? Each describes the matching of elements from one set with the elements in a second set. Consider Tables 1–3, which list values for the cube, square, and square root, respectively.

TABLE 1		TABLE 2		TABLE 3	
Domain (number)	**Range (cube)**	**Domain (number)**	**Range (square)**	**Domain (number)**	**Range (square root)**

Tables 1 and 2 specify functions, but Table 3 does not. Why not? The definition of the term *function* will explain.

DEFINITION 1   **Rule Form of the Definition of a Function**

A **function** is a rule that produces a correspondence between two sets of elements such that to each element in the first set there corresponds *one and only one* element in the second set.

The first set is called the **domain,** and the set of all corresponding elements in the second set is called the **range.**

Tables 1 and 2 specify functions, since to each domain value there corresponds exactly one range value (for example, the cube of -2 is -8 and no other number). On the other hand, Table 3 does not specify a function, since to at least one domain value there corresponds more than one range value (for example, to the domain value 9 there corresponds -3 and 3, both square roots of 9).

EXPLORE-DISCUSS 1   Consider the set of students enrolled in a college and the set of faculty members of that college. Define a correspondence between the two sets by saying that a student corresponds to a faculty member if the student is currently enrolled in a course taught by the faculty member. Is this correspondence a function? Discuss.

Since a function is a rule that pairs each element in the domain with a corresponding element in the range, this correspondence can be illustrated by using ordered pairs of elements, where the first component represents a domain element and the sec-

DEFINITION 3

The Symbol $f(x)$

The symbol $f(x)$ represents the real number in the range of the function f corresponding to the domain value x. Symbolically, $f:x \rightarrow f(x)$. The ordered pair $(x, f(x))$ belongs to the function f. If x is a real number that is not in the domain of f, then f is **not defined** at x and $f(x)$ **does not exist.**

EXAMPLE 4   **Evaluating Functions**

For

$$f(x) = \frac{15}{x - 3} \qquad g(x) = 16 + 3x - x^2 \qquad h(x) = \sqrt{25 - x^2}$$

find:

(A) $f(6)$ (B) $g(-7)$ (C) $h(10)$ (D) $f(0) + g(4) - h(-3)$

Solution   (A) $f(6)$ $\boxed{= \dfrac{15}{6 - 3}} = \dfrac{15}{3} = 5$

(B) $g(-7)$ $\boxed{= 16 + 3(-7) - (-7)^2} = 16 - 21 - 49 = -54$

(C) $h(10)$ $\boxed{= \sqrt{25 - 10^2}} = \sqrt{25 - 100} = \sqrt{-75}$

But $\sqrt{-75}$ is not a real number. Since we have agreed to restrict the domain of a function to values of x that produce real values for the function, 10 is not in the domain of h and $h(10)$ is not defined.

(D) $f(0) + g(4) - h(-3)$

$$\boxed{= \frac{15}{0 - 3} + [16 + 3(4) - 4^2] - \sqrt{25 - (-3)^2}}$$

$$= \frac{15}{-3} + 12 - \sqrt{16}$$

$$= -5 + 12 - 4 = 3$$

Matched Problem 4   Use the functions in Example 4 to find:

(A) $f(-2)$ (B) $g(6)$ (C) $h(-8)$ (D) $\dfrac{f(8)}{h(4)}$

EXAMPLE 5   **Finding Domains of Functions**

Find the domains of functions f, g, and h:

$$f(x) = \frac{15}{x-3} \qquad g(x) = 16 + 3x - x^2 \qquad h(x) = \sqrt{25 - x^2}$$

Solutions   **Domain of f**

The fraction $15/(x - 3)$ represents a real number for all replacements of x by real numbers except $x = 3$, since division by 0 is not defined. Thus, $f(3)$ does not exist, and the domain of f is the set of all real numbers except 3. We often indicate this by writing

$$f(x) = \frac{15}{x-3} \qquad x \neq 3$$

Domain of g

The domain is R, the set of all real numbers, since $16 + 3x - x^2$ represents a real number for all replacements of x by real numbers.

Domain of h

The domain is the set of all real numbers x such that $\sqrt{25 - x^2}$ is a real number— that is, such that $25 - x^2 \geq 0$. Solving $25 - x^2 = (5 - x)(5 + x) \geq 0$ by methods discussed in Section 2-8, we find

$$\text{Domain: } -5 \leq x \leq 5 \qquad \text{or} \qquad [-5, 5]$$

Matched Problem 5   Find the domains of functions F, G, and H:

$$F(x) = x^2 + 5x - 2 \qquad G(x) = \sqrt{\frac{x-2}{x+3}} \qquad H(x) = \frac{4}{x-2}$$

EXPLORE-DISCUSS 3   Let x and h be real numbers.

(A) If $f(x) = 4x + 3$, which of the following is true:
  (1) $f(x + h) = 4x + 3 + h$
  (2) $f(x + h) = 4x + 4h + 3$
  (3) $f(x + h) = 4x + 4h + 6$

(B) If $g(x) = x^2$, which of the following is true:
  (1) $g(x + h) = x^2 + h$
  (2) $g(x + h) = x^2 + h^2$
  (3) $g(x + h) = x^2 + 2hx + h^2$

(C) If $M(x) = x^2 + 4x + 3$, describe the operations that must be performed to evaluate $M(x + h)$.

In addition to evaluating functions at specific numbers, it is important to be able to evaluate functions at expressions that involve one or more variables. For example, the **difference quotient**

$$\frac{f(x + h) - f(x)}{h} \qquad x \text{ and } x + h \text{ in the domain of } f, h \neq 0$$

is studied extensively in a calculus course.

f EXAMPLE 6 **Evaluating and Simplifying a Difference Quotient**

For $f(x) = x^2 + 4x + 5$, find and simplify:

(A) $f(x + h)$ (B) $\dfrac{f(x + h) - f(x)}{h}, h \neq 0$

Solution   (A) To find $f(x + h)$, we replace x with $x + h$ everywhere it appears in the equation that defines f and simplify:

$$f(x + h) = (x + h)^2 + 4(x + h) + 5$$
$$= x^2 + 2xh + h^2 + 4x + 4h + 5$$

(B) Using the result of part A, we get

$$\frac{f(x + h) - f(x)}{h} = \frac{x^2 + 2xh + h^2 + 4x + 4h + 5 - (x^2 + 4x + 5)}{h}$$
$$= \frac{x^2 + 2xh + h^2 + 4x + 4h + 5 - x^2 - 4x - 5}{h}$$
$$= \frac{2xh + h^2 + 4h}{h} = \frac{h(2x + h + 4)}{h} = 2x + h + 4$$

Matched Problem 6   Repeat Example 6 for $f(x) = x^2 + 3x + 7$.

CAUTION   1. If f is a function, then the symbol $f(x + h)$ represents the value of f at the number $x + h$ and must be evaluated by replacing the independent variable in the equation that defines f with the expression $x + h$, as we did in Example 6. Do not confuse this notation with the familiar algebraic notation for multiplication:

$$f(x + h) \neq fx + fh \qquad f(x + h) \text{ is function notation.}$$
$$4(x + h) = 4x + 4h \qquad 4(x + h) \text{ is algebraic multiplication notation.}$$

2. There is another common incorrect interpretation of the symbol $f(x + h)$. If f is an arbitrary function, then

$$f(x + h) \neq f(x) + f(h)$$

It is possible to find some particular functions for which $f(x + h) = f(x) + f(h)$ is a true statement, but in general these two expressions are not equal.

• Application

EXAMPLE 7   **Construction**

A rectangular feeding pen for cattle is to be made with 100 meters of fencing.

(A) If x represents the width of the pen, express its area $A(x)$ in terms of x.
(B) What is the domain of the function A (determined by the physical restrictions)?

Solutions   (A) Draw a figure and label the sides.

Perimeter = 100 meters of fencing.
Half the perimeter = 50.
If x = Width, then $50 - x$ = Length.

x (Width)

$50 - x$ (Length)

$$A(x) = (\text{Width})(\text{Length}) = x(50 - x)$$

(B) To have a pen, x must be positive, but x must also be less than 50 (or the length will not exist). Thus,

Domain: $0 < x < 50$ Inequality notation

$(0, 50)$ Interval notation

Matched Problem 7   Rework Example 7 with the added assumption that a large barn is to be used as one side of the pen.

• A Brief History of the Function Concept

The history of the use of functions in mathematics illustrates the tendency of mathematicians to extend and generalize each concept. The word "function" appears to have been first used by Leibniz in 1694 to stand for any quantity associated with a curve. By 1718, Johann Bernoulli considered a function any expression made up of constants and a variable. Later in the same century, Euler came to regard a function as any equation made up of constants and variables. Euler made extensive use of the extremely important notation $f(x)$, although its origin is generally attributed to Clairaut (1734).

The form of the definition of function that has been used until well into the twentieth century (many texts still contain this definition) was formulated by Dirichlet (1805–1859). He stated that, if two variables x and y are so related that for each value of x there corresponds exactly one value of y, then y is said to be a (single-valued) function of x. He called x, the variable to which values are assigned at will, the independent variable, and y, the variable whose values depend on the values assigned to x, the dependent variable. He called the values assumed by x the domain of the function, and the corresponding values assumed by y the range of the function.

Now, since set concepts permeate almost all mathematics, we have the more general definition of function presented in this section in terms of sets of ordered pairs of elements.

Answers to Matched Problems

1. (A) S does not define a function
   (B) T defines a function with domain $\{-2, -1, 0, 1, 2\}$ and range $\{0, 1, 2\}$
2. (A) Does not define a function   (B) Defines a function
3. $x \geq -5$ Inequality notation
   $[-5, \infty)$ Interval notation
4. (A) -3 (B) -2 (C) Does not exist   (D) 1
5. Domain of F: all real numbers
   Domain of G: $x < -3$ or $x \geq 2$ Inequality notation
      $(-\infty, -3) \cup [2, \infty)$ Interval notation
   Domain of H: All real numbers except 2
6. (A) $x^2 + 2xh + h^2 + 3x + 3h + 7$ (B) $2x + h + 3$
7. (A) $A(x) = x(100 - 2x)$ (B) Domain: $0 < x < 50$ Inequality notation
      $(0, 50)$ Interval notation

EXERCISE 3-3

A

Indicate whether each table in Problems 1–6 defines a function.

1. Domain   Range

2. Domain   Range

3. Domain   Range

4. Domain   Range

5. Domain   Range

6. Domain   Range

Indicate whether each set in Problems 7–12 defines a function. Find the domain and range of each function.

7. $\{(2, 4), (3, 6), (4, 8), (5, 10)\}$

8. $\{(-1, 4), (0, 3), (1, 2), (2, 1)\}$

9. $\{(10, -10), (5, -5), (0, 0), (5, 5), (10, 10)\}$

10. $\{(-10, 10), (-5, 5), (0, 0), (5, 5), (10, 10)\}$

11. $\{(0, 1), (1, 1), (2, 1), (3, 2), (4, 2), (5, 2)\}$

12. $\{(1, 1), (2, 1), (3, 1), (1, 2), (2, 2), (3, 2)\}$

Indicate whether each graph in Problems 13–18 is the graph of a function.

13.

14.

15.

16.

17.

18.

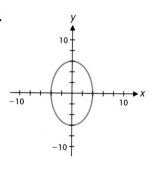

Problems 19–28 refer to the functions

$$f(x) = 2x + 6 \qquad g(t) = 3 - 2t$$
$$F(m) = 2m^2 + 3m - 1 \qquad G(u) = u^2 + u - 2$$

Evaluate as indicated.

19. $f(-1)$ 　　　　　 **20.** $g(6)$

21. $G(-2)$ 　　　　　 **22.** $F(-3)$

23. $F(-1) + f(3)$ 　　 **24.** $G(2) - g(-3)$

25. $2F(-2) - G(-1)$ 　 **26.** $3G(-2) + 2F(-1)$

27. $\dfrac{f(0) \cdot g(-2)}{F(-3)}$ 　 **28.** $\dfrac{g(4) \cdot f(2)}{G(1)}$

In Problems 29–32, use the following graph of a function f to determine x or y to the nearest integer, as indicated. Some problems may have more than one answer.

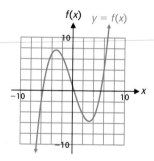

29. $y = f(-4)$ 　　　　 **30.** $y = f(4)$

31. $4 = f(x)$ 　　　　 **32.** $-2 = f(x)$

B

Determine which of the equations in Problems 33–42 define a function with independent variable x. For those that do, find the domain. For those that do not, find a value of x to which there corresponds more than one value of y.

33. $y - x^2 = 1$ 　　　　 **34.** $y^2 - x = 1$

35. $2x^3 + y^2 = 4$ 　　　 **36.** $3x^2 + y^3 = 8$

37. $x^3 - y = 2$

38. $x^3 + |y| = 6$

39. $xy = 0$

40. $xy = 1$

41. $x^2 + xy = 1$

42. $x^2 + xy = 0$

In Problems 43–56, find the domain of the indicated function.

43. $f(x) = \sqrt{x - 2}$

44. $g(t) = \sqrt[3]{5 - t}$

45. $h(u) = \sqrt[3]{u + 9}$

46. $k(s) = \sqrt{-1 - s}$

47. $F(w) = \dfrac{2w + 3}{w - 4}$

48. $G(m) = \dfrac{1 - 2m}{m^2 + 3}$

49. $H(n) = \dfrac{3n + 7}{n^2 + n - 2}$

50. $K(v) = \dfrac{2v^2 - 9}{v^2 - v - 6}$

51. $I(y) = \sqrt{y^2 - 2y - 3}$

52. $J(z) = \sqrt{z^2 - 2z + 5}$

53. $f(x) = \sqrt{\dfrac{x^2 - 4x + 5}{x^2 + 2x + 3}}$

54. $g(t) = \sqrt{\dfrac{t - 1}{5 - t}}$

55. $F(y) = \sqrt{\dfrac{y - 3}{y + 2}}$

56. $G(z) = \sqrt{\dfrac{z^2 + 4z + 6}{z^2 + 4z + 3}}$

The verbal statement "function f multiplies the square of the domain element by 3 and then subtracts 7 from the result" and the algebraic statement "$f(x) = 3x^2 - 7$" define the same function. In Problems 57–60, translate each verbal definition of a function into an algebraic definition.

57. Function g subtracts 5 from twice the cube of the domain element.

58. Function f multiplies the domain element by -3 and adds 4 to the result.

59. Function G multiplies the square root of the domain element by 2 and subtracts the square of the domain element from the result.

60. Function F multiplies the cube of the domain element by -8 and adds three times the square root of 3 to the result.

In Problems 61–64, translate each algebraic definition of the function into a verbal definition.

61. $f(x) = 2x - 3$

62. $g(x) = -2x + 7$

63. $F(x) = 3x^3 - 2\sqrt{x}$

64. $G(x) = 4\sqrt{x} - x^2$

65. If $F(s) = 3s + 15$, find: $\dfrac{F(2 + h) - F(2)}{h}$

66. If $K(r) = 7 - 4r$, find: $\dfrac{K(1 + h) - K(1)}{h}$

67. If $g(x) = 2 - x^2$, find: $\dfrac{g(3 + h) - g(3)}{h}$

68. If $P(m) = 2m^2 + 3$, find: $\dfrac{P(2 + h) - P(2)}{h}$

69. If $L(w) = -2w^2 + 3w - 1$, find: $\dfrac{L(-2 + h) - L(-2)}{h}$

70. If $D(p) = -3p^2 - 4p + 9$, find: $\dfrac{D(-1 + h) - D(-1)}{h}$

71. Find $f(x)$, given that
$$f(x + h) = 3(x + h)^2 - 5(x + h) + 9$$

72. Find $g(w)$, given that
$$g(w + h) = -4(w + h)^3 + 7(w + h) - 5$$

73. Find $m(t)$, given that
$$m(t + h) = -2(t + h)^2 - 5\sqrt{t + h} - 2$$

74. Find $s(z)$, given that
$$s(z + h) = 3(z + h) + 9\sqrt{z + h} + 1$$

C _____

In Problems 75–82, find and simplify:

(A) $\dfrac{f(x + h) - f(x)}{h}$ (B) $\dfrac{f(x) - f(a)}{x - a}$

75. $f(x) = 4x - 7$

76. $f(x) = -5x + 2$

77. $f(x) = 2x^2 - 4$

78. $f(x) = 5 - 3x^2$

79. $f(x) = -4x^2 + 3x - 2$

80. $f(x) = 3x^2 - 5x - 9$

81. $f(x) = x^3 - 2x$

82. $f(x) = x^2 - x^3$

83. The area of a rectangle is 64 square inches. Express the perimeter $P(w)$ as a function of the width w and state the domain.

84. The perimeter of a rectangle is 50 inches. Express the area $A(w)$ as a function of the width w and state the domain.

85. The altitude of a right triangle is 5 meters. Express the hypotenuse $h(b)$ as a function of the base b and state the domain.

86. The altitude of a right triangle is 4 meters. Express the base $b(h)$ as a function of the hypotenuse h and state the domain.

_____

APPLICATIONS

Most of the applications in this section are calculus-related. That is, similar problems will appear in a calculus course, but additional analysis of the functions will be required.

87. Cost Function. The fixed costs per day for a doughnut shop are $300, and the variable costs are $1.75 per dozen doughnuts produced. If x dozen doughnuts are produced daily, express the daily cost $C(x)$ as a function of x.

88. Cost Function. The fixed costs per day for a ski manufacturer are $3,750, and the variable costs are $68 per pair of skis produced. If x pairs of skis are produced daily, express the daily cost $C(x)$ as a function of x.

89. Physics—Rate. The distance in feet that an object falls in a vacuum is given by $s(t) = 16t^2$, where t is time in seconds. Find:

(A) $s(0)$, $s(1)$, $s(2)$, $s(3)$

(B) $\dfrac{s(2 + h) - s(2)}{h}$

(C) What happens in part B when h tends to 0? Interpret physically.

90. Physics—Rate. An automobile starts from rest and travels along a straight and level road. The distance in feet traveled by the automobile is given by $s(t) = 10t^2$, where t is time in seconds. Find:

(A) $s(8)$, $s(9)$, $s(10)$, $s(11)$

(B) $\dfrac{s(11 + h) - s(11)}{h}$

(C) What happens in part B as h tends to 0? Interpret physically.

91. Manufacturing. A candy box is to be made out of a piece of cardboard that measures 8 by 12 inches. Squares, x inches on a side, will be cut from each corner, and then the ends and sides will be folded down (see the figure). Find a formula for the volume of the box $V(x)$ in terms of x. From practical considerations, what is the domain of the function V?

92. Construction. A rancher has 20 miles of fencing to fence a rectangular piece of grazing land along a straight river. If no fence is required along the river and the sides perpendicular to the river are x miles long, find a formula for the area $A(x)$ of the rectangle in terms of x. From practical considerations, what is the domain of the function A?

★ 93. Construction. The manager of an animal clinic wants to construct a kennel with four individual pens, as indicated in the figure. State law requires that each pen have a gate 3 feet wide and an area of 50 square feet. If x is the width of one pen, express the total amount of fencing $F(x)$ (excluding the gates) required for the construction of the kennel as a function of x. Complete the following table [round values of $F(x)$ to one decimal place]:

x	4	5	6	7
$F(x)$				

Figure for 93

★ 94. Architecture. An architect wants to design a window with an area of 24 square feet in the shape of a rectangle surmounted by a semicircle, as indicated in the figure. If x is the width of the window, express the perimeter $P(x)$ of the window as a function of x. Complete the table below [round each value of $P(x)$ to one decimal place]:

x	4	5	6	7
$P(x)$				

★ 95. Construction. A freshwater pipeline is to be run from a source on the edge of a lake to a small resort community on an island 8 miles offshore, as indicated in the figure. It costs $10,000 per mile to lay the pipe on land and $15,000 per mile to lay the pipe in the lake. Express the total cost $C(x)$ of constructing the pipeline as a function of x. From practical considerations, what is the domain of the function C?

★ 96. Weather. An observation balloon is released at a point 10 miles from the station that receives its signal and rises vertically, as indicated in the figure. Express the distance $d(h)$ between the balloon and the receiving station as a function of the altitude h of the balloon.

Figure for 96

★★ **97. Operational Costs.** The cost per hour for fuel for running a train is $v^2/5$ dollars, where v is the speed in miles per hour. (Note that cost goes up as the square of the speed.) Other costs, including labor, are $400 per hour. Express the total cost of a 500-mile trip as a function of the speed v.

★★ **98. Operational Costs.** Refer to Problem 97. If it takes t hours for the train to complete the 500-mile trip, express the total cost as a function of t.

SECTION 3-4 Graphing Functions

- Basic Concepts
- Linear Functions
- Quadratic Functions
- Piecewise-Defined Functions
- The Greatest Integer Function

In this section we take another look at the graphs of linear equations, this time using the function concepts introduced in the preceding section. We also develop procedures for graphing functions defined by quadratic equations and functions formed by "piecing" together two or more other functions. We begin by discussing some general concepts related to the graphs of functions.

• Basic Concepts

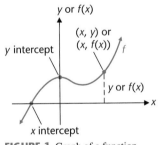

FIGURE 1 Graph of a function.

Each function that has a real number domain and range has a graph—the graph of the ordered pairs of real numbers that constitute the function. When functions are graphed, domain values usually are associated with the horizontal axis and range values with the vertical axis. Thus, the **graph of a function f** is the same as the graph of the equation

$$y = f(x)$$

where x is the independent variable and the abscissa of a point on the graph of f. The variables y and $f(x)$ are dependent variables, and either is the ordinate of a point on the graph of f (see Fig. 1)

The abscissa of a point where the graph of a function intersects the x axis is called an **x intercept** or **zero** of the function. The x intercept is also a real solution or **root** of the equation $f(x) = 0$. The ordinate of a point where the graph of a function crosses the y axis is called the **y intercept** of the function. The y intercept is given by $f(0)$, provided 0 is in the domain of f. Note that a function can have more than one x intercept but can never have more than one y intercept—a consequence of the vertical line test discussed in the preceding section.

The domain of a function is the set of all the x coordinates of points on the graph of the function, and the range is the set of all the y coordinates. It is instructive to view the domain and range as subsets of the coordinate axes as in Figure 2 on page 224. Note the effective use of interval notation in describing the domain and range of the functions in this figure. In Figure 2(a) a solid dot is used to indicate that a point is on the graph of the function and in Figure 2(b) an open dot to indicate that a point is not on the graph of the function. An open or solid dot at the end of a graph indicates that the graph terminates there, while an arrowhead indicates that the graph continues beyond the portion shown with no significant changes in shape [see Fig. 2(b)].

FIGURE 2 Domain and range.

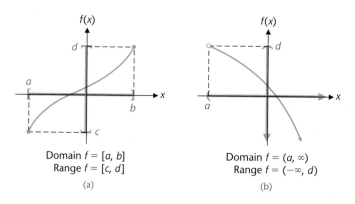

Domain $f = [a, b]$
Range $f = [c, d]$

(a)

Domain $f = (a, \infty)$
Range $f = (-\infty, d)$

(b)

EXAMPLE 1 | **Finding the Domain and Range from a Graph**

Find the domain and range of the function f in Figure 3:

Solution | The dots at each end of the graph of f indicate that the graph terminates at these points. Thus, the x coordinates of the points on the graph are between -3 and 6. The open dot at $(-3, 4)$ indicates that -3 is not in the domain of f, while the closed dot at $(6, -3)$ indicates that 6 is in the domain of f. That is,

$$\text{Domain: } -3 < x \le 6 \quad \text{or} \quad (-3, 6]$$

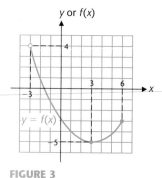

FIGURE 3

The y coordinates are between -5 and 4, and, as before, the open dot at $(-3, 4)$ indicates that 4 is not in the range of f and the closed dot at $(3, -5)$ indicates that -5 is in the range of f. Thus,

$$\text{Range: } -5 \le y < 4 \quad \text{or} \quad [-5, 4)$$

Matched Problem 1 | Find the domain and range of the function f given by the graph in Figure 4.

FIGURE 4

In general, you will not be expected to be able to determine the range of every function defined by an equation you will encounter in this course, but for certain basic functions collected in our library of elementary functions, it is important to know both the domain and range of each, and these will be discussed as these elementary functions are introduced. One of the primary goals of this course is to provide you with a library of basic mathematical functions, including their graphs and other important properties. These can then be used to analyze graphs and properties of a wide variety of more complex functions that arise quite naturally in important applications. In this section we begin this process by introducing some of the language commonly used to describe the behavior of a graph.

We now take a look at increasing and decreasing properties of functions. Intuitively, a function is increasing over an interval I in its domain if its graph rises as the independent variable increases over I. A function is decreasing over I if its graph falls as the independent variable increases over I (Fig. 5).

FIGURE 5 Increasing, decreasing, and constant functions.

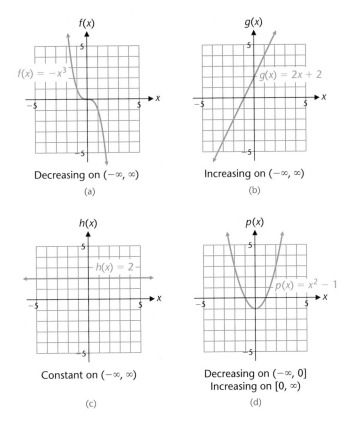

More formally, we define increasing, decreasing, and constant functions as follows:

DEFINITION 1   **Increasing, Decreasing, and Constant Functions**

Let I be an interval in the domain of a function f. Then:

1. f **is increasing** on I if $f(b) > f(a)$ whenever $b > a$ in I.

2. f **is decreasing** on I if $f(b) < f(a)$ whenever $b > a$ in I.

3. f **is constant** on I if $f(a) = f(b)$ for all a and b in I.

• **Linear Functions**   We now apply the general concepts discussed above to a specific class of functions known as *linear functions*.

DEFINITION 2   **Linear Function**

A function f is a **linear function** if

$$f(x) = mx + b \qquad m \neq 0$$

where m and b are real numbers.

Graphing a linear function is equivalent to graphing the equation

$$y = mx + b$$

which we recognize as the equation of a line with slope m and y intercept b. Since the expression $mx + b$ represents a real number for all real number replacements of x, the domain of a linear function is the set of all real numbers. The restriction $m \neq 0$ in the definition of a linear function implies that the graph is not a horizontal line. Hence, the range of a linear function is also the set of all real numbers.

Graph of $f(x) = mx + b$, $m \neq 0$

The graph of a linear function f is a nonvertical and nonhorizontal straight line with slope m and y intercept b.

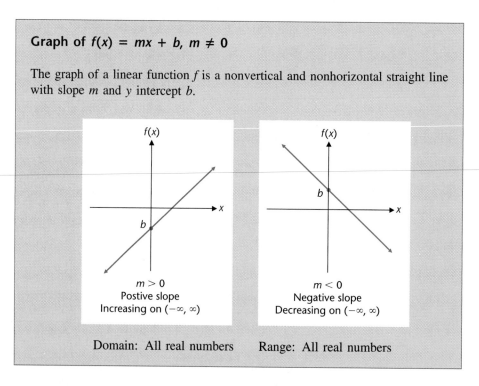

$m > 0$	$m < 0$
Postive slope	Negative slope
Increasing on $(-\infty, \infty)$	Decreasing on $(-\infty, \infty)$

Domain: All real numbers   Range: All real numbers

Notice that two types of lines are not the graphs of linear functions. A vertical line with equation $x = a$ does not pass the vertical line test and cannot define a function. A horizontal line with equation $y = b$ does pass the vertical line test and does define a function. However, a function of the form

$$f(x) = b \qquad \text{Constant function}$$

is called a **constant function,** not a linear function.

EXPLORE-DISCUSS 1   (A) Is it possible for a linear function to have two x intercepts? No x intercepts? If either of your answers is yes, give an example.

(B) Is it possible for a linear function to have two y intercepts? No y intercept? If either of your answers is yes, give an example.

(C) Discuss the possible number of x and y intercepts for a constant function.

EXAMPLE 2   **Graphing a Linear Function**

Find the slope and intercepts, and then sketch the graph of the linear function defined by

$$f(x) = -\tfrac{2}{3}x + 4$$

Check with a graphing utility.

Solution   The y intercept is $f(0) = 4$, and the slope is $-\tfrac{2}{3}$. To find the x intercept, we solve the equation $f(x) = 0$ for x:

$$f(x) = 0$$
$$-\tfrac{2}{3}x + 4 = 0$$
$$-\tfrac{2}{3}x = -4$$
$$x = (-\tfrac{3}{2})(-4) = 6 \quad \text{x intercept}$$

The graph of f is shown in Figure 6.

To find the y intercept with a graphing utility, simply evaluate the function at $x = 0$ [Fig. 7(a)]. Most graphing utilities have a built-in procedure for approximating x intercepts, usually called *root* or *zero* [Fig. 7(b)].

FIGURE 6

FIGURE 7

y intercept
(a)

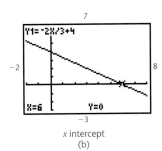

x intercept
(b)

Matched Problem 2   Find the slope and intercepts, and then sketch the graph of the linear function defined by

$$f(x) = \tfrac{3}{2}x - 6$$

● **Quadratic Functions**

Just as we used the first-degree polynomial $mx + b$, $m \neq 0$, to define a linear function, we use the second-degree polynomial $ax^2 + bx + c$, $a \neq 0$, to define a *quadratic function.*

DEFINITION 3

Quadratic Function

A function f is a **quadratic function** if

$$f(x) = ax^2 + bx + c \qquad a \neq 0 \tag{1}$$

where a, b, and c are real numbers.

The graphs of three quadratic functions are shown in Figure 8. The graph of a quadratic function is called a **parabola.**

FIGURE 8 Graphs of quadratic functions.

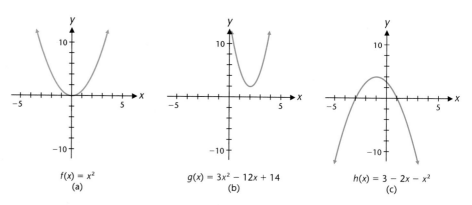

$f(x) = x^2$
(a)

$g(x) = 3x^2 - 12x + 14$
(b)

$h(x) = 3 - 2x - x^2$
(c)

Since the expression $ax^2 + bx + c$ represents a real number for all real number replacements of x:

The domain of a quadratic function is the set of all real numbers.

The range of a quadratic function and many important features of its graph can be determined by first transforming equation (1) by completing the square into the form

$$f(x) = a(x - h)^2 + k \tag{2}$$

A brief review of completing the square, which we discussed in Section 2-6, might prove helpful at this point. We illustrate this method through an example and then generalize the results.

Consider the quadratic function given by

$$f(x) = 2x^2 - 8x + 4 \tag{3}$$

We start by transforming equation (3) into form (2) by completing the square as follows:

$$f(x) = 2x^2 - 8x + 4$$

$$= 2(x^2 - 4x) + 4 \qquad \text{Factor the coefficient of } x^2 \text{ out of the first two terms.}$$

$$= 2(x^2 - 4x + ?) + 4$$

$$= 2(x^2 - 4x + 4) + 4 - 8 \qquad \text{We add 4 to complete the square inside the parentheses. But because of the 2 outside the parentheses, we have actually added 8, so we must subtract 8.}$$

$$= 2(x - 2)^2 - 4 \qquad \text{The transformation is complete.}$$

Thus,

$$f(x) = 2(x - 2)^2 - 4 \qquad (4)$$

If $x = 2$, then $2(x - 2)^2 = 0$ and $f(2) = -4$. For any other value of x, the positive number $2(x - 2)^2$ is added to -4, thus making $f(x)$ larger. Therefore,

$$f(2) = -4$$

is the *minimum value* of $f(x)$ for all x—a very important result! Furthermore, if we choose any two x values that are equidistant from the vertical line $x = 2$, we will obtain the same value for the function. For example, $x = 1$ and $x = 3$ are each one unit from $x = 2$, and the corresponding functional values are

$$f(1) = 2(-1)^2 - 4 = -2$$
$$f(3) = 2(1)^2 - 4 = -2$$

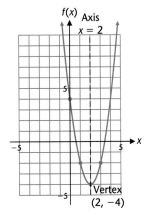

FIGURE 9 Graph of $f(x) = 2(x - 2)^2 - 4$.

Thus, the vertical line $x = 2$ is a line of symmetry. That is, if the graph is drawn on a piece of paper and the paper is folded along the line $x = 2$, then the two sides of the parabola will match exactly. All these results are illustrated by graphing equations (3) or (4) and the line $x = 2$ in the same coordinate system. (Fig. 9).

From the above discussion, we see that as x moves from left to right, $f(x)$ is decreasing on $(-\infty, 2]$ and increasing on $[2, \infty)$. Furthermore, $f(x)$ can assume any values greater than or equal to -4, but no values less than -4. Thus,

$$\text{Range of } f\!: y \geq -4 \qquad \text{or} \qquad [-4, \infty)$$

In general, the graph of a quadratic function is a parabola with line of symmetry parallel to the vertical axis. The lowest or highest point on the parabola, whichever exists, is called the **vertex.** The maximum or minimum value of a quadratic function always occurs at the vertex of the parabola. The line of symmetry through the vertex is called the **axis** of the parabola. In the above example, $x = 2$ is the axis of the parabola and $(2, -4)$ is its vertex.

Note the important results we have obtained by transforming equation (3) into equation (4):

The vertex of the parabola

The axis of the parabola

The minimum value of $f(x)$

The range of the function f

Now, let us explore the effect of changing the constants a, h, and k on the graph of $y = a(x - h)^2 + k$.

EXPLORE-DISCUSS 2   Explore the effect of changing the constants a, h, and k on the graph of $f(x) = a(x - h)^2 + k$.

(A) Let $a = 1$ and $h = 5$. Graph function f for $k = -4$, 0, and 3 simultaneously in the same coordinate system. Explain the effect of changing k on the graph of f.

(B) Let $a = 1$ and $k = 2$. Graph function f for $h = -4$, 0, 5 simultaneously in the same coordinate system. Explain the effect of changing h on the graph of f.

(C) Let $h = 5$ and $k = -2$. Graph function f for $a = 0.25$, 1, and 3 simultaneously in the same coordinate system. Graph function f for $a = 1$, -1, and -0.25 simultaneously in the same coordinate system.

(D) Discuss parts A–C using a graphing utility and a standard viewing window.

The above discussion is generalized for all quadratic functions in the following box:

Properties of a Quadratic Function and Its Graph

Given a quadratic function and the form obtained by completing the square

$$f(x) = ax^2 + bx + c = a(x - h)^2 + k \qquad a \neq 0$$

we summarize general properties as follows:

1. The graph of f is a parabola:

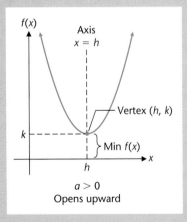

$a > 0$
Opens upward

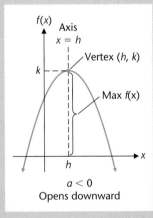

$a < 0$
Opens downward

2. Vertex: (h, k) (Parabola rises on one side of the vertex and falls on the other.)

3. Axis (of symmetry): $x = h$ (Parallel to y axis)

> **4.** $f(h) = k$ is the minimum if $a > 0$ and the maximum if $a < 0$.
>
> **5.** Domain: All real numbers
>
> Range: $(-\infty, k]$ if $a < 0$ or $[k, \infty)$ if $a > 0$

EXAMPLE 3   **Graph of a Quadratic Function**

Graph, finding the vertex, axis, maximum or minimum of $f(x)$, intervals where f is increasing or decreasing, and range.

$$f(x) = -0.5x^2 - x + 2$$

Solution   Complete the square:

$$
\begin{aligned}
f(x) &= -0.5x^2 - x + 2 \\
&= -0.5(x^2 + 2x + \;?) + 2 \\
&= -0.5(x^2 + 2x + 1) + 2 + 0.5 \\
&= -0.5(x + 1)^2 + 2.5
\end{aligned}
$$

From this last form we see that $h = -1$ and $k = 2.5$. Thus, the vertex is $(-1, 2.5)$, the axis of symmetry is $x = -1$, and the maximum value is $f(-1) = 2.5$. To graph f, locate the axis and vertex; then plot several points on either side of the axis (Fig. 10).

FIGURE 10

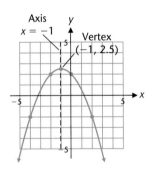

x	$f(x)$
-4	-2
-2	2
-1	2.5
0	2
2	-2

From the graph we see that f is increasing on $(-\infty, -1]$ and decreasing on $[-1, \infty)$. Also, $y = f(x)$ can be any number less than or equal to 2.5. Thus, the range of f is $y \leq 2.5$ or $(-\infty, 2.5]$.

Matched Problem 3   Graph, finding the vertex, axis, maximum or minimum of $f(x)$, intervals where f is increasing or decreasing, and range.

$$f(x) = -x^2 + 4x - 4$$

• Piecewise-Defined Functions

The **absolute value function** can be defined using the definition of absolute value from Section 2-4:

$$f(x) = |x| = \begin{cases} -x & \text{if } x < 0 \\ x & \text{if } x \geq 0 \end{cases}$$

Notice that this function is defined by different formulas for different parts of its domain. Functions whose definitions involve more than one formula are called **piecewise-defined functions.** As the next example illustrates, piecewise-defined functions occur naturally in many applications.

EXAMPLE 4 **Rental Charges**

A car rental agency charges $0.25 per mile if the total mileage does not exceed 100. If the total mileage exceeds 100, the agency charges $0.25 per mile for the first 100 miles plus $0.15 per mile for the additional mileage. If x represents the number of miles a rented vehicle is driven, express the mileage charge $C(x)$ as a function of x. Find $C(50)$ and $C(150)$, and graph C.

Solution   If $0 \leq x \leq 100$, then

$$C(x) = 0.25x$$

If $x > 100$, then

$$
\begin{array}{cc}
\text{Charge for the} & \text{Charge for the} \\
\text{first 100 miles} & \text{additional mileage}
\end{array}
$$

$$
\begin{aligned}
C(x) &= \quad 0.25(100) \quad + \quad 0.15(x - 100) \\
&= \qquad 25 \qquad + \quad 0.15x - 15 \\
&= 10 + 0.15x
\end{aligned}
$$

Thus, we see that C is a piecewise-defined function:

$$C(x) = \begin{cases} 0.25x & \text{if } 0 \leq x \leq 100 \\ 10 + 0.15x & \text{if } x > 100 \end{cases}$$

Piecewise-defined functions are evaluated by first determining which rule applies and then using the appropriate rule to find the value of the function. For example, to evaluate $C(50)$, we use the first rule and obtain

$$C(50) = 0.25(50) = \$12.50 \quad x = 50 \text{ satisfies } 0 \leq x \leq 100$$

To evaluate $C(150)$, we use the second rule and obtain

$$C(150) = 10 + 0.15(150) = \$32.50 \quad x = 150 \text{ satisfies } x > 100$$

To graph C, we graph each rule in the definition for the indicated values of x (Fig. 11).

FIGURE 11

x	$y = 0.25x$
50	12.5
100	25

x	$y = 10 + 0.15x$
100	25
150	32.5

Notice that the two formulas produce the same value at $x = 100$ and that the graph of C contains no breaks. Informally, a graph (or portion of a graph) is said to be **continuous** if it contains no breaks or gaps. (A formal presentation of continuity may be found in calculus texts.)

Matched Problem 4   Refer to Example 4. Find $C(x)$ if the agency charges $0.30 per mile when the total mileage does not exceed 75, and $0.30 per mile for the first 75 miles plus $0.20 per mile for additional mileage when the total mileage exceeds 75. Find $C(50)$ and $C(100)$, and graph C.

EXAMPLE 5   **Graphing a Function Involving Absolute Value**

Graph the function f given by

$$f(x) = x + \frac{x}{|x|}$$

and find its domain and range.
Check with a graphing utility.

Solution   We use the piecewise definition of $|x|$ to find a piecewise definition of f that does not involve $|x|$.
If $x < 0$, then $|x| = -x$ and

$$f(x) = x + \frac{x}{|x|} = x + \frac{x}{-x} = x - 1$$

If $x = 0$, then f is not defined, since division by 0 is not permissible.
If $x > 0$, then $|x| = x$ and

$$f(x) = x + \frac{x}{|x|} = x + \frac{x}{x} = x + 1$$

FIGURE 12

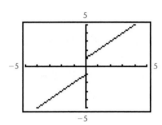

FIGURE 13

Thus, a piecewise definition for f is

$$f(x) = \begin{cases} x - 1 & \text{if } x < 0 \\ x + 1 & \text{if } x > 0 \end{cases}$$

Domain: $x \neq 0$ or $(-\infty, 0) \cup (0, \infty)$

We use this definition to graph f as shown in Figure 12. Examining this graph, we see that $y = f(x)$ can be any number less than -1 or any number greater than 1. Thus,

Range: $y < -1$ or $y > 1$ or $(-\infty, -1) \cup (1, \infty)$

Notice that we used open dots in the figure at $(0, -1)$ and $(0, 1)$ to indicate that these points do not belong to the graph of f. Because of the break in the graph at $x = 0$, we say that f is **discontinuous** at $x = 0$.

The check is shown in Figure 13. Most graphing utilities denote the absolute value function by abs(x)—check your manual.

Matched Problem 5   Graph the function f given by

$$f(x) = -\frac{2x}{|x|} - x$$

and find its domain and range.

• **The Greatest Integer Function**

We conclude this section with a discussion of an interesting and useful function called the *greatest integer function*.

The **greatest integer** of a real number x, denoted by $[\![x]\!]$, is the integer n such that $n \leq x < n + 1$; that is, $[\![x]\!]$ is the largest integer less than or equal to x. For example,

$$[\![3.45]\!] = 3 \qquad [\![-2.13]\!] = -3 \quad \text{Not } -2$$
$$[\![7]\!] = 7 \qquad [\![-8]\!] = -8$$
$$[\![0]\!] = 0$$

FIGURE 14 Greatest integer function.

The **greatest integer function** f is defined by the equation $f(x) = [\![x]\!]$. A piecewise definition of f for $-2 \leq x < 3$ is shown at the top of the next page and a sketch of the graph of f for $-5 \leq x \leq 5$ is shown in Figure 14. Since the domain of f is all real numbers, the piecewise definition continues indefinitely in both directions, as does the stairstep pattern in the figure. Thus, the range of f is the set of all integers. The greatest integer function is an example of a more general class of functions called **step functions**.

$$f(x) = [\![x]\!] = \begin{cases} \vdots \\ -2 & \text{if } -2 \leq x < -1 \\ -1 & \text{if } -1 \leq x < 0 \\ 0 & \text{if } 0 \leq x < 1 \\ 1 & \text{if } 1 \leq x < 2 \\ 2 & \text{if } 2 \leq x < 3 \\ \vdots \end{cases}$$

Notice in Figure 14 that at each integer value of x there is a break in the graph, and between integer values of x there is no break. Thus, the greatest integer function is discontinuous at each integer n and continuous on each interval of the form $[n, n + 1)$.

EXPLORE-DISCUSS 3   Most graphing utilities denote the greatest integer function as int (x), although not all define it the same way we have here. Graph $y = $ int (x) for $-5 \leq x \leq 5$ and $-5 \leq y \leq 5$ and discuss any differences between your graph and Figure 14. If your graphing utility supports both a connected mode and a dot mode for graphing functions (consult your manual), which mode is preferable for this graph?

EXAMPLE 6   **Computer Science**

Let

$$f(x) = \frac{[\![10x + 0.5]\!]}{10}$$

Find:

(A) $f(6)$ (B) $f(1.8)$ (C) $f(3.24)$ (D) $f(4.582)$ (E) $f(-2.68)$

What operation does this function perform?

Solutions   (A) $f(6) = \dfrac{[\![60.5]\!]}{10} = \dfrac{60}{10} = 6$

(B) $f(1.8) = \dfrac{[\![18.5]\!]}{10} = \dfrac{18}{10} = 1.8$

(C) $f(3.24) = \dfrac{[\![32.9]\!]}{10} = \dfrac{32}{10} = 3.2$

(D) $f(4.582) = \dfrac{[\![46.32]\!]}{10} = \dfrac{46}{10} = 4.6$

(E) $f(-2.68) = \dfrac{[\![-26.3]\!]}{10} = \dfrac{-27}{10} = -2.7$

Comparing the values of x and $f(x)$ in Table 1 in the margin, we conclude that this function rounds decimal fractions to the nearest tenth.

Matched Problem 6   Let $f(x) = [\![x + 0.5]\!]$. Find:

(A) $f(6)$ (B) $f(1.8)$ (C) $f(3.24)$ (D) $f(-4.3)$ (E) $f(-2.69)$

What operation does this function perform?

TABLE 1

x	$f(x)$
6	6
1.8	1.8
3.24	3.2
4.582	4.6
−2.68	−2.7

Answers to Matched Problems

1. Domain: $-4 < x < 5$
   or $(-4, 5)$
   Range: $-4 < y \le 3$
   or $(-4, 3]$

2. y intercept: $f(0) = -6$
   x intercept: 4
   Slope: $\frac{3}{2}$

3. Axis: $x = 2$
   Vertex: $(2, f(2)) = (2, 0)$
   Max $f(x)$: $f(2) = 0$
   Increasing: $(-\infty, 2]$
   Decreasing: $[2, \infty)$
   Range: $(-\infty, f(2)] = (-\infty, 0]$

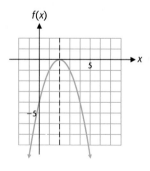

4. $C(x) = \begin{cases} 0.3x & \text{if } 0 \le x \le 75 \\ 7.5 + 0.2x & \text{if } x > 75 \end{cases}$
   $C(50) = \$15;\ C(100) = \$27.50$

5. $f(x) = \begin{cases} 2 - x & \text{if } x < 0 \\ -2 - x & \text{if } x > 0 \end{cases}$
   Domain: $x \ne 0$ or $(-\infty, 0) \cup (0, \infty)$
   Range: $(-\infty, -2) \cup (2, \infty)$

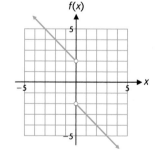

6. (A) 6   (B) 2   (C) 3   (D) −4
   (E) −3;
   f rounds decimal fractions to the nearest integer.

EXERCISE **3-4**

A

Problems 1–6 refer to functions f, g, h, k, p, and q given by the following graphs. (Assume the graphs continue as indicated beyond the parts shown.)

1. For the function f, find:
  (A) Domain   (B) Range
  (C) x intercepts   (D) y intercept
  (E) Intervals over which f is increasing
  (F) Intervals over which f is decreasing
  (G) Intervals over which f is constant
  (H) Any points of discontinuity

2. Repeat Problem 1 for the function g.

3. Repeat Problem 1 for the function h.

4. Repeat Problem 1 for the function k.

5. Repeat Problem 1 for the function p.

6. Repeat Problem 1 for the function q.

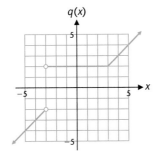

Problems 7–12 describe the graph of a continuous function f over the interval [−5, 5]. Sketch the graph of a function that is consistent with the given information.

7. The function f is increasing on [−5, −2], constant on [−2, 2], and decreasing on [2, 5].

8. The function f is decreasing on [−5, −2], constant on [−2, 2], and increasing on [2, 5].

9. The function f is decreasing on [−5, −2], constant on [−2, 2], and decreasing on [2, 5].

10. The function f is increasing on [−5, −2], constant on [−2, 2], and increasing on [2, 5].

11. The function f is decreasing on [−5, −2], increasing on [−2, 2], and decreasing on [2, 5].

12. The function f is increasing on [−5, −2], decreasing on [−2, 2], and increasing on [2, 5].

In Problems 13–16, find the slope and intercepts, and then sketch the graph.

13. $f(x) = 2x + 4$ **14.** $f(x) = 3x - 3$

15. $f(x) = -\frac{1}{2}x - \frac{5}{3}$ **16.** $f(x) = -\frac{3}{4}x + \frac{6}{5}$

In Problems 17 and 18, find a linear function f satisfying the given conditions.

17. $f(-2) = 7$ and $f(4) = -2$

18. $f(-3) = -2$ and $f(5) = 4$

B

In Problems 19–22, graph, finding the axis, vertex, maximum or minimum, and range.

19. $f(x) = (x - 3)^2 + 2$

20. $f(x) = \frac{1}{2}(x + 2)^2 - 4$

21. $f(x) = -(x + 3)^2 - 2$

22. $f(x) = -(x - 2)^2 + 4$

In Problems 23–26, graph, finding the axis, vertex, x intercepts, and y intercept.

23. $f(x) = x^2 - 4x - 5$ **24.** $f(x) = x^2 - 6x + 5$

25. $f(x) = -x^2 + 6x$ **26.** $f(x) = -x^2 + 2x + 8$

In Problems 27–30, graph, finding the axis, vertex, intervals over which f is increasing, and intervals over which f is decreasing.

27. $f(x) = x^2 + 6x + 11$ **28.** $f(x) = x^2 - 8x + 14$

29. $f(x) = -x^2 + 6x - 6$ **30.** $f(x) = -x^2 - 10x - 24$

In Problems 31–38, graph, finding the domain, range, and any points of discontinuity.

31. $f(x) = \begin{cases} x + 1 & \text{if } -1 \le x < 0 \\ -x + 1 & \text{if } 0 \le x \le 1 \end{cases}$

32. $f(x) = \begin{cases} x & \text{if } -2 \le x < 1 \\ -x + 2 & \text{if } 1 \le x \le 2 \end{cases}$

33. $f(x) = \begin{cases} -2 & \text{if } -3 \le x < -1 \\ 4 & \text{if } -1 < x \le 2 \end{cases}$

34. $f(x) = \begin{cases} 1 & \text{if } -2 \le x < 2 \\ -3 & \text{if } 2 < x \le 5 \end{cases}$

35. $f(x) = \begin{cases} x + 2 & \text{if } x < -1 \\ x - 2 & \text{if } x \ge -1 \end{cases}$

36. $f(x) = \begin{cases} -1 - x & \text{if } x \le 2 \\ 5 - x & \text{if } x > 2 \end{cases}$

37. $g(x) = \begin{cases} x^2 + 1 & \text{if } x < 0 \\ -x^2 - 1 & \text{if } x > 0 \end{cases}$

38. $h(x) = \begin{cases} -x^2 - 2 & \text{if } x < 0 \\ x^2 + 2 & \text{if } x > 0 \end{cases}$

C

In Problems 39–44, graph, finding the axis, vertex, maximum or minimum of f(x), range, intercepts, intervals over which f is increasing, and intervals over which f is decreasing.

39. $f(x) = \frac{1}{2}x^2 + 2x + 3$ **40.** $f(x) = 2x^2 - 12x + 14$

41. $f(x) = 4x^2 - 12x + 9$ **42.** $f(x) = -\frac{1}{2}x^2 + 4x - 10$

43. $f(x) = -2x^2 - 8x - 2$ **44.** $f(x) = -4x^2 - 4x - 1$

In Problems 45–50, find a piecewise definition of f that does not involve the absolute value function (see Example 5). Sketch the graph, and find the domain, range, and any points of discontinuity.

 Check your graphs in Problems 45–50 by graphing the given definition of f on a graphing utility.

45. $f(x) = \dfrac{|x|}{x}$ **46.** $f(x) = x|x|$

47. $f(x) = x + \dfrac{|x - 1|}{x - 1}$ **48.** $f(x) = x + 2\dfrac{|x + 1|}{x + 1}$

49. $f(x) = |x| + |x - 2|$ **50.** $f(x) = |x| - |x - 3|$

In Problems 51–56, write a piecewise definition for f (see the discussion of Fig. 14 in this section) and sketch the graph of f. Include sufficient intervals to clearly illustrate both the definition and the graph. Find the domain, range, and any points of discontinuity.

 Check your graphs in Problems 51–56 by graphing the given definition of f on a graphing utility.

51. $f(x) = [\![x/2]\!]$ **52.** $f(x) = [\![x/3]\!]$

53. $f(x) = [\![3x]\!]$ **54.** $f(x) = [\![2x]\!]$

55. $f(x) = x - [\![x]\!]$ **56.** $f(x) = [\![x]\!] - x$

57. Given that f is a quadratic function with min $f(x) = f(2) = 4$, find the axis, vertex, range, and x intercepts.

58. Given that f is a quadratic function with max $f(x) = f(-3) = -5$, find the axis, vertex, range, and x intercepts.

59. The function f is continuous and increasing on the interval $[1, 9]$ with $f(1) = -5$ and $f(9) = 4$.
(A) Sketch a graph of f that is consistent with the given information.
(B) How many times does your graph cross the x axis? Could the graph cross more times? Fewer times? Support your conclusions with additional sketches and/or verbal arguments.

60. Repeat Problem 59 if the function does not have to be continuous.

61. The function f is continuous on the interval $[-5, 5]$ with $f(-5) = -4$, $f(1) = 3$, and $f(5) = -2$.
(A) Sketch a graph of f that is consistent with the given information.
(B) How many times does your graph cross the x axis? Could the graph cross more times? Fewer times? Support your conclusions with additional sketches and/or verbal arguments.

62. Repeat Problem 61 if f is continuous on $[-8, 8]$ with $f(-8) = -6$, $f(-4) = 3$, $f(3) = -2$, and $f(8) = 5$.

∫ Problems 63–66 are calculus-related. In geometry, a line that intersects a circle in two distinct points is called a secant line, as shown in figure (a). In calculus, the line through the points $(x_1, f(x_1))$ and $(x_2, f(x_2))$ is called a **secant line** for the graph of the function f, as shown in figure (b).

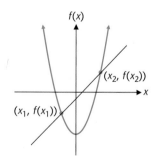

Secant line for a circle
(a)

Secant line for the graph of a function
(b)

In Problems 63 and 64, find the equation of the secant line through the indicated points on the graph of f. Graph f and the secant line on the same coordinate system.

63. $f(x) = x^2 - 4; (-1, -3), (3, 5)$

64. $f(x) = 9 - x^2; (-2, 5), (4, -7)$

65. Let $f(x) = x^2 - 3x + 5$. If h is a nonzero real number, then $(2, f(2))$ and $(2 + h, f(2 + h))$ are two distinct points on the graph of f.
(A) Find the slope of the secant line through these two points.
(B) Evaluate the slope of the secant line for $h = 1$, $h = 0.1$, $h = 0.01$, and $h = 0.001$. What value does the slope seem to be approaching?

66. Repeat Problem 65 for $f(x) = x^2 + 2x - 6$.

 Problems 67–74 require the use of a graphing utility.

In Problems 67–72, first graph functions f and g in the same viewing window, then graph m(x) and n(x) in their own viewing windows:

$$m(x) = 0.5[f(x) + g(x) + |f(x) - g(x)|]$$

$$n(x) = 0.5[f(x) + g(x) - |f(x) - g(x)|]$$

67. $f(x) = -2x, g(x) = 0.5x$

68. $f(x) = 3x + 1, g(x) = -0.5x - 4$

69. $f(x) = 5 - 0.2x^2, g(x) = 0.3x^2 - 4$

70. $f(x) = 0.15x^2 - 5, g(x) = 5 - 1.5|x|$

71. $f(x) = 0.2x^2 - 0.4x - 5, g(x) = 0.3x - 3$

72. $f(x) = 8 + 1.5x - 0.4x^2, g(x) = -0.2x + 5$

73. How would you characterize the relationship between f, g, and m in Problems 67–72? [*Hint:* See Problem 89 in Exercise 2-4.]

74. How would you characterize the relationship between f, g, and n in Problems 67–72? [*Hint:* See Problem 90 in Exercise 2-4.]

APPLICATIONS

75. Tire Mileage. An automobile tire manufacturer collected the data in Table 1 relating tire pressure x, in pounds per square inch (lb/in.^{2}) and mileage, in thousands of miles.

TABLE 1

x	28	30	32	34	36
Mileage	45	52	55	51	47

A mathematical model for this data is given by

$$f(x) = -0.518x^2 + 33.3x - 481$$

(A) Complete Table 2. Round values of $f(x)$ to one decimal place.

TABLE 2

x	28	30	32	34	36
Mileage	45	52	55	51	47
$f(x)$					

(B) Sketch the graph of f and the mileage data on the same axes.
(C) Use values of the modeling function rounded to two decimal places to estimate the mileage for a tire pressure of 31 lb/in.^{2}. For 35 lb/in.^{2}.
(D) Write a brief description of the relationship between tire pressure and mileage.

76. Automobile Production. Table 3 lists General Motors' total U.S. vehicle production in millions of units from 1989 to 1993.

TABLE 3

Year	89	90	91	92	93
Production	4.7	4.1	3.5	3.7	5.0

A mathematical model for GM's production data is given by

$$f(x) = 0.33x^2 - 1.3x + 4.8$$

where $x = 0$ corresponds to 1989.

(A) Complete Table 4. Round values of x to one decimal place.

TABLE 4

x	0	1	2	3	4
Production	4.7	4.1	3.5	3.7	5.0
$f(x)$					

(B) Sketch the graph of f and the production data on the same axes.
(C) Use values of the modeling function f rounded to two decimal places to estimate the production in 1994. In 1995.
(D) Write a brief verbal description of GM's production from 1989 to 1993.

77. Physics—Spring Stretch. Hooke's law states that the relationship between the stretch s of a spring and the weight w causing the stretch is linear (a principle upon which all spring scales are constructed). A 10-pound weight stretches a spring 1 inch, while with no weight the stretch of the spring is 0.

(A) Find a linear function $f: s = f(w) = mw + b$ that represents this relationship. [*Hint:* Both points (10, 1) and (0, 0) are on the graph of f.]

(B) Find $f(15)$ and $f(30)$—that is, the stretch of the spring for 15-pound and 30-pound weights.

(C) What is the slope of the graph of f? (The slope indicates the increase in stretch for each pound increase in weight.)

(D) Graph f for $0 \le w \le 40$.

78. Business—Depreciation. An electronic computer was purchased by a company for $20,000 and is assumed to have a salvage value of $2,000 after 10 years. Its value is depreciated linearly from $20,000 to $2,000.

(A) Find the linear function $f: V = f(t)$ that relates value V in dollars to time t in years.

(B) Find $f(4)$ and $f(8)$, the values of the computer after 4 and 8 years, respectively.

(C) Find the slope of the graph of f. (The slope indicates the decrease in value per year.)

(D) Graph f for $0 \le t \le 10$.

★ **79. Sales Commissions.** An appliance salesperson receives a base salary of $200 a week and a commission of 4% on all sales over $3,000 during the week. In addition, if the weekly sales are $8,000 or more, the salesperson receives a $100 bonus. If x represents weekly sales (in dollars), express the weekly earnings $E(x)$ as a function of x, and sketch its graph. Identify any points of discontinuity. Find $E(5,750)$ and $E(9,200)$.

★ **80. Service Charges.** On weekends and holidays, an emergency plumbing repair service charges $2.00 per minute for the first 30 minutes of a service call and $1.00 per minute for each additional minute. If x represents the duration of a service call in minutes, express the total service charge $S(x)$ as a function of x, and sketch its graph. Identify any points of discontinuity. Find $S(25)$ and $S(45)$.

81. Construction. A rectangular dog pen is to be made with 100 feet of fence wire.

(A) If x represents the width of the pen, express its area $A(x)$ in terms of x.

(B) Considering the physical limitations, what is the domain of the function A?

(C) Graph the function for this domain.

(D) Determine the dimensions of the rectangle that will make the area maximum.

82. Construction. Rework Problem 81 with the added assumption that an existing property fence will be used for one side of the pen. (Let x = Width; see the figure.)

Existing fence

83. Computer Science. Let $f(x) = 10[\![0.5 + x/10]\!]$. Evaluate f at 4, -4, 6, -6, 24, 25, 247, -243, -245, and -246. What operation does this function perform?

84. Computer Science. Let $f(x) = 100[\![0.5 + x/100]\!]$. Evaluate f at 40, -40, 60, -60, 740, 750, 7,551, -601, -649, and -651. What operation does this function perform?

★ **85. Computer Science.** Use the greatest integer function to define a function f that rounds real numbers to the nearest hundredth.

★ **86. Computer Science.** Use the greatest integer function to define a function f that rounds real numbers to the nearest thousandth.

87. Delivery Charges. A nationwide package delivery service charges $15 for overnight delivery of packages weighing 1 pound or less. Each additional pound (or fraction thereof) costs an additional $3. Let $C(x)$ be the charge for overnight delivery of a package weighing x pounds.

(A) Write a piecewise definition of C for $0 < x \le 6$, and sketch the graph of C by hand.

(B) Can the function f defined by $f(x) = 15 + 3[\![x]\!]$ be used to compute the delivery charges for all x, $0 < x \le 6$? Justify your answer.

88. Telephone Charges. Calls to 900 numbers are charged to the caller. A 900 number hot line for tips and hints for video games charges $4 for the first minute of the call and $2 for each additional minute (or fraction thereof). Let $C(x)$ be the charge for a call lasting x minutes.

(A) Write a piecewise definition of C for $0 < x \le 6$, and sketch the graph of C by hand.

(B) Can the function f defined by $f(x) = 4 + 2[\![x]\!]$ be used to compute the charges for all x, $0 < x \le 6$? Justify your answer.

★★ **89. Car Rental.** A car rental agency rents 300 cars a day at a rate of $40 per day. For each $1 increase in rate, five fewer cars are rented. At what rate should the cars be rented to produce the maximum income? What is the maximum income?

★★ **90. Rental Income.** A 400-room hotel in Las Vegas is filled to capacity every night at $70 a room. For each $1 increase in rent, four fewer rooms are rented. If each rented room costs $10 to service per day, how much should the management charge for each room to maximize profit? What is the maximum profit?

★★ **91. Physics.** A stunt driver is planning to jump a motorcycle from one ramp to another as illustrated in the figure. The ramps are 10 feet high, and the distance between the ramps is 80 feet. The trajectory of the cycle through the air is given by the graph of

$$f(x) = \frac{1}{4}x - \left(\frac{16}{v^2}\right)x^2$$

where v is the velocity of the cycle in feet per second as it leaves the ramp.

(A) How fast must the cycle be traveling when it leaves the ramp in order to follow the trajectory illustrated in the figure?

(B) What is the maximum height of the cycle above the ground as it follows this trajectory?

★★ **92. Physics.** The trajectory of a circus performer shot from a cannon is given by the graph of the function

$$f(x) = x - \frac{1}{100}x^2$$

Both the cannon and the net are 10 feet high (see the figure).

(A) How far from the muzzle of the cannon should the center of the net be placed so that the performer lands in the center of the net?

(B) What is the maximum height of the performer above the ground?

SECTION 3-5 Combining Functions

- Operations on Functions
- Composition
- Elementary Functions
- Vertical and Horizontal Shifts
- Reflections, Expansions, and Contractions

If two functions f and g are both defined at a real number x, and if $f(x)$ and $g(x)$ are both real numbers, then it is possible to perform real number operations such as addition, subtraction, multiplication, or division with $f(x)$ and $g(x)$. Furthermore, if $g(x)$ is a number in the domain of f, then it is also possible to evaluate f at $g(x)$. In this section we see how operations on the values of functions can be used to define operations on the functions themselves. We also investigate the graphic implications of some of these operations.

• Operations on Functions

The functions f and g given by

$$f(x) = 2x + 3 \quad \text{and} \quad g(x) = x^2 - 4$$

are defined for all real numbers. Thus, for any real x we can perform the following operations:

$$f(x) + g(x) = 2x + 3 + x^2 - 4 = x^2 + 2x - 1$$
$$f(x) - g(x) = 2x + 3 - (x^2 - 4) = -x^2 + 2x + 7$$
$$f(x)g(x) = (2x + 3)(x^2 - 4) = 2x^3 + 3x^2 - 8x - 12$$

For $x \neq \pm 2$ we can also form the quotient

$$\frac{f(x)}{g(x)} = \frac{2x + 3}{x^2 - 4} \qquad x \neq \pm 2$$

Notice that the result of each operation is a new function. Thus, we have

$$(f + g)(x) = f(x) + g(x) = x^2 + 2x - 1 \qquad \text{Sum}$$
$$(f - g)(x) = f(x) - g(x) = -x^2 + 2x + 7 \qquad \text{Difference}$$
$$(fg)(x) = f(x)g(x) = 2x^3 + 3x^2 - 8x - 12 \qquad \text{Product}$$
$$\left(\frac{f}{g}\right)(x) = \frac{f(x)}{g(x)} = \frac{2x + 3}{x^2 - 4} \qquad x \neq \pm 2 \qquad \text{Quotient}$$

Notice that the sum, difference, and product functions are defined for all values of x, as were f and g, but the domain of the quotient function must be restricted to exclude those values where $g(x) = 0$.

DEFINITION 1   **Operations on Functions**

The **sum, difference, product,** and **quotient** of the functions f and g are the functions defined by

$$(f + g)(x) = f(x) + g(x) \qquad \text{Sum function}$$
$$(f - g)(x) = f(x) - g(x) \qquad \text{Difference function}$$
$$(fg)(x) = f(x)g(x) \qquad \text{Product function}$$
$$\left(\frac{f}{g}\right)(x) = \frac{f(x)}{g(x)} \qquad g(x) \neq 0 \quad \text{Quotient function}$$

Each function is defined on the intersection of the domains of f and g, with the exception that the values of x where $g(x) = 0$ must be excluded from the domain of the quotient function.

EXAMPLE 1   **Finding the Sum, Difference, Product, and Quotient Functions**

Let $f(x) = \sqrt{4 - x}$ and $g(x) = \sqrt{3 + x}$. Find the functions $f + g, f - g, fg$, and f/g, and find their domains.

Solution

$$(f + g)(x) = f(x) + g(x) = \sqrt{4 - x} + \sqrt{3 + x}$$
$$(f - g)(x) = f(x) - g(x) = \sqrt{4 - x} - \sqrt{3 + x}$$
$$(fg)(x) = f(x)g(x) = \sqrt{4 - x}\sqrt{3 + x}$$
$$= \sqrt{(4 - x)(3 + x)}$$
$$= \sqrt{12 + x - x^2}$$

Check your graphs in Problems 35–40 by graphing f, g, and the line y = x in a squared viewing window on a graphing utility.

35. $f(x) = 3x + 6; g(x) = \frac{1}{3}x - 2$

36. $f(x) = -\frac{1}{2}x + 2; g(x) = -2x + 4$

37. $f(x) = 4 + x^2, x \geq 0; g(x) = \sqrt{x - 4}$

38. $f(x) = \sqrt{x + 2}; g(x) = x^2 - 2, x \geq 0$

39. $f(x) = -\sqrt{x - 2}; g(x) = x^2 + 2, x \leq 0$

40. $f(x) = 6 - x^2, x \leq 0; g(x) = -\sqrt{6 - x}$

The functions in Problems 41–60 are one-to-one. Find f^{-1}.

41. $f(x) = \frac{1}{5}x$

42. $f(x) = 4x$

43. $f(x) = 2x + 7$

44. $f(x) = 0.25x + 2.25$

45. $f(x) = 0.2x + 0.4$

46. $f(x) = 7 - 8x$

47. $f(x) = 3 - \dfrac{2}{x}$

48. $f(x) = 5 + \dfrac{4}{x}$

49. $f(x) = \dfrac{2x}{x + 1}$

50. $f(x) = \dfrac{4x}{2 - x}$

51. $f(x) = \dfrac{0.2x - 0.4}{0.1x + 0.5}$

52. $f(x) = \dfrac{x - 0.2}{x + 0.5}$

53. $f(x) = 8x^3 - 5$

54. $f(x) = 2x^5 + 9$

55. $f(x) = 2 + \sqrt[5]{3x - 7}$

56. $f(x) = -1 + \sqrt[3]{4 - 5x}$

57. $f(x) = 2\sqrt{9 - x}$

58. $f(x) = 3\sqrt{x - 4}$

59. $f(x) = 2 + \sqrt{3 - x}$

60. $f(x) = 4 - \sqrt{x + 5}$

61. How are the *x* and *y* intercepts of a function and its inverse related?

62. Does a constant function have an inverse? Explain.

C

The functions in Problems 63–66 are one-to-one. Find f^{-1}.

63. $f(x) = (x - 1)^2 + 2, x \geq 1$

64. $f(x) = 3 - (x - 5)^2, x \leq 5$

65. $f(x) = x^2 + 2x - 2, x \leq -1$

66. $f(x) = x^2 + 8x + 7, x \geq -4$

The graph of each function in Problems 67–70 is one-quarter of the graph of the circle with radius 3 and center (0, 0). Find f^{-1}, find the domain and range of f^{-1}, and sketch the graphs of f and f^{-1} in the same coordinate system.

67. $f(x) = -\sqrt{9 - x^2}, 0 \leq x \leq 3$

68. $f(x) = \sqrt{9 - x^2}, 0 \leq x \leq 3$

69. $f(x) = \sqrt{9 - x^2}, -3 \leq x \leq 0$

70. $f(x) = -\sqrt{9 - x^2}, -3 \leq x \leq 0$

The graph of each function in Problems 71–74 is one-quarter of the graph of the circle with radius 1 and center (0, 1). Find f^{-1}, find the domain and range of f^{-1}, and sketch the graphs of f and f^{-1} in the same coordinate system.

71. $f(x) = 1 + \sqrt{1 - x^2}, 0 \leq x \leq 1$

72. $f(x) = 1 - \sqrt{1 - x^2}, 0 \leq x \leq 1$

73. $f(x) = 1 - \sqrt{1 - x^2}, -1 \leq x \leq 0$

74. $f(x) = 1 + \sqrt{1 - x^2}, -1 \leq x \leq 0$

75. Find $f^{-1}(x)$ for $f(x) = ax + b, a \neq 0$.

76. Find $f^{-1}(x)$ for $f(x) = \sqrt{a^2 - x^2}, a > 0, 0 \leq x \leq a$.

77. Refer to Problem 75. For which *a* and *b* is *f* its own inverse?

78. How could you recognize the graph of a function that is its own inverse?

79. Show that the line through the points (a, b) and (b, a), $a \neq b$, is perpendicular to the line $y = x$ (see the figure).

80. Show that the point $((a + b)/2, (a + b)/2)$ bisects the line segment from (a, b) to $(b, a), a \neq b$ (see the figure).

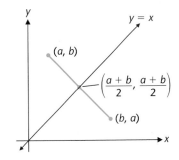

In Problems 81–84, the function f is not one-to-one. Find the inverses of the functions formed by restricting the domain of f as indicated.

 Check Problems 81–84 by graphing f, g, and the line y = x in a squared viewing window on a graphing utility. [Hint: To restrict the graph of y = f(x) to an interval of the form $a \leq x \leq b$, enter $y = f(x)/((a \leq x)*(x \leq b))$.]

81. $f(x) = (2 - x)^2$:
    (A) $x \leq 2$ (B) $x \geq 2$

82. $f(x) = (1 + x)^2$:
    (A) $x \leq -1$ (B) $x \geq -1$

83. $f(x) = \sqrt{4x - x^2}$:
    (A) $0 \leq x \leq 2$ (B) $2 \leq x \leq 4$

84. $f(x) = \sqrt{6x - x^2}$:
    (A) $0 \leq x \leq 3$ (B) $3 \leq x \leq 6$

CHAPTER 3 GROUP ACTIVITY Mathematical Modeling in Business*

This group activity is concerned with analyzing a basic model for manufacturing and selling a product by using tables of data and linear regression to determine appropriate values for the constants a, b, m, and n in the following functions:

TABLE 1 Business Modeling Functions

Function	Definition	Interpretation
Price–demand	$p(x) = m - nx$	x is the number of items that can be sold at $\$p$ per item
Cost	$C(x) = a + bx$	Total cost of producing x items
Revenue	$R(x) = xp$ $= x(m - nx)$	Total revenue from the sale of x items
Profit	$P(x) = R(x) - C(x)$	Total profit from the sale of x items

A manufacturing company manufactures and sells mountain bikes. The management would like to have price–demand and cost functions for break-even and profit–loss analysis. Price–demand and cost functions may be established by collecting appropriate data at different levels of output, and then finding a model in the form of a basic elementary function (from our library of elementary functions) that "closely fits" the collected data. The financial department, using statistical techniques, arrived at the price–demand and cost data in Tables 2 and 3, where p is the wholesale price of a bike for a demand of x thousand bikes and C is the cost, in thousands of dollars, of producing and selling x thousand bikes.

TABLE 2 Price–Demand

x (thousand)	p($)
7	530
13	360
19	270
25	130

TABLE 3 Cost

x (thousand)	C (thousand $)
5	2,100
12	2,940
19	3,500
26	3,920

(A) Building a Mathematical Model for Price–Demand.   Plot the data in Table 2 and observe that the relationship between p and x is almost linear. After observing a relationship between variables, analysts often try to model the relationship in terms of a basic function, from a portfolio of elementary functions, which "best fits" the data.

1. **Linear regression lines** are frequently used to model linear phenomena. This is a process of fitting to a set of data a straight line that minimizes the sum of the squares of the distances of all the points in the graph of the data to the line by using the **method of least squares.** Many graphing utilities have this routine built in. Read your user's manual for your particular graphing utility, and discuss among the members of the

*This group project may be done without the use of a graphing utility, but significant additional insight into mathematical modeling will be gained if one is available.

group how this is done. After obtaining the linear regression line for the data in Table 2, graph the line and the data in the same viewing window.

2. The linear regression line found in part 1 is a mathematical model for the price–demand function and is given by

$$p(x) = 666.5 - 21.5x \quad \text{Price–demand function}$$

Graph the data points from Table 2 and the price–demand function in the same rectangular coordinate system.

3. The linear regression line defines a linear price–demand function. Interpret the slope of the function. Discuss its domain and range. Using the mathematical model, determine the price for a demand of 10,000 bikes. For a demand of 20,000 bikes.

(B) Building a Mathematical Model for Cost. Plot the data in Table 3 in a rectangular coordinate system. Which type of function appears to best fit the data?

1. Fit a linear regression line to the data in Table 3. Then plot the data points and the line in the same viewing window.

2. The linear regression line found in part 1 is a mathematical model for the cost function and is given by

$$C(x) = 86x + 1{,}782 \quad \text{Cost function}$$

Graph the data points from Table 3 and the cost function in the same rectangular coordinate system.

3. Interpret the slope and the y intercept of the cost function. Discuss its domain and range. Using the mathematical model, determine the cost for an output and sales of 10,000 bikes. For an output and sales of 20,000 bikes.

(C) Break-Even and Profit–Loss Analysis. Write an equation for the revenue function and state its domain. Write the equation for the profit function and state its domain.

1. Graph the revenue function and the cost function simultaneously in the same rectangular coordinate system. Algebraically determine at what outputs (to the nearest unit) the company breaks even. Determine where costs exceed revenues and revenues exceed costs.

2. Graph the revenue function and the cost function simultaneously in the same viewing window. Graphically determine at what outputs (to the nearest unit) the company breaks even and where costs exceed revenues and revenues exceed costs.

3. Graph the profit function in a rectangular coordinate system. Algebraically determine at what outputs (to the nearest unit) the company breaks even. Determine where profits occur and where losses occur. At what output and price will a maximum profit occur? Do the maximum revenue and maximum profit occur for the same output? Discuss.

4. Graph the profit function in a graphing utility. Graphically determine at what outputs (to the nearest unit) the company breaks even and where losses occur and profits occur. At what output and price will a maximum profit occur? Do the maximum revenue and maximum profit occur for the same output? Discuss.

Chapter 3 Review

3-1   BASIC TOOLS; CIRCLES

A **Cartesian** or **rectangular coordinate system** is formed by the intersection of a horizontal real number line and a vertical real number line at their origins. These lines are called the **coordinate axes.** The **horizontal axis** is often referred to as the *x* **axis** and the **vertical axis** as the *y* **axis.** These axes divide the plane into four **quadrants.** Each point in the plane corresponds to its **coordinates**—an ordered pair (a, b) determined by passing horizontal and vertical lines through the point. The **abscissa** or *x* **coordinate** a is the coordinate of the intersection of the vertical line with the horizontal axis, and the **ordinate** or *y* **coordinate** b is the coordinate of the intersection of the horizontal line with the vertical axis. The point $(0, 0)$ is called the **origin.** The **solution set** of an equation in two variables is the set of all ordered pairs of real numbers that make the equation a true statement. The **graph of an equation in two variables** is the graph of its solution set formed using **point-by-point plotting** or with the aid of a **graphing utility.**

The **distance between the two points** $P_1(x_1, y_1)$ and $P_2(x_2, y_2)$ is

$$d(P_1, P_2) = \sqrt{(x_2 - x_1)^2 + (y_2 - y_1)^2}$$

The **standard equations for a circle** are

$$(x - h)^2 + (y - k)^2 = r^2 \qquad \text{Radius: } r > 0,$$
$$\text{Center: } (h, k)$$
$$x^2 + y^2 = r^2 \qquad \text{Radius: } r > 0,$$
$$\text{Center: } (0, 0)$$

Using a **squared window** will improve the appearance of a circle on a graphing utility.

3-2   STRAIGHT LINES

The **standard form** for the equation of a line is $Ax + By = C$, where A, B, and C are constants, A and B not both 0. The *y* **intercept** is the ordinate of the point where the graph crosses the *y* axis, and the *x* **intercept** is the abscissa of the point where the graph crosses the *x* axis. The **slope** of the line through the points (x_1, y_1) and (x_2, y_2) is

$$m = \frac{y_2 - y_1}{x_2 - x_1} \qquad \text{if } x_1 \neq x_2$$

The slope is not defined for a vertical line where $x_1 = x_2$. Two lines with slopes m_1 and m_2 are **parallel** if and only if $m_1 = m_2$ and **perpendicular** if and only if $m_1 m_2 = -1$.

Equations of a Line

Standard form	$Ax + By = C$	A and B not both 0
Slope–intercept form	$y = mx + b$	Slope: m; y intercept: b
Point–slope form	$y - y_1 = m(x - x_1)$	Slope: m; Point: (x_1, y_1)
Horizontal line	$y = b$	Slope: 0
Vertical line	$x = a$	Slope: Undefined

3-3   FUNCTIONS

A **function** is a **rule** that produces a correspondence between two sets of elements such that to each element in the first set there corresponds one and only one element in the second set. The first set is called the **domain** and the set of all corresponding elements in the second set is called the **range.** Equivalently, a **function** is a **set of ordered pairs** with the property that no two ordered pairs have the same first component and different second components. The **domain** is the set of all first components, and the **range** is the set of all second components. An **equation** in two variables **defines a function** if to each value of the **independent variable,** the placeholder for domain values, there corresponds exactly one value of the **dependent variable,** the placeholder for range values. A **vertical line** will intersect the graph of a function in at most one point. Unless otherwise specified, the **domain of a function defined by an equation** is assumed to be the set of all real number replacements for the independent variable that produce real values for the dependent variable. The symbol $f(x)$ represents the real number in the range of the function f corresponding to the domain value x. Equivalently, the ordered pair $(x, f(x))$ belongs to the function f.

3-4   GRAPHING FUNCTIONS

The **graph of a function** f is the graph of the equation $y = f(x)$. The abscissa of any point where the graph of a function f crosses the *x* axis is called an *x* **intercept** of f. The ordinate of a point where the graph crosses the *y* axis is called the *y* **intercept.**

Let I be an interval in the domain of a function f. Then:

1. f **is increasing** on I if $f(b) > f(a)$ whenever $b > a$ in I.

2. f **is decreasing** on I if $f(b) < f(a)$ whenever $b > a$ in I.

3. f **is constant** on I if $f(a) = f(b)$ for all a and b in I.

A function f is a **linear function** if $f(x) = mx + b$, $m \neq 0$, and a **quadratic function** if $f(x) = ax^2 + bx + c$, $a \neq 0$. The graph of a linear function is a straight line that is neither horizontal nor vertical.

Properties of $f(x) = ax^2 + bx + c = a(x - h)^2 + k$, $a \neq 0$, and its graph:

1. The graph of f is a parabola:

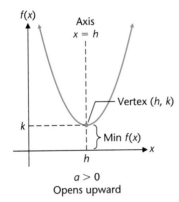

$a > 0$
Opens upward

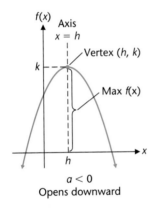

$a < 0$
Opens downward

2. Vertex: (h, k) (Parabola increases on one side of the vertex and decreases on the other.)

3. Axis (of symmetry): $x = h$ (Parallel to y axis)

4. $f(h) = k$ is the minimum if $a > 0$ and the maximum if $a < 0$.

5. Domain: All real numbers

Range: $(-\infty, k]$ if $a < 0$ or $[k, \infty)$ if $a > 0$

A **piecewise-defined function** is a function whose definition involves more than one formula. The graph of a function is **continuous** if it has no holes or breaks and **discontinuous** at any point where it has a hole or break. The **greatest integer function** is defined by

$$f(x) = [\![x]\!] = n \qquad \text{where } n \text{ is an integer}, n \leq x < n + 1$$

3-5 COMBINING FUNCTIONS

The **sum, difference, product,** and **quotient** of the functions f and g are defined by

$$(f + g)(x) = f(x) + g(x) \qquad (f - g)(x) = f(x) - g(x)$$

$$(fg)(x) = f(x)g(x) \qquad \left(\frac{f}{g}\right)(x) = \frac{f(x)}{g(x)} \qquad g(x) \neq 0$$

The **domain** of each function is the intersection of the domains of f and g, with the exception that values of x where $g(x) = 0$ must be excluded from the domain of f/g.

The **composition** of functions f and g is defined by $(f \circ g)(x) = f[g(x)]$. The **domain** of $f \circ g$ is the set of all real numbers x in the domain of g where $g(x)$ is in the domain of f. The domain of $f \circ g$ is always a subset of the domain of g.

Vertical Translation:

$y = f(x) + k$, $k > 0$	Shift the graph of $y = f(x)$ upward k units.
$y = f(x) - k$, $k > 0$	Shift the graph of $y = f(x)$ downward k units.

Horizontal Translation:

$y = f(x - h)$, $h > 0$	Shift the graph of $y = f(x)$ to the right h units.
$y = f(x + h)$, $h > 0$	Shift the graph of $y = f(x)$ to the left h units.

Reflection:

$y = -f(x)$	Reflect the graph of $y = f(x)$ in the x axis.

Expansion and Contraction:

$y = Cf(x)$, $C > 1$	Expand the graph of $y = f(x)$ by multiplying each ordinate value by C.
$y = Cf(x)$, $0 < C < 1$	Contract the graph of $y = f(x)$ by multiplying each ordinate value by C.

3-6 INVERSE FUNCTIONS

A function is **one-to-one** if no two ordered pairs in the function have the same second component and different first components. A **horizontal line** will intersect the graph of a one-to-one function in at most one point. A function that is increasing (decreasing) throughout its domain is one-to-one. The **inverse** of the one-to-one function f is the function f^{-1} formed by reversing all the ordered pairs in f. If f is not one-to-one, then f^{-1} **does not exist.**

Assuming that f^{-1} exists, then:

1. f^{-1} is one-to-one.

2. Domain of f^{-1} = Range of f.

3. Range of f^{-1} = Domain of f.

4. $x = f^{-1}(y)$ if and only if $y = f(x)$.

5. $f^{-1}[f(x)] = x$ for all x in the domain of f.

6. $f[f^{-1}(x)] = x$ for all x in the domain of f^{-1}.

7. To find f^{-1}, solve the equation $y = f(x)$ for x and then interchange x and y.

8. The graphs of $y = f(x)$ and $y = f^{-1}(x)$ are symmetric with respect to the line $y = x$.

Chapter 3 Review Exercise

Work through all the problems in this chapter review and check answers in the back of the book. Answers to all review problems are there, and following each answer is a number in italics indicating the section in which that type of problem is discussed. Where weaknesses show up, review appropriate sections in the text.

(A)

A

1. Given the points $A(-2, 3)$ and $B(4, 0)$, find:
  (A) Distance between A and B
  (B) Slope of the line through A and B
  (C) Slope of a line perpendicular to the line through A and B

2. Write the equation of a circle with radius $\sqrt{7}$ and center:
  (A) $(0, 0)$ (B) $(3, -2)$

3. Find the center and radius of the circle given by

$$(x + 3)^2 + (y - 2)^2 = 5$$

4. Graph $3x + 2y = 9$ and indicate its slope.

5. Write an equation of a line with x intercept 6 and y intercept 4. Write the final answer in the standard form $Ax + By = C$, where A, B, and C are integers.

6. Write the slope–intercept form of the equation of the line with slope $-\frac{2}{3}$ and y intercept 2.

7. Write the equations of the vertical and horizontal lines passing through the point $(-3, 4)$. What is the slope of each?

8. Indicate whether each set defines a function. Find the domain and range of each function.
  (A) $\{(1, 1), (2, 4), (3, 9)\}$
  (B) $\{(1, 1), (1, -1), (2, 2), (2, -2)\}$
  (C) $\{(-2, 2), (-1, 2), (0, 2), (1, 2), (2, 2)\}$

9. Indicate whether each graph specifies a function:

(B)

(C)

(D)

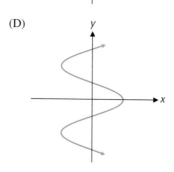

10. Which of the following equations define functions?
(A) $y = x$ (B) $y^2 = x$
(C) $y^3 = x$ (D) $|y| = x$

Problems 11–20 refer to the functions f, g, k, and m given by:

$$f(x) = 3x + 5 \qquad g(x) = 4 - x^2$$
$$k(x) = 5$$
$$m(x) = 2|x| - 1$$

Find the indicated quantities or expressions.

11. $f(2) + g(-2) + k(0)$

12. $\dfrac{m(-2) + 1}{g(2) + 4}$

13. $\dfrac{f(2 + h) - f(2)}{h}$

14. $\dfrac{g(a + h) - g(a)}{h}$

15. $(f + g)(x)$

16. $(f - g)(x)$

17. $(fg)(x)$

18. $\left(\dfrac{f}{g}\right)(x)$

19. $(f \circ g)(x)$

20. $(g \circ f)(x)$

21. Sketch a graph of each of the functions in parts A–D using the graph of function f in the figure.
(A) $y = -f(x)$ (B) $y = f(x) + 4$
(C) $y = f(x - 2)$ (D) $y = -f(x + 3) - 3$

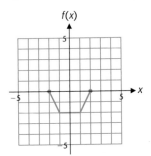

22. Match each equation with a graph of one of the functions f, g, m, or n in the figure. Each graph is a graph of one of the equations and is assumed to continue without bound beyond the portion shown.
(A) $y = (x - 2)^2 - 4$ (B) $y = -(x + 2)^2 + 4$
(C) $y = -(x - 2)^2 + 4$ (D) $y = (x + 2)^2 - 4$

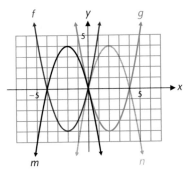

23. Referring to the graph of function f in the figure for Problem 22 and using known properties of quadratic functions, find each of the following to the nearest integer:
(A) Intercepts   (B) Vertex
(C) Maximum or minimum   (D) Range
(E) Increasing interval   (F) Decreasing interval

24. Find the maximum or minimum value of $f(x) = x^2 - 6x + 11$ without graphing. What are the coordinates of the vertex of the graph?

25. How are the graphs of the following related to the graph of $y = x^2$?
(A) $y = -x^2$ (B) $y = x^2 - 3$
(C) $y = (x + 3)^2$

B _____

Problems 26–32 refer to the function q given by the following graph. (Assume the graph continues as indicated beyond the part shown.)

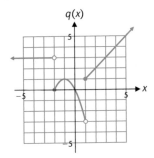

26. Find y to the nearest integer:
(A) $y = q(0)$ (B) $y = q(1)$
(C) $y = q(2)$ (D) $y = q(-2)$

27. Find x to the nearest integer:
(A) $q(x) = 0$ (B) $q(x) = 1$
(C) $q(x) = -3$ (D) $q(x) = 3$

28. Find the domain and range of q.

29. Find the intervals over which q is increasing.

30. Find the intervals over which q is decreasing.

31. Find the intervals over which q is constant.

32. Identify any points of discontinuity.

33. The function f multiplies the cube of the domain element by 4 and then subtracts the square root of the domain element. Write an algebraic definition of f.

34. Write a verbal description of the function $f(x) = 3x^2 + 4x - 6$.

35. (A) Find an equation of the line through $P(-4, 3)$ and $Q(0, -3)$. Write the final answer in the standard form $Ax + By = C$, where A, B, and C are integers with $A > 0$.
(B) Find $d(P, Q)$.

36. Write equations of the lines
   (A) Parallel to   (B) Perpendicular to
   the line $6x + 3y = 5$ and passing through the point $(-2, 1)$. Write the final answers in the slope–intercept form $y = mx + b$.

37. Discuss the graph of $4x^2 + 9y^2 = 36$ relative to symmetry with respect to the x axis, y axis and origin.

38. Find the domain of $g(x) = 1/\sqrt{3 - x}$

39. Graph $f(x) = x^2 - 6x + 5$. Show the axis of symmetry and vertex, and find the range, intercepts, and maximum or minimum value of $f(x)$.

40. Find the domain of $h(x) = 1/(4 - \sqrt{x})$.

41. Given $f(x) = \sqrt{x} - 8$ and $g(x) = |x|$:
   (A) Find $f \circ g$ and $g \circ f$.
   (B) Find the domains of $f \circ g$ and $g \circ f$.

42. Which of the following functions are one-to-one?
   (A) $f(x) = x^3$
   (B) $g(x) = (x - 2)^2$
   (C) $h(x) = 2x - 3$
   (D) $F(x) = (x + 3)^2, x \geq -3$

43. Given $f(x) = 3x - 7$:
   (A) Find $f^{-1}(x)$.
   (B) Find $f^{-1}(5)$.
   (C) Find $f^{-1}[f(x)]$.
   (D) Is f increasing, decreasing, or constant on $(-\infty, \infty)$?

 Check by graphing f, f^{-1}, and $y = x$ in a squared viewing window on a graphing utility.

44. Graph, finding the domain, range, and any points of discontinuity:

$$f(x) = \begin{cases} 2 - x & \text{if } -1 \leq x < 0 \\ x^2 & \text{if } 0 \leq x \leq 1 \end{cases}$$

45. The following graph is the result of applying a sequence of transformations to the graph of $y = x^2$. Describe the transformations verbally and write an equation for the given graph.

 Check by graphing your equation on a graphing utility.

46. The graph of $f(x) = |x|$ is expanded by a factor of 3, reflected in the x axis, and shifted 2 units to the right and 5

units up to form the graph of the function g. Find an equation for the function g and graph g.

47. Write an equation for the following graph in the form $y = a(x - h)^2 + k$, where a is either -1 or $+1$ and h and k are integers.

Check by graphing your equation on a graphing utility.

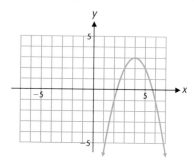

48. Graph:
   (A) $y = |x| - 2$ (B) $y = |x + 1|$
   (C) $y = \frac{1}{2}|x|$

49. Given $f(x) = \sqrt{x - 1}$:
   (A) Find $f^{-1}(x)$.
   (B) Find the domain and range of f and f^{-1}.
   (C) Graph f, f^{-1}, and $y = x$ on the same coordinate system.

 Check by graphing f, f^{-1}, and $y = x$ in a squared window on a graphing utility.

50. Find the equation of a circle that passes through the point $(-1, 4)$ with center at $(3, 0)$.

51. Find the center and radius of the circle given by $x^2 + y^2 + 4x - 6y = 3$.

52. Determine symmetry with respect to the x axis, y axis, and origin and graph $xy = 4$.

53. If the slope of a line is negative, is the linear function represented by the graph of the line increasing, decreasing, or constant on $(-\infty, \infty)$?

54. Given $f(x) = x^2 - 1, x \geq 0$:
   (A) Find the domain and range of f and f^{-1}.
   (B) Find $f^{-1}(x)$.
   (C) Find $f^{-1}(3)$.
   (D) Find $f^{-1}[f(4)]$.
   (E) Find $f^{-1}[f(x)]$.

Check by graphing f, f^{-1}, and $y = x$ in a squared window on a graphing utility.

C

55. The graph at the top of next page is the result of applying a sequence of transformations to the graph of $y = \sqrt[3]{x}$. Describe the transformations verbally, and write an equation for the given graph.

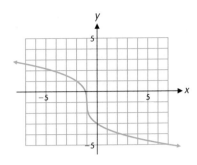

Check by graphing your equation on a graphing utility.

56. How is the graph of $f(x) = -(x - 2)^2 - 1$ related to the graph of $g(x) = x^2$?

57. Graph: $f(x) = -|x + 1| - 1$

58. Find the domain of $f(x) = \sqrt{25 - x^2}$.

59. Given $f(x) = x^2$ and $g(x) = \sqrt{1 - x}$, find each function and its domain.
(A) fg (B) f/g (C) $f \circ g$ (D) $g \circ f$

60. For the one-to-one function f given by

$$f(x) = \frac{x + 2}{x - 3}$$

(A) Find $f^{-1}(x)$.
(B) Find $f^{-1}(3)$.
(C) Find $f^{-1}[f(x)]$.

61. Find a piecewise definition of $f(x) = |x + 1| - |x - 1|$ that does not involve the absolute value function. Find the domain and range of f.

62. Find the equation of the set of points equidistant from $(3, 3)$ and $(6, 0)$. What is the name of the geometric figure formed by this set?

63. Prove that two nonvertical lines are parallel if and only if their slopes are the same.

64. Prove that the lines $mx - y = b$ and $x + my = c$ are perpendicular.

65. Graph:
(A) $f(x) = [|x|]$ (B) $g(x) = |[x]|$

66. Graph in a standard viewing window:

$$f(x) = 0.1(x - 2)^2 + \frac{|3x - 6|}{x - 2}$$

Assuming the graph continues as indicated beyond the part shown in this viewing window, find the domain, range, and any points of discontinuity. (Use the dot mode on your graphing utility, if it has one.)

67. A partial graph of the function f is shown in the figure. Complete the graph of f over the interval $[0, 5]$ given that:
(A) f is symmetric with respect to the y axis.
(B) f is symmetric with respect to the origin.

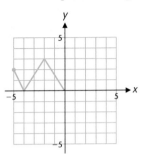

68. The function f is decreasing on $[-5, 5]$ with $f(-5) = 4$ and $f(5) = -3$.
(A) If f is continuous on $[-5, 5]$, how many times can the graph of f cross the x axis? Support your conclusion with examples and/or verbal arguments.
(B) Repeat part A if the function does not have to be continuous.

APPLICATIONS

69. Linear Depreciation. A computer system was purchased by a small company for $12,000 and is assumed to have a depreciated value of $2,000 after 8 years. If the value is depreciated linearly from $12,000 to $2,000:
(A) Find the linear equation that relates value V (in dollars) to time t (in years).
(B) What would be the depreciated value of the system after 5 years?

70. Business—Pricing. A sporting goods store sells tennis shorts that cost $30 for $48 and sunglasses that cost $20 for $32.
(A) If the markup policy of the store for items that cost over $10 is assumed to be linear and is reflected in the pricing of these two items, write an equation that expresses retail price R as a function of cost C.
(B) What should be the retail price of a pair of skis that cost $105?

★ **71. Income.** A salesperson receives a base salary of $200 per week and a commission of 10% on all sales over $3,000 during the week. If x represents the salesperson's weekly sales, express the total weekly earnings $E(x)$ as a function of x. Find $E(2,000)$ and $E(5,000)$.

72. Demand. Egg consumption has been decreasing for some time, presumably because of increasing awareness of the high cholesterol in egg yolks. Table 1 lists the annual per capita consumption of eggs in the United States (*source: Department of Agriculture*).

TABLE 1

Year	1970	1975	1980	1985	1990
Consumption	309	276	271	255	233

A mathematical model for this data is given by

$$f(x) = 303.4 - 3.46x$$

where $x = 0$ corresponds to 1970.
(A) Complete Table 2. Round values of $f(x)$ to the nearest integer.

TABLE 2

x	0	5	10	15	20
Consumption	309	276	271	255	233
$f(x)$					

(B) Graph $y = f(x)$ and the data in the table on the same set of axes.
(C) Use the modeling function f to estimate the per capita egg consumption in 1995. In 2000.
(D) Based on the information in the table, write a brief verbal description of egg consumption from 1970 to 1990.

73. Pricing. An office supply store sells ballpoint pens for $0.49 each. For an order of 3 dozen or more pens, the price per pen for all pens ordered is reduced to $0.44, and for an order of 6 dozen or more, the price per pen for all pens ordered is reduced to $0.39.
(A) If $C(x)$ is the total cost in dollars for an order of x pens, write a piecewise definition for C.
(B) Graph $y = C(x)$ for $0 \le x \le 108$, and identify any points of discontinuity.

74. Break-Even Analysis. A video production company is planning to produce an instructional videotape. The producer estimates that it will cost $84,000 to shoot the video and $15 per unit to copy and distribute the tape. The wholesale price of the tape is $50 per unit.
(A) Write the cost equation and the revenue equation, and graph both simultaneously in the rectangular coordinate system.

(B) Determine when $R = C$; then, with the aid of part A, determine when $R < C$ and $R > C$.

★ **75. Market Research.** If x units of a product are produced each week and sold for a price of p per unit, then the weekly demand, revenue, and cost equations are, respectively,

$$x = 500 - 10p$$
$$R(x) = 50x - \tfrac{1}{10}x^2$$
$$C(x) = 20x + 4{,}000$$

Express the weekly profit as a function of the price p.

★ **76. Architecture.** A circular arc forms the top of an entryway with 6-foot vertical sides 8 feet apart. If the top of the arc is 2 feet above the ends, what is the radius of the arc?

77. Construction. A farmer has 120 feet of fencing to be used in the construction of two identical rectangular pens sharing a common side (see the figure).

(A) Express the total area $A(x)$ enclosed by both pens as a function of the width x.
(B) From physical considerations, what is the domain of the function A?
(C) Find the dimensions of the pens that will make the total enclosed area maximum.

78. Computer Science. In computer programming, it is often necessary to check numbers for certain properties (even, odd, perfect square, etc.). The greatest integer function provides a convenient method for determining some of these properties. For example, the following function can be used to determine whether a number is the square of an integer:

$$f(x) = x - (\llbracket \sqrt{x} \rrbracket)^2$$

(A) Find $f(1)$.   (B) Find $f(2)$.   (C) Find $f(3)$.
(D) Find $f(4)$.   (E) Find $f(5)$.
(F) Find $f(n^2)$, where n is a positive integer.

POLYNOMIAL AND RATIONAL FUNCTIONS

CHAPTER 4

Recall that the zeros of a function f are the solutions or roots of the equation $f(x) = 0$, if any exist. We know how to find all real and imaginary zeros of linear and quadratic functions (see Table 1).

TABLE 1 **Zeros of Linear and Quadratic Functions**

Function	Form	Equation	Zeros/roots
Linear	$f(x) = ax + b, a \neq 0$	$ax + b = 0$	$x = -\dfrac{b}{a}$
Quadratic	$f(x) = ax^2 + bx + c, a \neq 0$	$ax^2 + bx + c = 0$	$x = \dfrac{-b \pm \sqrt{b^2 - 4ac}}{2a}$

Linear and quadratic functions are also called first- and second-degree *polynomial functions,* respectively. Thus, Table 1 contains formulas for the zeros of any first- or second-degree polynomial function. What about higher-degree polynomial functions such as

$$p(x) = 4x^3 - 2x^2 + 3x + 5 \qquad \text{Third degree}$$
$$q(x) = -2x^4 + 5x^2 - 6 \qquad \text{Fourth degree}$$
$$r(x) = x^5 - x^4 + x^3 - 10 \qquad \text{Fifth degree}$$

It turns out that there are direct, though complicated, methods for finding all zeros for any third- or fourth-degree polynomial function. However, the Frenchman Évariste Galois (1811–1832) proved at the age of 20 that for polynomial functions of degree greater than 4 there is no finite step-by-step process that will always yield all zeros.* This does not mean that we give up looking for zeros of higher-degree polynomial functions. It just means that we will have to use a variety of specialized methods, and sometimes we will have to approximate the zeros. The development of these methods is one of the primary objectives of this chapter.

We begin in Section 4-1 by discussing the graphic properties of polynomial functions. In Section 4-2 we develop tools for finding all the rational zeros of a polynomial equation with rational coefficients. In Section 4-3 we discuss methods for locating the real zeros of a polynomial with real coefficients. Once located, the real zeros are easily approximated with a graphing utility. Section 4-4 deals with rational functions and their graphs, and Section 4-5 discusses the decomposition of rational functions into simpler forms, an important tool for calculus.

*Galois' contribution, using the new concept of "group," was of the highest mathematical significance and originality. However, his contemporaries hardly read his papers, dismissing them as "almost unintelligible." At the age of 21, involved in political agitation, Galois met an untimely death in a duel. A short but fascinating account of Galois' tragic life can be found in E. T. Bell's *Men of Mathematics* (New York: Simon & Schuster, 1937), pp. 362–377.

SECTION **4-1**   **Polynomial Functions and Graphs**

- Polynomial Functions
- Polynomial Division
- Division Algorithm
- Remainder Theorem
- Graphing Polynomial Functions

• Polynomial
Functions

In Chapter 3 you were introduced to the basic functions

$$f(x) = b \qquad\qquad\text{Constant function}$$

$$f(x) = ax + b, \qquad a \neq 0 \qquad\text{Linear function}$$

$$f(x) = ax^2 + bx + c, \qquad a \neq 0 \quad\text{Quadratic function}$$

as well as some special cases of more complicated functions such as

$$f(x) = ax^3 + bx^2 + cx + d, \qquad a \neq 0 \quad\text{Cubic function}$$

Notice the evolving pattern going from the constant function to the cubic function—the terms in each equation are of the form ax^n, where n is a nonnegative integer and a is a real number. All these functions are special cases of the general class of functions called *polynomial functions*. The function

$$P(x) = a_n x^n + a_{n-1} x^{n-1} + \cdots + a_1 x + a_0 \qquad a_n \neq 0$$

is called an ***n*th-degree polynomial function.** We will also refer to $P(x)$ as a **polynomial of degree *n*** or, more simply, as a **polynomial.** The numbers $a_n, a_{n-1}, \ldots, a_1,$ a_0 are called the **coefficients of the function.** A nonzero constant function is a zero-degree polynomial, a linear function is a first-degree polynomial, and a quadratic function is a second-degree polynomial. The zero function $Q(x) = 0$ is also considered to be a polynomial but is not assigned a degree. The coefficients of a polynomial function may be complex numbers, or may be restricted to real numbers, rational numbers, or integers, depending on our interests. The domain of a polynomial function can be the set of complex numbers, the set of real numbers, or appropriate subsets of either, depending on our interests. In general, the context will dictate the choice of coefficients and domain.

The number r is said to be a **zero of the function P,** or a **zero of the polynomial $P(x)$,** or a **solution or root of the equation $P(x) = 0$,** if

$$P(r) = 0$$

A zero of a polynomial may or may not be the number 0. A zero of a polynomial is any number that makes the value of the polynomial 0. If the coefficients of a polynomial $P(x)$ are real numbers, then a real zero is simply an x intercept for the graph of $y = P(x)$. Consider the polynomial

$$P(x) = x^2 - 4x + 3$$

The graph of P is shown in Figure 1.

FIGURE 1 Zeros, roots, and x intercepts.

The x intercepts 1 and 3 are zeros of $P(x) = x^2 - 4x + 3$, since $P(1) = 0$ and $P(3) = 0$. The x intercepts 1 and 3 are also solutions or roots for the equation $x^2 - 4x + 3 = 0$.

In general:

Zeros and Roots

If the coefficients of a polynomial $P(x)$ are real, then the x intercepts of the graph of $y = P(x)$ are real **zeros** of P and $P(x)$ and real **solutions,** or **roots,** for the equation $P(x) = 0$.

- **Polynomial Division**

We can find quotients of polynomials by a long-division process similar to that used in arithmetic. An example will illustrate the process.

EXAMPLE 1   **Algebraic Long Division**

Divide $5 + 4x^3 - 3x$ by $2x - 3$.

Solution

$$
\begin{array}{r}
2x^2 + 3x + 3 \\
2x - 3 \overline{)\,4x^3 + 0x^2 - 3x + 5} \\
\underline{4x^3 - 6x^2} \\
6x^2 - 3x \\
\underline{6x^2 - 9x} \\
6x + 5 \\
\underline{6x - 9} \\
14 = R
\end{array}
$$

Remainder

Arrange the dividend and the divisor in descending powers of the variable. Insert, with 0 coefficients, any missing terms of degree less than 3.
Divide the first term of the divisor into the first term of the dividend.
Multiply the divisor by $2x^2$, line up like terms, subtract as in arithmetic, and bring down $-3x$. Repeat the process until the degree of the remainder is less than that of the divisor.

Thus,

$$\frac{4x^3 - 3x + 5}{2x - 3} = 2x^2 + 3x + 3 + \frac{14}{2x - 3}$$

Check

$$(2x - 3)\left[(2x^2 + 3x + 3) + \frac{14}{2x - 3}\right] = (2x - 3)(2x^2 + 3x + 3) + 14$$

$$= 4x^3 - 3x + 5$$

Matched Problem 1   Divide $6x^2 - 30 + 9x^3$ by $3x - 4$.

Being able to divide a polynomial $P(x)$ by a linear polynomial of the form $x - r$ quickly and accurately will be of great help in the search for zeros of higher-degree polynomial functions. This kind of division can be carried out more efficiently by a method called **synthetic division.** The method is most easily understood through an example. Let's start by dividing $P(x) = 2x^4 + 3x^3 - x - 5$ by $x + 2$, using ordinary long division. The critical parts of the process are indicated in color.

$$
\begin{array}{r}
2x^3 - 1x^2 + 2x - 5 \qquad \text{Quotient} \\
x + 2 \overline{)\; 2x^4 + 3x^3 + 0x^2 - 1x - 5} \quad \text{Dividend} \\
\underline{2x^4 + 4x^3} \\
-1x^3 + 0x^2 \\
\underline{-1x^3 - 2x^2} \\
2x^2 - 1x \\
\underline{2x^2 + 4x} \\
-5x - 5 \\
\underline{-5x - 10} \\
5 \qquad \text{Remainder}
\end{array}
$$

Divisor (labels the $x + 2$)

The numerals printed in color, which represent the essential part of the division process, are arranged more conveniently as follows:

Mechanically, we see that the second and third rows of numerals are generated as follows. The first coefficient, 2, of the dividend is brought down and multiplied by 2 from the divisor; and the product, 4, is placed under the second dividend coefficient, 3, and subtracted. The difference, -1, is again multiplied by the 2 from the divisor; and the product is placed under the third coefficient from the dividend and subtracted. This process is repeated until the remainder is reached. The process can be made a little faster, and less prone to sign errors, by changing $+2$ from the divisor to -2 and adding instead of subtracting. Thus

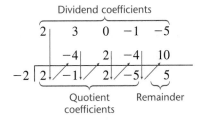

Key Steps in the Synthetic Division Process

To divide the polynomial $P(x)$ by $x - r$:

Step 1. Arrange the coefficients of $P(x)$ in order of descending powers of x. Write 0 as the coefficient for each missing power.

Step 2. After writing the divisor in the form $x - r$, use r to generate the second and third rows of numbers as follows. Bring down the first coefficient of the dividend and multiply it by r; then add the product to the second coefficient of the dividend. Multiply this sum by r, and add the product to the third coefficient of the dividend. Repeat the process until a product is added to the constant term of $P(x)$.

Step 3. The last number to the right in the third row of numbers is the remainder. The other numbers in the third row are the coefficients of the quotient, which is of degree 1 less than $P(x)$.

EXAMPLE 2   **Synthetic Division**

Use synthetic division to find the quotient and remainder resulting from dividing $P(x) = 4x^5 - 30x^3 - 50x - 2$ by $x + 3$. Write the answer in the form $Q(x) + R/(x - r)$, where R is a constant.

Solution   Since $x + 3 = x - (-3)$, we have $r = -3$, and

$$
\begin{array}{r|rrrrrr}
 & 4 & 0 & -30 & 0 & -50 & -2 \\
 &   & -12 & 36 & -18 & 54 & -12 \\
\hline
-3 & 4 & -12 & 6 & -18 & 4 & -14
\end{array}
$$

The quotient is $4x^4 - 12x^3 + 6x^2 - 18x + 4$ with a remainder of -14. Thus,

$$\frac{P(x)}{x + 3} = 4x^4 - 12x^3 + 6x^2 - 18x + 4 + \frac{-14}{x + 3}$$

Matched Problem 2   Repeat Example 2 with $P(x) = 3x^4 - 11x^3 - 18x + 8$ and divisor $x - 4$.

A calculator is a convenient tool for performing synthetic division. Any type of calculator can be used, although one with a memory will save some keystrokes. The flowchart in Figure 2 shows the repetitive steps in the synthetic division process, and Figure 3 illustrates the results of applying this process to Example 2 on a graphing calculator.

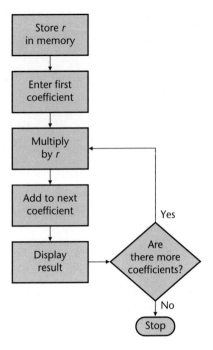

FIGURE 2 Synthetic division.

FIGURE 3

• **Division Algorithm**   If we divide $P(x) = 2x^4 - 5x^3 - 4x^2 + 13$ by $x - 3$, we obtain

$$\frac{2x^4 - 5x^3 - 4x^2 + 13}{x - 3} = 2x^3 + x^2 - x - 3 + \frac{4}{x - 3} \qquad x \neq 3$$

If we multiply both sides of this equation by $x - 3$, then we get

$$2x^4 - 5x^3 - 4x^2 + 13 = (x - 3)(2x^3 + x^2 - x - 3) + 4$$

This last equation is an identity in that the left side is equal to the right side for *all* replacements of x by real or imaginary numbers, including $x = 3$. This example suggests the important **division algorithm,** which we state as Theorem 1 without proof.

Theorem 1   **Division Algorithm**

For each polynomial $P(x)$ of degree greater than 0 and each number r, there exists a unique polynomial $Q(x)$ of degree 1 less than $P(x)$ and a unique number R such that

$$P(x) = (x - r)Q(x) + R$$

The polynomial $Q(x)$ is called the **quotient,** $x - r$ is the **divisor,** and R is the **remainder.** Note that R may be 0.

EXPLORE-DISCUSS 1   Let $P(x) = x^3 - 3x^2 - 2x + 8$.

(A) Evaluate $P(x)$ for
      (i) $x = -2$ (ii) $x = 1$ (iii) $x = 3$
(B) Use synthetic division to find the remainder when $P(x)$ is divided by
      (i) $x + 2$ (ii) $x - 1$ (iii) $x - 3$

What conclusion does a comparison of the results in parts A and B suggest?

● **Remainder**   We now use the division algorithm in Theorem 1 to prove the *remainder theorem*.
Theorem   The equation in Theorem 1,

$$P(x) = (x - r)Q(x) + R$$

is an identity; that is, it is true for all real or imaginary replacements for x. In particular, if we let $x = r$, then we observe a very interesting and useful relationship:

$$P(r) = (r - r)Q(r) + R$$
$$= 0 \cdot Q(r) + R$$
$$= 0 + R$$
$$= R$$

In words, the value of a polynomial $P(x)$ at $x = r$ is the same as the remainder R obtained when we divide $P(x)$ by $x - r$. We have proved the well-known remainder theorem:

Theorem 2   **Remainder Theorem**

If R is the remainder after dividing the polynomial $P(x)$ by $x - r$, then

$$P(r) = R$$

EXAMPLE 3   **Two Methods for Evaluating Polynomials**

If $P(x) = 4x^4 + 10x^3 + 19x + 5$, find $P(-3)$ by:

(A) Using the remainder theorem and synthetic division
(B) Evaluating $P(-3)$ directly

Solutions   (A) Use synthetic division to divide $P(x)$ by $x - (-3)$.

$$
\begin{array}{r}
4 \quad\ \ 10 \quad\ \ 0 \quad\ \ 19 \quad\ \ 5 \\
-12 \quad\ 6 \quad -18 \quad -3 \\
\hline
-3\,\lfloor\ 4 \quad -2 \quad\ \ 6 \quad\ \ 1 \quad\ \ 2 = R = P(-3)
\end{array}
$$

(B) $P(-3) = 4(-3)^4 + 10(-3)^3 + 19(-3) + 5$

$= 2$

Matched Problem 3   Repeat Example 3 for $P(x) = 3x^4 - 16x^2 - 3x + 7$ and $x = -2$.

• Graphing
Polynomial Functions

The shape of the graph of a polynomial function is connected to the degree of the polynomial. The shapes of odd-degree polynomial functions have something in common, and the shapes of even-degree polynomial functions have something in common. Figure 4 shows graphs of representative polynomial functions from degrees 1 to 6 and suggests some general properties of graphs of polynomial functions.

FIGURE 4 Graphs of polynomial functions.

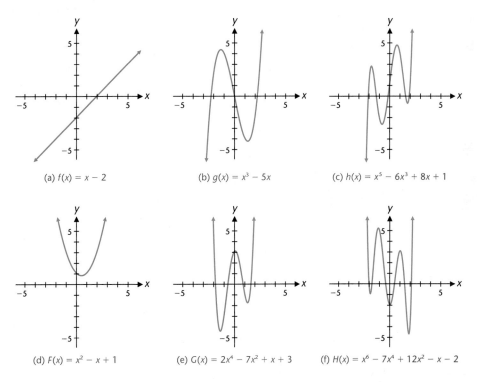

(a) $f(x) = x - 2$

(b) $g(x) = x^3 - 5x$

(c) $h(x) = x^5 - 6x^3 + 8x + 1$

(d) $F(x) = x^2 - x + 1$

(e) $G(x) = 2x^4 - 7x^2 + x + 3$

(f) $H(x) = x^6 - 7x^4 + 12x^2 - x - 2$

Notice that the odd-degree polynomial graphs start negative, end positive, and cross the x axis at least once. The even-degree polynomial graphs start positive, end positive, and may not cross the x axis at all. In all cases in Figure 4, the coefficient of the highest-degree term was chosen positive. If any leading coefficient had been chosen negative, then we would have a similar graph but reflected in the x axis.

The shape of the graph of a polynomial is also related to the shape of the graph of the highest-degree or **leading term** of the polynomial. Figure 5 compares the graph of one of the polynomials from Figure 4 with the graph of its leading term. Although quite dissimilar for points close to the origin, as we "zoom out" to points distant from the origin, the graphs become quite similar. The leading term in the polynomial dominates all other terms combined.

FIGURE 5 $p(x) = x^5$, $h(x) = x^5 - 6x^3 + 8x + 1$.

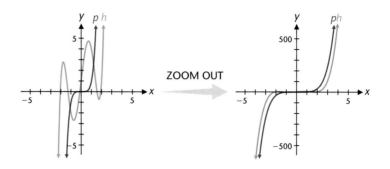

In general, the behavior of the graph of a polynomial function as x decreases without bound to the left or as x increases without bound to the right is determined by its leading term. We often use the symbols $-\infty$ and ∞ to help describe this left and right behavior.* The various possibilities are summarized in Theorem 3.

Theorem 3   **Left and Right Behavior of a Polynomial**

$$P(x) = a_n x^n + a_{n-1} x^{n-1} + \cdots + a_1 x + a_0, \qquad a_n \neq 0$$

1. $a_n > 0$ and n even
Graph of $P(x)$ increases without bound as x decreases to the left and as x increases to the right.

2. $a_n > 0$ and n odd
Graph of $P(x)$ decreases without bound as x decreases to the left and increases without bound as x increases to the right.

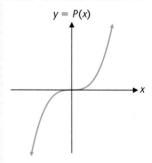

$$P(x) \rightarrow \begin{cases} \infty & \text{as } x \to -\infty \\ \infty & \text{as } x \to \infty \end{cases}$$

$$P(x) \rightarrow \begin{cases} -\infty & \text{as } x \to -\infty \\ \infty & \text{as } x \to \infty \end{cases}$$

*Remember, the symbol ∞ does not represent a real number. Earlier, we used ∞ to denote unbounded intervals, such as $[0, \infty)$. Now we are using it to describe quantities that are growing with no upper limit on their size.

3. $a_n < 0$ and n even
Graph of $P(x)$ decreases without bound as x decreases to the left and as x increases to the right.

4. $a_n < 0$ and n odd
Graph of $P(x)$ increases without bound as x decreases to the left and decreases without bound as x increases to the right.

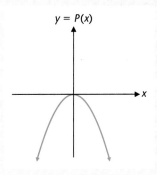

$$P(x) \to \begin{cases} -\infty & \text{as } x \to -\infty \\ -\infty & \text{as } x \to \infty \end{cases}$$

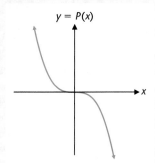

$$P(x) \to \begin{cases} \infty & \text{as } x \to -\infty \\ -\infty & \text{as } x \to \infty \end{cases}$$

Figure 4 gives examples of polynomial functions with graphs containing the maximum number of *turning points* possible for a polynomial of that degree. A **turning point** on a continuous graph is a point that separates an increasing portion from a decreasing portion. Listed in Theorem 4 below are useful properties of polynomial functions we accept without proof. Property 3 is discussed in detail later in this chapter. The other properties are established in calculus.

Theorem 4

Graph Properties of Polynomial Functions

Let P be an nth-degree polynomial function with real coefficients.

1. P is continuous for all real numbers.

2. The graph of P is a smooth curve.

3. The graph of P has at most n x intercepts.

4. P has at most $n - 1$ turning points.

EXPLORE-DISCUSS 2

(A) What is the least number of turning points an odd-degree polynomial function can have? An even-degree polynomial function?

(B) What is the maximum number of x intercepts the graph of a polynomial function of degree n can have?

(C) What is the maximum number of real solutions an nth-degree polynomial equation can have?

(D) What is the least number of x intercepts the graph of a polynomial function of odd degree can have? Of even degree?

(E) What is the least number of real solutions a polynomial equation of odd degree can have? Of even degree?

EXAMPLE 4   **Graphing a Polynomial**

Graph $P(x) = x^3 - 12x + 2$, $-4 \leq x \leq 4$. Find points by using synthetic division and the remainder theorem. How many x intercepts does the graph have? How many turning points? Describe the left and right behavior of $P(x)$.

Solution   We evaluate $P(x)$ from $x = -4$ to $x = 4$ for integer values of x. The process is speeded up by forming a synthetic division table. To simplify the form of the table, we dispense with writing the product of r with each coefficient in the quotient and perform the calculations mentally or with a calculator. A calculator becomes increasingly useful as the coefficients become more numerous or complicated. The table also provides other important information, as will be seen in subsequent sections.

	1	0	-12	2	
-4	1	-4	4	-14	$= P(-4)$
-3	1	-3	-3	11	$= P(-3)$
-2	1	-2	-8	18	$= P(-2)$
-1	1	-1	-11	13	$= P(-1)$
0	1	0	-12	2	$= P(0)$
1	1	1	-11	-9	$= P(1)$
2	1	2	-8	-14	$= P(2)$
3	1	3	-3	-7	$= P(3)$
4	1	4	4	18	$= P(4)$

Next we plot the points found in the table and connect them with a smooth curve (Fig. 6). As we draw this curve, we notice that the graph crosses the x axis three times and changes direction twice. The next two sections will address the question of determining precisely where a graph crosses the x axis. Precise determination of the location of turning points requires calculus techniques. Lacking this precise information, we simply change direction at $x = -2$ and $x = 2$.

The leading term of $P(x)$ is x^3. From case 2 in Theorem 3 we see that $P(x) \to -\infty$ as $x \to -\infty$ and $P(x) \to \infty$ as $x \to \infty$.

FIGURE 6 $P(x) = x^3 - 12x + 2$.

Matched Problem 4   Graph $P(x) = x^3 - 4x^2 - 4x + 16$, $-3 \leq x \leq 5$. Find points by using synthetic division and the remainder theorem. How many x intercepts does the graph have? How many turning points? Describe the left and right behavior of $P(x)$.

Remark. A graphing utility can quickly produce a table of values, without using synthetic division, and can graph a polynomial just as quickly. In Section 4-3 we will find that a synthetic division table is a valuable tool when used in conjunction with a graphing utility. Thus, students with graphing utilities should also learn to construct synthetic division tables. (See Table 2 in Section 4-3 for a more efficient way to construct a synthetic division table on a graphing utility.)

Answers to Matched Problems

1. $3x^2 + 6x + 8 + \dfrac{2}{3x - 4}$

2. $\dfrac{P(x)}{x - 4} = 3x^3 + x^2 + 4x - 2 + \dfrac{0}{x - 4} = 3x^3 + x^2 + 4x - 2$

3. $P(-2) = -3$ for both parts, as it should.

4.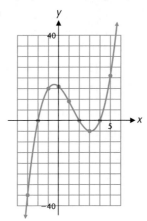

x	$P(x)$
-3	-35
-2	0
-1	15
0	16
1	9
2	0
3	-5
4	0
5	21

Three x intercepts and two turning points; $P(x) \to -\infty$ as $x \to -\infty$; $P(x) \to \infty$ as $x \to \infty$.

EXERCISE 4-1

A

In Problems 1–4, a is a positive real number. Match each function with one of graphs (a)–(d).

1. $f(x) = ax^3$ **2.** $g(x) = -ax^4$

3. $h(x) = ax^6$ **4.** $k(x) = -ax^5$

(c)   (d)

(a)

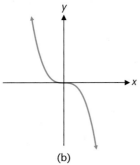

(b)

Problems 5–8 refer to the graphs of functions f, g, h, and k shown below.

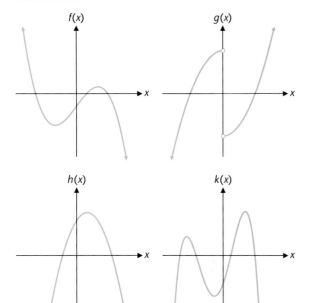

5. Which of these functions could be a second-degree polynomial?

6. Which of these functions could be a third-degree polynomial?

7. Which of these functions could be a fourth-degree polynomial?

8. Which of these functions is not a polynomial?

In Problems 9–16, divide, using algebraic long division. Write the quotient, and indicate the remainder.

9. $(a^2 - 4) \div (a + 2)$ **10.** $(a^2 + 4) \div (a + 2)$

11. $(b^2 - 6b - 9) \div (b - 3)$

12. $(b^2 - 6b + 9) \div (b - 3)$

13. $(3x + 2 + x^3 - x^2) \div (x - 1)$

14. $(2x^2 + 4x - x^3 - 8) \div (x - 2)$

15. $(1 + 8y^3 - 4y^2 - 2y) \div (2y + 1)$

16. $(3 + 8y^3 - 6y^2 - 7y) \div (2y - 3)$

In Problems 17–22, use synthetic division to write the quotient $P(x) \div (x - r)$ in the form $P(x)/(x - r) = Q(x) + R/(x - r)$, where R is a constant.

17. $(x^2 + 4x - 15) \div (x - 3)$

18. $(x^2 - 2x - 1) \div (x - 4)$ **19.** $(3x^2 - x - 7) \div (x + 2)$

20. $(4x^2 + 18x + 4) \div (x + 5)$

21. $(2x^3 + 3x^2 - 8x + 1) \div (x + 3)$

22. $(3x^3 - 4x^2 - 7x + 9) \div (x - 2)$

B _____

Use synthetic division and the remainder theorem in Problems 23–28.

23. Find $P(-5)$, given $P(x) = 2x^2 + 8x - 6$.

24. Find $P(2)$, given $P(x) = 3x^2 - 4x + 2$.

25. Find $P(-4)$, given $P(x) = 4x^3 + 12x^2 - 8x + 25$.

26. Find $P(-3)$, given $P(x) = 5x^3 + 14x^2 + 3x + 10$.

27. Find $P(3)$, given $P(x) = 2x^4 - 5x^3 + 2x^2 - 11x - 14$.

28. Find $P(-6)$, given $P(x) = 3x^4 + 18x^3 + x^2 + 4x - 7$.

In Problems 29–44, divide, using synthetic division. Write the quotient, and indicate the remainder. As coefficients get more involved, a calculator should prove helpful. Do not round off—all quantities are exact.

29. $(2x^5 - 5x^2 + 3) \div (x - 1)$

30. $(3x^4 - 2x - 5) \div (x + 1)$

31. $(x^4 - 16) \div (x + 4)$ **32.** $(x^5 - 32) \div (x - 2)$

33. $(4x^4 - 9x^3 - 8x^2 - 2x - 7) \div (x - 3)$

34. $(2x^4 + 6x^3 - 4x^2 - 5x + 7) \div (x + 3)$

35. $(x^6 + 7x^5 + 10x^4 - x^2 - 5x) \div (x + 5)$

36. $(x^6 + 6x^5 + 2x^4 + 12x^3 - 3x - 18) \div (x + 6)$

37. $(2x^4 + 9x^3 + 5x^2 - 4x + 3) \div (x + \frac{3}{2})$

38. $(2x^4 + 5x^3 + 5x + 8) \div (x + \frac{1}{2})$

39. $(3x^4 + 5x^3 - 5x^2 + 10x - 1) \div (x - \frac{1}{3})$

40. $(4x^4 - 11x^3 + 18x^2 - 5x + 4) \div (x - \frac{3}{4})$

41. $(5x^4 - 4x^3 + 2x - 5) \div (x - 0.2)$

42. $(3x^4 - 4x^2 + 5x + 8) \div (x + 0.8)$

43. $(5x^5 + 2x^4 + 4x^3 + 6x^2 - 6) \div (x + 0.6)$

44. $(10x^5 - 4x^4 + 2x^2 + 4x - 1) \div (x - 0.4)$

In Problems 45–52, graph each polynomial function using synthetic division and the remainder theorem. Then describe each graph verbally, including the number of x intercepts, the number of turning points, and the left and right behavior.

 *Check your work in Problems 45–52 by graphing on a graphing utility.

45. $P(x) = x^3 - 5x^2 + 2x + 8, -2 \leq x \leq 5$

46. $P(x) = x^3 + 2x^2 - 5x - 6, -4 \leq x \leq 3$

47. $P(x) = x^3 + 4x^2 - x - 4, -5 \leq x \leq 2$

48. $P(x) = x^3 - 2x^2 - 5x + 6, -3 \leq x \leq 4$

49. $P(x) = -x^3 + 2x^2 - 3, -2 \leq x \leq 3$

*Please note that use of a graphing utility is not required to complete these exercises. Checking them with a g.u. is optional.

50. $P(x) = -x^3 - x + 4,\ -2 \le x \le 2$

51. $P(x) = -x^3 + 3x^2 - 3x + 2,\ -1 \le x \le 3$

52. $P(x) = -x^3 + x^2 + 4x + 6,\ -3 \le x \le 4$

In Problems 53–56, either give an example of a polynomial with real coefficients that satisfies the given conditions or explain why such a polynomial cannot exist.

53. $P(x)$ is a third-degree polynomial with one x intercept.

54. $P(x)$ is a fourth-degree polynomial with no x intercepts.

55. $P(x)$ is a third-degree polynomial with no x intercepts.

56. $P(x)$ is a fourth-degree polynomial with no turning points.

C

In Problems 57–60, divide, using algebraic long division. Write the quotient, and indicate the remainder.

57. $(x^4 + x^3 + x^2 - x - 2) \div (x^2 - 1)$

58. $(x^4 + 2x^3 - 3x^2 - 8x - 4) \div (x^2 - 4)$

59. $(x^4 + x^3 - 7x^2 + 8x + 1) \div (x^2 + 3x - 2)$

60. $(x^4 - 7x^3 + 15x^2 - 9x + 1) \div (x^2 - 4x + 1)$

In Problems 61 and 62, divide, using synthetic division. Do not use a calculator.

61. $(x^4 + 2x^3 - 2x^2 + 2x - 3) \div (x - i)$

62. $(x^4 + 2x^3 - 2x^2 + 2x - 3) \div (x + i)$

63. Let $P(x) = x^2 + 2ix - 10$. Find:
    (A) $P(2 - i)$ (B) $P(5 - 5i)$
    (C) $P(3 - i)$ (D) $P(-3 - i)$

64. Let $P(x) = x^2 - 4ix - 13$. Find:
    (A) $P(5 + 6i)$ (B) $P(1 + 2i)$
    (C) $P(3 + 2i)$ (D) $P(-3 + 2i)$

In Problems 65–72, graph each polynomial function using synthetic division and the remainder theorem. Then describe each graph verbally, including the number of x intercepts, the number of turning points, and the left and right behavior.

Check your work in Problems 65–72 by graphing on a graphing utility.

65. $P(x) = x^4 - 2x^3 - 2x^2 + 8x - 8$

66. $P(x) = x^4 + x^3 - 3x^2 + 7x - 6$

67. $P(x) = x^4 + 4x^3 - x^2 - 10x - 8$

68. $P(x) = x^4 - 8x^2 - 4x + 10$

69. $P(x) = -x^4 + 2x^3 + 10x^2 - 10x - 9$

70. $P(x) = -x^4 - 5x^3 + x^2 + 20x + 5$

71. $P(x) = x^5 - 6x^4 + 4x^3 + 17x^2 - 5x - 7$

72. $P(x) = x^5 - 9x^3 + 4x^2 + 15x - 10$

73. (A) Divide $P(x) = a_2x^2 + a_1x + a_0$ by $x - r$, using both synthetic division and the long-division process, and compare the coefficients of the quotient and the remainder produced by each method.
    (B) Expand the expression representing the remainder. What do you observe?

74. Repeat Problem 73 for
$$P(x) = a_3x^3 + a_2x^2 + a_1x + a_0$$

75. Polynomials also can be evaluated conveniently using a "nested factoring" scheme. For example, the polynomial $P(x) = 2x^4 - 3x^3 + 2x^2 - 5x + 7$ can be written in a nested factored form as follows:

$$P(x) = 2x^4 - 3x^3 + 2x^2 - 5x + 7$$
$$= (2x - 3)x^3 + 2x^2 - 5x + 7$$
$$= [(2x - 3)x + 2]x^2 - 5x + 7$$
$$= \{[(2x - 3)x + 2]x - 5\}x + 7$$

Use the nested factored form to find $P(-2)$ and $P(1.7)$. [*Hint:* To evaluate $P(-2)$, store -2 in your calculator's memory and proceed from left to right recalling -2 as needed.]

76. Find $P(-2)$ and $P(1.3)$ for $P(x) = 3x^4 + x^3 - 10x^2 + 5x - 2$ using the nested factoring scheme presented in Problem 75.

SECTION 4-2 Finding Rational Zeros of Polynomials

- Factor Theorem
- Fundamental Theorem of Algebra
- Imaginary Zeros
- Rational Zeros

In this section we develop some important properties of polynomials with arbitrary coefficients. Then we consider the problem of finding all the rational zeros of a polynomial with rational coefficients. In some cases, this process will also enable us to find irrational or imaginary zeros.

● **Factor Theorem**   The division algorithm (Theorem 1 in Section 4-1)

$$P(x) = (x - r)Q(x) + R$$

may, because of the remainder theorem (Theorem 2 in Section 4-1), be written in a form where R is replaced by $P(r)$:

$$P(x) = (x - r)Q(x) + P(r)$$

It is easy to see that $x - r$ is a factor of $P(x)$ if and only if $P(r) = 0$; that is, if and only if r is a zero of the polynomial $P(x)$. This result is known as the **factor theorem:**

Theorem 1   **Factor Theorem**

If r is a zero of the polynomial $P(x)$, then $x - r$ is a factor of $P(x)$. Conversely, if $x - r$ is a factor of $P(x)$, then r is a zero of $P(x)$.

The relationship between zeros, roots, factors, and x intercepts is fundamental to the study of polynomials. With the addition of the factor theorem, we now know that the following statements are equivalent for any polynomial $P(x)$:

1. r is a root of the equation $P(x) = 0$.

2. r is a zero of $P(x)$.

3. $x - r$ is a factor of $P(x)$.

If, in addition, the coefficients of $P(x)$ are real numbers and r is a real number, then we can add a fourth statement to this list:

4. r is an x intercept of the graph of $P(x)$.

EXAMPLE 1   **Factors, Zeros, Roots, and Intercepts**

(A) Use the factor theorem to show that $x + 1$ is a factor of $P(x) = x^{25} + 1$.
(B) What are the zeros of $P(x) = 3(x - 5)(x + 2)(x - 3)$?
(C) What are the roots of $x^4 - 1 = 0$?
(D) What are the x intercepts of the graph of $P(x) = x^4 - 1$?

Solutions   (A) Since $x + 1 = x - (-1)$, we have $r = -1$ and

$$P(r) = P(-1) = (-1)^{25} + 1 = -1 + 1 = 0$$

Hence, -1 is a zero of $P(x) = x^{25} + 1$. By the factor theorem, $x - (-1) = x + 1$ is a factor of $x^{25} + 1$.

(B) Since $(x - 5)$, $(x + 2)$, and $(x - 3)$ are all factors of $P(x)$, 5, -2, and 3 are zeros of $P(x)$.

(C) Factoring the left side, we have

$$x^4 - 1 = 0$$

$$(x^2 - 1)(x^2 + 1) = 0$$

$$(x - 1)(x + 1)(x - i)(x + i) = 0$$

Thus, the roots of $x^4 - 1 = 0$ are 1, -1, i, and $-i$.

(D) From part C, the zeros of $P(x)$ are 1, -1, i, and $-i$. However, x intercepts must be real numbers. Thus the x intercepts of the graph of $P(x) = x^4 - 1$ are 1 and -1 (see Fig. 1).

FIGURE 1 x intercepts of $P(x) = x^4 - 1$.

Matched Problem 1   (A) Use the factor theorem to show that $x - 1$ is a factor of $P(x) = x^{54} - 1$.

(B) What are the zeros of

$$P(x) = 2(x + 3)(x + 7)(x - 8)(x + 1)?$$

(C) What are the roots of $x^2 + 4 = 0$?

(D) What are the x intercepts of $P(x) = x^2 + 4$?

• Fundamental Theorem of Algebra

Theorem 2, often referred to as the **fundamental theorem of algebra,** requires verification that is beyond the scope of this book, so we state it without proof.

Theorem 2   **Fundamental Theorem of Algebra**

Every polynomial $P(x)$ of degree $n > 0$ has at least one zero.

If $P(x) = a_n x^n + a_{n-1} x^{n-1} + \cdots + a_1 x + a_0$ is a polynomial of degree $n > 0$ with complex coefficients, then, according to Theorem 2, it has at least one zero, say r_1. According to the factor theorem, $x - r_1$ is a factor of $P(x)$. Thus,

$$P(x) = (x - r_1)Q_1(x)$$

where $Q_1(x)$ is a polynomial of degree $n - 1$. If $n - 1 = 0$, then $Q_1(x) = a_n$. If $n - 1 > 0$, then, by Theorem 2, $Q_1(x)$ has at least one zero, say r_2. And

$$Q_1(x) = (x - r_2)Q_2(x)$$

where $Q_2(x)$ is a polynomial of degree $n - 2$. Thus,

$$P(x) = (x - r_1)(x - r_2)Q_2(x)$$

If $n - 2 = 0$, then $Q_2(x) = a_n$. If $n - 2 > 0$, then $Q_2(x)$ has at least one zero, say r_3. And

$$Q_2(x) = (x - r_3)Q_3(x)$$

where $Q_3(x)$ is a polynomial of degree $n - 3$.

We continue in this way until $Q_k(x)$ is of degree zero—that is, until $k = n$. At this point, $Q_n(x) = a_n$, and we have

$$P(x) = (x - r_1)(x - r_2) \cdot \cdots \cdot (x - r_n)a_n$$

Thus, $r_1, r_2, \ldots, r_n$ are n zeros, not necessarily distinct, of $P(x)$. Is it possible for $P(x)$ to have more than these n zeros? Let's assume that r is a number different from the zeros above. Then

$$P(r) = a_n(r - r_1)(r - r_2) \cdot \cdots \cdot (r - r_n) \neq 0$$

since r is not equal to any of the zeros. Hence, r is not a zero, and we conclude that $r_1, r_2, \ldots, r_n$ are the only zeros of $P(x)$. We have just sketched a proof of Theorem 3.

Theorem 3	**n Zeros Theorem**
	Every polynomial $P(x)$ of degree $n > 0$ can be expressed as the product of n linear factors. Hence, $P(x)$ has exactly n zeros—not necessarily distinct.

Theorems 2 and 3 were first proved in 1797 by Carl Friedrich Gauss, one of the greatest mathematicians of all time, at the age of 20.

If $P(x)$ is represented as the product of linear factors and $x - r$ occurs m times, then r is called a **zero of multiplicity m.** For example, if

$$P(x) = 4(x - 5)^3(x + 1)^2(x - i)(x + i)$$

then this seventh-degree polynomial has seven zeros, not all distinct. That is, 5 is a zero of multiplicity 3, or a triple zero; -1 is a zero of multiplicity 2, or a double zero; and i and $-i$ are zeros of multiplicity 1, or simple zeros. Thus, this seventh-degree polynomial has exactly seven zeros if we count 5 and -1 with their respective multiplicities.

EXPLORE-DISCUSS 1   If r is a real zero of a polynomial $P(x)$ with real coefficients, then r is also an x intercept for the graph of $P(x)$. Discuss the difference between the graph of $P(x)$ at a real zero of odd multiplicity and at a real zero of even multiplicity.

EXAMPLE 2   **Factoring a Polynomial**

If -2 is a double zero of $P(x) = x^4 - 7x^2 + 4x + 20$, write $P(x)$ as a product of first-degree factors.

Solution   Since -2 is a double zero of $P(x)$, we can write

$$P(x) = (x + 2)^2 Q(x)$$
$$= (x^2 + 4x + 4)Q(x)$$

and find $Q(x)$ by dividing $P(x)$ by $x^2 + 4x + 4$. Carrying out the algebraic long division, we obtain

$$Q(x) = x^2 - 4x + 5$$

The zeros of $Q(x)$ are found, using the quadratic formula, to be $2 - i$ and $2 + i$. Thus, $P(x)$ written as a product of linear factors is

$$P(x) = (x + 2)^2 [x - (2 - i)][x - (2 + i)]$$

[*Note:* Any time $Q(x)$ is a quadratic polynomial, its zeros can be found using the quadratic formula.]

Matched Problem 2   If 3 is a double zero of $P(x) = x^4 - 12x^3 + 55x^2 - 114x + 90$, write $P(x)$ as a product of first-degree factors.

● **Imaginary Zeros**   Something interesting happens if we restrict the coefficients of a polynomial to real numbers. Let's use the quadratic formula to find the zeros of the polynomial

$$P(x) = x^2 - 6x + 13$$

To find the zeros of $P(x)$, we solve $P(x) = 0$:

$$x^2 - 6x + 13 = 0$$

$$x = \frac{6 \pm \sqrt{36 - 52}}{2}$$

$$= \frac{6 \pm \sqrt{-16}}{2} = \frac{6 \pm 4i}{2} = 3 \pm 2i$$

The zeros of $P(x)$ are $3 - 2i$ and $3 + 2i$, conjugate imaginary numbers (see Section 2-5). Also observe that the imaginary zeros in Example 2 are the conjugate imaginary numbers $2 - i$ and $2 + i$.

This is generalized in the following theorem:

Theorem 4 **Imaginary Zeros Theorem**

Imaginary zeros of polynomials with real coefficients, if they exist, occur in conjugate pairs.

As a consequence of Theorems 3 and 4, we also know (think this through) the following:

Theorem 5 **Real Zeros and Odd-Degree Polynomials**

A polynomial of odd degree with real coefficients always has at least one real zero.

EXPLORE-DISCUSS 2   (A) Let $P(x)$ be a third-degree polynomial with real coefficients. Indicate which of the following statements are true and which are false. Justify your conclusions.
  (i) $P(x)$ has at least one real zero.
  (ii) $P(x)$ has three zeros.
  (iii) $P(x)$ can have two real zeros and one imaginary zero.
  (B) Let $P(x)$ be a fourth-degree polynomial with real coefficients. Indicate which of the following statements are true and which are false. Justify your conclusions.
  (i) $P(x)$ has four zeros.
  (ii) $P(x)$ has at least two real zeros.
  (iii) If we know $P(x)$ has three real zeros, then the fourth zero must be real.

● **Rational Zeros**   First note that a polynomial with rational coefficients can always be written as a constant times a polynomial with integer coefficients. For example,

$$P(x) = \tfrac{1}{2}x^3 - \tfrac{2}{3}x^2 + \tfrac{7}{4}x + 5$$
$$= \tfrac{1}{12}(6x^3 - 8x^2 + 21x + 60)$$

Thus, it is sufficient to confine our attention to polynomials with integer coefficients. We introduce the rational zero theorem by examining the following quadratic polynomial whose zeros can be found easily by factoring:

$$P(x) = 6x^2 - 13x - 5 = (2x - 5)(3x + 1)$$

Zeros of $P(x)$: $\dfrac{5}{2}$ and $-\dfrac{1}{3} = \dfrac{-1}{3}$

Solution We start the process with a synthetic division table:

$$
\begin{array}{r|rrrrrl}
 & 1 & -2 & -10 & 40 & -90 & \\
\hline
3 & 1 & 1 & -7 & 19 & -33 & = P(3) \\
4 & 1 & 2 & -2 & 32 & 38 & = P(4)
\end{array}
$$

TABLE 1   Bisection Approximation

Sign change interval (a, b)	Midpoint m	Sign of P		
		P(a)	P(m)	P(b)
(3, 4)	3.5	−	−	+
(3.5, 4)	3.75	−	+	+
(3.5, 3.75)	3.625	−	+	+
(3.5, 3.625)	3.563	−	−	+
(3.563, 3.625)	We stop here	−		+

Since the sign of $P(x)$ changes at the endpoints of the interval (3.563, 3.625), we conclude that a real zero lies on this interval and is given by $r = 3.6$ to one decimal place accuracy (each endpoint rounds to 3.6).

Figure 3 illustrates the nested intervals produced by the bisection method in Table 1. Match each step in Table 1 with an interval in Figure 3. Note how each interval that contains a zero gets smaller and smaller and is contained in the preceding interval that contained the zero.

FIGURE 3 Nested intervals produced by the bisection method in Table 1.

If we had wanted two decimal place accuracy, we would use four decimal places for the values of x and continue the process in Table 1 until the endpoints of a sign change interval rounded to the same two decimal place number.

Matched Problem 3 Use the bisection method to approximate to one decimal place accuracy a zero on the interval $(-5, -4)$ for the polynomial in Example 3.

• Approximating Real Zeros Using a Graphing Utility

The bisection method is easy to understand but tedious to carry out, especially if the approximation must be accurate to more than two decimal places. Fortunately, this is the type of repetitive calculation that a graphing utility can be programmed to carry out. In fact, we have been using a graphing utility for some time now to find the zeros of a function (see Section 4 in "Graphs and Functions"). Now we will see how the upper and lower bound theorem can be used in conjunction with the zero approximation routine on a graphing utility to approximate all the real zeros of a polynomial.

EXAMPLE 4   **Approximating Real Zeros Using a Graphing Utility**

Given the polynomial $P(x) = x^5 + x - 1$:

(A) Form a synthetic division table to find upper and lower bounds for any real zeros, and locate real zeros between successive integers.

(B) Graph $P(x)$ in a graphing utility, and approximate any real zeros to four decimal places using a built-in zero approximation routine.

Solution   (A) Form a synthetic division table:

		1	0	0	0	1	−1	
	0	1	0	0	0	1	−1	} Real zero
UB	1	1	1	1	1	2	1	
LB	−1	1	−1	1	−1	2	−3	

From the table we see that all real zeros of $P(x)$ are between −1 and 1, and a real zero lies on the interval (0, 1).

(B) Enter $P(x)$ in a graphing utility, and set the window dimensions with the synthetic division table in part A as a guide. Figure 4(a) shows the graph of $P(x)$, and Figure 4(b) shows the zero approximation using a built-in routine.

FIGURE 4

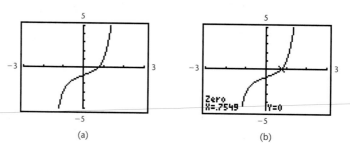

(a)   (b)

It is clear from the graph and the upper and lower bounds of the zeros found in part A that $P(x)$ has only one real zero, which is, to four decimal places, $x = 0.7549$.

Matched Problem 4   Given the polynomial $P(x) = x^5 - x^2 + 1$:

(A) Form a synthetic division table to find upper and lower bounds for any real zeros, and locate real zeros between successive integers.

(B) Graph $P(x)$ in a graphing utility and approximate any real zeros to four decimal places using a built-in zero approximation routine.

Earlier in this section and in Section 4.1 we saw that a calculator is a useful tool for constructing a synthetic division table. A graphing utility that can store and execute programs is even more useful. Table 2 shows a simple program for a graphing calculator that will perform synthetic division. The table also shows the output generated when we use this program to construct the synthetic division table in Example 2.

In Problems 23–30:

(A) Find the smallest positive integer and largest negative integer that, by Theorem 2, are upper and lower bounds, respectively, for the real zeros of P(x).

(B) Approximate the real zeros of each polynomial to two decimal places.

23. $P(x) = x^3 - 2x^2 + 3x - 8$

24. $P(x) = x^3 + 3x^2 + 4x + 5$

25. $P(x) = x^4 + x^3 - 5x^2 + 7x - 22$

26. $P(x) = x^4 - x^3 - 8x^2 - 12x - 25$

27. $P(x) = x^5 - 3x^3 - 4x + 4$

28. $P(x) = x^5 - x^4 - 2x^2 - 4x - 5$

29. $P(x) = x^5 + x^4 + 3x^3 + x^2 + 2x - 5$

30. $P(x) = x^5 - 2x^4 - 6x^2 - 9x + 10$

C

In Problems 31–34:

(A) Find the smallest positive integer and largest negative integer that, by Theorem 2, are upper and lower bounds, respectively, for the real zeros of P(x). Also note the location of any zeros between successive integers.

(B) Approximate to two decimal places the largest real zero of P(x) using the bisection method.

31. $P(x) = x^5 - 5x^4 + 5x^3 + 5x^2 - 10x + 5$

32. $P(x) = x^5 - 2x^4 - 7x^3 + 8x^2 + 12x - 5$

33. $P(x) = x^5 - 10x^3 + 9x + 10$

34. $P(x) = x^5 - 9x^3 + 4x^2 + 12x - 15$

In Problems 35–44:

(A) Find the smallest positive integer multiple of 10 and largest negative integer multiple of 10 that, by Theorem 2, are upper and lower bounds, respectively, for the real zeros of each polynomial.

(B) Approximate the real zeros of each polynomial to two decimal places.

35. $P(x) = x^3 - 24x^2 - 25x + 10$

36. $P(x) = x^3 - 37x^2 + 70x - 20$

37. $P(x) = x^4 + 12x^3 - 900x^2 + 5,000$

38. $P(x) = x^4 - 12x^3 - 425x^2 + 7,000$

39. $P(x) = x^4 - 100x^2 - 1,000x - 5,000$

40. $P(x) = x^4 - 5x^3 - 50x^2 - 500x + 7,000$

41. $P(x) = 4x^4 - 40x^3 - 1,475x^2 + 7,875x - 10,000$

42. $P(x) = 9x^4 + 120x^3 - 3,083x^2 - 25,674x - 48,400$

43. $P(x) = 0.01x^5 - 0.1x^4 - 12x^3 + 9,000$

44. $P(x) = 0.1x^5 + 0.7x^4 - 18.775x^3 - 340x^2 - 1,645x - 2,450$

APPLICATIONS

Express the solutions to Problems 45–50 as the roots of a polynomial equation of the form P(x) = 0 and approximate these solutions to one decimal place. Use a graphing utility, if available; otherwise, use the bisection method.

★ **45. Geometry.** Find all points on the graph of $y = x^2$ that are 1 unit away from the point (1, 2). [Hint: Use the distance-between-two-points formula from Section 3-1.]

★ **46. Geometry.** Find all points on the graph of $y = x^2$ that are 1 unit away from the point (2, 1).

★ **47. Manufacturing.** A box is to be made out of a piece of cardboard that measures 18 by 24 inches. Squares, x inches on a side, will be cut from each corner, and then the ends and sides will be folded up (see the figure). Find the value of x that would result in a box with a volume of 600 cubic inches.

★ **48. Manufacturing.** A box with a hinged lid is to be made out of a piece of cardboard that measures 20 by 40 inches. Six squares, x inches on a side, will be cut from each corner and the middle, and then the ends and sides will be folded up to form the box and its lid (see the figure). Find the value of x that would result in a box with a volume of 500 cubic inches.

★ **49. Construction.** A propane gas tank is in the shape of a right circular cylinder with a hemisphere at each end (see the figure). If the overall length of the tank is 10 feet and the volume is 20π cubic feet, find the common radius of the hemispheres and the cylinder.

★ **50. Shipping.** A shipping box is reinforced with steel bands in all three directions (see the figure). A total of 20.5 feet of steel tape is to be used, with 6 inches of waste because of a 2-inch overlap in each direction. If the box has a square base and a volume of 2 cubic feet, find its dimensions.

SECTION 4-4   Rational Functions

- Rational Functions
- Vertical and Horizontal Asymptotes
- Graphing Rational Functions

• **Rational Functions**

Just as rational numbers are defined in terms of quotients of integers, rational functions are defined in terms of quotients of polynomials. The following equations define rational functions:

$$f(x) = \frac{x - 1}{x^2 - x - 6} \qquad g(x) = \frac{1}{x} \qquad h(x) = \frac{x^3 - 1}{x}$$

$$p(x) = 2x^2 - 3 \qquad q(x) = 3 \qquad r(x) = 0$$

In general, a function f is a **rational function** if

$$f(x) = \frac{n(x)}{d(x)} \qquad d(x) \neq 0$$

where $n(x)$ and $d(x)$ are polynomials. The **domain of** f is the set of all real numbers x such that $d(x) \neq 0$.

If $x = a$ and $d(a) = 0$, then f is not defined at $x = a$ and there can be no point on the graph of f with abscissa $x = a$. Remember, division by 0 is never allowed. It can be shown that:

> If $f(x) = n(x)/d(x)$ and $d(a) = 0$, then f is discontinuous at $x = a$ and the graph of f has a hole or break at $x = a$.

If $x = a$ is in the domain of $f(x)$ and $n(a) = 0$, then the graph of f crosses the x axis at $x = a$. Thus:

> If $f(x) = n(x)/d(x)$, $n(a) = 0$, and $d(a) \neq 0$, then $x = a$ is an x intercept for the graph of f.

What happens if both $n(a) = 0$ and $d(a) = 0$? In this case, we know that $x - a$ is a factor of both $n(x)$ and $d(x)$, and thus, $f(x)$ is not in lowest terms (see Section 1-4).

> Unless specifically stated to the contrary, we assume that all the rational functions we consider are reduced to lowest terms.

EXAMPLE 1   **Finding the Domain and x Intercepts for a Rational Function**

Find the domain and x intercepts for $f(x) = \dfrac{2x^2 - 2x - 4}{x^2 - 9}$

Solution $$f(x) = \frac{n(x)}{d(x)} = \frac{2x^2 - 2x - 4}{x^2 - 9} = \frac{2(x - 2)(x + 1)}{(x - 3)(x + 3)}$$

Since $d(3) = 0$ and $d(-3) = 0$, the domain of f is

$$x \neq \pm 3 \qquad \text{or} \qquad (-\infty, -3) \cup (-3, 3) \cup (3, \infty)$$

Since $n(2) = 0$ and $n(-1) = 0$, the graph of f crosses the x axis at $x = 2$ and $x = -1$.

Matched Problem 1   Find the domain and x intercepts for: $f(x) = \dfrac{3x^2 - 12}{x^2 + 2x - 3}$

● **Vertical and Horizontal Asymptotes**   Even though a rational function f may be discontinuous at $x = a$ (no graph for $x = a$), it is still useful to know what happens to the graph of f when x is close to a. For example, consider the very simple rational function f defined by

$$f(x) = \frac{1}{x}$$

It is clear that the function f is discontinuous at $x = 0$. But what happens to $f(x)$ when x approaches 0 from either side of 0? A numerical approach will give us an idea of what happens to $f(x)$ when x gets close to 0. From Table 1, we see that as x approaches 0 from the right, $1/x$ gets larger and larger—that is, $1/x$ increases without bound. We write this symbolically* as

$$\frac{1}{x} \to \infty \qquad \text{as} \qquad x \to 0^{+}$$

TABLE 1   Behavior of $1/x$ as $x \to 0^{+}$

x	1	0.1	0.01	0.001	0.0001	0.000 01	0.000 001	...	x approaches 0 from the right ($x \to 0^{+}$)
$1/x$	1	10	100	1,000	10,000	100,000	1,000,000	...	$1/x$ increases without bound ($1/x \to \infty$)

*Remember, the symbol ∞ does not represent a real number. In this context, ∞ is used to indicate that the values of $1/x$ increase without bound. That is, $1/x$ exceeds any given number N no matter how large N is chosen.

If x approaches 0 from the left, then x and $1/x$ are both negative and the values of $1/x$ decrease without bound (see Table 2). This is denoted as

$$\frac{1}{x} \to -\infty \qquad \text{as} \qquad x \to 0^-$$

TABLE 2 Behavior of $1/x$ as $x \to 0^-$

x	-1	-0.1	-0.01	-0.001	-0.0001	$-0.000\ 01$	$-0.000\ 001$	$\ldots$	x approaches 0 from the left $(x \to 0^-)$
$1/x$	-1	-10	-100	$-1,000$	$-10,000$	$-100,000$	$-1,000,000$	$\ldots$	$1/x$ decreases without bound $(1/x \to -\infty)$

The graph of $f(x) = 1/x$ for $-1 \le x \le 1$, $x \ne 0$, is shown in Figure 1. The behavior of f as x approaches 0 from the right is illustrated on the graph by drawing a curve that becomes almost vertical and placing an arrow on the curve to indicate that the values of $1/x$ continue to increase without bound as x approaches 0 from the right. The behavior as x approaches 0 from the left is illustrated in a similar manner.

FIGURE 1 $f(x) = \dfrac{1}{x}$ near $x = 0$.

EXPLORE-DISCUSS 1   Construct tables similar to Tables 1 and 2 for $g(x) = 1/x^2$, and discuss the behavior of the graph of $g(x)$ near $x = 0$.

The preceding discussion suggests that vertical asymptotes are associated with the zeros of the denominator of a rational function. Using the same kind of reasoning, we state the following general method of locating vertical asymptotes for rational functions.

Theorem 1   **Vertical Asymptotes and Rational Functions**

Let f be a rational function defined by

$$f(x) = \frac{n(x)}{d(x)}$$

where $n(x)$ and $d(x)$ are polynomials. If a is a real number such that $d(a) = 0$ and $n(a) \neq 0$, then the line $x = a$ is a vertical asymptote of the graph of $y = f(x)$.

Now we look at the behavior of $f(x) = 1/x$ as $|x|$ gets very large—that is, as $x \to \infty$ and as $x \to -\infty$. Consider Tables 3 and 4. As x increases without bound, $1/x$ is positive and approaches 0 from above. As x decreases without bound, $1/x$ is negative and approaches 0 from below. For our purposes, it is not necessary to distinguish between $1/x$ approaching 0 from above and from below. Thus, we will describe this behavior by writing

$$\frac{1}{x} \to 0 \qquad \text{as} \qquad x \to \infty \text{ and as } x \to -\infty$$

TABLE 3   Behavior of $1/x$ as $x \to \infty$

x	1	10	100	1,000	10,000	100,000	1,000,000	...	x increases without bound ($x \to \infty$)
$1/x$	1	0.1	0.01	0.001	0.0001	0.000 01	0.000 001	...	$1/x$ approaches 0 ($1/x \to 0$)

TABLE 4   Behavior of $1/x$ as $x \to -\infty$

x	-1	-10	-100	$-1,000$	$-10,000$	$-100,000$	$-1,000,000$	...	x decreases without bound ($x \to -\infty$)
$1/x$	-1	-0.1	-0.01	-0.001	-0.0001	$-0.000\ 01$	$-0.000\ 001$	...	$1/x$ approaches 0 ($1/x \to 0$)

The completed graph of $f(x) = 1/x$ is shown in Figure 2. Notice that the behavior as $x \to \infty$ and as $x \to -\infty$ is illustrated by drawing a curve that is almost horizontal and adding arrows at the ends. The curve in Figure 2 is an example of a plane curve called a **hyperbola,** and the coordinate axes are called **asymptotes** for this curve. The y axis is a *vertical asymptote* for $1/x$, and the x axis is a *horizontal asymptote* for $1/x$.

FIGURE 2 $f(x) = \dfrac{1}{x}, x \neq 0.$

DEFINITION 1   **Horizontal and Vertical Asymptotes**

The line $x = a$ is a **vertical asymptote** for the graph of $y = f(x)$ if $f(x)$ either increases or decreases without bound as x approaches a from the right or from the left. Symbolically,

$$f(x) \to \infty \quad \text{or} \quad f(x) \to -\infty \quad \text{as} \quad x \to a^+ \quad \text{or} \quad x \to a^-$$

The line $y = b$ is a **horizontal asymptote** for the graph of $y = f(x)$ if $f(x)$ approaches b as x increases without bound or as x decreases without bound. Symbolically,

$$f(x) \to b \quad \text{as} \quad x \to \infty \quad \text{or} \quad x \to -\infty$$

EXPLORE-DISCUSS 2   Construct tables similar to Tables 3 and 4 for each of the following functions and discuss the behavior of each as $x \to \infty$ and as $x \to -\infty$:

(A) $f(x) = \dfrac{3x}{x^2 + 1}$ (B) $g(x) = \dfrac{3x^2}{x^2 + 1}$ (C) $h(x) = \dfrac{3x^3}{x^2 + 1}$

In Section 4-1, we saw that the behavior of a polynomial

$$P(x) = a_n x^n + \ldots + a_1 x + a_0$$

as $x \to \pm\infty$ is determined by its leading term, $a_n x^n$. In a similar manner, the behavior of a rational function is determined by the ratio of the leading terms of its numerator and denominator; that is, the graphs of

$$f(x) = \frac{a_m x^m + \cdots + a_1 x + a_0}{b_n x^n + \cdots + b_1 x + b_0} \quad \text{and} \quad g(x) = \frac{a_m x^m}{b_n x^n}$$

exhibit the same behavior as $x \to \infty$ and as $x \to -\infty$ (see Fig. 3).

FIGURE 3 Graphs of rational functions as $x \to \pm\infty$.

$f(x) = \dfrac{2x + 7}{x^2 + x + 3}$

$g(x) = \dfrac{2x}{x^2} = \dfrac{2}{x}$

(a)

$f(x) = \dfrac{2x^2 + 4}{x^2 + x + 3}$

$g(x) = \dfrac{2x^2}{x^2} = 2$

(b)

$f(x) = \dfrac{2x^3 + 3}{x^2 + x + 3}$

$g(x) = \dfrac{2x^3}{x^2} = 2x$

(c)

In Figure 3(a), the degree of the numerator is less than the degree of the denominator and the x axis is a horizontal asymptote. In Figure 3(b), the degree of the numerator equals the degree of the denominator and the line $y = 2$ is a horizontal asymptote. In Figure 3(c), the degree of the numerator is greater than the degree of the denominator and there are no horizontal asymptotes. These ideas are generalized in Theorem 2 to provide a simple way to locate the horizontal asymptotes of any rational function.

Theorem 2

Horizontal Asymptotes and Rational Functions

Let f be a rational function defined by the quotient of two polynomials as follows:

$$f(x) = \frac{a_m x^m + \cdots + a_1 x + a_0}{b_n x^n + \cdots + b_1 x + b_0}$$

1. For $m < n$, the line $y = 0$ (the x axis) is a horizontal asymptote.

2. For $m = n$, the line $y = a_m/b_n$ is a horizontal asymptote.

3. For $m > n$, the graph will increase or decrease without bound, depending on m, n, a_m, and b_n, and there are no horizontal asymptotes.

EXAMPLE 2 **Finding Vertical and Horizontal Asymptotes for a Rational Function**

Find all vertical and horizontal asymptotes for

$$f(x) = \frac{n(x)}{d(x)} = \frac{2x^2 - 2x - 4}{x^2 - 9}$$

Solution   Since $d(x) = x^2 - 9 = (x - 3)(x + 3)$, the graph of $f(x)$ has vertical asymptotes at $x = 3$ and $x = -3$ (Theorem 1). Since $n(x)$ and $d(x)$ have the same degree, the line

$$y \quad \boxed{= \frac{a_2}{b_2}} \quad = \frac{2}{1} = 2 \quad a_2 = 2,\ b_2 = 1$$

is a horizontal asymptote (Theorem 2, part 2).

Matched Problem 2   Find all vertical and horizontal asymptotes for

$$f(x) = \frac{3x^2 - 12}{x^2 + 2x - 3}$$

• Graphing Rational Functions

We now use the techniques for locating asymptotes, along with other graphing aids discussed in the text, to graph several rational functions. First, we outline a systematic approach to the problem of graphing rational functions:

Graphing a Rational Function: $f(x) = n(x)/d(x)$

Step 1. Intercepts. Find the real solutions of the equation $n(x) = 0$ and use these solutions to plot any x intercepts of the graph of f. Evaluate $f(0)$, if it exists, and plot the y intercept.

Step 2. Vertical Asymptotes. Find the real solutions of the equation $d(x) = 0$ and use these solutions to determine the domain of f, the points of discontinuity, and the vertical asymptotes. Sketch any vertical asymptotes as dashed lines.

Step 3. Sign Chart. Construct a sign chart for f and use it to determine the behavior of the graph near each vertical asymptote.

Step 4. Horizontal Asymptotes. Determine whether there is a horizontal asymptote and if so, sketch it as a dashed line.

Step 5. Complete the Sketch. Complete the sketch of the graph by plotting additional points and joining these points with a smooth continuous curve over each interval in the domain of f. Do not cross any points of discontinuity.

EXAMPLE 3 Graphing a Rational Function

Graph: $y = f(x) = \dfrac{2x}{x - 3}$

Solution

$$f(x) = \frac{2x}{x - 3} = \frac{n(x)}{d(x)}$$

Step 1. Intercepts. Find real zeros of $n(x) = 2x$ and find $f(0)$:

$$2x = 0$$

$$x = 0 \quad \text{x intercept}$$

$$f(0) = 0 \quad \text{y intercept}$$

The graph crosses the coordinate axes only at the origin. Plot this intercept, as shown in Figure 4.

FIGURE 4

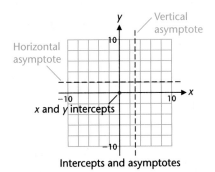

Intercepts and asymptotes

Step 2. *Vertical Asymptotes.* Find real zeros of $d(x) = x - 3$:

$$x - 3 = 0$$

$$x = 3$$

The domain of f is $(-\infty, 3) \cup (3, \infty)$, f is discontinuous at $x = 3$, and the graph has a vertical asymptote at $x = 3$. Sketch this asymptote, as shown in Figure 4.

Step 3. *Sign Chart.* Construct a sign chart for $f(x)$ (review Section 2-8), as shown in the margin. Since $x = 3$ is a vertical asymptote and $f(x) < 0$ for $0 < x < 3$,

$$f(x) \to -\infty \qquad \text{as} \qquad x \to 3^-$$

Since $x = 3$ is a vertical asymptote and $f(x) > 0$ for $x > 3$,

$$f(x) \to \infty \qquad \text{as} \qquad x \to 3^+$$

Notice how much information is contained in the sign chart for f. The solid dot determines the x intercept; the open dot determines the domain, the point of discontinuity, and the vertical asymptote; and the signs of $f(x)$ determine the behavior of the graph at the vertical asymptote.

Step 4. *Horizontal Asymptote.* Since $n(x)$ and $d(x)$ have the same degree, the line $y = 2$ is a horizontal asymptote. Sketch this asymptote, as shown in Figure 4.

Step 5. *Complete the Sketch.* By plotting a few additional points, we obtain the graph in Figure 5. Notice that the graph is a smooth continuous curve over the interval $(-\infty, 3)$ and over the interval $(3, \infty)$. As expected, there is a break in the graph at $x = 3$.

Test number	-1	1	4
Value of f	$\frac{1}{2}$	-1	8
Sign of f	$+$	$-$	$+$

FIGURE 5

As you gain experience in graphing, many of the steps in Example 3 can be done mentally (or on scratch paper) and the process can be speeded up considerably.

Matched Problem 3   Proceed as in Example 3 and graph: $y = f(x) = \dfrac{3x}{x + 2}$

Remark: Refer to Example 3. When we graph $f(x) = 2x/(x - 3)$ on a graphing utility [Fig. 6(a)], it appears that the graphing utility has also drawn the vertical asymptote, but this is not the case. Most graphing utilities, when set in *connected mode,* calculate points on a graph and connect these points with line segments. The last point plotted to the left of the asymptote and the first plotted to the right of the asymptote will usually have very large y coordinates. If these y coordinates have opposite sign, then the graphing utility may connect the two points with a nearly vertical line segment, which gives the appearance of an asymptote. If you wish, you can set the calculator in *dot mode* to plot the points without the connecting line segments (Fig. 6(b)].

FIGURE 6 Graphing utility graphs of $f(x) = \dfrac{2x}{x - 3}$.

(a) Connected mode

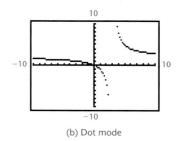

(b) Dot mode

In the remaining examples we will just list the results of each step in the graphing strategy and omit the computational details.

EXAMPLE 4   **Graphing a Rational Function**

Graph: $y = f(x) = \dfrac{x^2 - 6x + 9}{x^2 + x - 2}$

Solution

$$f(x) = \frac{x^2 - 6x + 9}{x^2 + x - 2} = \frac{(x - 3)^2}{(x + 2)(x - 1)}$$

x intercept: $x = 3$

y intercept: $y = f(0) = -\frac{9}{2} = -4.5$

Domain: $(-\infty, -2) \cup (-2, 1) \cup (1, \infty)$

Points of discontinuity: $x = -2$ and $x = 1$

Vertical asymptotes: $x = -2$ and $x = 1$

Test number	−3	0	2	4
Value of f	9	$-\frac{9}{2}$	$\frac{1}{4}$	$\frac{1}{18}$
Sign of f	+	−	+	+

$$f(x) \to \infty \qquad \text{as} \qquad x \to -2^-$$
$$f(x) \to -\infty \qquad \text{as} \qquad x \to -2^+$$
$$f(x) \to -\infty \qquad \text{as} \qquad x \to 1^-$$
$$f(x) \to \infty \qquad \text{as} \qquad x \to 1^+$$

Horizontal asymptote: $y = 1$

Sketch in the intercepts and asymptotes (Fig. 7), then sketch the graph of f (Fig. 8).

Intercepts and asymptotes

FIGURE 7

$$f(x) = \frac{x^2 - 6x + 9}{x^2 + x - 2}$$

FIGURE 8

Matched Problem 4   Graph: $y = f(x) = \dfrac{x^2}{x^2 - 7x + 10}$

CAUTION   The graph of a function cannot cross a vertical asymptote, but the same statement is not true for horizontal asymptotes. The graph in Example 4 clearly shows that **the graph of a function can cross a horizontal asymptote.** The definition of a horizontal asymptote requires $f(x)$ to approach b as x increases or decreases without bound, but it does not preclude the possibility that $f(x) = b$ for one or more values of x. In fact, using the cosine function from trigonometry, it is possible to construct a function whose graph crosses a horizontal asymptote an infinite number of times (see Fig. 9).

FIGURE 9 Multiple intersections of a graph and a horizontal asymptote.

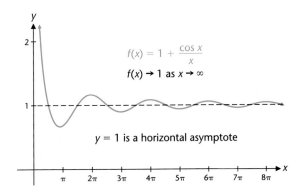

$$f(x) = 1 + \frac{\cos x}{x}$$

$f(x) \to 1$ as $x \to \infty$

$y = 1$ is a horizontal asymptote

EXAMPLE 5   **Graphing a Rational Function**

Graph: $y = f(x) = \dfrac{x^2 - 3x - 4}{x - 2}$

Solution

$$f(x) = \frac{x^2 - 3x - 4}{x - 2} = \frac{(x + 1)(x - 4)}{x - 2}$$

x intercepts: $x = -1$ and $x = 4$

y intercept: $y = f(0) = 2$

Domain: $(-\infty, 2) \cup (2, \infty)$

Points of discontinuity: $x = 2$

Vertical asymptote: $x = 2$

$f(x) \to \infty$ as $x \to 2^-$

$f(x) \to -\infty$ as $x \to 2^+$

No horizontal asymptote

Test number	-2	0	3	5
Value of f	$-\frac{3}{2}$	2	-4	2
Sign of f	$-$	$+$	$-$	$+$

Even though the graph of f does not have a horizontal asymptote, we can still gain some useful information about the behavior of the graph as $x \to -\infty$ and as $x \to \infty$ if we first perform a long division:

$$
\begin{array}{r}
x - 1 \\
x - 2 \overline{)\,x^2 - 3x - 4} \\
\underline{x^2 - 2x} \\
-x - 4 \\
\underline{-x + 2} \\
-6
\end{array}
$$

Quotient

Remainder

Thus,

$$f(x) = \frac{x^2 - 3x - 4}{x - 2} = x - 1 - \frac{6}{x - 2}$$

As $x \to -\infty$ or $x \to \infty$, $6/(x - 2) \to 0$ and the graph of f approaches the line $y = x - 1$. This line is called an **oblique asymptote** for the graph of f. The asymptotes and intercepts are sketched in Figure 10, and the graph of f is sketched in Figure 11.

Intercepts and asymptotes

FIGURE 10

$$f(x) = \frac{x^2 - 3x - 4}{x - 2}$$

FIGURE 11

Generalizing the results of Example 5, we have Theorem 3.

Theorem 3

Oblique Asymptotes and Rational Functions

If $f(x) = n(x)/d(x)$, where $n(x)$ and $d(x)$ are polynomials and the degree of $n(x)$ is 1 more than the degree of $d(x)$, then $f(x)$ can be expressed in the form

$$f(x) = mx + b + \frac{r(x)}{d(x)}$$

where the degree of $r(x)$ is less than the degree of $d(x)$. The line

$$y = mx + b$$

is an oblique asymptote for the graph of f. That is,

$$[f(x) - (mx + b)] \to 0 \quad \text{as} \quad x \to -\infty \quad \text{or} \quad x \to \infty$$

Matched Problem 5   Graph, including any oblique asymptotes: $y = f(x) = \dfrac{x^2 + 5}{x + 1}$

Answers to Matched Problems

1. Domain: $(-\infty, -3) \cup (-3, 1) \cup (1, \infty)$; x intercepts: $x = -2$, $x = 2$
2. Vertical asymptotes: $x = -3$, $x = 1$; horizontal asymptote: $y = 3$

3.

$$f(x) = \frac{3x}{x + 2}$$

4.

$$f(x) = \frac{x^2}{x^2 - 7x + 10}$$

5.

$y = x - 1$

$$f(x) = \frac{x^2 + 5}{x + 1}$$

EXERCISE 4-4

A

In Problems 1–4, match each graph with one of the following functions:

$$f(x) = \frac{2x - 4}{x + 2} \qquad g(x) = \frac{2x + 4}{2 - x}$$

$$h(x) = \frac{2x + 4}{x - 2} \qquad k(x) = \frac{4 - 2x}{x + 2}$$

1.

2.

3.

4.

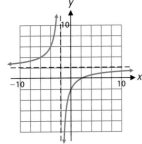

In Problems 5–12, find the domain and x intercepts. Do not graph.

5. $f(x) = \dfrac{2x - 4}{x + 1}$

6. $g(x) = \dfrac{3x + 6}{x - 1}$

7. $h(x) = \dfrac{x^2 - 1}{x^2 - 16}$

8. $k(x) = \dfrac{x^2 - 36}{x^2 - 25}$

9. $r(x) = \dfrac{x^2 - x - 6}{x^2 - x - 12}$

10. $s(x) = \dfrac{x^2 + x - 12}{x^2 + x - 6}$

11. $F(x) = \dfrac{x}{x^2 + 4}$

12. $G(x) = \dfrac{x^2}{x^2 + 16}$

In Problems 13–20, find all vertical and horizontal asymptotes. Do not graph.

13. $f(x) = \dfrac{2x}{x - 4}$

14. $h(x) = \dfrac{3x}{x + 5}$

15. $s(x) = \dfrac{2x^2 + 3x}{3x^2 - 48}$

16. $r(x) = \dfrac{5x^2 - 7x}{2x^2 - 50}$

17. $p(x) = \dfrac{2x}{x^4 + 1}$

18. $q(x) = \dfrac{5x^4}{2x^2 + 3x - 2}$

19. $t(x) = \dfrac{6x^4}{3x^2 - 2x - 5}$

20. $g(x) = \dfrac{3x}{x^4 + 2x^2 + 1}$

B

In Problems 21–40, use the graphing strategy outlined in the text to sketch the graph of each function.

 Check Problems 21–40 on a graphing utility.

21. $f(x) = \dfrac{1}{x - 4}$

22. $g(x) = \dfrac{1}{x + 3}$

23. $f(x) = \dfrac{x}{x + 1}$

24. $f(x) = \dfrac{3x}{x - 3}$

25. $h(x) = \dfrac{x}{2x - 2}$

26. $p(x) = \dfrac{3x}{4x + 4}$

27. $f(x) = \dfrac{2x - 4}{x + 3}$

28. $f(x) = \dfrac{3x + 3}{2 - x}$

29. $g(x) = \dfrac{1 - x^2}{x^2}$

30. $f(x) = \dfrac{x^2 + 1}{x^2}$

31. $f(x) = \dfrac{9}{x^2 - 9}$

32. $g(x) = \dfrac{6}{x^2 - x - 6}$

33. $f(x) = \dfrac{x}{x^2 - 1}$

34. $p(x) = \dfrac{x}{1 - x^2}$

35. $g(x) = \dfrac{2}{x^2 + 1}$

36. $f(x) = \dfrac{x}{x^2 + 1}$

37. $f(x) = \dfrac{12x^2}{(3x + 5)^2}$

38. $f(x) = \dfrac{7x^2}{(2x - 3)^2}$

39. $f(x) = \dfrac{x^2 - 1}{x^2 + 7x + 10}$

40. $f(x) = \dfrac{x^2 + 6x + 8}{x^2 - x - 2}$

41. If $f(x) = n(x)/d(x)$, where $n(x)$ and $d(x)$ are quadratic functions, what is the maximum number of x intercepts $f(x)$ can have? What is the minimum number? Illustrate both cases with examples.

42. If $f(x) = n(x)/d(x)$, where $n(x)$ and $d(x)$ are quadratic functions, what is the maximum number of vertical asymptotes $f(x)$ can have? What is the minimum number? Illustrate both cases with examples.

In Problems 43–48, find all vertical, horizontal, and oblique asymptotes. Do not graph.

43. $f(x) = \dfrac{2x^2}{x - 1}$

44. $g(x) = \dfrac{3x^2}{x + 2}$

45. $p(x) = \dfrac{x^3}{x^2 + 1}$

46. $q(x) = \dfrac{x^5}{x^3 - 8}$

47. $r(x) = \dfrac{2x^2 - 3x + 5}{x}$

48. $s(x) = \dfrac{-3x^2 + 5x + 9}{x}$

 In Problems 49–52, use a graphing utility to investigate the behavior of each function as $x \to \infty$ and as $x \to -\infty$, and to find any horizontal asymptotes.

49. $f(x) = \dfrac{5x}{\sqrt{x^2 + 1}}$

50. $f(x) = \dfrac{2x}{\sqrt{x^2 - 1}}$

51. $f(x) = \dfrac{4\sqrt{x^2 - 4}}{x}$

52. $f(x) = \dfrac{3\sqrt{x^2 + 1}}{x - 1}$

C

In Problems 53–58, use the graphing strategy outlined in the text to sketch the graph of each function. Include any oblique asymptotes.

 Check Problems 53–58 with a graphing utility.

53. $f(x) = \dfrac{x^2 + 1}{x}$

54. $g(x) = \dfrac{x^2 - 1}{x}$

55. $k(x) = \dfrac{x^2 - 4x + 3}{2x - 4}$

56. $h(x) = \dfrac{x^2 + x - 2}{2x - 4}$

57. $F(x) = \dfrac{8 - x^3}{4x^2}$

58. $G(x) = \dfrac{x^4 + 1}{x^3}$

If $f(x) = n(x)/d(x)$, where the degree of $n(x)$ is greater than the degree of $d(x)$, then long division can be used to write $f(x) = p(x) + q(x)/d(x)$, where $p(x)$ and $q(x)$ are polynomials with the degree of $q(x)$ less than the degree of $d(x)$. In Problems 59–62, perform the long division and discuss the relationship between the graphs of $f(x)$ and $p(x)$ as $x \to \infty$ and as $x \to -\infty$.

59. $f(x) = \dfrac{x^4}{x^2 + 1}$

60. $f(x) = \dfrac{x^5}{x^2 + 1}$

61. $f(x) = \dfrac{x^5}{x^2 - 1}$

62. $f(x) = \dfrac{x^5}{x^3 - 1}$

 In calculus, it is often necessary to consider rational functions that are not in lowest terms, such as the functions given in Problems 63–66. For each function, state the domain, reduce the function to lowest terms, and sketch its graph. Remember to exclude from the graph any points with x values that are not in the domain.

63. $f(x) = \dfrac{x^2 - 4}{x - 2}$

64. $g(x) = \dfrac{x^2 - 1}{x + 1}$

65. $r(x) = \dfrac{x + 2}{x^2 - 4}$

66. $s(x) = \dfrac{x - 1}{x^2 - 1}$

APPLICATIONS

67. Employee Training. A company producing electronic components used in television sets has established that on the average, a new employee can assemble $N(t)$ components per day after t days of on-the-job training, as given by

$$N(t) = \frac{50t}{t + 4} \qquad t \geq 0$$

Sketch the graph of N, including any vertical or horizontal asymptotes. What does N approach as $t \to \infty$?

68. Physiology. In a study on the speed of muscle contraction in frogs under various loads, researchers W. O. Fems and J. Marsh found that the speed of contraction decreases with increasing loads. More precisely, they found that the relationship between speed of contraction S (in centimeters per second) and load w (in grams) is given approximately by

$$S(w) = \frac{26 + 0.06w}{w} \qquad w \geq 5$$

Sketch the graph of S, including any vertical or horizontal asymptotes. What does S approach as $w \to \infty$?

69. Retention. An experiment on retention is conducted in a psychology class. Each student in the class is given 1 day to memorize the same list of 40 special characters. The lists are turned in at the end of the day, and for each succeeding day for 20 days each student is asked to turn in a list of as many of the symbols as can be recalled. Averages are taken, and it is found that a good approximation of the average number of symbols, $N(t)$, retained after t days is given by

$$N(t) = \frac{5t + 30}{t} \qquad t \geq 1$$

Sketch the graph of N, including any vertical or horizontal asymptotes. What does N approach as $t \to \infty$?

70. Learning Theory. In 1917, L. L. Thurstone, a pioneer in quantitative learning theory, proposed the function

$$f(x) = \frac{a(x + c)}{(x + c) + b}$$

to describe the number of successful acts per unit time that a person could accomplish after x practice sessions. Suppose that for a particular person enrolling in a typing class,

$$f(x) = \frac{50(x + 1)}{x + 5} \qquad x \geq 0$$

where $f(x)$ is the number of words per minute the person is able to type after x weeks of lessons. Sketch the graph of f, including any vertical or horizontal asymptotes. What does f approach as $x \to \infty$?

Using calculus techniques, it can be shown that the minimum value of a function of the form

$$g(x) = ax + b + \frac{c}{x} \qquad a > 0, c > 0, x > 0$$

is min $g(x) = g(\sqrt{c/a})$. Use this fact in Problems 71–74.

★ **71. Replacement Time.** A desktop office copier has an initial price of $2,500. A maintenance/service contract costs $200 for the first year and increases $50 per year thereafter. It can be shown that the total cost of the copier after n years is given by

$$C(n) = 2,500 + 175n + 25n^2$$

The average cost per year for n years is $\overline{C}(n) = C(n)/n$.
(A) Find the rational function $\overline{C}$.
(B) When is the average cost per year minimum? (This is frequently referred to as the *replacement time* for this piece of equipment.)
(C) Sketch the graph of $\overline{C}$, including any asymptotes.

★ **72. Average Cost.** The total cost of producing x units of a certain product is given by

$$C(x) = \tfrac{1}{5}x^2 + 2x + 2{,}000$$

The average cost per unit for producing x units is $\overline{C}(x) = C(x)/x$.
(A) Find the rational function $\overline{C}$.
(B) At what production level will the average cost per unit be minimal?
(C) Sketch the graph of $\overline{C}$, including any asymptotes.

★ **73. Construction.** A rectangular dog pen is to be made to enclose an area of 225 square feet.
(A) If x represents the width of the pen, express the total length $L(x)$ of the fencing material required for the pen in terms of x.
(B) Considering the physical limitations, what is the domain of the function L?

(C) Find the dimensions of the pen that will require the least amount of fencing material.
(D) Graph the function L, including any asymptotes.

★ **74. Construction.** Rework Problem 73 with the added assumption that the pen is to be divided into two sections, as shown in the figure.

SECTION 4-5 Partial Fractions

- Basic Theorems
- Partial Fraction Decomposition

You have now had considerable experience combining two or more rational expressions into a single rational expression. For example, problems such as

$$\frac{2}{x+5} + \frac{3}{x-4} = \frac{2(x-4) + 3(x+5)}{(x+5)(x-4)} = \frac{5x+7}{(x+5)(x-4)}$$

should seem routine. Frequently in more advanced courses, particularly in calculus, it is advantageous to be able to reverse this process—that is, to be able to express a rational expression as the sum of two or more simpler rational expressions called **partial fractions.** As is often the case with reverse processes, the process of decomposing a rational expression into partial fractions is more difficult than combining rational expressions. Basic to the process is the factoring of polynomials, so the topics discussed earlier in this chapter can be put to effective use.

We confine our attention to rational expressions of the form $P(x)/D(x)$, where $P(x)$ and $D(x)$ are polynomials with real coefficients and no common factors. In addition, we assume that the degree of $P(x)$ is less than the degree of $D(x)$. If the degree of $P(x)$ is greater than or equal to that of $D(x)$, we have only to divide $P(x)$ by $D(x)$ to obtain

$$\frac{P(x)}{D(x)} = Q(x) + \frac{R(x)}{D(x)}$$

where the degree of $R(x)$ is less than that of $D(x)$. For example,

$$\frac{x^4 - 3x^3 + 2x^2 - 5x + 1}{x^2 - 2x + 1} = x^2 - x - 1 + \frac{-6x + 2}{x^2 - 2x + 1}$$

If the degree of $P(x)$ is less than that of $D(x)$, then $P(x)/D(x)$ is called a **proper fraction.**

• Basic Theorems   Our task now is to establish a systematic way to decompose a proper fraction into the sum of two or more partial fractions. The following three theorems take care of the problem completely. Theorems 1 and 3 are stated without proof.

Theorem 1   **Equal Polynomials**

Two polynomials are equal to each other if and only if the coefficients of terms of like degree are equal.

For example, if

Equate the constant terms.

$$(A + 2B)x + B = 5x - 3$$

Equate the coefficients of x.

then

$$B = -3 \quad \text{Substitute } B = -3 \text{ into the second equation to solve for } A.$$

$$A + 2B = 5$$

$$A + 2(-3) = 5$$

$$A = 11$$

EXPLORE-DISCUSS 1   If

$$x + 5 = A(x + 1) + B(x - 3) \tag{1}$$

is a polynomial identity (that is, both sides represent the same polynomial), then equating coefficients produces the system

$$1 = A + \ \ B \qquad \text{Equating coefficients of } x$$

$$5 = A - 3B \qquad \text{Equating constant terms}$$

(A) Solve this system (see Section 2-2).
(B) For an alternate method of solution, substitute $x = 3$ in (1) to find A and then substitute $x = -1$ in (1) to find B. Explain why method B is valid.

Theorem 2   **Linear and Quadratic Factor Theorem**

For a polynomial with real coefficients, there always exists a complete factoring involving only linear and/or quadratic factors with real coefficients where the linear and quadratic factors are prime relative to the real numbers.

That Theorem 2 is true can be seen as follows: From earlier theorems in this chapter, we know that an nth-degree polynomial $P(x)$ has n zeros and n linear factors. The real zeros of $P(x)$ correspond to linear factors of the form $(x - r)$, where r is a real number. Since $P(x)$ has real coefficients, the imaginary zeros occur in conjugate pairs. Thus, the imaginary zeros correspond to pairs of factors of the form $[x - (a + bi)]$ and $[x - (a - bi)]$, where a and b are real numbers. Multiplying these two imaginary factors, we have

$$[x - (a + bi)][x - (a - bi)] = x^2 - 2ax + a^2 + b^2$$

This quadratic polynomial with real coefficients is a factor of $P(x)$. Thus, $P(x)$ can be factored into a product of linear factors and quadratic factors, all with real coefficients.

● **Partial Fraction Decomposition**

We are now ready to state Theorem 3, which forms the basis for partial fraction decomposition.

Theorem 3   **Partial Fraction Decomposition**

Any proper fraction $P(x)/D(x)$ reduced to lowest terms can be decomposed into the sum of partial fractions as follows:

1. If $D(x)$ has a nonrepeating linear factor of the form $ax + b$, then the partial fraction decomposition of $P(x)/D(x)$ contains a term of the form

$$\frac{A}{ax + b} \qquad A \text{ a constant}$$

2. If $D(x)$ has a k-repeating linear factor of the form $(ax + b)^k$, then the partial fraction decomposition of $P(x)/D(x)$ contains k terms of the form

$$\frac{A_1}{ax + b} + \frac{A_2}{(ax + b)^2} + \cdots + \frac{A_k}{(ax + b)^k} \qquad A_1, A_2, \ldots, A_k \text{ constants}$$

3. If $D(x)$ has a nonrepeating quadratic factor of the form $ax^2 + bx + c$, which is prime relative to the real numbers, then the partial fraction decomposition of $P(x)/D(x)$ contains a term of the form

$$\frac{Ax + B}{ax^2 + bx + c} \qquad A, B \text{ constants}$$

4. If $D(x)$ has a k-repeating quadratic factor of the form $(ax^2 + bx + c)^k$, where $ax^2 + bx + c$ is prime relative to the real numbers, then the partial fraction decomposition of $P(x)/D(x)$ contains k terms of the form

$$\frac{A_1x + B_1}{ax^2 + bx + c} + \frac{A_2x + B_2}{(ax^2 + bx + c)^2} + \cdots + \frac{A_kx + B_k}{(ax^2 + bx + c)^k}$$

$$A_1, \ldots, A_k, \quad B_1, \ldots, B_k \text{ constants}$$

Let's see how the theorem is used to obtain partial fraction decompositions in several examples.

EXAMPLE 1   **Nonrepeating Linear Factors**

Decompose into partial fractions: $\dfrac{5x + 7}{x^2 + 2x - 3}$

Solution   We first try to factor the denominator. If it can't be factored in the real numbers, then we can't go any further. In this example, the denominator factors, so we apply part 1 from Theorem 3:

$$\frac{5x + 7}{(x - 1)(x + 3)} = \frac{A}{x - 1} + \frac{B}{x + 3} \tag{2}$$

To find the constants A and B, we combine the fractions on the right side of equation (2) to obtain

$$\frac{5x + 7}{(x - 1)(x + 3)} = \frac{A(x + 3) + B(x - 1)}{(x - 1)(x + 3)}$$

Since these fractions have the same denominator, their numerators must be equal. Thus,

$$5x + 7 = A(x + 3) + B(x - 1) \tag{3}$$

We could multiply the right side and find A and B by using Theorem 1, but in this case it is easier to take advantage of the fact that equation (3) is an identity—that is, it must hold for all values of x. In particular, we note that if we let $x = 1$, then the second term of the right side drops out and we can solve for A:

$$5 \cdot 1 + 7 = A(1 + 3) + B(1 - 1)$$

$$12 = 4A$$

$$A = 3$$

Similarly, if we let $x = -3$, the first term drops out and we find

$$-8 = -4B$$

$$B = 2$$

Hence,

$$\frac{5x + 7}{x^2 + 2x - 3} = \frac{3}{x - 1} + \frac{2}{x + 3} \tag{4}$$

as can easily be checked by adding the two fractions on the right.

Matched Problem 1 Decompose into partial fractions: $\dfrac{7x + 6}{x^2 + x - 6}$

 EXPLORE-DISCUSS 2 A graphing utility can also be used to check a partial fraction decomposition. To check Example 1, we graph the left and right sides of (4) in a graphing utility (Fig. 1). Discuss how the Trace feature on the graphing utility can be used to check that the graphing utility is displaying two identical graphs.

FIGURE 1

EXAMPLE 2 **Repeating Linear Factors**

Decompose into partial fractions: $\dfrac{6x^2 - 14x - 27}{(x + 2)(x - 3)^2}$

Solution Using parts 1 and 2 from Theorem 3, we write

$$\frac{6x^2 - 14x - 27}{(x + 2)(x - 3)^2} = \frac{A}{x + 2} + \frac{B}{x - 3} + \frac{C}{(x - 3)^2}$$

$$= \frac{A(x - 3)^2 + B(x + 2)(x - 3) + C(x + 2)}{(x + 2)(x - 3)^2}$$

Thus, for all x,

$$6x^2 - 14x - 27 = A(x - 3)^2 + B(x + 2)(x - 3) + C(x + 2)$$

If $x = 3$, then   If $x = -2$, then

$$-15 = 5C \qquad\qquad 25 = 25A$$

$$C = -3 \qquad\qquad A = 1$$

There are no other values of x that will cause terms on the right to drop out. Since any value of x can be substituted to produce an equation relating A, B, and C, we let $x = 0$ and obtain

$$-27 = 9A - 6B + 2C \quad \text{Substitute } A = 1 \text{ and } C = -3.$$

$$-27 = 9 - 6B - 6$$

$$B = 5$$

Thus,

$$\frac{6x^2 - 14x - 27}{(x + 2)(x - 3)^2} = \frac{1}{x + 2} + \frac{5}{x - 3} - \frac{3}{(x - 3)^2}$$

Matched Problem 2   Decompose into partial fractions: $\dfrac{x^2 + 11x + 15}{(x - 1)(x + 2)^2}$

EXAMPLE 3   **Nonrepeating Linear and Quadratic Factors**

Decompose into partial fractions: $\dfrac{5x^2 - 8x + 5}{(x - 2)(x^2 - x + 1)}$

Solution   First, we see that the quadratic in the denominator can't be factored further in the real numbers. Then, we use parts 1 and 3 from Theorem 3 to write

$$\frac{5x^2 - 8x + 5}{(x - 2)(x^2 - x + 1)} = \frac{A}{x - 2} + \frac{Bx + C}{x^2 - x + 1}$$

$$= \frac{A(x^2 - x + 1) + (Bx + C)(x - 2)}{(x - 2)(x^2 - x + 1)}$$

Thus, for all x,

$$5x^2 - 8x + 5 = A(x^2 - x + 1) + (Bx + C)(x - 2)$$

If $x = 2$, then

$$9 = 3A$$

$$A = 3$$

If $x = 0$, then, using $A = 3$, we have

$$5 = 3 - 2C$$

$$C = -1$$

If $x = 1$, then, using $A = 3$ and $C = -1$, we have

$$2 = 3 + (B - 1)(-1)$$

$$B = 2$$

Hence,

$$\frac{5x^2 - 8x + 5}{(x - 2)(x^2 - x + 1)} = \frac{3}{x - 2} + \frac{2x - 1}{x^2 - x + 1}$$

Matched Problem 3   Decompose into partial fractions: $\dfrac{7x^2 - 11x + 6}{(x - 1)(2x^2 - 3x + 2)}$

EXAMPLE 4   **Repeating Quadratic Factors**

Decompose into partial fractions: $\dfrac{x^3 - 4x^2 + 9x - 5}{(x^2 - 2x + 3)^2}$

Solution   Since $x^2 - 2x + 3$ can't be factored further in the real numbers, we proceed to use part 4 from Theorem 3 to write

$$\frac{x^3 - 4x^2 + 9x - 5}{(x^2 - 2x + 3)^2} = \frac{Ax + B}{x^2 - 2x + 3} + \frac{Cx + D}{(x^2 - 2x + 3)^2}$$

$$= \frac{(Ax + B)(x^2 - 2x + 3) + Cx + D}{(x^2 - 2x + 3)^2}$$

Thus, for all x,

$$x^3 - 4x^2 + 9x - 5 = (Ax + B)(x^2 - 2x + 3) + Cx + D$$

Since the substitution of carefully chosen values of x doesn't lead to the immediate determination of A, B, C, or D, we multiply and rearrange the right side to obtain

$$x^3 - 4x^2 + 9x - 5 = Ax^3 + (B - 2A)x^2 + (3A - 2B + C)x + (3B + D)$$

Now we use Theorem 1 to equate coefficients of terms of like degree:

$$A = 1$$
$$B - 2A = -4$$
$$3A - 2B + C = 9$$
$$3B + D = -5$$

$$\underset{Ax^3 \; + \; (B - 2A)x^2 \; + \; (3A - 2B + C)x \; + \; (3B + D)}{\overset{1x^3 \qquad\quad -4x^2 \qquad\qquad +9x \qquad\qquad -5}{\downarrow \qquad\qquad \downarrow \qquad\qquad\quad \downarrow \qquad\qquad\quad \downarrow}}$$

From these equations we easily find that $A = 1$, $B = -2$, $C = 2$, and $D = 1$. Now we can write

$$\frac{x^3 - 4x^2 + 9x - 5}{(x^2 - 2x + 3)^2} = \frac{x - 2}{x^2 - 2x + 3} + \frac{2x + 1}{(x^2 - 2x + 3)^2}$$

Matched Problem 4   Decompose into partial fractions: $\dfrac{3x^3 - 6x^2 + 7x - 2}{(x^2 - 2x + 2)^2}$

Answers to Matched Problems

1. $\dfrac{4}{x - 2} + \dfrac{3}{x + 3}$ **2.** $\dfrac{3}{x - 1} - \dfrac{2}{x + 2} + \dfrac{1}{(x + 2)^2}$ **3.** $\dfrac{2}{x - 1} + \dfrac{3x - 2}{2x^2 - 3x + 2}$

4. $\dfrac{3x}{x^2 - 2x + 2} + \dfrac{x - 2}{(x^2 - 2x + 2)^2}$

EXERCISE 4-5

A

In Problems 1–4, find constants A and B so that the indicated equation is a polynomial identity.

1. $5x - 10 = A(x + 4) + B(x - 6)$

2. $3x + 18 = A(x - 1) + B(x + 2)$

3. $-9x + 5 = A(2x + 5) + B(3x + 2)$

4. $-32x - 13 = A(4x - 7) + B(2x + 8)$

In Problems 5–10, find constants A, B, C, and D so that the right side is equal to the left.

5. $\dfrac{5x^2 + 7x + 6}{(x - 1)(x + 2)^2} = \dfrac{A}{x - 1} + \dfrac{B}{x + 2} + \dfrac{C}{(x + 2)^2}$

6. $\dfrac{x^2 + 5x - 12}{(x + 1)(x - 3)^2} = \dfrac{A}{x + 1} + \dfrac{B}{x - 3} + \dfrac{C}{(x - 3)^2}$

7. $\dfrac{x^2 + 5x + 3}{x(x^2 + 1)} = \dfrac{A}{x} + \dfrac{Bx + C}{x^2 + 1}$

8. $\dfrac{3x^2 - 2x - 16}{x(x^2 + 4)} = \dfrac{A}{x} + \dfrac{Bx + C}{x^2 + 4}$

9. $\dfrac{2x^3 + x^2 - 2}{(x^2 + x + 2)^2} = \dfrac{Ax + B}{x^2 + x + 2} + \dfrac{Cx + D}{(x^2 + x + 2)^2}$

10. $\dfrac{4x^2 - 2x + 13}{(x^2 - x + 4)^2} = \dfrac{Ax + B}{x^2 - x + 4} + \dfrac{Cx + D}{(x^2 - x + 4)^2}$

B

In Problems 11–22, decompose into partial fractions.

11. $\dfrac{3x - 40}{x^2 - x - 12}$ **12.** $\dfrac{4x - 43}{x^2 + 3x - 10}$

13. $\dfrac{8x - 22}{6x^2 + 17x - 14}$ **14.** $\dfrac{x - 27}{10x^2 - 13x - 3}$

15. $\dfrac{6x^2 - 14x + 4}{4x^3 - 4x^2 + x}$ **16.** $\dfrac{7x - 8}{9x^3 - 12x^2 + 4x}$

17. $\dfrac{10x^2 + 4x + 3}{2x^3 + x^2 + x}$ **18.** $\dfrac{9x^2 + 14}{x^3 + x^2 + 2x}$

19. $\dfrac{4x^3 - 5x^2 + 6x - 5}{x^4 + 2x^2 + 1}$ **20.** $\dfrac{2x^3 - x^2 + 11x}{x^4 + 8x^2 + 16}$

21. $\dfrac{4x^2 - 5x + 3}{x^3 + x + 2}$ **22.** $\dfrac{5x^2 - 8x + 12}{x^3 - x^2 - 4}$

C

In Problems 23–26, decompose into a sum of a polynomial and partial fractions.

23. $\dfrac{x^5 - 4x^2 - 16x + 40}{x^4 - 2x^3 - 8x + 16}$ **24.** $\dfrac{2x^5 + 2x^4 - 6x^3}{x^4 + 4x + 3}$

25. $\dfrac{2x^5 - 7x^2 + 20x + 21}{x^4 - 4x^2 - 12x - 9}$ **26.** $\dfrac{x^5 + 2x^4 + 3x^2 - 14x}{x^4 - x^2 - 4x - 4}$

In Problems 27–30, a and b are real constants. Decompose into partial fractions.

27. $\dfrac{x}{(x + a)^2}, a \neq 0$

28. $\dfrac{1}{x(x + a)^2}, a \neq 0$

29. $\dfrac{1}{(x - a)(x - b)}, a \neq b$

30. $\dfrac{x}{(x - a)(x - b)}, a \neq b$

CHAPTER 4 GROUP ACTIVITY Interpolating Polynomials

Given two points in the plane, we can use the point-slope form of the equation of a line to find a polynomial whose graph passes through these two points. How can we proceed if we are given more than two points? For example, how can we find the equation of a polynomial $P(x)$ whose graph passes through the points listed in Table 1 and graphed in Figure 1?

TABLE 1

x	1	2	3	4
$P(x)$	1	3	-3	1

FIGURE 1

The key to solving this problem is to write the unknown polynomial $P(x)$ in the following special form:

$$P(x) = a_0 + a_1(x - 1) + a_2(x - 1)(x - 2) + a_3(x - 1)(x - 2)(x - 3) \tag{1}$$

Since the graph of $P(x)$ is to pass through each point in Table 1, we can substitute each value of x in (1) to determine the coefficients a_0, a_1, a_2, and a_3. First we evaluate (1) at $x = 1$ to determine a_0:

$$1 = P(1)$$
$$= a_0 \qquad \text{All other terms in (1) are 0 when } x = 1.$$

Using this value for a_0 in (1) and evaluating at $x = 2$, we have

$$3 = P(2) = 1 + a_1(1) \qquad \text{All other terms are 0.}$$
$$2 = a_1$$

Continuing in this manner, we have

$$-3 = P(3) = 1 + 2(2) + a_2(2)(1)$$
$$-8 = 2a_2$$
$$-4 = a_2$$

$$1 = P(4) = 1 + 2(3) - 4(3)(2) + a_3(3)(2)(1)$$

$$18 = 6a_3$$

$$3 = a_3$$

We have now evaluated all the coefficients in (1) and can write

$$P(x) = 1 + 2(x - 1) - 4(x - 1)(x - 2) + 3(x - 1)(x - 2)(x - 3) \tag{2}$$

If we expand the products in (2) and collect like terms, we can express $P(x)$ in the more conventional form (verify this)

$$P(x) = 3x^3 - 22x^2 + 47x - 27$$

(A) To check these calculations, evaluate $P(x)$ at $x = 1, 2, 3,$ and 4 and compare the results with Table 1. Then add the graph of $P(x)$ to Figure 1.

(B) Write a verbal description of the special form of $P(x)$ in (1).

In general, given a set of $n + 1$ points:

x	x_0	x_1	$\cdots$	x_n
y	y_0	y_1	$\cdots$	y_n

the **interpolating polynomial** for these points is the polynomial $P(x)$ of degree less than or equal to n that satisfies $P(x_k) = y_k$ for $k = 0, 1, \ldots, n$. The **general form** of the interpolating polynomial is

$$P(x) = a_0 + a_1(x - x_0) + a_2(x - x_0)(x - x_1) + \cdots + a_n(x - x_0)(x - x_1) \cdot \cdots \cdot (x - x_{n-1})$$

(C) Summarize the procedure for using the points in the table to find the coefficients in the general form.

(D) Give an example to show that the interpolating polynomial can have degree strictly less than n.

(E) Could there be two different polynomials of degree less than or equal to n whose graph passes through the given $n + 1$ points? Justify your answer.

(F) Find the interpolating polynomial for each of Tables 2 and 3. Check your answers by evaluating the polynomial, and illustrate by graphing the points in the table and the polynomial on the same axes.

TABLE 2

x	-1	0	1	2
y	5	3	3	11

TABLE 3

x	-2	-1	0	1	2
y	-3	0	5	0	-3

A surprisingly short program on a graphing utility can be used to calculate the coefficients in the general form of an interpolating polynomial. Table 4 shows such a program for a Texas Instruments graphing calculator and the output generated when we use the program to find the coefficients of the interpolating polynomial for Table 1.

TABLE 4   Interpolating Polynomial Coefficients on a Graphing Utility

Program INTERP*	Output

``
L2→L3
dimL L3→M
For(I,2,M,1)
For(J,M,I,-1)
(L3(J)-L3(J-1))/(L1(J)-L1(J-I+1))→L3(J)
End
End
Disp L3
``
	``
	{1,2,3,4}→L1
	           {1 2 3 4}
	{1,3, -3,1}→L2
	           {1 3 -3 1}
	INTERP
	           {1 2 -4 3}
	               Done
	``

(G) If you have a TI-85 or TI-86 graphing calculator, enter INTERP in your calculator exactly as shown in Table 4. To use the program, enter the x values in L1 and the corresponding y values in L2 (see the output in Table 4) and then execute the program. If you have some other graphing utility that can store and execute programs, consult your manual and modify the statements in INTERP so that the program works on your graphing utility. Use INTERP to check your answers to part F.

*Available for download at *www.mhhe.com/barnett*.

Chapter 4 Review

In this chapter, unless indicated otherwise, the coefficients of the **nth-degree polynomial function** $P(x) = a_n x^n + a_{n-1} x^{n-1} + \cdots + a_1 x + a_0$ are complex numbers and the domain is the set of complex numbers. The number r is said to be a **zero of the function P,** or a **zero of the polynomial $P(x)$,** or a **solution or root of the equation $P(x) = 0$,** if $P(r) = 0$. If the coefficients of $P(x)$ are real numbers, then the x intercepts of the graph of $y = P(x)$ are real **zeros** of P and $P(x)$ and real **solutions** or **roots** for the equation $P(x) = 0$.

4-1   POLYNOMIAL FUNCTIONS AND GRAPHS

Synthetic division is an efficient method for dividing polynomials by linear terms of the form $x - r$ that is well-suited to calculator use.

Let $P(x)$ be a polynomial of degree greater than 0 and let r be a real number. Then we have the following important theorems:

Division Algorithm. $P(x) = (x - r)Q(x) + R$, where $x - r$ is the **divisor;** $Q(x)$, a unique polynomial of degree 1 less than $P(x)$, is the **quotient;** and R, a unique real number, is the **remainder.**

Remainder Theorem. $P(r) = R$.

The left and right behavior of an nth-degree polynomial $P(x)$ with real coefficients is determined by its highest degree or **leading term.** As $x \to \pm\infty$, $a_n x^n$ and $P(x)$ both approach $\pm\infty$, depending on n and the sign of a_n. A **turning point** on a continuous graph is a point that separates an increasing portion from a decreasing portion. Important graph properties are:

1. P is continuous for all real numbers.

2. The graph of P is a smooth curve.

3. The graph of P has at most n x intercepts.

4. P has at most $n - 1$ turning points.

4-2   FINDING RATIONAL ZEROS OF POLYNOMIALS

If $P(x)$ is a polynomial of degree $n > 0$, then we have the following important theorems:

Factor Theorem. The number r is a zero of $P(x)$ if and only if $(x - r)$ is a factor of $P(x)$.

Fundamental Theorem of Algebra. $P(x)$ has at least one zero.

n Zeros Theorem. $P(x)$ can be expressed as a product of n linear factors and has n zeros, not necessarily distinct.

If $P(x)$ is represented as the product of linear factors and $x - r$ occurs m times, then r is called a **zero of multiplicity m.**

Imaginary Zeros Theorem. If $P(x)$ has real coefficients, then imaginary zeros of $P(x)$, if they exist, must occur in conjugate pairs.

Real Zeros and Odd-Degree Polynomials. If $P(x)$ has real coefficients and is of odd degree, then $P(x)$ always has at least one real zero.

Rational Zero Theorem. If the rational number b/c, in lowest terms, is a zero of the polynomial

$$P(x) = a_n x^n + a_{n-1} x^{n-1} + \ldots + a_1 x + a_0,$$

$a_n \neq 0$ with integer coefficients, then b must be an integer factor of a_0 and c must be an integer factor of a_n.

Strategy for Finding Rational Zeros

Assume that $P(x)$ is a polynomial with integer coefficients and is of degree greater than 2.

Step 1. List the possible rational zeros of $P(x)$ using the rational zero theorem (Theorem 6).

Step 2. Construct a synthetic division table. If a rational zero r is found, stop, write

$$P(x) = (x - r)Q(x)$$

and immediately proceed to find the rational zeros for $Q(x)$, the **reduced polynomial** relative to $P(x)$. If the degree of $Q(x)$ is greater than 2, return to step 1 using $Q(x)$ in place of $P(x)$. If $Q(x)$ is quadratic, find all its zeros using standard methods for solving quadratic equations.

4-3   APPROXIMATING REAL ZEROS OF POLYNOMIALS

The following theorems are important tools for locating the real zeros of a polynomial with real coefficients. Once located, a graphing utility can be used to approximate the zeros.

Location Theorem. If f is continuous on an interval I, a and b are two numbers in I, and $f(a)$ and $f(b)$ are of opposite sign, then there is at least one x intercept between a and b.

Upper and Lower Bounds of Real Zeros. If $a_n > 0$ and $P(x)$ is divided by $x - r$ using synthetic division:

1. If $r > 0$ and all numbers in the quotient row of the synthetic division, including the remainder, are nonnegative, then r is greater than or equal to the largest zero of $P(x)$ and is called an **upper bound** of the real zeros of $P(x)$.

2. If $r < 0$ and all numbers in the quotient row of the synthetic division, including the remainder, alternate in sign,

then r is less than or equal to the smallest zero of $P(x)$ and is called a **lower bound** of the real zeros of $P(x)$.

4-4   RATIONAL FUNCTIONS

A function of the form $f(x) = n(x)/d(x)$, where $n(x)$ and $d(x)$ are polynomials, is a **rational function.** The line $x = a$ is a **vertical asymptote** for the graph of $y = f(x)$ if $f(x) \to \infty$ or $f(x) \to -\infty$ as $x \to a^+$ or $x \to a^-$. If $d(a) = 0$ and $n(a) \neq 0$, then the line $x = a$ is a vertical asymptote. The line $y = b$ is a **horizontal asymptote** for the graph of $y = f(x)$ if $f(x) \to b$ as $x \to \infty$ or $x \to -\infty$. The line $y = mx + b$ is an **oblique asymptote** if $[f(x) - (mx + b)] \to 0$ as $x \to \infty$ or $x \to -\infty$.

Let

$$f(x) = \frac{a_m x^m + \cdots + a_1 x + a_0}{b_n x^n + \cdots + b_1 x + b_0}, \qquad a_m, b_n \neq 0$$

The behavior of the graph of f as $x \to \infty$ or $x \to -\infty$ is determined by the ratio of the leading terms of the numerator and denominator, $a_m x^m / b_n x^n$

1. If $m < n$, then the x axis is a horizontal asymptote.

2. If $m = n$, then the line $y = a_m/b_n$ is a horizontal asymptote.

3. If $m > n$, then there are no horizontal asymptotes.

Graphing a Rational Function: $f(x) = n(x)/d(x)$

Step 1. Intercepts. Find the real solutions of the equation $n(x) = 0$ and use these solutions to plot any x intercepts of the graph of f. Evaluate $f(0)$, if it exists, and plot the y intercept.

Step 2. Vertical Asymptotes. Find the real solutions of the equation $d(x) = 0$ and use these solutions to determine the domain of f, the points of discontinuity, and the vertical asymptotes. Sketch any vertical asymptotes as dashed lines.

Step 3. Sign Chart. Construct a sign chart for f and use it to determine the behavior of the graph near each vertical asymptote.

Step 4. Horizontal Asymptotes. Determine whether there is a horizontal asymptote and if so, sketch it as a dashed line.

Step 5. Complete the Sketch. Complete the sketch of the graph by plotting additional points and joining these points with a smooth continuous curve over each interval in the domain of f. (Do not cross any points of discontinuity.)

4-5 PARTIAL FRACTIONS

A rational function $P(x)/D(x)$ often can be decomposed into a sum of simpler rational functions called **partial fractions.** If the degree of $P(x)$ is less than the degree of $D(x)$, then $P(x)/D(x)$ is called a **proper fraction.** We have the following important theorems:

Equal Polynomials. Two polynomials are equal to each other if and only if the coefficients of terms of like degree are equal.

Linear and Quadratic Factor Theorem. A polynomial with real coefficients can be factored into a product of linear and/or quadratic factors with real coefficients where the linear and quadratic factors are prime relative to the real numbers.

Partial Fraction Decomposition. Any proper fraction $P(x)/D(x)$ reduced to lowest terms can be decomposed into the sum of partial fractions as follows:

1. If $D(x)$ has a nonrepeating linear factor of the form $ax + b$, then the partial fraction decomposition of $P(x)/D(x)$ contains a term of the form

$$\frac{A}{ax + b} \qquad A \text{ a constant}$$

2. If $D(x)$ has a k-repeating linear factor of the form $(ax + b)^k$, then the partial fraction decomposition of $P(x)/D(x)$ contains k terms of the form

$$\frac{A_1}{ax + b} + \frac{A_2}{(ax + b)^2} + \cdots + \frac{A_k}{(ax + b)^k}$$

$$A_1, A_2, \ldots, A_k \text{ constants}$$

3. If $D(x)$ has a nonrepeating quadratic factor of the form $ax^2 + bx + c$, which is prime relative to the real numbers, then the partial fraction decomposition of $P(x)/D(x)$ contains a term of the form

$$\frac{Ax + B}{ax^2 + bx + c} \qquad A, B \text{ constants}$$

4. If $D(x)$ has a k-repeating quadratic factor of the form $(ax^2 + bx + c)^k$, where $ax^2 + bx + c$ is prime relative to the real numbers, then the partial fraction decomposition of $P(x)/D(x)$ contains k terms of the form

$$\frac{A_1x + B_1}{ax^2 + bx + c} + \frac{A_2x + B_2}{(ax^2 + bx + c)^2} + \cdots + \frac{A_kx + B_k}{(ax^2 + bx + c)^k}$$

$$A_1, \ldots, A_k, \quad B_1, \ldots, B_k \text{ constants}$$

Chapter 4 Review Exercise

Work through all the problems in this chapter review, and check answers in the back of the book. Answers to all review problems are there, and following each answer is a number in italics indicating the section in which that type of problem is discussed. Where weaknesses show up, review appropriate sections in the text.

A

1. Use synthetic division to divide $P(x) = 2x^3 + 3x^2 - 1$ by $D(x) = x + 2$, and write the answer in the form $P(x) = D(x)Q(x) + R$.

2. If $P(x) = x^5 - 4x^4 + 9x^2 - 8$, find $P(3)$ using the remainder theorem and synthetic division.

3. What are the zeros of $P(x) = 3(x - 2)(x + 4)(x + 1)$?

4. If $P(x) = x^2 - 2x + 2$ and $P(1 + i) = 0$, find another zero of $P(x)$.

5. Let $P(x)$ be the polynomial whose graph is shown in the figure.
    (A) Assuming that $P(x)$ has integer zeros and leading coefficient 1, find the lowest-degree equation that could produce this graph.

(B) Describe the left and right behavior of $P(x)$.

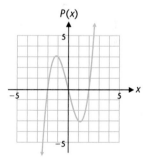

6. According to the upper and lower bound theorem, which of the following are upper or lower bounds of the zeros of $P(x) = x^3 - 4x^2 + 2$?

$$-2, -1, 3, 4$$

7. How do you know that $P(x) = 2x^3 - 3x^2 + x - 5$ has at least one real zero between 1 and 2?

8. Write the possible rational zeros for $P(x) = x^3 - 4x^2 + x + 6$.

9. Find all rational zeros for $P(x) = x^3 - 4x^2 + x + 6$.

10. Find the domain and x intercept(s) for:

(A) $f(x) = \dfrac{2x - 3}{x + 4}$ (B) $g(x) = \dfrac{3x}{x^2 - x - 6}$

11. Find the horizontal and vertical asymptotes for the functions in Problem 10.

12. Decompose into partial fractions: $\dfrac{7x - 11}{(x - 3)(x + 2)}$

B

13. Let $P(x) = x^3 - 3x^2 - 3x + 4$.
   (A) Graph $P(x)$ and describe the graph verbally, including the number of x intercepts, the number of turning points, and the left and right behavior.
   (B) Use the bisection method to approximate the largest x intercept to one decimal place.

14. If $P(x) = 8x^4 - 14x^3 - 13x^2 - 4x + 7$, find $Q(x)$ and R such that $P(x) = (x - \frac{1}{4})Q(x) + R$. What is $P(\frac{1}{4})$?

15. If $P(x) = 4x^3 - 8x^2 - 3x - 3$, find $P(-\frac{1}{2})$ using the remainder theorem and synthetic division.

16. Use the quadratic formula and the factor theorem to factor $P(x) = x^2 - 2x - 1$.

17. Is $x + 1$ a factor of $P(x) = 9x^{26} - 11x^{17} + 8x^{11} - 5x^4 - 7$? Explain, without dividing or using synthetic division.

18. Determine all rational zeros of $P(x) = 2x^3 - 3x^2 - 18x - 8$.

19. Factor the polynomial in Problem 18 into linear factors.

20. Find all rational zeros of $P(x) = x^3 - 3x^2 + 5$.

21. Find all zeros (rational, irrational, and imaginary) exactly for $P(x) = 2x^4 - x^3 + 2x - 1$.

22. Factor the polynomial in Problem 21 into linear factors.

23. Solve $2x^3 + 3x^2 \leq 11x + 6$. Write the answer in inequality and interval notation.

24. Let $P(x) = x^4 - 2x^3 - 30x^2 - 25$.
   (A) Find the smallest positive and the largest negative integers that, by Theorem 2 in Section 4-3, are upper and lower bounds, respectively, for the real zeros of $P(x)$.
   (B) Use the bisection method to approximate the largest real zero of $P(x)$ to two decimal places.
   (C) Use a graphing utility to approximate the real zeros of $P(x)$ to two decimal places.

25. Let $f(x) = \dfrac{x - 1}{2x + 2}$.
   (A) Find the domain and the intercepts for f.
   (B) Find the vertical and horizontal asymptotes for f.
   (C) Sketch a graph of f. Draw vertical and horizontal asymptotes with dashed lines.

26. Decompose into partial fractions: $\dfrac{-x^2 + 3x + 4}{x(x - 2)^2}$

27. Decompose into partial fractions: $\dfrac{8x^2 - 10x + 9}{2x^3 - 3x^2 + 3x}$

C

28. Use synthetic division to divide $P(x) = x^3 + 3x + 2$ by $[x - (1 + i)]$, and write the answer in the form $P(x) = D(x)Q(x) + R$.

29. Find a polynomial of lowest degree with leading coefficient 1 that has zeros $-\frac{1}{2}$ (multiplicity 2), -3, and 1 (multiplicity 3). (Leave the answer in factored form.) What is the degree of the polynomial?

30. Repeat Problem 29 for a polynomial $P(x)$ with zeros -5, $2 - 3i$, and $2 + 3i$.

31. Find all zeros (rational, irrational, and imaginary) exactly for $P(x) = 2x^5 - 5x^4 - 8x^3 + 21x^2 - 4$.

32. Factor the polynomial in Problem 31 into linear factors.

33. Solve

$$\frac{4x^2 + 4x - 3}{2x^3 + 3x^2 - 11x - 6} \geq 0$$

Write the answer in both inequality and interval notation.

34. What is the minimal degree of a polynomial $P(x)$, given that $P(-1) = -4$, $P(0) = 2$, $P(1) = -5$, and $P(2) = 3$? Justify your conclusion.

35. If $P(x)$ is a cubic polynomial with integer coefficients and if $1 + 2i$ is a zero of $P(x)$, can $P(x)$ have an irrational zero? Explain.

36. The solutions to the equation $x^3 - 27 = 0$ are the cube roots of 27.
   (A) How many cube roots of 27 are there?
   (B) 3 is obviously a cube root of 27; find all others.

37. Let $P(x) = x^4 + 2x^3 - 500x^2 - 4,000$.
   (A) Find the smallest positive integer multiple of 10 and the largest negative integer multiple of 10 that, by Theorem 2 in Section 4-3, are upper and lower bounds, respectively, for the real zeros of $P(x)$.
   (B) Approximate the real zeros of $P(x)$ to two decimal places.

38. Graph

$$f(x) = \frac{x^2 + 2x + 3}{x + 1}$$

Indicate any vertical, horizontal, or oblique asymptotes with dashed lines.

39. Use a graphing utility to find any horizontal asymptotes for

$$f(x) = \frac{2x}{\sqrt{x^2 + 3x + 4}}$$

40. Decompose into partial fractions:

$$\frac{5x^2 + 2x + 9}{x^4 - 3x^3 + x^2 - 3x}$$

APPLICATIONS

Express the solutions to each problem as the roots of a polynomial equation of the form P(x) = 0. Find rational solutions exactly and irrational solutions to one decimal place. Use a graphing utility or bisection only if necessary.

41. Architecture. An entryway is formed by placing a rectangular door inside an arch in the shape of the parabola with graph $y = 16 - x^2$, x and y in feet (see the figure). If the area of the door is 48 square feet, find the dimensions of the door.

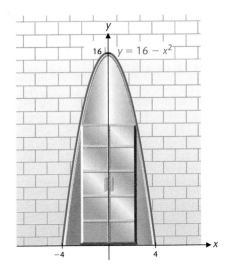

42. Construction. A grain silo is formed by attaching a hemisphere to the top of a right circular cylinder (see the figure). If the cylinder is 18 feet high and the volume of the silo is 486π cubic feet, find the common radius of the cylinder and the hemisphere.

18 feet

★ **43. Manufacturing.** A box is to be made out of a piece of cardboard that measures 15 by 20 inches. Squares, x inches on a side, will be cut from each corner, and then the ends and sides will be folded up (see the figure). Find the value of x that would result in a box with a volume of 300 cubic inches.

20 in.

15 in.

★ **44. Geometry.** Find all points on the graph of $y = x^2$ that are 3 units from the point (1, 4).

EXPONENTIAL AND LOGARITHMIC FUNCTIONS

CHAPTER 5

Most of the functions we have considered so far have been polynomial and rational functions, with a few others involving roots or powers of polynomial or rational functions. The general class of functions defined by means of the algebraic operations of addition, subtraction, multiplication, division, and the taking of powers and roots on variables and constants are called *algebraic functions*.

In this chapter we define and investigate the properties of two new and important types of functions called *exponential functions* and *logarithmic functions*. These functions are not algebraic, but are members of another class of functions called *transcendental functions*. The exponential functions and logarithmic functions are used in describing and solving a wide variety of real-world problems, including growth of populations of people, animals, and bacteria; radioactive decay; growth of money at compound interest; absorption of light as it passes through air, water, or glass; and magnitudes of sounds and earthquakes. We consider applications in these areas plus many more in the sections that follow.

SECTION **5-1**   **Exponential Functions**

- Exponential Functions
- Basic Exponential Graphs
- Additional Exponential Properties
- Applications

In this section we define exponential functions, look at some of their important properties—including graphs—and consider several significant applications.

- **Exponential Functions**

Let's start by noting that the functions f and g given by

$$f(x) = 2^x \qquad \text{and} \qquad g(x) = x^2$$

are not the same function. Whether a variable appears as an exponent with a constant base or as a base with a constant exponent makes a big difference. The function g is a quadratic function, which we have already discussed. The function f is a new type of function called an *exponential function*.

Many students, if asked to graph an exponential function such as $f(x) = 2^x$, would not hesitate at all. They would likely make up a table by assigning integers to x, plot the resulting points, and then join these points with a smooth curve, as shown in Figure 1.

FIGURE 1 $f(x) = 2^x$.

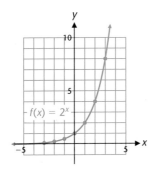

x	$f(x)$
-3	$\frac{1}{8}$
-2	$\frac{1}{4}$
-1	$\frac{1}{2}$
0	1
1	2
2	4
3	8

The only catch is that we have not defined 2^x for all real numbers x. We know what 2^5, 2^{-3}, $2^{2/3}$, $2^{-3/5}$, $2^{1.4}$, and $2^{-3.15}$ mean because we have defined 2^p for any rational number p, but what does

$$2^{\sqrt{2}}$$

mean? The question is not easy to answer at this time. In fact, a precise definition of $2^{\sqrt{2}}$ must wait for more advanced courses, where we can show that, if b is a positive real number and x is any real number, then

$$b^x$$

names a real number, and the graph of $f(x) = 2^x$ is as indicated in Figure 1. We also can show that for x irrational, b^x can be approximated as closely as we like by using rational number approximations for x. Since $\sqrt{2} = 1.414213\ldots$, for example, the sequence

$$2^{1.4},\ 2^{1.41},\ 2^{1.414},\ldots$$

approximates $2^{\sqrt{2}}$, and as we use more decimal places, the approximation improves.

DEFINITION 1

Exponential Function

The equation

$$f(x) = b^x \qquad b > 0,\ b \neq 1$$

defines an **exponential function** for each different constant b, called the **base**. The independent variable x may assume any real value.

Thus, the **domain of f** is the set of all real numbers, and it can be shown that the **range of f** is the set of all positive real numbers. We require the base b to be positive to avoid imaginary numbers such as $(-2)^{1/2}$.

• Basic Exponential Graphs

EXPLORE-DISCUSS 1   Compare the graphs of $f(x) = 3^x$ and $g(x) = 2^x$ by plotting both functions on the same coordinate system. Find all points of intersection of the graphs. For which values of x is the graph of f above the graph of g? Below the graph of g? Are the graphs of f and g close together as $x \to \infty$? As $x \to -\infty$? Discuss.

It is useful to compare the graphs of $y = 2^x$ and $y = (\frac{1}{2})^x = 2^{-x}$ by plotting both on the same coordinate system, as shown in Figure 2(a). The graph of

$$f(x) = b^x \qquad b > 1 \qquad [\text{Fig. 2(b)}]$$

looks very much like the graph of the particular case $y = 2^x$, and the graph of

$$f(x) = b^x \qquad 0 < b < 1 \qquad [\text{Fig. 2(b)}]$$

FIGURE 2 Basic exponential graphs.

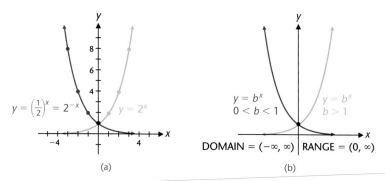

(a)

(b)

looks very much like the graph of $y = (\frac{1}{2})^x$. Note in both cases that the x axis is a *horizontal asymptote* for the graph.

The graphs in Figure 2 suggest the following important general properties of exponential functions, which we state without proof:

Basic Properties of the Graph of $f(x) = b^x$, $b > 0$, $b \neq 1$

1. All graphs pass through the point $(0, 1)$. $b^0 = 1$ for any permissible base b.

2. All graphs are continuous, with no holes or jumps.

3. The x axis is a horizontal asymptote.

4. If $b > 1$, then b^x increases as x increases.

5. If $0 < b < 1$, then b^x decreases as x increases.

6. The function f is one-to-one.

Property 6 implies that an exponential function has an inverse, called a *logarithmic function,* which we will discuss in Section 5-3.

A calculator may be used to create an accurate table of values from which the graph of an exponential function is drawn. Example 1 illustrates the process. (Of course, we may bypass the creation of a table of values with a graphing utility, which graphs the function directly.)

EXAMPLE 1   **Graphing Multiples of Exponential Functions**

Use integer values of x from -3 to 3 to construct a table of values for $y = \frac{1}{2}(4^x)$, and then graph this function.

Solution   Use a calculator to create the table of values shown below. Then plot the points, and join these points with a smooth curve (see Fig. 3).

FIGURE 3 $y = \frac{1}{2}(4^x)$.

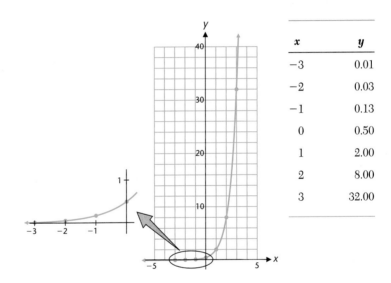

x	y
-3	0.01
-2	0.03
-1	0.13
0	0.50
1	2.00
2	8.00
3	32.00

Matched Problem 1   Repeat Example 1 for $y = \frac{1}{2}(\frac{1}{4})^x = \frac{1}{2}(4^{-x})$.

• Additional Exponential Properties   Exponential functions, whose domains include irrational numbers, obey the familiar laws of exponents we discussed earlier for rational exponents. We summarize these exponent laws here and add two other important and useful properties.

Exponential Function Properties

For a and b positive, $a \neq 1$, $b \neq 1$, and x and y real:

1. Exponent laws:

$$a^x a^y = a^{x+y} \qquad (a^x)^y = a^{xy} \qquad (ab)^x = a^x b^x$$

$$\left(\frac{a}{b}\right)^x = \frac{a^x}{b^x} \qquad \frac{a^x}{a^y} = a^{x-y} \qquad \frac{2^{5x}}{2^{7x}} \; \boxed{= 2^{5x-7x}} \; = 2^{-2x}$$

2. $a^x = a^y$ if and only if $x = y$.   If $6^{4x} = 6^{2x+4}$, then $4x = 2x + 4$, and $x = 2$.

3. For $x \neq 0$, then $a^x = b^x$ if and only if $a = b$.   If $a^4 = 3^4$, then $a = 3$.

EXAMPLE 2   **Using Exponential Function Properties**

Solve $4^{x-3} = 8$ for x.

Solution   Express both sides in terms of the same base, and use property 2 to equate exponents.

$$4^{x-3} = 8$$
$$(2^2)^{x-3} = 2^3 \qquad \text{Express 4 and 8 as powers of 2.}$$
$$2^{2x-6} = 2^3 \qquad (a^x)^y = a^{xy}$$
$$2x - 6 = 3 \qquad \text{Property 2}$$
$$2x = 9$$
$$x = \tfrac{9}{2}$$

Check $4^{(9/2)-3} = 4^{3/2} = (\sqrt{4})^3 = 2^3 \overset{\checkmark}{=} 8$

Matched Problem 2   Solve $27^{x+1} = 9$ for x.

● **Applications**   We now consider three applications that utilize exponential functions in their analysis: population growth, radioactive decay, and compound interest. Population growth and compound interest are examples of exponential growth, while radioactive decay is an example of negative exponential growth.

Our first example involves the growth of populations, such as people, animals, insects, and bacteria. Populations tend to grow exponentially and at different rates. A convenient and easily understood measure of growth rate is the **doubling time**—that is, the time it takes for a population to double. Over short periods of time the **doubling time growth model** is often used to model population growth:

$$P = P_0 2^{t/d}$$

where P = Population at time t
       P_0 = Population at time $t = 0$
       d = Doubling time

Note that when $t = d$,

$$P = P_0 2^{d/d} = P_0 2$$

and the population is double the original, as it should be. We use this model to solve a population growth problem in Example 3.

EXAMPLE 3   **Population Growth**

Mexico has a population of around 100 million people, and it is estimated that the population will double in 21 years. If population growth continues at the same rate, what will be the population:

(A) 15 years from now?   (B) 30 years from now?

Calculate answers to 3 significant digits.

Solutions   We use the doubling time growth model:

$$P = P_0 2^{t/d}$$

P (millions)

FIGURE 4 $P = 100(2^{t/21})$

Substituting $P_0 = 100$ and $d = 21$, we obtain

$$P = 100(2^{t/21})   \text{See Figure 4.}$$

(A) Find P when $t = 15$ years:

$$P = 100(2^{15/21})$$

$$\approx 164 \text{ million people}   \text{Use a calculator.}$$

(B) Find P when $t = 30$ years:

$$P = 100(2^{30/21})$$

$$\approx 269 \text{ million people}   \text{Use a calculator.}$$

Matched Problem 3   The bacterium *Escherichia coli* (*E. coli*) is found naturally in the intestines of many mammals. In a particular laboratory experiment, the doubling time for *E. coli* is found to be 25 minutes. If the experiment starts with a population of 1,000 *E. coli* and there is no change in the doubling time, how many bacteria will be present:

(A) In 10 minutes?   (B) In 5 hours?

Write answers to 3 significant digits.

Our second application involves radioactive decay, which is often referred to as negative growth. Radioactive materials are used extensively in medical diagnosis and therapy, as power sources in satellites, and as power sources in many countries. If we start with an amount A_0 of a particular radioactive isotope, the amount declines exponentially in time. The rate of decay varies from isotope to isotope. A convenient and easily understood measure of the rate of decay is the **half-life** of the isotope—that is, the time it takes for half of a particular material to decay. In this section we use the following **half-life decay model:**

$$A = A_0 \left(\tfrac{1}{2}\right)^{t/h}$$
$$= A_0 2^{-t/h}$$

where A = Amount at time t
   A_0 = Amount at time $t = 0$
   h = Half-life

Note that when $t = h$,

$$A = A_0 2^{-h/h} = A_0 2^{-1} = \frac{A_0}{2}$$

and the amount of isotope is half the original amount, as it should be.

EXAMPLE 4   **Radioactive Decay**

The radioactive isotope gallium 67 (^{67}Ga), used in the diagnosis of malignant tumors, has a biological half-life of 46.5 hours. If we start with 100 milligrams of the isotope, how many milligrams will be left after:

(A) 24 hours?   (B) 1 week?

Compute answers to 3 significant digits.

Solutions   We use the half-life decay model:

$$A = A_0 \left(\tfrac{1}{2}\right)^{t/h} = A_0 2^{-t/h}$$

Using $A_0 = 100$ and $h = 46.5$, we obtain

$$A = 100(2^{-t/46.5}) \quad \text{See Figure 5.}$$

A (milligrams)

Hours

FIGURE 5 $A = 100(2^{-t/46.5})$.

(A) Find A when $t = 24$ hours:

$$A = 100(2^{-24/46.5})$$

$$= 69.9 \text{ milligrams} \quad \text{Use a calculator.}$$

(B) Find A when $t = 168$ hours (1 week = 168 hours):

$$A = 100(2^{-168/46.5})$$

$$= 8.17 \text{ milligrams} \quad \text{Use a calculator.}$$

Matched Problem 4 Radioactive gold 198 (^{198}Au), used in imaging the structure of the liver, has a half-life of 2.67 days. If we start with 50 milligrams of the isotope, how many milligrams will be left after:

(A) $\frac{1}{2}$ day?   (B) 1 week?

Compute answers to 3 significant digits.

Our third application deals with the growth of money at compound interest. This topic is important to most people and is fundamental to many topics in the mathematics of finance.

The fee paid to use another's money is called **interest.** It is usually computed as a percent, called the **interest rate,** of the principal over a given period of time. If, at the end of a payment period, the interest due is reinvested at the same rate, then the interest earned as well as the principal will earn interest during the next payment period. Interest paid on interest reinvested is called **compound interest.**

Suppose you deposit $1,000 in a savings and loan that pays 8% compounded semiannually. How much will the savings and loan owe you at the end of 2 years? Compounded semiannually means that interest is paid to your account at the end of each 6-month period, and the interest will in turn earn interest. The **interest rate per period** is the annual rate, 8% = 0.08, divided by the number of compounding periods per year, 2. If we let A_1, A_2, A_3, and A_4 represent the new amounts due at the end of the first, second, third, and fourth periods, respectively, then

$$A_1 = \$1,000 + \$1,000\left(\frac{0.08}{2}\right)$$

$$= \$1,000(1 + 0.04) \qquad P\left(1 + \frac{r}{n}\right)$$

$$A_2 = A_1(1 + 0.04)$$

$$= [\$1,000(1 + 0.04)](1 + 0.04)$$

$$= \$1,000(1 + 0.04)^2 \qquad P\left(1 + \frac{r}{n}\right)^2$$

$$A_3 = A_2(1 + 0.04)$$

$$= [\$1,000(1 + 0.04)^2](1 + 0.04)$$

$$= \$1,000(1 + 0.04)^3 \qquad P\left(1 + \frac{r}{n}\right)^3$$

$$A_4 = A_3(1 + 0.04)$$
$$= [\$1,000(1 + 0.04)^3](1 + 0.04)$$
$$= \$1,000(1 + 0.04)^4 \qquad P\left(1 + \frac{r}{n}\right)^4$$

What do you think the savings and loan will owe you at the end of 6 years? If you guessed

$$A = \$1,000(1 + 0.04)^{12}$$

you have observed a pattern that is generalized in the following compound interest formula:

Compound Interest

If a **principal** P is invested at an annual **rate** r compounded n times a year, then the **amount** A in the account at the end of t years is given by

$$A = P\left(1 + \frac{r}{n}\right)^{nt}$$

The annual rate r is expressed in decimal form.

Since the principal P represents the initial amount in the account and A represents the amount in the account t years later, we also call P the **present value** of the account and A the **future value** of the account.

EXAMPLE 5   **Compound Interest**

If you deposit $5,000 in an account paying 9% compounded daily, how much will you have in the account in 5 years? Compute the answer to the nearest cent.

Solution   We use the compound interest formula as follows:

$$A = P\left(1 + \frac{r}{n}\right)^{nt}$$

$$= 5,000\left(1 + \frac{0.09}{365}\right)^{(365)(5)}$$

$$= \$7,841.12 \qquad \text{Use a calculator.}$$

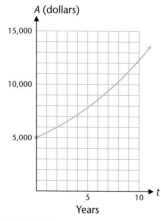

The graph of

$$A = 5,000\left(1 + \frac{0.09}{365}\right)^{365t}$$

is shown in Figure 6.

FIGURE 6

Matched Problem 5 If $1,000 is invested in an account paying 10% compounded monthly, how much will be in the account at the end of 10 years? Compute the answer to the nearest cent.

 EXAMPLE 6 **Visualizing Investments with a Graphing Utility**

Use a graphing utility to compare the growth of an investment of $1,000 at 10% compounded monthly with an investment of $2,000 at 5% compounded monthly. When do the two investments have the same value?

Solution We use the compound interest formula to express the future value y_1 of the first investment by $y_1 = 1,000(1 + 0.10/12)^{12x}$, and the future value y_2 of the second investment by $y_2 = 2,000(1 + 0.05/12)^{12x}$, where x is time in years. We graph both functions and use the intersection routine of the graphing utility to conclude that the investments have the same value when $x \approx 14$ years, as shown in Figure 7. After that time the $1,000 investment has the greater value.

FIGURE 7

 Matched Problem 6 Use a graphing utility to determine when an investment of $5,000 at 6% compounded quarterly has the same value as an investment of $4,000 at 10% compounded daily.

Answers to Matched Problems

1. $y = \frac{1}{2}(4^{-x})$

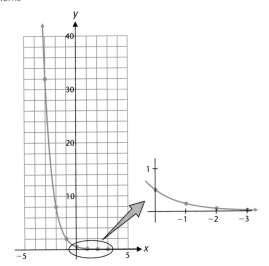

x	y
-3	32.00
-2	8.00
-1	2.00
0	0.50
1	0.13
2	0.03
3	0.01

2. $x = -\frac{1}{3}$
3. (A) 1,320   (B) 4,100,000 = 4.10 × 10^6
4. (A) 43.9 mg   (B) 8.12 mg
5. $2,707.04
6. 5 years, 6 months

EXERCISE 5-1

A

In Problems 1–10, construct a table of values for integer values of x over the indicated interval and then graph the function.

1. $y = 3^x; [-3, 3]$
2. $y = 5^x; [-2, 2]$

3. $y = (\frac{1}{3})^x = 3^{-x}; [-3, 3]$
4. $y = (\frac{1}{5})^x = 5^{-x}; [-2, 2]$

5. $g(x) = -3^{-x}; [-3, 3]$
6. $f(x) = -5^x; [-2, 2]$

7. $h(x) = 5(3^x); [-3, 3]$
8. $f(x) = 4(5^x); [-2, 2]$

9. $y = 3^{x+3} - 5; [-6, 0]$
10. $y = 5^{x+2} + 4; [-4, 0]$

In Problems 11–22, simplify.

11. $2^{5x+1}2^{3-2x}$
12. $5^{2x-3}5^{6-4x}$
13. $\dfrac{3^{2y-x}}{3^{3x-5y}}$

14. $\dfrac{7^{-3y-x}}{7^{-4x-6y}}$
15. $(4^x)^3$
16. $(x^3)^4$

17. $(10^2)^x(10^x)^2$
18. $2^{x^3}2^{x^2}$
19. $\left(\dfrac{5^x}{4^y}\right)^3$

20. $\left(\dfrac{3^{2x}}{7^{3y}}\right)^2$
21. $\left(\dfrac{a^{-2}b^3c^{-1}}{a^{-3}b^2c^{-3}}\right)^2$
22. $\left(\dfrac{a^{-4}b^3c^2}{a^{-6}b^2c^{-1}}\right)^3$

B

In Problems 23–34, solve for x.

23. $3^{2x-5} = 3^{4x+2}$
24. $4^{4x+1} = 4^{2x-2}$

25. $10^{x^2+2} = 10^{2x+2}$
26. $5^{x^2-5} = 5^{3x+5}$

27. $(2x + 1)^3 = 8$
28. $(2x - 1)^5 = -32$

29. $5^{3x} = 25^{x+3}$
30. $4^{5x+1} = 16^{2x-1}$

31. $4^{2x+2} = 8^{x+2}$
32. $100^{2x+4} = 1,000^{x+4}$

33. $100^{x^2} = 10^{5x-3}$
34. $3^{x^2} = 9^{x+4}$

35. Find all real numbers a such that $a^2 = a^{-2}$. Explain why this does not violate the second exponential function property in the box on page 358.

36. Find real numbers a and b such that $a \neq b$ but $a^4 = b^4$. Explain why this does not violate the third exponential function property in the box on page 358.

Graph each function in Problems 37–46 by constructing a table of values.

 *Check Problems 37–46 with a graphing utility.

37. $G(t) = 3^{t/100}$
38. $f(t) = 2^{t/10}$
39. $y = 11(3^{-x/2})$

*Please note that use of graphing utility is not required to complete these exercises. Checking them with a g.u. is optional.

40. $y = 7(2^{-2x})$
41. $g(x) = 2^{-|x|}$
42. $f(x) = 2^{|x|}$

43. $y = 1,000(1.08)^x$
44. $y = 100(1.03)^x$

45. $y = 2^{-x^2}$
46. $y = 3^{-x^2}$

C

In Problems 47–50, simplify.

47. $(6^x + 6^{-x})(6^x - 6^{-x})$
48. $(3^x - 3^{-x})(3^x + 3^{-x})$

49. $(6^x + 6^{-x})^2 - (6^x - 6^{-x})^2$
50. $(3^x - 3^{-x})^2 + (3^x + 3^{-x})^2$

Graph each function in Problems 51–54 by constructing a table of values.

 Check Problems 51–54 with a graphing utility.

51. $m(x) = x(3^{-x})$
52. $h(x) = x(2^x)$

53. $f(x) = \dfrac{2^x + 2^{-x}}{2}$
54. $g(x) = \dfrac{3^x + 3^{-x}}{2}$

 In Problems 55–58:

(A) Approximate the real zeros of each function to two decimal places.

(B) Investigate the behavior of each function as $x \to \infty$ and $x \to -\infty$ and find any horizontal asymptotes.

55. $f(x) = 3^x - 5$
56. $f(x) = 4 + 2^{-x}$

57. $f(x) = 1 + x + 10^x$
58. $f(x) = 8 - x^2 + 2^{-x}$

APPLICATIONS

59. Gaming. A person bets on red and black on a roulette wheel using a *Martingale strategy*. That is, a \$2 bet is placed on red, and the bet is doubled each time until a win occurs. The process is then repeated. If black occurs n times in a row, then $L = 2^n$ dollars is lost on the nth bet. Graph this function for $1 \le n \le 10$. Even though the function is defined only for positive integers, points on this type of graph are usually joined with a smooth curve as a visual aid.

60. Bacterial Growth. If bacteria in a certain culture double every $\frac{1}{2}$ hour, write an equation that gives the number of bacteria N in the culture after t hours, assuming the culture has 100 bacteria at the start. Graph the equation for $0 \le t \le 5$.

61. Population Growth. Because of its short life span and frequent breeding, the fruit fly *Drosophila* is used in some genetic studies. Raymond Pearl of Johns Hopkins University,

for example, studied 300 successive generations of descendants of a single pair of *Drosophila* flies. In a laboratory situation with ample food supply and space, the doubling time for a particular population is 2.4 days. If we start with 5 male and 5 female flies, how many flies should we expect to have in:
(A) 1 week?   (B) 2 weeks?

62. Population Growth. If Kenya has a population of about 30,000,000 people and a doubling time of 19 years and if the growth continues at the same rate, find the population in:
(A) 10 years   (B) 30 years
Compute answers to 2 significant digits.

63. Insecticides. The use of the insecticide DDT is no longer allowed in many countries because of its long-term adverse effects. If a farmer uses 25 pounds of active DDT, assuming its half-life is 12 years, how much will still be active after:
(A) 5 years?   (B) 20 years?
Compute answers to 2 significant digits.

64. Radioactive Tracers. The radioactive isotope technetium 99m (^{99m}Tc) is used in imaging the brain. The isotope has a half-life of 6 hours. If 12 milligrams are used, how much will be present after:
(A) 3 hours?   (B) 24 hours?
Compute answers to 3 significant digits.

65. Finance. Suppose $4,000 is invested at 11% compounded weekly. How much money will be in the account in:
(A) $\frac{1}{2}$ year?   (B) 10 years?
Compute answers to the nearest cent.

66. Finance. Suppose $2,500 is invested at 7% compounded quarterly. How much money will be in the account in:
(A) $\frac{3}{4}$ year?   (B) 15 years?
Compute answers to the nearest cent.

★ **67. Finance.** A couple just had a new child. How much should they invest now at 8.25% compounded daily in order to have $40,000 for the child's education 17 years from now? Compute the answer to the nearest dollar.

★ **68. Finance.** A person wishes to have $15,000 cash for a new car 5 years from now. How much should be placed in an account now if the account pays 9.75% compounded weekly? Compute the answer to the nearest dollar.

★ **69. Finance.** Will an investment of $10,000 at 8.9% compounded daily ever be worth more at the end of a quarter than an investment of $10,000 at 9% compounded quarterly? Explain.

★ **70. Finance.** A sum of $5,000 is invested at 13% compounded semiannually. Suppose that a second investment of $5,000 is made at interest rate *r* compounded daily. For which values of *r*, to the nearest tenth of a percent, is the second investment better than the first? Discuss.

SECTION 5-2   The Exponential Function with Base *e*

- Base *e* Exponential Function
- Growth and Decay Applications Revisited
- Continuous Compound Interest
- A Comparison of Exponential Growth Phenomena

Until now the number π has probably been the most important irrational number you have encountered. In this section we will introduce another irrational number, *e*, that is just as important in mathematics and its applications.

• Base *e* Exponential Function   The following expression is important to the study of calculus and, as we will see later in this section, also is closely related to the compound interest formula discussed in the preceding section:

$$\left(1 + \frac{1}{m}\right)^m$$

(A) Calculate the value of $[1 + (1/m)]^m$ for $m = 1,2,3,4,5$. Are the values increasing or decreasing as m gets larger?

(B) What is the smallest positive integer m such that $[1 + (1/m)]^m$ is greater than 2.5? Greater than 2.7? Greater than 2.9?

TABLE 1

m	$\left(1 + \dfrac{1}{m}\right)^m$
1	2
10	2.59374 . . .
100	2.70481 . . .
1,000	2.71692 . . .
10,000	2.71814 . . .
100,000	2.71827 . . .
1,000,000	2.71828 . . .
⋮	⋮

Interestingly, by calculating the value of the expression for larger and larger values of m (see Table 1), it appears that $[1 + (1/m)]^m$ approaches a number close to 2.7183. In a calculus course we can show that as m increases without bound, the value of $[1 + (1/m)]^m$ approaches an irrational number that we call e. Just as irrational numbers such as π and $\sqrt{2}$ have unending, nonrepeating decimal representations (see Section 1-1), e also has an unending, nonrepeating decimal representation. To 12 decimal places,

$$e = 2.718\ 281\ 828\ 459$$

Exactly who discovered e is still being debated. It is named after the great Swiss mathematician Leonhard Euler (1707–1783), who computed e to 23 decimal places using $[1 + (1/m)]^m$.

The constant e turns out to be an ideal base for an exponential function because in calculus and higher mathematics many operations take on their simplest form using this base. This is why you will see e used extensively in expressions and formulas that model real-world phenomena.

DEFINITION 1   **Exponential Function with Base e**

For x a real number, the equation

$$f(x) = e^x$$

defines the **exponential function with base e.**

The exponential function with base e is used so frequently that it is often referred to as *the* exponential function. The graphs of $y = e^x$ and $y = e^{-x}$ are shown in Figure 1.

FIGURE 1 Exponential functions with base e.

2.

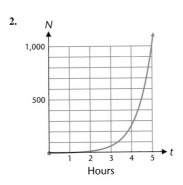

3. 119.4 mg

4. Annually: $176.23; quarterly: $180.61; continuously: $182.21

EXERCISE **5-2**

A

In Problems 1–6, construct a table of values for integer values of x over the indicated interval and then graph the function.

Check Problems 1–6 with a graphing utility.

1. $y = -e^x$; $[-3, 3]$ **2.** $y = -e^{-x}$; $[-3, 3]$

3. $y = 10e^{0.2x}$; $[-5, 5]$ **4.** $y = 100e^{0.1x}$; $[-5, 5]$

5. $f(t) = 100e^{-0.1t}$; $[-5, 5]$ **6.** $g(t) = 10e^{-0.2t}$; $[-5, 5]$

In Problems 7–12, simplify.

7. $e^{2x}e^{-3x}$ **8.** $(e^{-x})^4$ **9.** $(e^x)^3$

10. $e^{-4x}e^{6x}$ **11.** $\dfrac{e^{5x}}{e^{2x+1}}$ **12.** $\dfrac{e^{4-3x}}{e^{2-5x}}$

13. (A) Explain what is wrong with the following reasoning about the expression $[1 + (1/m)]^m$: As m gets large, $1 + (1/m)$ approaches 1 because $1/m$ approaches 0, and 1 raised to any power is 1, so $[1 + (1/m)]^m$ approaches 1.

     (B) Which number does $[1 + (1/m)]^m$ approach as $m \to \infty$?

14. (A) Explain what is wrong with the following reasoning about the expression $[1 + (1/m)]^m$: If $b > 1$, then the exponential function $b^x \to \infty$ as $x \to \infty$, and $1+(1/m)$ is greater than 1, so $[1 + (1/m)]^m$ approaches infinity as $m \to \infty$.

     (B) Which number does $[1 + (1/m)]^m$ approach as $m \to \infty$?

B

In Problems 15–22, graph each function by constructing a table of values.

Check Problems 15–22 with a graphing utility.

15. $y = 2 + e^{x-2}$ **16.** $y = -3 + e^{1+x}$

17. $y = e^{-|x|}$ **18.** $y = e^{|x|}$

19. $M(x) = e^{x/2} + e^{-x/2}$ **20.** $C(x) = \dfrac{e^x + e^{-x}}{2}$

21. $N = \dfrac{200}{1 + 3e^{-t}}$ **22.** $N = \dfrac{100}{1 + e^{-t}}$

In Problems 23–28, simplify.

23. $\dfrac{-2x^3e^{-2x} - 3x^2e^{-2x}}{x^6}$ **24.** $\dfrac{5x^4e^{5x} - 4x^3e^{5x}}{x^8}$

25. $(e^x + e^{-x})^2 + (e^x - e^{-x})^2$

26. $e^x(e^{-x} + 1) - e^{-x}(e^x + 1)$

27. $\dfrac{e^{-x}(e^x - e^{-x}) + e^{-x}(e^x + e^{-x})}{e^{-2x}}$

28. $\dfrac{e^x(e^x + e^{-x}) - (e^x - e^{-x})e^x}{e^{2x}}$

In Problems 29–32, solve each equation. [Remember: $e^{-x} \neq 0$.]

29. $2xe^{-x} = 0$ **30.** $(x - 3)e^x = 0$

31. $x^2e^x - 5xe^x = 0$ **32.** $3xe^{-x} + x^2e^{-x} = 0$

C

One of the most important functions in statistics is the **normal probability density function**

$$f(x) = \frac{1}{\sigma\sqrt{2\pi}}e^{-(x-\mu)^2/2\sigma^2}$$

where μ is the **mean** and σ is the **standard deviation**. The graph of this function is the "bell-shaped" curve that instructors refer to when they say that they are grading on a curve.

Graph the related functions given in Problems 33 and 34.

33. $f(x) = e^{-x^2}$ **34.** $g(x) = \dfrac{1}{\sqrt{\pi}}e^{-x^2/2}$

35. Given $f(s) = (1 + s)^{1/s}$, $s \neq 0$:
  (A) Complete the tables below to four decimal places.
  (B) What does $(1 + s)^{1/s}$ seem to tend to as s approaches 0?

s	$f(s)$	s	$f(s)$
-0.5	4.0000	0.5	2.2500
-0.2	3.0518	0.2	2.4883
-0.1		0.1	
-0.01		0.01	
-0.001		0.001	
-0.0001		0.0001	

36. Refer to Problem 35. Graph $f(s) = (1 + s)^{1/s}$ for s in $[-0.5, 0) \cup (0, 0.5]$.

Problems 37–40 require the use of a graphing utility.

*It is common practice in many applications of mathematics to approximate nonpolynomial functions with appropriately selected polynomials. For example, the polynomials in Problems 37–40, called **Taylor polynomials,** can be used to approximate the exponential function $f(x) = e^x$. To illustrate this approximation graphically, in each problem graph $f(x) = e^x$ and the indicated polynomial in the same viewing window, $-4 \leq x \leq 4$ and $-5 \leq y \leq 50$.*

37. $P_1(x) = 1 + x + \frac{1}{2}x^2$

38. $P_2(x) = 1 + x + \frac{1}{2}x^2 + \frac{1}{6}x^3$

39. $P_3(x) = 1 + x + \frac{1}{2}x^2 + \frac{1}{6}x^3 + \frac{1}{24}x^4$

40. $P_4(x) = 1 + x + \frac{1}{2}x^2 + \frac{1}{6}x^3 + \frac{1}{24}x^4 + \frac{1}{120}x^5$

41. Investigate the behavior of the functions $f_1(x) = x/e^x$, $f_2(x) = x^2/e^x$, and $f_3(x) = x^3/e^x$ as $x \to \infty$ and as $x \to -\infty$, and find any horizontal asymptotes. Generalize to functions of the form $f_n(x) = x^n/e^x$, where n is any positive integer.

42. Investigate the behavior of the functions $g_1(x) = xe^x$, $g_2(x) = x^2e^x$, and $g_3(x) = x^3e^x$ as $x \to \infty$ and as $x \to -\infty$, and find any horizontal asymptotes. Generalize to functions of the form $g_n(x) = x^ne^x$, where n is any positive integer.

APPLICATIONS

43. Population Growth. If the world population is about 6 billion people now and if the population grows continuously at an annual rate of 1.7%, what will the population be in 10 years? Compute the answer to 2 significant digits.

44. Population Growth. If the population in Mexico is around 100 million people now and if the population grows continuously at an annual rate of 2.3%, what will the population be in 8 years? Compute the answer to 2 significant digits.

45. Population Growth. In 1996 the population of Russia was 148 million and the population of Nigeria was 104 million. If the populations of Russia and Nigeria grow continuously at annual rates of -0.62% and 3.0%, respectively, when will Nigeria have a greater population than Russia?

46. Population Growth. In 1996 the population of Germany was 84 million and the population of Egypt was 64 million. If the populations of Germany and Egypt grow continuously at annual rates of -0.15% and 1.9%, respectively, when will Egypt have a greater population than Germany?

47. Space Science. Radioactive isotopes, as well as solar cells, are used to supply power to space vehicles. The isotopes gradually lose power because of radioactive decay. On a particular space vehicle the nuclear energy source has a power output of P watts after t days of use as given by

$$P = 75e^{-0.0035t}$$

Graph this function for $0 \leq t \leq 100$.

48. Earth Science. The atmospheric pressure P, in pounds per square inch, decreases exponentially with altitude h, in miles above sea level, as given by

$$P = 14.7e^{-0.21h}$$

Graph this function for $0 \leq h \leq 10$.

49. Marine Biology. Marine life is dependent upon the microscopic plant life that exists in the *photic zone*, a zone that goes to a depth where about 1% of the surface light still remains. Light intensity I relative to depth d, in feet, for one of the clearest bodies of water in the world, the Sargasso Sea in the West Indies, can be approximated by

$$I = I_0e^{-0.00942d}$$

where I_0 is the intensity of light at the surface. What percentage of the surface light will reach a depth of:
  (A) 50 feet?   (B) 100 feet?

50. Marine Biology. Refer to Problem 49. In some waters with a great deal of sediment, the photic zone may go down only 15 to 20 feet. In some murky harbors, the intensity of light d feet below the surface is given approximately by

$$I = I_0e^{-0.23d}$$

What percentage of the surface light will reach a depth of:
  (A) 10 feet?   (B) 20 feet?

51. Money Growth. If you invest $5,250 in an account paying 11.38% compounded continuously, how much money will be in the account at the end of:
  (A) 6.25 years?   (B) 17 years?

52. Money Growth. If you invest $7,500 in an account paying

8.35% compounded continuously, how much money will be in the account at the end of:
(A) 5.5 years?   (B) 12 years?

53. Money Growth. *Barron's,* a national business and financial weekly, published the following "Top Savings Deposit Yields" for $2\frac{1}{2}$-year certificate of deposit accounts:

Gill Saving	8.30% (CC)
Richardson Savings and Loan	8.40% (CQ)
USA Savings	8.25% (CD)

where CC represents compounded continuously, CQ compounded quarterly, and CD compounded daily. Compute the value of $1,000 invested in each account at the end of $2\frac{1}{2}$ years.

54. Money Growth. Refer to Problem 53. In another issue of *Barron's,* 1-year certificate of deposit accounts included:

Alamo Savings	8.25% (CQ)
Lamar Savings	8.05% (CC)

Compute the value of $10,000 invested in each account at the end of 1 year.

★ **55. Present Value.** A promissory note will pay $30,000 at maturity 10 years from now. How much should you be willing to pay for the note now if the note gains value at a rate of 9% compounded continuously?

★ **56. Present Value.** A promissory note will pay $50,000 at maturity $5\frac{1}{2}$ years from now. How much should you be willing to pay for the note now if the note gains value at a rate of 10% compounded continuously?

57. AIDS Epidemic. In June of 1996 the World Health Organization estimated that 7.7 million cases of AIDS (acquired immunodeficiency syndrome) had occurred worldwide since the beginning of the epidemic. Assuming that the disease spreads continuously at an annual rate of 17%, estimate the total number of AIDS cases which will have occurred by June of the year:
(A) 2000   (B) 2004

58. AIDS Epidemic. In June 1996 the World Health Organization estimated that 28 million people worldwide had been infected with HIV (human immunodeficiency virus) since the beginning of the AIDS epidemic. Assuming that HIV infection spreads continuously at an annual rate of 19%, estimate the total number of people who will have been infected with HIV by June of the year:
(A) 2000   (B) 2004

★ **59. Learning Curve.** People assigned to assemble circuit boards for a computer manufacturing company undergo on-the-job training. From past experience it was found that the learning curve for the average employee is given by

$$N = 40(1 - e^{-0.12t})$$

where N is the number of boards assembled per day after t days of training. Graph this function for $0 \le t \le 30$. What is the maximum number of boards an average employee can be expected to produce in 1 day?

★ **60. Advertising.** A company is trying to expose a new product to as many people as possible through television advertising in a large metropolitan area with 2 million possible viewers. A model for the number of people N, in millions, who are aware of the product after t days of advertising was found to be

$$N = 2(1 - e^{-0.037t})$$

Graph this function for $0 \le t \le 50$. What value does N tend to as t increases without bound?

61. Newton's Law of Cooling. This law states that the rate at which an object cools is proportional to the difference in temperature between the object and its surrounding medium. The temperature T of the object t hours later is given by

$$T = T_m + (T_0 - T_m)e^{-kt}$$

where T_m is the temperature of the surrounding medium and T_0 is the temperature of the object at $t = 0$. Suppose a bottle of wine at a room temperature of 72°F is placed in the refrigerator to cool before a dinner party. If the temperature in the refrigerator is kept at 40°F and $k = 0.4$, find the temperature of the wine, to the nearest degree, after 3 hours. (In Exercise 5-5 we will find out how to determine k.)

62. Newton's Law of Cooling. Refer to Problem 61. What is the temperature, to the nearest degree, of the wine after 5 hours in the refrigerator?

★ **63. Photography.** An electronic flash unit for a camera is activated when a capacitor is discharged through a filament of wire. After the flash is triggered, and the capacitor is discharged, the circuit (see the figure) is connected and the battery pack generates a current to recharge the capacitor. The time it takes for the capacitor to recharge is called the *recycle time.* For a particular flash unit using a 12-volt battery pack, the charge q, in coulombs, on the capacitor t seconds after recharging has started is given by

$$q = 0.0009(1 - e^{-0.2t})$$

Graph this function for $0 \le t \le 10$. Estimate the maximum charge on the capacitor.

★ **64. Medicine.** An electronic heart pacemaker utilizes the same type of circuit as the flash unit in Problem 63, but it is designed so that the capacitor discharges 72 times a minute. For a particular pacemaker, the charge on the capacitor t seconds after it starts recharging is given by

$$q = 0.000\,008(1 - e^{-2t})$$

Graph this function for $0 \leq t \leq \frac{5}{6}$. Estimate the maximum charge on the capacitor.

★ **65. Wildlife Management.** A herd of 20 white-tailed deer is introduced to a coastal island where there had been no deer before. Their population is predicted to increase according to the logistic curve

$$N = \frac{100}{1 + 4e^{-0.14t}}$$

where N is the number of deer expected in the herd after t years. Graph the equation for $0 \leq t \leq 30$. Estimate the herd size that the island can support.

★ **66. Training.** A trainee is hired by a computer manufacturing company to learn to test a particular model of a personal computer after it comes off the assembly line. The learning curve for an average trainee is given by

$$N = \frac{200}{4 + 21e^{-0.1t}}$$

where N is the number of computers tested per day after t days on the job. Graph the equation for $0 \leq t \leq 50$. What is the maximum number of computers that an average tester can be expected to test per day after training?

★ **67. Catenary.** A free-hanging power line between two supporting towers looks something like a parabola, but it is actually a curve called a **catenary.** A catenary has an equation of the form

$$y = \frac{e^{mx} + e^{-mx}}{2m}$$

where m is a constant. Graph the equation for $m = 0.25$.

★ **68. Catenary.** Graph the equation in Problem 67 for $m = 0.4$.

SECTION 5-3   Logarithmic Functions

- Definition of Logarithmic Function
- From Logarithmic Form to Exponential Form, and Vice Versa
- Properties of Logarithmic Functions

• Definition of Logarithmic Function

We now define a new class of functions, called **logarithmic functions,** as inverses of exponential functions. Since exponential functions are one-to-one, their inverses exist. Here you will see why we placed special emphasis on the general concept of inverse functions in Section 3-6. If you know quite a bit about a function, then, based on a knowledge of inverses in general, you will automatically know quite a bit about its inverse. For example, the graph of f^{-1} is the graph of f reflected across the line $y = x$, and the domain and range of f^{-1} are, respectively, the range and domain of f.

If we start with the exponential function,

$$f: \quad y = 2^x$$

and interchange the variables x and y, we obtain the inverse of f:

$$f^{-1}: \quad x = 2^y$$

The graphs of f, f^{-1}, and the line $y = x$ are shown in Figure 1. This new function is given the name **logarithmic function with base 2.** Since we cannot solve the equation $x = 2^y$ for y using the algebra properties discussed so far, we introduce a new symbol to represent this inverse function:

$$y = \log_2 x \quad \text{Read "log to the base 2 of } x.\text{"}$$

Thus,

$$y = \log_2 x \qquad \text{is equivalent to} \qquad x = 2^y$$

that is, $\log_2 x$ is the exponent to which 2 must be raised to obtain x. Symbolically, $x = 2^y = 2^{\log_2 x}$.

FIGURE 1 Logarithmic function with base 2.

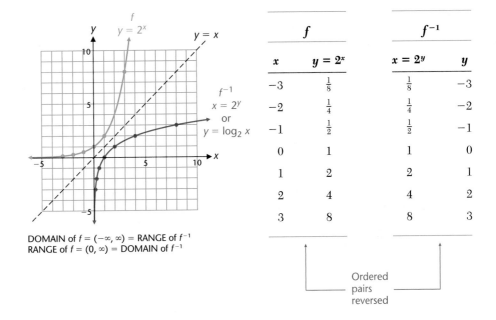

	f		f^{-1}	
x	$y = 2^x$		$x = 2^y$	y
-3	$\frac{1}{8}$		$\frac{1}{8}$	-3
-2	$\frac{1}{4}$		$\frac{1}{4}$	-2
-1	$\frac{1}{2}$		$\frac{1}{2}$	-1
0	1		1	0
1	2		2	1
2	4		4	2
3	8		8	3

Ordered pairs reversed

DOMAIN of $f = (-\infty, \infty) =$ RANGE of f^{-1}
RANGE of $f = (0, \infty) =$ DOMAIN of f^{-1}

In general, we define the **logarithmic function with base b** to be the inverse of the exponential function with base b ($b > 0$, $b \neq 1$).

DEFINITION 1

Definition of Logarithmic Function

For $b > 0$ and $b \neq 1$,

Logarithmic form   Exponential form
$$y = \log_b x \qquad \text{is equivalent to} \qquad x = b^y$$

The log to the base b of x is the exponent to which b must be raised to obtain x.

$$y = \log_{10} x \qquad \text{is equivalent to} \qquad x = 10^y$$

$$y = \log_e x \qquad \text{is equivalent to} \qquad x = e^y$$

Remember: A logarithm is an exponent.

It is very important to remember that $y = \log_b x$ and $x = b^y$ define the same function, and as such can be used interchangeably.

Since the domain of an exponential function includes all real numbers and its range is the set of positive real numbers, the **domain** of a logarithmic function is the set of all positive real numbers and its **range** is the set of all real numbers. Thus, $\log_{10} 3$ is defined, but $\log_{10} 0$ and $\log_{10} (-5)$ are not defined. That is, 3 is a logarithmic domain value, but 0 and -5 are not. Typical logarithmic curves are shown in Figure 2.

FIGURE 2 Typical logarithmic graphs.

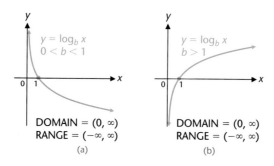

<table>
<tr><td>DOMAIN = $(0, \infty)$
RANGE = $(-\infty, \infty)$
(a)</td><td>DOMAIN = $(0, \infty)$
RANGE = $(-\infty, \infty)$
(b)</td></tr>
</table>

EXPLORE-DISCUSS 1   For the exponential function $f = \{(x, y) \mid y = (\frac{2}{3})^x\}$, graph f, f^{-1}, and $y = x$ on the same coordinate system. Discuss the domains and ranges of f and its inverse. By what other name is f^{-1} known?

• From Logarithmic Form to Exponential Form, and Vice Versa

We now look into the matter of converting logarithmic forms to equivalent exponential forms, and vice versa.

EXAMPLE 1   **Logarithmic–Exponential Conversions**

Change each logarithmic form to an equivalent exponential form:

(A) $\log_2 8 = 3$ (B) $\log_{25} 5 = \frac{1}{2}$ (C) $\log_2 (\frac{1}{4}) = -2$

Solutions   (A) $\log_2 8 = 3$ is equivalent to $8 = 2^3$
(B) $\log_{25} 5 = \frac{1}{2}$ is equivalent to $5 = 25^{1/2}$
(C) $\log_2 (\frac{1}{4}) = -2$ is equivalent to $\frac{1}{4} = 2^{-2}$

Matched Problem 1   Change each logarithmic form to an equivalent exponential form:

(A) $\log_3 27 = 3$ (B) $\log_{36} 6 = \frac{1}{2}$ (C) $\log_3 (\frac{1}{9}) = -2$

EXAMPLE 2   **Logarithmic–Exponential Conversions**

Change each exponential form to an equivalent logarithmic form:

(A) $49 = 7^2$ (B) $3 = \sqrt{9}$ (C) $\frac{1}{5} = 5^{-1}$

EXERCISE 5-4

A

In Problems 1–8, evaluate to four decimal places.

1. log 27,593

2. log 0.000 539

3. log 0.004 321

4. log 120,564

5. ln 0.000 654

6. ln 0.023 198

7. ln 456.76

8. ln 132.43

In Problems 9–16, evaluate x to 4 significant digits, given:

9. $\ln x = 3.4797$

10. $\ln x = -0.2985$

11. $\ln x = -2.2643$

12. $\ln x = 6.8236$

13. $\log x = -1.2543$

14. $\log x = -2.6123$

15. $\log x = 3.5324$

16. $\log x = 2.5017$

B

In Problems 17–24, evaluate to three decimal places.

17. $\dfrac{\log 3.215}{\log 2.569}$

18. $\log \dfrac{3.215}{2.569}$

19. $\ln \dfrac{0.5545}{0.0545}$

20. $\dfrac{\ln 0.5545}{\ln 0.0545}$

21. $\dfrac{\ln 0.6}{0.01}$

22. $\dfrac{\log 0.7}{0.005}$

23. $\dfrac{\log 300}{\log 3}$

24. $\dfrac{\ln 300}{\ln 3}$

In Problems 25–32, evaluate x to 5 significant digits.

25. $x = \ln (3.4562 \times 10^{15})$

26. $x = \ln (4.3931 \times 10^{-11})$

27. $x = \log (6.7744 \times 10^{-13})$

28. $x = \log (5.1212 \times 10^{14})$

29. $\log x = 21.667\ 503$

30. $\ln x = 14.561\ 094$

31. $\ln x = -11.112\ 445$

32. $\log x = -15.599\ 943$

Graph each function in Problems 33–40.

Check Problems 33–40 with a graphing utility.

33. $y = \ln x$

34. $y = -\ln x$

35. $y = |\ln x|$

36. $y = \ln |x|$

37. $y = 2 \ln (x + 2)$

38. $y = 2 \ln x + 2$

39. $y = 4 \ln x - 3$

40. $y = 4 \ln (x - 3)$

C

41. Find the fallacy:

$$1 < 3$$
$$\frac{1}{27} < \frac{3}{27} \qquad \text{Divide both sides by 27.}$$
$$\frac{1}{27} < \frac{1}{9}$$
$$(\tfrac{1}{3})^3 < (\tfrac{1}{3})^2$$
$$\log (\tfrac{1}{3})^3 < \log (\tfrac{1}{3})^2$$
$$3 \log \tfrac{1}{3} < 2 \log \tfrac{1}{3}$$
$$3 < 2 \qquad \text{Divide both sides by } \log \tfrac{1}{3}.$$

42. Find the fallacy:

$$3 > 2$$
$$3 \log \tfrac{1}{2} > 2 \log \tfrac{1}{2} \qquad \text{Multiply both sides by } \log \tfrac{1}{2}.$$
$$\log (\tfrac{1}{2})^3 > \log (\tfrac{1}{2})^2$$
$$(\tfrac{1}{2})^3 > (\tfrac{1}{2})^2$$
$$\tfrac{1}{8} > \tfrac{1}{4}$$
$$1 > 2 \qquad \text{Multiply both sides by 8.}$$

43. The function $f(x) = \log x$ increases extremely slowly as $x \to \infty$, but the composite function $g(x) = \log(\log x)$ increases still more slowly.
   (A) Illustrate this fact by computing the values of both functions for several large values of x.
   (B) Determine the domain and range of the function g.
   (C) Discuss the graphs of both functions.

44. The function $f(x) = \ln x$ increases extremely slowly as $x \to \infty$, but the composite function $g(x) = \ln(\ln x)$ increases still more slowly.
   (A) Illustrate this fact by computing the values of both functions for several large values of x.
   (B) Determine the domain and range of the function g.
   (C) Discuss the graphs of both functions.

In Problems 45–48, use a graphing utility to find the coordinates of all points of intersection to two decimal places.

45. $f(x) = \ln x, g(x) = 0.1x - 0.2$

46. $f(x) = \log x, g(x) = 4 - x^2$

47. $f(x) = \ln x, g(x) = x^{1/3}$

48. $f(x) = 3 \ln(x-2), g(x) = 4e^{-x}$

Problems 49–52 require the use of a graphing utility.

*The polynomials in Problems 49–52, called **Taylor polynomials,** can be used to approximate the function $g(x) = \ln(1 + x)$. To illustrate this approximation graphically, in each problem, graph $g(x) = \ln(1 + x)$ and the indicated polynomial in the same viewing window, $-1 \le x \le 3$ and $-2 \le y \le 2$.*

49. $P_1(x) = x - \frac{1}{2}x^2$

50. $P_2(x) = x - \frac{1}{2}x^2 + \frac{1}{3}x^3$

51. $P_3(x) = x - \frac{1}{2}x^2 + \frac{1}{3}x^3 - \frac{1}{4}x^4$

52. $P_4(x) = x - \frac{1}{2}x^2 + \frac{1}{3}x^3 - \frac{1}{4}x^4 + \frac{1}{5}x^5$

APPLICATIONS

53. Sound. What is the decibel level of:
  (A) The threshold of hearing, 1.0×10^{-12} watt per square meter?
  (B) The threshold of pain, 1.0 watt per square meter?
  Compute answers to 2 significant digits.

54. Sound. What is the decibel level of:
  (A) A normal conversation, 3.2×10^{-6} watt per square meter?
  (B) A jet plane with an afterburner, 8.3×10^2 watts per square meter?
  Compute answers to 2 significant digits.

55. Sound. If the intensity of a sound from one source is 1,000 times that of another, how much more is the decibel level of the louder sound than the quieter one?

56. Sound. If the intensity of a sound from one source is 10,000 times that of another, how much more is the decibel level of the louder sound than the quieter one?

57. Earthquakes. The largest recorded earthquake to date was in Colombia in 1906, with an energy release of 1.99×10^{17} joules. What was its magnitude on the Richter scale? Compute the answer to one decimal place.

58. Earthquakes. Anchorage, Alaska, had a major earthquake in 1964 that released 7.08×10^{16} joules of energy. What was its magnitude on the Richter scale? Compute the answer to one decimal place.

★★ **59. Earthquakes.** The 1933 Long Beach, California, earthquake had a Richter scale reading of 6.3, and the 1964 Anchorage, Alaska, earthquake had a Richter scale reading of 8.3. How many times more powerful was the Anchorage earthquake than the Long Beach earthquake?

★★ **60. Earthquakes.** Generally, an earthquake requires a magnitude of over 5.6 on the Richter scale to inflict serious damage. How many times more powerful than this was the great 1906 Colombia earthquake, which registered a magnitude of 8.6 on the Richter scale?

61. Space Vehicles. A new solid-fuel rocket has a weight ratio $W_t/W_b = 19.8$ and an exhaust velocity $c = 2.57$ kilometers per second. What is its velocity at burnout? Compute the answer to two decimal places.

62. Space Vehicles. A liquid-fuel rocket has a weight ratio $W_t/W_b = 6.2$ and an exhaust velocity $c = 5.2$ kilometers per second. What is its velocity at burnout? Compute the answer to two decimal places.

63. Chemistry. The hydrogen ion concentration of a substance is related to its acidity and basicity. Because hydrogen ion concentrations vary over a very wide range, logarithms are used to create a compressed **pH scale,** which is defined as follows:

$$pH = -\log[H^+]$$

where $[H^+]$ is the hydrogen ion concentration, in moles per liter. Pure water has a pH of 7, which means it is neutral. Substances with a pH less than 7 are acidic, and those with a pH greater than 7 are basic. Compute the pH of each substance listed, given the indicated hydrogen ion concentration.
  (A) Seawater, 4.63×10^{-9}
  (B) Vinegar, 9.32×10^{-4}
  Also, indicate whether it is acidic or basic. Compute answers to one decimal place.

64. Chemistry. Refer to Problem 63. Compute the pH of each substance below, given the indicated hydrogen ion concentration. Also, indicate whether it is acidic or basic. Compute answers to one decimal place.
  (A) Milk, 2.83×10^{-7}
  (B) Garden mulch, 3.78×10^{-6}

★ **65. Ecology.** Refer to Problem 63. Many lakes in Canada and the United States will no longer sustain some forms of wildlife because of the increase in acidity of the water from acid rain and snow caused by sulfur dioxide emissions from industry. If the pH of a sample of rainwater is 5.2, what is its hydrogen ion concentration in moles per liter? Compute the answer to two significant digits.

★ **66. Ecology.** Refer to Problem 63. If normal rainwater has a pH of 5.7, what is its hydrogen ion concentration in moles per liter? Compute the answer to two significant digits.

CHAPTER 5 GROUP ACTIVITY   Growth of Increasing Functions

The exponential function 2^x and the logarithmic function $\log_2 x$ are both increasing functions: their graphs rise as x increases. Moreover, they increase without bound. But they grow in strikingly different ways. The exponential function increases extremely rapidly, the logarithmic function extremely slowly. By computing values of functions and studying their graphs, we can investigate the manner in which various increasing functions grow.

(A) The functions $f(x) = 2^x$ and $g(x) = x^3$ are both increasing, but for large values of x, f increases much faster than g does. Illustrate this fact by computing values of both functions at several large values of x. Graph both functions and determine the number of intersection points.

(B) The functions $h(x) = \ln x$ and $k(x) = x^{1/4}$ are both increasing for $x > 0$, but for large values of x, h increases much more slowly than k does. Illustrate this fact by computing values of both functions at several large values of x. Graph both functions and determine the number of intersection points.

(C) The following 12 functions all increase without bound for $x > 1$. Arrange them in order, starting with the function that increases most slowly and ending with the function that increases most rapidly: $\sqrt{x}$, 2^x, $\log x$, e^{e^x}, x^6, $x \ln x$, e^x, $x^{0.1}$, $\ln(\ln x)$, e^{x^2}, $2x$, $\ln x$.

Chapter 5 Review

5-1   EXPONENTIAL FUNCTIONS

The equation $f(x) = b^x$, $b > 0$, $b \neq 1$, defines an **exponential function** with **base** b. The domain of f is $(-\infty, \infty)$ and the range is $(0, \infty)$. The **graph** of an exponential function is a continuous curve that always passes through the point $(0, 1)$ and has the x axis as a horizontal asymptote. If $b > 1$, then b^x increases as x increases, and if $0 < b < 1$, then b^x decreases as x increases. The function f is one-to-one and has an inverse. We have the following **exponential function properties:**

1. $a^x a^y = a^{x+y}$ $(a^x)^y = a^{xy}$ $(ab)^x = a^x b^x$

   $\left(\dfrac{a}{b}\right)^x = \dfrac{a^x}{b^x}$ $\dfrac{a^x}{a^y} = a^{x-y}$

2. $a^x = a^y$ if and only if $x = y$.

3. For $x \neq 0$, then $a^x = b^x$ if and only if $a = b$.

Exponential functions are used to describe various types of **growth.**

1. **Population growth** can be modeled by using the **doubling time growth model** $P = P_0 2^{t/d}$, where P is population at time t, P_0 is the population at time $t = 0$, and d is the **doubling time**—the time it takes for the population to double.

2. **Radioactive decay** can be modeled by using the **half-life decay model** $A = A_0 (\tfrac{1}{2})^{t/h} = A_0 2^{-t/h}$, where A is the amount at time t, A_0 is the amount at time $t = 0$, and h is the **half-life**—the time it takes for half the material to decay.

3. The growth of money in an account paying **compound interest** is described by $A = P(1 + r/n)^{nt}$, where P is the **principal**, r is the annual **rate**, n is the number of compounding periods in one year, and A is the **amount** in the account after t years. We also call P the **present value** and A the **future value** of the account.

5-2   THE EXPONENTIAL FUNCTION WITH BASE e

As m approaches ∞, the expression $(1 + 1/m)^m$ approaches the irrational number $e \approx 2.718\ 281\ 828\ 459$. The function $f(x) = e^x$ is called the **exponential function with base e.** Exponential functions with base e are used to model a variety of different types of exponential growth and decay, including growth of money in accounts that pay **continuous compound interest.** If a principal P is invested at an annual rate r compounded continuously, then the amount A in the account after t years is given by $A = Pe^{rt}$.

5-3   LOGARITHMIC FUNCTIONS

The **logarithmic function with base b** is defined to be the inverse of the exponential function with base b and is denoted by $y = \log_b x$. Thus, $y = \log_b x$ if and only if $x = b^y$, $b > 0$, $b \neq 1$. The domain of a logarithmic function is $(0, \infty)$ and the range is $(-\infty, \infty)$. The graph of a logarithmic function is a continuous curve that always passes through the point $(1, 0)$ and has the y

axis as a vertical asymptote. We have the following **properties of logarithmic functions:**

1. $\log_b 1 = 0$

2. $\log_b b = 1$

3. $\log_b b^x = x$

4. $b^{\log_b x} = x, x > 0$

5. $\log_b MN = \log_b M + \log_b N$

6. $\log_b \dfrac{M}{N} = \log_b M - \log_b N$

7. $\log_b M^p = p \log_b M$

8. $\log_b M = \log_b N$ if and only if $M = N$

5-4 COMMON AND NATURAL LOGARITHMS

Logarithms to the base 10 are called **common logarithms** and are denoted by $\log x$. Logarithms to the base e are called **natural logarithms** and are denoted by $\ln x$. Thus, $\log x = y$ is equivalent to $x = 10^y$, and $\ln x = y$ is equivalent to $x = e^y$.

The following applications involve logarithms:

1. The **decibel** is defined by $D = 10 \log (I/I_0)$, where D is the **decibel level** of the sound, I is the **intensity** of the sound, and $I_0 = 10^{-12}$ watt per square meter is a standardized sound level.

2. The **magnitude** M of an earthquake on the **Richter scale** is given by $M = \frac{2}{3} \log (E/E_0)$, where E is the energy released by the earthquake and $E_0 = 10^{4.40}$ joules is a standardized energy level.

3. The **velocity** v of a rocket at burnout is given by the **rocket equation** $v = c \ln (W_t/W_b)$, where c is the exhaust velocity, W_t is the takeoff weight, and W_b is the burnout weight.

5-5 EXPONENTIAL AND LOGARITHMIC EQUATIONS

Various techniques for solving **exponential equations,** such as $2^{3x-2} = 5$, and **logarithmic equations,** such as $\log (x + 3) + \log x = 1$, are illustrated by examples. The **change-of-base formula,** $\log_b N = (\log_a N)/(\log_a b)$, relates logarithms to two different bases and can be used, along with a calculator, to evaluate logarithms to bases other than e or 10.

Chapter 5 Review Exercise

Work through all the problems in this chapter review and check answers in the back of the book. Answers to all review problems are there, and following each answer is a number in italics indicating the section in which that type of problem is discussed. Where weaknesses show up, review appropriate sections in the text.

A

1. Write in logarithmic form using base 10: $m = 10^n$

2. Write in logarithmic form using base e: $x = e^y$

Write Problems 3 and 4 in exponential form.

3. $\log x = y$ 4. $\ln y = x$

In Problems 5 and 6, simplify.

5. $\dfrac{7^{x+2}}{7^{2-x}}$ 6. $\left(\dfrac{e^x}{e^{-x}}\right)^x$

Solve Problems 7–9 for x exactly. Do not use a calculator or table.

7. $\log_2 x = 3$ 8. $\log_x 25 = 2$ 9. $\log_3 27 = x$

Solve Problems 10–13 for x to 3 significant digits.

10. $10^x = 17.5$ 11. $e^x = 143{,}000$

12. $\ln x = -0.015\,73$ 13. $\log x = 2.013$

B

Solve Problems 14–24 for x exactly. Do not use a calculator or table.

14. $\ln (2x - 1) = \ln (x + 3)$

15. $\log (x^2 - 3) = 2 \log (x - 1)$

16. $e^{x^2-3} = e^{2x}$ 17. $4^{x-1} = 2^{1-x}$

18. $2x^2 e^{-x} = 18e^{-x}$ 19. $\log_{1/4} 16 = x$

20. $\log_x 9 = -2$ 21. $\log_{16} x = \frac{3}{2}$

22. $\log_x e^5 = 5$ 23. $10^{\log_{10} x} = 33$

24. $\ln x = 0$

Evaluate Problems 25–28 to 4 significant digits using a calculator.

25. $\ln \pi$ 26. $\log (-e)$

27. $\pi^{\ln 2}$ 28. $\dfrac{e^\pi + e^{-\pi}}{2}$

Solve Problems 29–38 for x to 3 significant digits.

29. $x = 2(10^{1.32})$ 30. $x = \log_5 23$

31. $\ln x = -3.218$ 32. $x = \log (2.156 \times 10^{-7})$

33. $x = \dfrac{\ln 4}{\ln 2.31}$ 34. $25 = 5(2^x)$

SYSTEMS OF EQUATIONS AND INEQUALITIES

CHAPTER 6

In this chapter we move from the standard methods of solving two linear equations with two variables to a method that can be used to solve linear systems with any number of variables and equations. This method is well-suited to computer use in the solution of larger systems. We also discuss systems involving nonlinear equations and systems of linear inequalities. Finally, we introduce a relatively new and powerful mathematical tool called *linear programming.* Many applications in mathematics involve systems of equations or inequalities. The mathematical techniques discussed in this chapter are applied to a variety of interesting and significant applications.

SECTION 6-1 Systems of Linear Equations and Augmented Matrices

- Substitution: A Brief Review
- Graphing
- Elimination by Addition
- Matrices
- Solving Linear Systems Using Augmented Matrices

In this section we will continue our discussion, began in Section 2-2, of systems involving two linear equations and two variables:

$$ax + by = h$$
$$cx + dy = k \qquad (1)$$

where x and y are variables, a, b, c, and d are real numbers called the **coefficients** of x and y, and h and k are real numbers called the **constant terms** in the equations. Recall that a pair of numbers $x = x_0$ and $y = y_0$, also written as an ordered pair (x_0, y_0), is a **solution** of this system if each equation is satisfied by the pair. The set of all such ordered pairs of numbers is called the **solution set** for the system.

In Section 2-2, we used substitution to solve system (1). After briefly reviewing this method, we will explore the relationship between the graph of the equations in a system and the solution of the system and review the standard *elimination-by-addition method.* Then we will introduce augmented matrices to transform the elimination-by-addition method into a solution process that is well-suited for computer use in the solution of linear systems involving large numbers of equations and variables.

To keep the introduction to augmented matrices in this section as simple as possible, we restrict the discussion to two equations with two variables. In the next section, the solution process is generalized and applied to larger linear systems.

- **Substitution: A Brief Review**
To review some of the basic concepts introduced in Section 2-2, consider the following simple example. At a computer fair, student tickets cost $2 and general admission tickets cost $3. If a total of 7 tickets are purchased for a total cost of $18, how many of each type were purchased?

Let

$$x = \text{Number of student tickets}$$
$$y = \text{Number of general admission tickets}$$

Then

$$x + y = 7 \qquad \text{Total number of tickets purchased}$$

$$2x + 3y = 18 \qquad \text{Total purchase cost}$$

To solve this system by substitution, we solve the first equation for y in terms of x and substitute in the second equation:

$$x + y = 7 \qquad\qquad 2x + 3y = 18$$

$$y = 7 - x \qquad 2x + 3(7 - x) = 18$$

$$-x + 21 = 18$$

$$x = 3$$

Now, replace x with 3 in $y = 7 - x$:

$$y = 7 - x$$

$$= 7 - 3$$

$$y = 4$$

Thus the solution is 3 student tickets and 4 general admission tickets. You should check this result in each of the original equations.

● **Graphing**   Recall that the graph of a linear equation is the line consisting of all ordered pairs that satisfy the equation. If we graph both equations in system (1) in the same coordinate system, then the coordinates of any points that the lines have in common must be solutions to the system. Example 1 illustrates this process for the ticket problem discussed above.

EXAMPLE 1   **Solving a System by Graphing**

Solve the ticket problem by graphing: $\begin{aligned} x + y &= 7 \\ 2x + 3y &= 18 \end{aligned}$

Solution   From Figure 1 we see that:

FIGURE 1

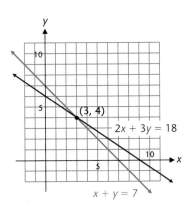

$x = 3$ Student tickets

$y = 4$ General admission tickets

Check $x + y = 7 \qquad\qquad 2x + 3y = 18$

$3 + 4 \overset{?}{=} 7 \qquad\quad 2(3) + 3(4) \overset{?}{=} 18$

$7 \overset{\checkmark}{=} 7 \qquad\qquad\qquad 18 \overset{\checkmark}{=} 18$

Matched Problem 1   Solve by graphing and check: $x - y = 3$
$$x + 2y = -3$$

It is clear that the preceding example has exactly one solution, since the lines have exactly one point of intersection. In general, lines in a rectangular coordinate system are related to each other in one of three ways illustrated in the next example.

EXAMPLE 2   **Solving Three Important Types of Systems by Graphing**

Solve each of the following systems by graphing:

(A) $2x - 3y = 2$ (B) $4x + 6y = 12$ (C) $2x - 3y = -6$
$\quad\ x + 2y = 8$ $\quad\ \ 2x + 3y = -6$ $\quad -x + \frac{3}{2}y = 3$

Solutions   (A)

(B)

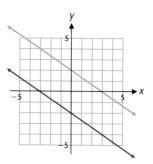

Lines intersect at one point only.
Exactly one solution: $x = 4, y = 2$

Lines are parallel (each has slope $-\frac{2}{3}$). No solution.

(C)

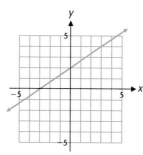

Lines coincide.
Infinitely many solutions.

Matched Problem 2   Solve each of the following systems by graphing:

(A) $2x + 3y = 12$ (B) $\quad\ x - 3y = -3$ (C) $2x - 3y = 12$
$\quad\ x - 3y = -3$ $\quad -2x + 6y = 12$ $\quad -x + \frac{3}{2}y = -6$

We now define some terms that can be used to describe the different types of solutions to systems of equations illustrated in Example 2.

Systems of Linear Equations: Basic Terms

A system of linear equations is **consistent** if it has one or more solutions and **inconsistent** if no solutions exist. Furthermore, a consistent system is said to be **independent** if it has exactly one solution (often referred to as the **unique solution**) and **dependent** if it has more than one solution.

Referring to the three systems in Example 2, the system in part A is consistent and independent, with the unique solution $x = 4$ and $y = 2$. The system in part B is inconsistent, with no solution. And the system in part C is consistent and dependent, with an infinite number of solutions: all the points on the two coinciding lines.

EXPLORE-DISCUSS 1

(A) Can a consistent and dependent system have exactly two solutions? Exactly three solutions? Explain.

(B) Solve each of the systems in Example 2 by the substitution method. Based on your results, describe how you might recognize a dependent system or an inconsistent system when using substitution.

By geometrically interpreting a system of two linear equations in two variables, we gain useful information about what to expect in the way of solutions to the system. In general, any two lines in a rectangular coordinate plane must intersect in exactly one point, be parallel, or coincide (have identical graphs). Thus, the systems in Example 2 illustrate the only three possible types of solutions for systems of two linear equations in two variables. These ideas are summarized in Theorem 1.

Theorem 1   **Possible Solutions to a Linear System**

The linear system

$$ax + by = h$$
$$cx + dy = k$$

must have:

1. Exactly one solution   Consistent and independent
   or
2. No solution   Inconsistent
   or
3. Infinitely many solutions   Consistent and dependent

There are no other possibilities.

One drawback of finding a solution by graphing is the inaccuracy of hand-drawn graphs. Graphic solutions performed on a graphing utility, however, provide both a useful geometric interpretation and an accurate approximation of the solution to a system of linear equations in two variables.

EXAMPLE 3   **Solving a System Using a Graphing Utility**

Solve to two decimal places using a graphing utility: $5x - 3y = 13$
$2x + 4y = 15$

Solution   First solve each equation for y:

$$5x - 3y = 13 \qquad\qquad 2x + 4y = 15$$
$$-3y = -5x + 13 \qquad\qquad 4y = -2x + 15$$
$$y = \tfrac{5}{3}x - \tfrac{13}{3} \qquad\qquad y = -0.5x + 3.75$$

Next, enter each equation in a graphing utility [Fig. 2(a)], graph in an appropriate viewing window, and approximate the intersection point [Fig. 2(b)].

FIGURE 2

(a) Equation definitions

(b) Intersection point

Rounding the values in Figure 2(b) to two decimal places, we see that the solution is

$$x = 3.73 \text{ and } y = 1.88 \qquad \text{or} \qquad (3.73, 1.88)$$

Check

$$5x - 3y = 13 \qquad\qquad 2x + 4y = 15$$
$$5(3.73) - 3(1.88) \stackrel{?}{=} 13 \qquad 2(3.73) + 4(1.88) \stackrel{?}{=} 15$$
$$13.01 \stackrel{\checkmark}{\approx} 13 \qquad\qquad 14.98 \stackrel{\checkmark}{\approx} 15$$

The checks are not exact because the values of x and y are approximations.

Matched Problem 3   Solve to two decimal places using a graphing utility: $2x - 5y = -25$
$4x + 3y = 5$

Graphic methods help us visualize a system and its solutions, frequently reveal relationships that might otherwise be hidden, and, with the assistance of a graphing utility, provide very accurate approximations to solutions.

The **position** of an element in a matrix is the row and column containing the element. This is usually denoted using **double subscript notation** a_{ij}, where i is the row and j is the column containing the element a_{ij}, as illustrated below:

$$A = \begin{bmatrix} 1 & 5 & -3 \\ 6 & 0 & -4 \end{bmatrix} \qquad \begin{matrix} a_{11} = 1, a_{12} = 5, a_{13} = -3 \\ a_{21} = 6, a_{22} = 0, a_{23} = -4 \end{matrix}$$

Note that a_{12} is read "a sub one two," not "a sub twelve." The elements $a_{11} = 1$ and $a_{22} = 0$ make up the *principal diagonal* of A. In general, the **principal diagonal** of a matrix A consists of the elements $a_{11}, a_{22}, a_{33}, \ldots$.

FIGURE 3 Matrix notation on a graphing utility.

Remark. Most graphing utilities are capable of storing and manipulating matrices. Figure 3 shows matrix A displayed in the editing screen of a particular graphing calculator. The size of the matrix is given at the top of the screen, and the position of the currently selected element is given at the bottom. Notice that a comma is used in the notation for the position. This is common practice on graphing utilities but not in mathematical literature.

The coefficients and constant terms in a system of linear equations can be used to form several matrices of interest to our work. Related to the system

$$\begin{matrix} 2x - 3y = & 5 \\ x + 2y = & -3 \end{matrix} \qquad (3)$$

are the following matrices:

Coefficient matrix	Constant matrix	Augmented coefficient matrix
$\begin{bmatrix} 2 & -3 \\ 1 & 2 \end{bmatrix}$	$\begin{bmatrix} 5 \\ -3 \end{bmatrix}$	$\left[\begin{array}{cc|c} 2 & -3 & 5 \\ 1 & 2 & -3 \end{array}\right]$

The augmented coefficient matrix will be used in this section. The other matrices will be used in later sections. The augmented coefficient matrix contains the essential parts of the system—both the coefficients and the constants. The vertical bar is included only as a visual aid to help us separate the coefficients from the constant terms. (Matrices entered and displayed on a graphing calculator or computer will not display this line.)

For ease of generalization to the larger systems in the following sections, we are now going to change the notation for the variables in system (3) to a subscript form (we would soon run out of letters, but we will not run out of subscripts). That is, in place of x and y, we will use x_1 and x_2, respectively, and (3) will be written as

$$\begin{matrix} 2x_1 - 3x_2 = & 5 \\ x_1 + 2x_2 = & -3 \end{matrix}$$

In general, associated with each linear system of the form

$$\begin{matrix} a_{11}x_1 + a_{12}x_2 = k_1 \\ a_{21}x_1 + a_{22}x_2 = k_2 \end{matrix} \qquad (4)$$

where x_1 and x_2 are variables, is the **augmented matrix** of the system:

This matrix contains the essential parts of system (4). Our objective is to learn how to manipulate augmented matrices in such a way that a solution to system (4) will result, if a solution exists.

In our earlier discussion of using elimination by addition, we said that two systems were equivalent if they had the same solution. And we used the operations in Theorem 2 to transform a system into an equivalent system. Paralleling this approach, we now say that two augmented matrices are **row-equivalent,** denoted by the symbol $\sim$ between the two matrices, if they are augmented matrices of equivalent systems of equations. And we use the operations listed in Theorem 3 below to transform augmented matrices into row-equivalent matrices. Note that Theorem 3 is a direct consequence of Theorem 2.

Theorem 3

Elementary Row Operations Producing Row-Equivalent Matrices

An augmented matrix is transformed into a row-equivalent matrix if any of the following **row operations** is performed:

1. Two rows are interchanged ($R_i \leftrightarrow R_j$).

2. A row is multiplied by a nonzero constant ($kR_i \to R_i$).

3. A constant multiple of one row is added to another row ($kR_j + R_i \to R_i$).

[*Note:* The arrow means "replaces."]

● **Solving Linear Systems Using Augmented Matrices**

The use of Theorem 3 in solving systems in the form of (3) is best illustrated by examples.

EXAMPLE 6   **Solving a System Using Augmented Matrix Methods**

Solve, using augmented matrix methods:

$$3x_1 + 4x_2 = 1$$
$$x_1 - 2x_2 = 7 \tag{5}$$

Solution   We start by writing the augmented matrix corresponding to system (5):

$$\begin{bmatrix} 3 & 4 & | & 1 \\ 1 & -2 & | & 7 \end{bmatrix} \tag{6}$$

Our objective is to use row operations from Theorem 3 to try to transform matrix (6) into the form

$$\begin{bmatrix} 1 & 0 & | & m \\ 0 & 1 & | & n \end{bmatrix} \tag{7}$$

where m and n are real numbers. The solution to system (5) will then be obvious, since matrix (7) will be the augmented matrix of the following system:

$$x_1 = m \qquad x_1 + 0x_2 = m$$

$$x_2 = n \qquad 0x_1 + x_2 = n$$

We now proceed to use row operations to transform (6) into form (7).

Step 1. To get a 1 in the upper left corner, we interchange rows 1 and 2—Theorem 3, part 1:

$$\begin{bmatrix} 3 & 4 & | & 1 \\ 1 & -2 & | & 7 \end{bmatrix} \underset{R_1 \leftrightarrow R_2}{\sim} \begin{bmatrix} 1 & -2 & | & 7 \\ 3 & 4 & | & 1 \end{bmatrix} \quad \begin{array}{l} \text{Now you see why we} \\ \text{wanted Theorem 2, part 1.} \end{array}$$

Step 2. To get a 0 in the lower left corner, we multiply R_1 by -3 and add to R_2—Theorem 3, part 3. This changes R_2 but not R_1. Some people find it useful to write $(-3)R_1$ outside the matrix to help reduce errors in arithmetic:

$$\begin{bmatrix} 1 & -2 & | & 7 \\ 3 & 4 & | & 1 \end{bmatrix} \underset{(-3)R_1 + R_2 \to R_2}{\sim} \begin{bmatrix} 1 & -2 & | & 7 \\ 0 & 10 & | & -20 \end{bmatrix}$$

$$-3 \qquad 6 \qquad -21 \leftarrow \text{-----}$$

Step 3. To get a 1 in the second row, second column, we multiply R_2 by $\frac{1}{10}$—Theorem 3, part 2:

$$\begin{bmatrix} 1 & -2 & | & 7 \\ 0 & 10 & | & -20 \end{bmatrix} \underset{\frac{1}{10}R_2 \to R_2}{\sim} \begin{bmatrix} 1 & -2 & | & 7 \\ 0 & 1 & | & -2 \end{bmatrix}$$

Step 4. To get a 0 in the first row, second column, we multiply R_2 by 2 and add the result to R_1—Theorem 3, part 3. This changes R_1 but not R_2.

$$0 \qquad 2 \qquad -4 \leftarrow \text{---}$$

$$\begin{bmatrix} 1 & -2 & | & 7 \\ 0 & 1 & | & -2 \end{bmatrix} \underset{2R_2 + R_1 \to R_1}{\sim} \begin{bmatrix} 1 & 0 & | & 3 \\ 0 & 1 & | & -2 \end{bmatrix}$$

We have accomplished our objective! The last matrix is the augmented matrix for the system

$$x_1 = 3$$

$$x_2 = -2 \tag{8}$$

Since system (8) is equivalent to the original system (5), we have solved system (5). That is, $x_1 = 3$ and $x_2 = -2$.

Check

$$3x_1 + 4x_2 = 1 \qquad x_1 - 2x_2 = 7$$
$$3(3) + 4(-2) \overset{?}{=} 1 \qquad 3 - 2(-2) \overset{?}{=} 7$$
$$1 \overset{\checkmark}{=} 1 \qquad 7 \overset{\checkmark}{=} 7$$

The above process is written more compactly as follows:

Step 1:
Need a 1 here
$$\begin{bmatrix} 3 & 4 & | & 1 \\ 1 & -2 & | & 7 \end{bmatrix} \quad R_1 \leftrightarrow R_2$$

Step 2:
Need a 0 here
$$\sim \begin{bmatrix} 1 & -2 & | & 7 \\ 3 & 4 & | & 1 \end{bmatrix} \quad (-3)R_1 + R_2 \to R_2$$
$$\qquad -3 \quad 6 \quad -21 \leftarrow\!-\!-\!-\!\rfloor$$

Step 3:
Need a 1 here
$$\sim \begin{bmatrix} 1 & -2 & | & 7 \\ 0 & 10 & | & -20 \end{bmatrix} \quad \tfrac{1}{10}R_2 \to R_2$$
$$\qquad 0 \quad 2 \quad -4 \leftarrow\!-\!-\!\rceil$$

Step 4:
Need a 0 here
$$\sim \begin{bmatrix} 1 & -2 & | & 7 \\ 0 & 1 & | & -2 \end{bmatrix} \quad 2R_2 + R_1 \to R_1$$

$$\sim \begin{bmatrix} 1 & 0 & | & 3 \\ 0 & 1 & | & -2 \end{bmatrix}$$

Therefore, $x_1 = 3$ and $x_2 = -2$.

Matched Problem 6 Solve, using augmented matrix methods: $2x_1 - x_2 = -7$
$x_1 + 2x_2 = 4$

EXPLORE-DISCUSS 2 The summary at the end of Example 6 shows five augmented coefficient matrices. Write the linear system that each matrix represents, solve each system graphically, and discuss the relationship between these solutions.

EXAMPLE 7 **Solving a System Using Augmented Matrix Methods**

Solve, using augmented matrix methods: $2x_1 - 3x_2 = 7$
$3x_1 + 4x_2 = 2$

Solution

Step 1: Need a 1 here
$$\begin{bmatrix} 2 & -3 & | & 7 \\ 3 & 4 & | & 2 \end{bmatrix} \quad \tfrac{1}{2}R_1 \to R_1$$

Step 2: Need a 0 here
$$\sim \begin{bmatrix} 1 & -\tfrac{3}{2} & | & \tfrac{7}{2} \\ 3 & 4 & | & 2 \end{bmatrix} \quad (-3)R_1 + R_2 \to R_2$$

$$\begin{array}{ccc} -3 & \tfrac{9}{2} & -\tfrac{21}{2} \end{array}$$

Step 3: Need a 1 here
$$\sim \begin{bmatrix} 1 & -\tfrac{3}{2} & | & \tfrac{7}{2} \\ 0 & \tfrac{17}{2} & | & -\tfrac{17}{2} \end{bmatrix} \quad \tfrac{2}{17}R_2 \to R_2$$

$$\begin{array}{ccc} 0 & \tfrac{3}{2} & -\tfrac{3}{2} \end{array}$$

Step 4: Need a 0 here
$$\sim \begin{bmatrix} 1 & -\tfrac{3}{2} & | & \tfrac{7}{2} \\ 0 & 1 & | & -1 \end{bmatrix} \quad \tfrac{3}{2}R_2 + R_1 \to R_1$$

$$\sim \begin{bmatrix} 1 & 0 & | & 2 \\ 0 & 1 & | & -1 \end{bmatrix}$$

Thus, $x_1 = 2$ and $x_2 = -1$. You should check this solution in the original system.

Matched Problem 7
Solve, using augmented matrix methods: $\begin{aligned} 5x_1 - 2x_2 &= 12 \\ 2x_1 + 3x_2 &= 1 \end{aligned}$

EXAMPLE 8 **Solving a System Using Augmented Matrix Methods**

Solve, using augmented matrix methods:
$$\begin{aligned} 2x_1 - x_2 &= 4 \\ -6x_1 + 3x_2 &= -12 \end{aligned} \tag{9}$$

Solution
$$\begin{bmatrix} 2 & -1 & | & 4 \\ -6 & 3 & | & -12 \end{bmatrix} \quad \begin{array}{l} \tfrac{1}{2}R_1 \to R_1 \text{ (This produces a 1 in the upper left corner.)} \\ \tfrac{1}{3}R_2 \to R_2 \text{ (This simplifies } R_2.) \end{array}$$

$$\sim \begin{bmatrix} 1 & -\tfrac{1}{2} & | & 2 \\ -2 & 1 & | & -4 \end{bmatrix} \quad 2R_1 + R_2 \to R_2 \text{ (This produces a 0 in the lower left corner.)}$$

$$\begin{array}{ccc} 2 & -1 & 4 \end{array}$$

$$\sim \begin{bmatrix} 1 & -\tfrac{1}{2} & | & 2 \\ 0 & 0 & | & 0 \end{bmatrix}$$

The last matrix corresponds to the system

$$x_1 - \tfrac{1}{2}x_2 = 2 \qquad x_1 - \tfrac{1}{2}x_2 = 2$$
$$0 = 0 \qquad 0x_1 + 0x_2 = 0$$

Thus, $x_1 = \tfrac{1}{2}x_2 + 2$. Hence, for any real number t, if $x_2 = t$, then $x_1 = \tfrac{1}{2}t + 2$. That is, the solution set is described by

$$(\tfrac{1}{2}t + 2, t) \qquad t \text{ a real number} \tag{10}$$

For example, if $t = 6$, then $(5, 6)$ is a particular solution; if $t = -2$, then $(1, -2)$ is another particular solution; and so on. Geometrically, the graphs of the two original equations coincide and there are infinitely many solutions.

In general, if we end up with a row of 0's in an augmented matrix for a two-equation–two-variable system, the system is dependent and there are infinitely many solutions.

Check   The following is a check that (10) provides a solution for system (9) for any real number t:

$$2x_1 - x_2 = 4 \qquad\qquad -6x_1 + 3x_2 = -12$$
$$2(\tfrac{1}{2}t + 2) - t \overset{?}{=} 4 \qquad -6(\tfrac{1}{2}t + 2) + 3t \overset{?}{=} -12$$
$$t + 4 - t \overset{?}{=} 4 \qquad\qquad -3t - 12 + 3t \overset{?}{=} -12$$
$$4 \overset{\checkmark}{=} 4 \qquad\qquad\qquad -12 \overset{\checkmark}{=} -12$$

Matched Problem 8   Solve, using augmented matrix methods: $-2x_1 + 6x_2 = 6$
$$3x_1 - 9x_2 = -9$$

EXPLORE-DISCUSS 3   Most graphing utilities can perform row operations. Figure 4 shows the solution to Example 8 on a particular graphing calculator. Consult your manual to see how to perform row operations, and solve Matched Problem 8 on your graphing utility.

FIGURE 4 Performing row operations on a graphing utility.

```
[A]
    [[2  -1  4   ]
     [-6  3  -12]]
*row(1/2,[A],1)→
[A]
    [[1   -.5  2  ]
     [-6   3  -12]]
```

```
*row(1/3,[A],2)→
[A]
    [[1   -.5  2 ]
     [-2   1   -4]]
*row+(2,[A],1,2)
→[A]
    [[1  -.5  2]
     [0   0   0]]
```

EXAMPLE 9   **Solving a System Using Augmented Matrix Methods**

Solve, using augmented matrix methods: $2x_1 + 6x_2 = -3$
$$x_1 + 3x_2 = 2$$

Solution

$$\begin{bmatrix} 2 & 6 & | & -3 \\ 1 & 3 & | & 2 \end{bmatrix} \quad R_1 \leftrightarrow R_2$$

$$\sim \begin{bmatrix} 1 & 3 & | & 2 \\ 2 & 6 & | & -3 \end{bmatrix} \quad (-2)R_1 + R_2 \rightarrow R_2$$
$$\begin{matrix} -2 & -6 & -4 \end{matrix}$$

$$\sim \begin{bmatrix} 1 & 3 & | & 2 \\ 0 & 0 & | & -7 \end{bmatrix} \quad \text{\small R_2 implies the contradiction: } 0 = -7$$

This is the augmented matrix of the system

$$x_1 + 3x_2 = 2 \qquad \text{\small $x_1 + 3x_2 = 2$}$$
$$0 = -7 \qquad \text{\small $0x_1 + 0x_2 = -7$}$$

The second equation is not satisfied by any ordered pair of real numbers. Hence, the original system is inconsistent and has no solution. Otherwise, we have proved that $0 = -7$!

Thus, if we obtain all 0's to the left of the vertical bar and a nonzero number to the right of the bar in a row of an augmented matrix, then the system is inconsistent and there are no solutions.

Matched Problem 9 Solve, using augmented matrix methods: $\begin{aligned} 2x_1 - x_2 &= 3 \\ 4x_1 - 2x_2 &= -1 \end{aligned}$

Summary

For m, n, p real numbers: $p \neq 0$:

Form 1: A Unique Solution (Consistent and Independent)	**Form 2: Infinitely Many Solutions (Consistent and Dependent)**	**Form 3: No Solution (Inconsistent)**
$\begin{bmatrix} 1 & 0 & \| & m \\ 0 & 1 & \| & n \end{bmatrix}$	$\begin{bmatrix} 1 & m & \| & n \\ 0 & 0 & \| & 0 \end{bmatrix}$	$\begin{bmatrix} 1 & m & \| & n \\ 0 & 0 & \| & p \end{bmatrix}$

The process of solving systems of equations described in this section is referred to as *Gauss–Jordan elimination*. We will use this method to solve larger-scale systems in the next section, including systems where the number of equations and the number of variables are not the same.

Answers to Matched Problems

1.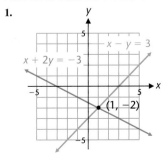

$x = 1$, $y = -2$

Check:
$$x - y = 3$$
$$1 - (-2) \overset{?}{=} 3$$
$$3 \overset{\checkmark}{=} 3$$
$$x + 2y = -3$$
$$1 + 2(-2) \overset{?}{=} -3$$
$$-3 \overset{\checkmark}{=} -3$$

2. (A) (3, 2), or $x = 3$ and $y = 2$ (B) No solution   (C) Infinite number of solutions
3. $(-1.92, 4.23)$, or $x = -1.92$ and $y = 4.23$ 4. $(-1, 3)$, or $x = -1$ and $y = 3$
5. 840 pounds of regular mix, 780 pounds of deluxe mix
6. $x_1 = -2, x_2 = 3$ 7. $x_1 = 2, x_2 = -1$
8. The system is dependent. For t any real number, $x_2 = t$, $x_1 = 3t - 3$ is a solution.
9. Inconsistent—no solution

EXERCISE **6-1**

A

Match each system in Problems 1–4 with one of the following graphs, and use the graph to solve the system.

(a)

(b)

(c)

(d)

1. $2x - 4y = 8$
   $x - 2y = 0$

2. $x + y = 3$
   $x - 2y = 0$

3. $2x - y = 5$
   $3x + 2y = -3$

4. $4x - 2y = 10$
   $2x - y = 5$

Solve Problems 5–10 by graphing.

5. $x + y = 7$
   $x - y = 3$

6. $x - y = 2$
   $x + y = 4$

7. $3x - 2y = 12$
   $7x + 2y = 8$

8. $3x - y = 2$
   $x + 2y = 10$

9. $3u + 5v = 15$
   $6u + 10v = -30$

10. $m + 2n = 4$
    $2m + 4n = -8$

Solve Problems 11–14 using elimination by addition.

11. $2x + 3y = 1$
    $3x - y = 7$

12. $2m - n = 10$
    $m - 2n = -4$

13. $4x - 3y = 15$
    $3x + 4y = 5$

14. $5x + 2y = 1$
    $2x - 3y = -11$

Problems 15–24 refer to the following matrices:

$$A = \begin{bmatrix} 3 & -2 & 0 \\ 4 & 1 & -6 \end{bmatrix} \qquad B = \begin{bmatrix} -2 & 8 & 0 \\ -3 & 6 & 9 \\ 4 & 2 & 0 \end{bmatrix}$$

$$C = \begin{bmatrix} 3 & -2 & 0 \end{bmatrix} \qquad D = \begin{bmatrix} -4 \\ 7 \end{bmatrix}$$

15. What is the size of A? Of C?

16. What is the size of B? Of D?

17. Identify all row matrices.

18. Identify all column matrices.

19. Identify all square matrices.

20. How many additional rows would matrix A need to be a square matrix?

21. For matrix A, find a_{12} and a_{23}.

22. For matrix A, find a_{21} and a_{13}.

23. Find the elements on the principal diagonal of matrix B.

24. Find the elements on the principal diagonal of matrix A.

Perform each of the row operations indicated in Problems 25–36 on the following matrix:

$$\begin{bmatrix} 1 & -3 & 2 \\ 4 & -6 & -8 \end{bmatrix}$$

25. $R_1 \leftrightarrow R_2$ 26. $\frac{1}{2}R_2 \to R_2$ 27. $-4R_1 \to R_1$

28. $-2R_1 \to R_1$ 29. $2R_2 \to R_2$ 30. $-1R_2 \to R_2$

31. $(-4)R_1 + R_2 \to R_2$ 32. $(-\frac{1}{2})R_2 + R_1 \to R_1$

33. $(-2)R_1 + R_2 \to R_2$ 34. $(-3)R_1 + R_2 \to R_2$

35. $(-1)R_1 + R_2 \rightarrow R_2$ **36.** $1R_1 + R_2 \rightarrow R_2$

Solve Problems 37 and 38 using augmented matrix methods. Write the linear system represented by each augmented matrix in your solution, and solve each of these systems graphically. Discuss the relationship between the solutions of these systems.

37. $x_1 + x_2 = 7$
$x_1 - x_2 = 1$

38. $x_1 + x_2 = 5$
$x_1 - x_2 = -3$

B

Solve Problems 39–50 using augmented matrix methods:

39. $x_1 - 4x_2 = -2$
$-2x_1 + x_2 = -3$

40. $x_1 - 3x_2 = -5$
$-3x_1 - x_2 = 5$

41. $3x_1 - x_2 = 2$
$x_1 + 2x_2 = 10$

42. $2x_1 + x_2 = 0$
$x_1 - 2x_2 = -5$

43. $x_1 + 2x_2 = 4$
$2x_1 + 4x_2 = -8$

44. $2x_1 - 3x_2 = -2$
$-4x_1 + 6x_2 = 7$

45. $2x_1 + x_2 = 6$
$x_1 - x_2 = -3$

46. $3x_1 - x_2 = -5$
$x_1 + 3x_2 = 5$

47. $3x_1 - 6x_2 = -9$
$-2x_1 + 4x_2 = 6$

48. $2x_1 - 4x_2 = -2$
$-3x_1 + 6x_2 = 3$

49. $4x_1 - 2x_2 = 2$
$-6x_1 + 3x_2 = -3$

50. $-6x_1 + 2x_2 = 4$
$3x_1 - x_2 = -2$

In Problems 51–54 use an intersection routine on a graphing utility to approximate the solution of each system to two decimal places.

51. $2x - 3y = -5$
$3x + 4y = 13$

52. $7x - 3y = 20$
$5x + 2y = 8$

53. $3.5x - 2.4y = 0.1$
$2.6x - 1.7y = -0.2$

54. $5.4x + 4.2y = -12.9$
$3.7x + 6.4y = -4.5$

C

55. The coefficients of the three systems below are very similar. One might guess that the solution sets to the three systems would also be nearly identical. Develop evidence for or against this guess by considering graphs of the systems and solutions obtained using elimination by addition.

(A) $4x + 5y = 4$
$9x + 11y = 4$

(B) $4x + 5y = 4$
$8x + 11y = 4$

(C) $4x + 5y = 4$
$8x + 10y = 4$

56. Repeat Problem 55 for the following systems.

(A) $5x - 6y = -10$
$11x - 13y = -20$

(B) $5x - 6y = -10$
$10x - 13y = -20$

(C) $5x - 6y = -10$
$10x - 12y = -20$

Solve Problems 57–60 using augmented matrix methods. Use a graphing utility to perform the row operations.

57. $0.8x_1 + 2.88x_2 = 4$
$1.25x_1 + 4.34x_2 = 5$

58. $2.7x_1 - 15.12x_2 = 27$
$3.25x_1 - 18.52x_2 = 33$

59. $4.8x_1 - 40.32x_2 = 295.2$
$-3.75x_1 + 28.7x_2 = -211.2$

60. $5.7x_1 - 8.55x_2 = -35.91$
$4.5x_1 + 5.73x_2 = 76.17$

APPLICATIONS

61. Puzzle. A friend of yours came out of the post office having spent $19.50 on 32¢ and 23¢ stamps. If she bought 75 stamps in all, how many of each type did she buy?

62. Puzzle. A parking meter contains only nickels and dimes worth $6.05. If there are 89 coins in all, how many of each type are there?

63. Investments. Bond A pays 6% compounded annually and bond B pays 9% compounded annually. If a $200,000 investment in a combination of the two bonds returns $14,775 annually, how much is invested in each bond?

64. Investments. Past history indicates that mutual fund A will earn 14.6% annually and mutual fund B will earn 9.8% annually. How should an investment be divided between the two funds to produce an expected return of 11%?

65. Chemistry. A chemist has two solutions of sulfuric acid: a 20% solution and an 80% solution. How much of each should be used to obtain 100 liters of a 62% solution?

66. Chemistry. A chemist has two solutions: one containing 40% alcohol and another containing 70% alcohol. How much of each should be used to obtain 80 liters of a 49% solution?

67. Nutrition. Animals in an experiment are to be kept on a strict diet. Each animal is to receive, among other things, 54 grams of protein and 24 grams of fat. The laboratory technician is able to purchase two food mixes of the following compositions: Mix A has 15% protein and 10% fat; mix B has 30% protein and 5% fat. How many grams of each mix should be used to obtain the right diet for a single animal?

68. Nutrition—Plants. A fruit grower can use two types of fertilizer in his orange grove, brand A and brand B. Each bag of brand A contains 9 pounds of nitrogen and 5 pounds of phosphoric acid. Each bag of brand B contains 8 pounds of nitrogen and 6 pounds of phosphoric acid. Tests indicate that the grove needs 770 pounds of nitrogen and 490 pounds of phosphoric acid. How many bags of each brand should be used to provide the required amounts of nitrogen and phosphoric acid?

69. Delivery Charges. United Express, a nationwide package delivery service, charges a base price for overnight delivery of packages weighing 1 pound or less and a surcharge for each additional pound (or fraction thereof). A customer is billed $27.75 for shipping a 5-pound package and $64.50 for shipping a 20-pound package. Find the base price and the surcharge for each additional pound.

70. Delivery Charges. Refer to Problem 69. Federated Shipping, a competing overnight delivery service, informs the customer in Problem 69 that it would ship the 5-pound package for $29.95 and the 20-pound package for $59.20.
(A) If Federated Shipping computes its cost in the same manner as United Express, find the base price and the surcharge for Federated Shipping.
(B) Devise a simple rule that the customer can use to choose the cheaper of the two services for each package shipped. Justify your answer.

71. Resource Allocation. A coffee manufacturer uses Colombian and Brazilian coffee beans to produce two blends, robust and mild. A pound of the robust blend requires 12 ounces of Colombian beans and 4 ounces of Brazilian beans. A pound of the mild blend requires 6 ounces of Colombian beans and 10 ounces of Brazilian beans. Coffee is shipped in 132-pound burlap bags. The company has 50 bags of Colombian beans and 40 bags of Brazilian beans on hand. How many pounds of each blend should it produce in order to use all the available beans?

72. Resource Allocation. Refer to Problem 71.
(A) If the company decides to discontinue production of the robust blend and only produce the mild blend, how many pounds of the mild blend can it produce and how many beans of each type will it use? Are there any beans that are not used?
(B) Repeat part A if the company decides to discontinue production of the mild blend and only produce the robust blend.

SECTION **6-2** Gauss–Jordan Elimination

- Reduced Matrices
- Solving Systems by Gauss–Jordan Elimination
- Application

Now that you have had some experience with row operations on simple augmented matrices, we will consider systems involving more than two variables. In addition, we will not require that a system have the same number of equations as variables. It turns out that the results for two-variable–two-equation linear systems, stated in Theorem 1 in Section 6-1, actually hold for linear systems of any size.

Possible Solutions to a Linear System

It can be shown that any linear system must have exactly one solution, no solution, or an infinite number of solutions, regardless of the number of equations or the number of variables in the system. The terms *unique, consistent, inconsistent, dependent,* and *independent* are used to describe these solutions, just as they are for systems with two variables.

● **Reduced Matrices**   In the last section we used row operations to transform the augmented coefficient matrix for a system of two equations in two variables

$$\begin{bmatrix} a_{11} & a_{12} & \bigm| & k_1 \\ a_{21} & a_{22} & \bigm| & k_2 \end{bmatrix} \qquad \begin{aligned} a_{11}x_1 + a_{12}x_2 &= k_1 \\ a_{21}x_1 + a_{22}x_2 &= k_2 \end{aligned}$$

into one of the following simplified forms:

$$
\begin{array}{ccc}
\text{Form 1} & \text{Form 2} & \text{Form 3} \\
\left[\begin{array}{cc|c} 1 & 0 & m \\ 0 & 1 & n \end{array}\right] & \left[\begin{array}{cc|c} 1 & m & n \\ 0 & 0 & 0 \end{array}\right] & \left[\begin{array}{cc|c} 1 & m & n \\ 0 & 0 & p \end{array}\right]
\end{array}
\tag{1}
$$

where m, n, and p are real numbers, $p \neq 0$. Each of these reduced forms represents a system that has a different type of solution set, and no two of these forms are row-equivalent. Thus, we consider each of these to be a different simplified form. Now we want to consider larger systems with more variables and more equations.

EXPLORE-DISCUSS 1   Forms 1, 2, and 3 above represent systems that have, respectively, a unique solution, an infinite number of solutions, and no solution. Discuss the number of solutions for the systems of three equations in three variables represented by the following augmented coefficient matrices.

$$
\text{(A)} \left[\begin{array}{ccc|c} 1 & 1 & 1 & 2 \\ 0 & 0 & 0 & 3 \\ 0 & 0 & 0 & 0 \end{array}\right] \quad
\text{(B)} \left[\begin{array}{ccc|c} 1 & 1 & 1 & 2 \\ 0 & 0 & 0 & 0 \\ 0 & 0 & 0 & 0 \end{array}\right] \quad
\text{(C)} \left[\begin{array}{ccc|c} 1 & 0 & 0 & 2 \\ 0 & 1 & 0 & 3 \\ 0 & 0 & 1 & 4 \end{array}\right]
$$

Since there is no upper limit on the number of variables or the number of equations in a linear system, it is not feasible to explicitly list all possible "simplified forms" for larger systems, as we did for systems of two equations in two variables. Instead, we state a general definition of a simplified form called a *reduced matrix* that can be applied to all matrices and systems, regardless of size.

DEFINITION 1   **Reduced Matrix**

A matrix is in **reduced form** if:

1. Each row consisting entirely of 0's is below any row having at least one nonzero element.

2. The leftmost nonzero element in each row is 1.

3. The column containing the leftmost 1 of a given row has 0's above and below the 1.

4. The leftmost 1 in any row is to the right of the leftmost 1 in the preceding row.

EXAMPLE 1   **Reduced Forms**

The matrices below are not in reduced form. Indicate which condition in the definition is violated for each matrix. State the row operation(s) required to transform the matrix to reduced form, and find the reduced form.

$$\text{(A)} \begin{bmatrix} 0 & 1 & -2 \\ 1 & 0 & 3 \end{bmatrix} \qquad \text{(B)} \begin{bmatrix} 1 & 2 & -2 & 3 \\ 0 & 0 & 1 & -1 \end{bmatrix}$$

$$\text{(C)} \begin{bmatrix} 1 & 0 & -3 \\ 0 & 0 & 0 \\ 0 & 1 & -2 \end{bmatrix} \qquad \text{(D)} \begin{bmatrix} 1 & 0 & 0 & -1 \\ 0 & 2 & 0 & 3 \\ 0 & 0 & 1 & -5 \end{bmatrix}$$

Solutions   (A) Condition 4 is violated: The leftmost 1 in row 2 is not to the right of the left-most 1 in row 1. Perform the row operation $R_1 \leftrightarrow R_2$ to obtain the reduced form:

$$\begin{bmatrix} 1 & 0 & 3 \\ 0 & 1 & -2 \end{bmatrix}$$

(B) Condition 3 is violated: The column containing the leftmost 1 in row 2 does not have a zero above the 1. Perform the row operation $2R_2 + R_1 \to R_1$ to obtain the reduced form:

$$\begin{bmatrix} 1 & 2 & 0 & 1 \\ 0 & 0 & 1 & -1 \end{bmatrix}$$

(C) Condition 1 is violated: The second row contains all zeros, and it is not below any row having at least one nonzero element. Perform the row operation $R_2 \leftrightarrow R_3$ to obtain the reduced form:

$$\begin{bmatrix} 1 & 0 & -3 \\ 0 & 1 & -2 \\ 0 & 0 & 0 \end{bmatrix}$$

(D) Condition 2 is violated: The leftmost nonzero element in row 2 is not a 1. Perform the row operation $\frac{1}{2}R_2 \to R_2$ to obtain the reduced form:

$$\begin{bmatrix} 1 & 0 & 0 & -1 \\ 0 & 1 & 0 & \frac{3}{2} \\ 0 & 0 & 1 & -5 \end{bmatrix}$$

Matched Problem 1   The matrices below are not in reduced form. Indicate which condition in the definition is violated for each matrix. State the row operation(s) required to transform the matrix to reduced form and find the reduced form.

$$\text{(A)} \begin{bmatrix} 1 & 0 & 2 \\ 0 & 3 & -6 \end{bmatrix} \qquad \text{(B)} \begin{bmatrix} 1 & 5 & 4 & 3 \\ 0 & 1 & 2 & -1 \\ 0 & 0 & 0 & 0 \end{bmatrix}$$

$$\text{(C)} \begin{bmatrix} 0 & 1 & 0 & -3 \\ 1 & 0 & 0 & 0 \\ 0 & 0 & 1 & 2 \end{bmatrix} \qquad \text{(D)} \begin{bmatrix} 1 & 2 & 0 & 3 \\ 0 & 0 & 0 & 0 \\ 0 & 0 & 1 & 4 \end{bmatrix}$$

• Solving Systems by Gauss–Jordan Elimination

We are now ready to outline the Gauss–Jordan elimination method for solving systems of linear equations. The method systematically transforms an augmented matrix into a reduced form. The system corresponding to a reduced augmented coefficient matrix is called a **reduced system.** As we will see, reduced systems are easy to solve.

The Gauss–Jordan elimination method is named after the German mathematician Carl Friedrich Gauss (1777–1885) and the German geodesist Wilhelm Jordan (1842–1899). Gauss, one of the greatest mathematicians of all time, used a method of solving systems of equations that was later generalized by Jordan to solve problems in large-scale surveying.

EXAMPLE 2 **Solving a System Using Gauss–Jordan Elimination**

Solve by Gauss–Jordan elimination:
$$\begin{aligned} 2x_1 - 2x_2 + x_3 &= 3 \\ 3x_1 + x_2 - x_3 &= 7 \\ x_1 - 3x_2 + 2x_3 &= 0 \end{aligned}$$

Solution Write the augmented matrix and follow the steps indicated at the right to produce a reduced form.

Need a 1 here.
$$\begin{bmatrix} 2 & -2 & 1 & | & 3 \\ 3 & 1 & -1 & | & 7 \\ 1 & -3 & 2 & | & 0 \end{bmatrix} \quad R_1 \leftrightarrow R_3$$

Step 1: Choose the leftmost nonzero column and get a 1 at the top.

Need 0's here.
$$\sim \begin{bmatrix} 1 & -3 & 2 & | & 0 \\ 3 & 1 & -1 & | & 7 \\ 2 & -2 & 1 & | & 3 \end{bmatrix} \quad \begin{array}{l} (-3)R_1 + R_2 \to R_2 \\ (-2)R_1 + R_3 \to R_3 \end{array}$$

Step 2: Use multiples of the row containing the 1 from step 1 to get zeros in all remaining places in the column containing this 1.

Need a 1 here.
$$\sim \begin{bmatrix} 1 & -3 & 2 & | & 0 \\ 0 & 10 & -7 & | & 7 \\ 0 & 4 & -3 & | & 3 \end{bmatrix} \quad 0.1R_2 \to R_2$$

Step 3: Repeat step 1 with the *submatrix* formed by (mentally) deleting the top row.

Need 0's here.
$$\sim \begin{bmatrix} 1 & -3 & 2 & | & 0 \\ 0 & 1 & -0.7 & | & 0.7 \\ 0 & 4 & -3 & | & 3 \end{bmatrix} \quad \begin{array}{l} 3R_2 + R_1 \to R_1 \\ \\ (-4)R_2 + R_3 \to R_3 \end{array}$$

Step 4: Repeat step 2 with the *entire matrix.*

Need a 1 here.
$$\sim \begin{bmatrix} 1 & 0 & -0.1 & | & 2.1 \\ 0 & 1 & -0.7 & | & 0.7 \\ 0 & 0 & -0.2 & | & 0.2 \end{bmatrix} \quad (-5)R_3 \to R_3$$

Step 3: Repeat step 1 with the *submatrix* formed by (mentally) deleting the top two rows.

Need 0's here.
$$\sim \begin{bmatrix} 1 & 0 & -0.1 & | & 2.1 \\ 0 & 1 & -0.7 & | & 0.7 \\ 0 & 0 & 1 & | & -1 \end{bmatrix} \quad \begin{array}{l} 0.1R_3 + R_1 \to R_1 \\ 0.7R_3 + R_2 \to R_2 \end{array}$$

Step 4: Repeat step 2 with the *entire matrix.*

$$\sim \begin{bmatrix} 1 & 0 & 0 & 2 \\ 0 & 1 & 0 & 0 \\ 0 & 0 & 1 & -1 \end{bmatrix}$$

The matrix is now in reduced form, and we can proceed to solve the corresponding reduced system.

$$x_1 \qquad = 2$$
$$x_2 \quad = 0$$
$$x_3 = -1$$

The solution to this system is $x_1 = 2$, $x_2 = 0$, $x_3 = -1$. You should check this solution in the original system.

Gauss–Jordan Elimination

Step 1. Choose the leftmost nonzero column and use appropriate row operations to get a 1 at the top.

Step 2. Use multiples of the row containing the 1 from step 1 to get zeros in all remaining places in the column containing this 1.

Step 3. Repeat step 1 with the **submatrix** formed by (mentally) deleting the row used in step 2 and all rows above this row.

Step 4. Repeat step 2 with the **entire matrix,** including the mentally deleted rows. Continue this process until it is impossible to go further.

[*Note:* If at any point in this process we obtain a row with all zeros to the left of the vertical line and a nonzero number to the right, we can stop, since we will have a contradiction: $0 = n$, $n \neq 0$. We can then conclude that the system has no solution.]

Remarks
1. Even though each matrix has a unique reduced form, the sequence of steps (algorithm) presented here for transforming a matrix into a reduced form is not unique. That is, other sequences of steps (using row operations) can produce a reduced matrix. (For example, it is possible to use row operations in such a way that computations involving fractions are minimized.) But we emphasize again that we are not interested in the most efficient hand methods for transforming small matrices into reduced forms. Our main interest is in giving you a little experience with a method that is suitable for solving large-scale systems on a computer or graphing utility.

2. Most graphing utilities have the ability to find reduced forms, either directly or with some programming. Figure 1 illustrates the solution of Example 2 on a graphing calculator that has a built-in routine for finding reduced forms. Notice that in row 2 and column 4 of the reduced form the graphing calculator has displayed the very small number $-3.5E{-}13$ instead of the exact value 0. This is a common occurrence on a graphing calculator and causes no problems. Just replace any very small numbers displayed in scientific notation with 0.

FIGURE 1 Gauss–Jordan elimination on a graphing calculator.

```
A
      [[2 -2 1  3]
       [3 1  -1 7]
       [1 -3 2  0]]
rref A
      [[1 0 0 2        ]
       [0 1 0 -3.5E-13]
       [0 0 1 -1       ]]
```

Matched Problem 2   Solve by Gauss–Jordan elimination:
$$3x_1 + x_2 - 2x_3 = 2$$
$$x_1 - 2x_2 + x_3 = 3$$
$$2x_1 - x_2 - 3x_3 = 3$$

EXAMPLE 3   **Solving a System Using Gauss–Jordan Elimination**

Solve by Gauss–Jordan elimination:
$$2x_1 - 4x_2 + x_3 = -4$$
$$4x_1 - 8x_2 + 7x_3 = 2$$
$$-2x_1 + 4x_2 - 3x_3 = 5$$

Solution

$$\begin{bmatrix} 2 & -4 & 1 & | & -4 \\ 4 & -8 & 7 & | & 2 \\ -2 & 4 & -3 & | & 5 \end{bmatrix} \quad 0.5R_1 \to R_1$$

$$\sim \begin{bmatrix} 1 & -2 & 0.5 & | & -2 \\ 4 & -8 & 7 & | & 2 \\ -2 & 4 & -3 & | & 5 \end{bmatrix} \quad \begin{matrix} \\ (-4)R_1 + R_2 \to R_2 \\ 2R_1 + R_3 \to R_3 \end{matrix}$$

$$\sim \begin{bmatrix} 1 & -2 & 0.5 & | & -2 \\ 0 & 0 & 5 & | & 10 \\ 0 & 0 & -2 & | & 1 \end{bmatrix} \quad \begin{matrix} 0.2R_2 \to R_2 & \text{Note that column 3 is the} \\ & \text{leftmost nonzero column} \\ & \text{in this submatrix.} \end{matrix}$$

$$\sim \begin{bmatrix} 1 & -2 & 0.5 & | & -2 \\ 0 & 0 & 1 & | & 2 \\ 0 & 0 & -2 & | & 1 \end{bmatrix} \quad \begin{matrix} (-0.5)R_2 + R_1 \to R_1 \\ \\ 2R_2 + R_3 \to R_3 \end{matrix}$$

$$\sim \begin{bmatrix} 1 & -2 & 0 & | & -3 \\ 0 & 0 & 1 & | & 2 \\ 0 & 0 & 0 & | & 5 \end{bmatrix} \quad \begin{matrix} \text{We stop the Gauss–Jordan} \\ \text{elimination, even though the} \\ \text{matrix is not in reduced form,} \\ \text{since the last row produces a} \\ \text{contradiction.} \end{matrix}$$

The system is inconsistent and has no solution.

Matched Problem 3   Solve by Gauss–Jordan elimination:
$$2x_1 - 4x_2 - x_3 = -8$$
$$4x_1 - 8x_2 + 3x_3 = 4$$
$$-2x_1 + 4x_2 + x_3 = 11$$

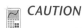 *CAUTION*   Figure 2 shows the solution to Example 3 on a graphing calculator with a built-in reduced form routine. Notice that the graphing calculator does not stop when a contradiction first occurs, as we did in the solution to Example 3, but continues on to find the reduced form. Nevertheless, the last row in the reduced form still produces a contradiction, indicating that the system has no solution.

FIGURE 2 Recognizing contradictions on a graphing calculator.

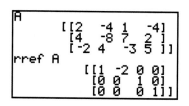

EXAMPLE 4   **Solving a System Using Gauss–Jordan Elimination**

Solve by Gauss–Jordan elimination:

$$3x_1 + 6x_2 - 9x_3 = 15$$
$$2x_1 + 4x_2 - 6x_3 = 10$$
$$-2x_1 - 3x_2 + 4x_3 = -6$$

Solution

$$\begin{bmatrix} 3 & 6 & -9 & | & 15 \\ 2 & 4 & -6 & | & 10 \\ -2 & -3 & 4 & | & -6 \end{bmatrix} \quad \tfrac{1}{3}R_1 \to R_1$$

$$\sim \begin{bmatrix} 1 & 2 & -3 & | & 5 \\ 2 & 4 & -6 & | & 10 \\ -2 & -3 & 4 & | & -6 \end{bmatrix} \quad \begin{array}{l}(-2)R_1 + R_2 \to R_2 \\ 2R_1 + R_3 \to R_3 \end{array}$$

$$\sim \begin{bmatrix} 1 & 2 & -3 & | & 5 \\ 0 & 0 & 0 & | & 0 \\ 0 & 1 & -2 & | & 4 \end{bmatrix} \quad \begin{array}{l} R_2 \leftrightarrow R_3 \quad \text{Note that we must interchange} \\ \text{rows 2 and 3 to obtain a nonzero} \\ \text{entry at the top of the second} \\ \text{column of this submatrix.} \end{array}$$

$$\sim \begin{bmatrix} 1 & 2 & -3 & | & 5 \\ 0 & 1 & -2 & | & 4 \\ 0 & 0 & 0 & | & 0 \end{bmatrix} \quad (-2)R_2 + R_1 \to R_1$$

$$\sim \begin{bmatrix} 1 & 0 & 1 & | & -3 \\ 0 & 1 & -2 & | & 4 \\ 0 & 0 & 0 & | & 0 \end{bmatrix} \quad \begin{array}{l}\text{This matrix is now in reduced form.} \\ \text{Write the corresponding reduced} \\ \text{system and solve.} \end{array}$$

$$\begin{aligned} x_1 \phantom{{}+{}} + x_3 &= -3 \\ x_2 - 2x_3 &= 4 \end{aligned} \quad \begin{array}{l}\text{We discard the equation corresponding} \\ \text{to the third (all 0) row in the reduced} \\ \text{form, since it is satisfied by all values of} \\ x_1, x_2, \text{ and } x_3. \end{array}$$

Note that the leftmost variable in each equation appears in one and only one equation. We solve for the leftmost variables x_1 and x_2 in terms of the remaining variable x_3:

$$x_1 = -x_3 - 3$$
$$x_2 = 2x_3 + 4$$

This dependent system has an infinite number of solutions. We will use a parameter to represent all the solutions. If we let $x_3 = t$, then for any real number t,

$$x_1 = -t - 3$$
$$x_2 = 2t + 4$$
$$x_3 = t$$

You should check that $(-t - 3, 2t + 4, t)$ is a solution of the original system for any real number t. Some particular solutions are

$t = 0$	$t = -2$	$t = 3.5$
$(-3, 4, 0)$	$(-1, 0, -2)$	$(-6.5, 11, 3.5)$

Matched Problem 4   Solve by Gauss–Jordan elimination:

$$2x_1 - 2x_2 - 4x_3 = -2$$
$$3x_1 - 3x_2 - 6x_3 = -3$$
$$-2x_1 + 3x_2 + x_3 = 7$$

In general,

> If the number of leftmost 1's in a reduced augmented coefficient matrix is less than the number of variables in the system and there are no contradictions, then the system is dependent and has infinitely many solutions.

There are many different ways to use the reduced augmented coefficient matrix to describe the infinite number of solutions of a dependent system. We will always proceed as follows: Solve each equation in a reduced system for its leftmost variable and then introduce a different parameter for each remaining variable. As the solution to Example 4 illustrates, this method produces a concise and useful representation of the solutions to a dependent system. Example 5 illustrates a dependent system where two parameters are required to describe the solution.

EXPLORE-DISCUSS 2   Explain why the definition of reduced form ensures that each leftmost variable in a reduced system appears in one and only one equation and no equation contains more than one leftmost variable. Discuss methods for determining if a consistent system is independent or dependent by examining the reduced form.

EXAMPLE 5   **Solving a System Using Gauss–Jordan Elimination**

Solve by Gauss–Jordan elimination:

$$\begin{aligned} x_1 + 2x_2 + 4x_3 + x_4 - x_5 &= 1 \\ 2x_1 + 4x_2 + 8x_3 + 3x_4 - 4x_5 &= 2 \\ x_1 + 3x_2 + 7x_3 + 3x_5 &= -2 \end{aligned}$$

Solution

$$\begin{bmatrix} 1 & 2 & 4 & 1 & -1 & | & 1 \\ 2 & 4 & 8 & 3 & -4 & | & 2 \\ 1 & 3 & 7 & 0 & 3 & | & -2 \end{bmatrix} \quad \begin{matrix} \\ (-2)R_1 + R_2 \rightarrow R_2 \\ (-1)R_1 + R_3 \rightarrow R_3 \end{matrix}$$

$$\sim \begin{bmatrix} 1 & 2 & 4 & 1 & -1 & | & 1 \\ 0 & 0 & 0 & 1 & -2 & | & 0 \\ 0 & 1 & 3 & -1 & 4 & | & -3 \end{bmatrix} \quad R_2 \leftrightarrow R_3$$

$$\sim \begin{bmatrix} 1 & 2 & 4 & 1 & -1 & | & 1 \\ 0 & 1 & 3 & -1 & 4 & | & -3 \\ 0 & 0 & 0 & 1 & -2 & | & 0 \end{bmatrix} \quad (-2)R_2 + R_1 \rightarrow R_1$$

$$\sim \begin{bmatrix} 1 & 0 & -2 & 3 & -9 & | & 7 \\ 0 & 1 & 3 & -1 & 4 & | & -3 \\ 0 & 0 & 0 & 1 & -2 & | & 0 \end{bmatrix} \quad \begin{matrix} (-3)R_3 + R_1 \rightarrow R_1 \\ R_3 + R_2 \rightarrow R_2 \end{matrix}$$

$$\sim \begin{bmatrix} 1 & 0 & -2 & 0 & -3 & | & 7 \\ 0 & 1 & 3 & 0 & 2 & | & -3 \\ 0 & 0 & 0 & 1 & -2 & | & 0 \end{bmatrix} \quad \text{Matrix is in reduced form.}$$

$$\begin{aligned} x_1 - 2x_3 - 3x_5 &= 7 \\ x_2 + 3x_3 + 2x_5 &= -3 \\ x_4 - 2x_5 &= 0 \end{aligned}$$

Solve for the leftmost variables x_1, x_2, and x_4 in terms of the remaining variables x_3 and x_5:

$$\begin{aligned} x_1 &= 2x_3 + 3x_5 + 7 \\ x_2 &= -3x_3 - 2x_5 - 3 \\ x_4 &= 2x_5 \end{aligned}$$

If we let $x_3 = s$ and $x_5 = t$, then for any real numbers s and t,

$$\begin{aligned} x_1 &= 2s + 3t + 7 \\ x_2 &= -3s - 2t - 3 \\ x_3 &= s \\ x_4 &= 2t \\ x_5 &= t \end{aligned}$$

is a solution. The check is left for you to perform.

Matched Problem 5 Solve by Gauss–Jordan elimination:

$$\begin{aligned} x_1 - x_2 + 2x_3 \qquad\;\; - 2x_5 &= 3 \\ -2x_1 + 2x_2 - 4x_3 - x_4 + x_5 &= -5 \\ 3x_1 - 3x_2 + 7x_3 + x_4 - 4x_5 &= 6 \end{aligned}$$

• Application We now consider an application that involves a dependent system of equations.

EXAMPLE 6 **Purchasing**

A chemical manufacturer wants to purchase a fleet of 24 railroad tank cars with a combined carrying capacity of 250,000 gallons. Tank cars with three different carrying capacities are available: 6,000 gallons, 8,000 gallons, and 18,000 gallons. How many of each type of tank car should be purchased?

Solution Let

$$x_1 = \text{Number of 6,000-gallon tank cars}$$
$$x_2 = \text{Number of 8,000-gallon tank cars}$$
$$x_3 = \text{Number of 18,000-gallon tank cars}$$

Then

$$\begin{aligned} x_1 + x_2 + x_3 &= 24 && \text{Total number of tank cars} \\ 6{,}000x_1 + 8{,}000x_2 + 18{,}000x_3 &= 250{,}000 && \text{Total carrying capacity} \end{aligned}$$

Now we can form the augmented matrix of the system and solve by using Gauss–Jordan elimination:

$$\begin{bmatrix} 1 & 1 & 1 & | & 24 \\ 6{,}000 & 8{,}000 & 18{,}000 & | & 250{,}000 \end{bmatrix} \qquad \tfrac{1}{1{,}000}R_2 \to R_2 \text{ (simplify } R_2)$$

$$\sim \begin{bmatrix} 1 & 1 & 1 & | & 24 \\ 6 & 8 & 18 & | & 250 \end{bmatrix} \qquad (-6)R_1 + R_2 \to R_2$$

$$\sim \begin{bmatrix} 1 & 1 & 1 & | & 24 \\ 0 & 2 & 12 & | & 106 \end{bmatrix} \qquad \tfrac{1}{2}R_2 \to R_2$$

$$\sim \begin{bmatrix} 1 & 1 & 1 & | & 24 \\ 0 & 1 & 6 & | & 53 \end{bmatrix} \qquad (-1)R_2 + R_1 \to R_1$$

$$\sim \begin{bmatrix} 1 & 0 & -5 & | & -29 \\ 0 & 1 & 6 & | & 53 \end{bmatrix} \qquad \text{Matrix is in reduced form}$$

$$\begin{aligned} x_1 - 5x_3 &= -29 && \text{or} && x_1 = 5x_3 - 29 \\ x_2 + 6x_3 &= 53 && \text{or} && x_2 = -6x_3 + 53 \end{aligned}$$

Let $x_3 = t$. Then for t any real number,

$$x_1 = 5t - 29$$
$$x_2 = -6t + 53$$
$$x_3 = t$$

is a solution—or is it? Since the variables in this system represent the number of tank cars purchased, the values of x_1, x_2, and x_3 must be nonnegative integers. Thus, the third equation requires that t must be a nonnegative integer. The first equation requires that $5t - 29 \geq 0$, so t must be at least 6. The middle equation requires that $-6t + 53 \geq 0$, so t can be no larger than 8. Thus, 6, 7, and 8 are the only possible values for t. There are only three possible combinations that meet the company's specifications of 24 tank cars with a total carrying capacity of 250,000 gallons, as shown in Table 1:

TABLE 1

t	6,000-gallon tank cars x_1	8,000-gallon tank cars x_2	18,000-gallon tank cars x_3
6	1	17	6
7	6	11	7
8	11	5	8

The final choice would probably be influenced by other factors. For example, the company might want to minimize the cost of the 24 tank cars.

Matched Problem 6   A commuter airline wants to purchase a fleet of 30 airplanes with a combined carrying capacity of 960 passengers. The three available types of planes carry 18, 24, and 42 passengers, respectively. How many of each type of plane should be purchased?

Answers to Matched Problems

1. (A) Condition 2 is violated: The 3 in row 2 and column 2 should be a 1. Perform the operation $\frac{1}{3}R_2 \to R_2$ to obtain:

$$\begin{bmatrix} 1 & 0 & | & 2 \\ 0 & 1 & | & -2 \end{bmatrix}$$

(B) Condition 3 is violated: The 5 in row 1 and column 2 should be a 0. Perform the operation $(-5)R_2 + R_1 \to R_1$ to obtain:

$$\begin{bmatrix} 1 & 0 & -6 & | & 8 \\ 0 & 1 & 2 & | & -1 \\ 0 & 0 & 0 & | & 0 \end{bmatrix}$$

(C) Condition 4 is violated: The leftmost 1 in the second row is not to the right of the leftmost 1 in the first row. Perform the operation $R_1 \leftrightarrow R_2$ to obtain:

$$\begin{bmatrix} 1 & 0 & 0 & | & 0 \\ 0 & 1 & 0 & | & -3 \\ 0 & 0 & 1 & | & 2 \end{bmatrix}$$

(D) Condition 1 is violated: The all-zero second row should be at the bottom. Perform the operation $R_2 \leftrightarrow R_3$ to obtain:

$$\begin{bmatrix} 1 & 2 & 0 & | & 3 \\ 0 & 0 & 1 & | & 4 \\ 0 & 0 & 0 & | & 0 \end{bmatrix}$$

2. $x_1 = 1, x_2 = -1, x_3 = 0$
3. Inconsistent; no solution
4. $x_1 = 5t + 4, x_2 = 3t + 5, x_3 = t, t$ any real number
5. $x_1 = s + 7, x_2 = s, x_3 = t - 2, x_4 = -3t - 1, x_5 = t, s$ and t any real numbers
6.

t	18-passenger planes x_1	24-passenger planes x_2	42-passenger planes x_3
14	2	14	14
15	5	10	15
16	8	6	16
17	11	2	17

EXERCISE 6-2

A

In Problems 1–8, indicate whether each matrix is in reduced form.

1. $\begin{bmatrix} 1 & 0 & | & -1 \\ 0 & 2 & | & 6 \end{bmatrix}$ 2. $\begin{bmatrix} 1 & 0 & | & 5 \\ 0 & 1 & | & -3 \end{bmatrix}$

3. $\begin{bmatrix} 0 & 1 & -2 & | & 0 \\ 0 & 0 & 0 & | & 1 \\ 0 & 0 & 0 & | & 0 \end{bmatrix}$ 4. $\begin{bmatrix} 1 & -1 & 4 & | & 0 \\ 0 & 0 & 0 & | & 0 \\ 0 & 0 & 0 & | & 1 \end{bmatrix}$

5. $\begin{bmatrix} 0 & 0 & 1 & | & 2 \\ 0 & 1 & 0 & | & -5 \\ 1 & 0 & 0 & | & 4 \end{bmatrix}$ 6. $\begin{bmatrix} 1 & -2 & 4 & | & 1 \\ 0 & 0 & 1 & | & -3 \\ 0 & 0 & 0 & | & 0 \end{bmatrix}$

7. $\begin{bmatrix} 0 & 1 & 6 & 0 & | & -8 \\ 0 & 0 & 0 & 1 & | & 1 \end{bmatrix}$ 8. $\begin{bmatrix} 0 & 0 & 1 & | & 0 \\ 0 & 0 & 0 & | & 0 \end{bmatrix}$

In Problems 9–16, write the linear system corresponding to each reduced augmented matrix and solve.

9. $\begin{bmatrix} 1 & 0 & 0 & | & -2 \\ 0 & 1 & 0 & | & 3 \\ 0 & 0 & 1 & | & 0 \end{bmatrix}$ 10. $\begin{bmatrix} 1 & 0 & 0 & 0 & | & -2 \\ 0 & 1 & 0 & 0 & | & 0 \\ 0 & 0 & 1 & 0 & | & 1 \\ 0 & 0 & 0 & 1 & | & 3 \end{bmatrix}$

11. $\begin{bmatrix} 1 & 0 & -2 & | & 3 \\ 0 & 1 & 1 & | & -5 \\ 0 & 0 & 0 & | & 0 \end{bmatrix}$ 12. $\begin{bmatrix} 1 & -2 & 0 & | & -3 \\ 0 & 0 & 1 & | & 5 \\ 0 & 0 & 0 & | & 0 \end{bmatrix}$

13. $\begin{bmatrix} 1 & 0 & | & 0 \\ 0 & 1 & | & 0 \\ 0 & 0 & | & 1 \end{bmatrix}$ 14. $\begin{bmatrix} 1 & 0 & | & 5 \\ 0 & 1 & | & -3 \\ 0 & 0 & | & 0 \end{bmatrix}$

15. $\begin{bmatrix} 1 & -2 & 0 & -3 & | & -5 \\ 0 & 0 & 1 & 3 & | & 2 \end{bmatrix}$ 16. $\begin{bmatrix} 1 & 0 & -2 & 3 & | & 4 \\ 0 & 1 & -1 & 2 & | & -1 \end{bmatrix}$

B

Use row operations to change each matrix in Problems 17–22 to reduced form.

17. $\begin{bmatrix} 1 & 2 & | & -1 \\ 0 & 1 & | & 3 \end{bmatrix}$ 18. $\begin{bmatrix} 1 & 3 & | & 1 \\ 0 & 2 & | & -4 \end{bmatrix}$

19. $\begin{bmatrix} 1 & 0 & -3 & | & 1 \\ 0 & 1 & 2 & | & 0 \\ 0 & 0 & 3 & | & -6 \end{bmatrix}$ 20. $\begin{bmatrix} 1 & 0 & 4 & | & 0 \\ 0 & 1 & -3 & | & -1 \\ 0 & 0 & -2 & | & 2 \end{bmatrix}$

21. $\begin{bmatrix} 1 & 2 & -2 & | & -1 \\ 0 & 3 & -6 & | & 1 \\ 0 & -1 & 2 & | & -\frac{1}{3} \end{bmatrix}$ **22.** $\begin{bmatrix} 0 & -2 & 8 & | & 1 \\ 2 & -2 & 6 & | & -4 \\ 0 & -1 & 4 & | & \frac{1}{2} \end{bmatrix}$

Solve Problems 23–42 using Gauss–Jordan elimination.

23. $2x_1 + 4x_2 - 10x_3 = -2$
$3x_1 + 9x_2 - 21x_3 = 0$
$x_1 + 5x_2 - 12x_3 = 1$

24. $3x_1 + 5x_2 - x_3 = -7$
$x_1 + x_2 + x_3 = -1$
$2x_1 + 11x_3 = 7$

25. $3x_1 + 8x_2 - x_3 = -18$
$2x_1 + x_2 + 5x_3 = 8$
$2x_1 + 4x_2 + 2x_3 = -4$

26. $2x_1 + 7x_2 + 15x_3 = -12$
$4x_1 + 7x_2 + 13x_3 = -10$
$3x_1 + 6x_2 + 12x_3 = -9$

27. $2x_1 - x_2 - 3x_3 = 8$
$x_1 - 2x_2 = 7$

28. $2x_1 + 4x_2 - 6x_3 = 10$
$3x_1 + 3x_2 - 3x_3 = 6$

29. $2x_1 - x_2 = 0$
$3x_1 + 2x_2 = 7$
$x_1 - x_2 = -1$

30. $2x_1 - x_2 = 0$
$3x_1 + 2x_2 = 7$
$x_1 - x_2 = -2$

31. $3x_1 - 4x_2 - x_3 = 1$
$2x_1 - 3x_2 + x_3 = 1$
$x_1 - 2x_2 + 3x_3 = 2$

32. $3x_1 + 7x_2 - x_3 = 11$
$x_1 + 2x_2 - x_3 = 3$
$2x_1 + 4x_2 - 2x_3 = 10$

33. $-2x_1 + x_2 + 3x_3 = -7$
$x_1 - 4x_2 + 2x_3 = 0$
$x_1 - 3x_2 + x_3 = 1$

34. $2x_1 + 5x_2 + 4x_3 = -7$
$-4x_1 - 5x_2 + 2x_3 = 9$
$-2x_1 - x_2 + 4x_3 = 3$

35. $2x_1 - 2x_2 - 4x_3 = -2$
$-3x_1 + 3x_2 + 6x_3 = 3$

36. $2x_1 + 8x_2 - 6x_3 = 4$
$-3x_1 - 12x_2 + 9x_3 = -6$

37. $4x_1 - x_2 + 2x_3 = 3$
$-4x_1 + x_2 - 3x_3 = -10$
$8x_1 - 2x_2 + 9x_3 = -1$

38. $4x_1 - 2x_2 + 2x_3 = 5$
$-6x_1 + 3x_2 - 3x_3 = -2$
$10x_1 - 5x_2 + 9x_3 = 4$

39. $2x_1 - 5x_2 - 3x_3 = 7$
$-4x_1 + 10x_2 + 2x_3 = 6$
$6x_1 - 15x_2 - x_3 = -19$

40. $-4x_1 + 8x_2 + 10x_3 = -6$
$6x_1 - 12x_2 - 15x_3 = 9$
$-8x_1 + 14x_2 + 19x_3 = -8$

41. $5x_1 - 3x_2 + 2x_3 = 13$
$2x_1 - x_2 - 3x_3 = 1$
$4x_1 - 2x_2 + 4x_3 = 12$

42. $4x_1 - 2x_2 + 3x_3 = 3$
$3x_1 - x_2 - 2x_3 = -10$
$2x_1 + 4x_2 - x_3 = -1$

43. Consider a consistent system of three linear equations in three variables. Discuss the nature of the solution set for the system if the reduced form of the augmented coefficient matrix has
(A) One leftmost 1
(B) Two leftmost 1's
(C) Three leftmost 1's
(D) Four leftmost 1's

44. Consider a system of three linear equations in three variables. Give examples of two reduced forms that are not row equivalent if the system is
(A) Consistent and dependent
(B) Inconsistent

C

Solve Problems 45–50 using Gauss–Jordan elimination.

45. $x_1 + 2x_2 - 4x_3 - x_4 = 7$
$2x_1 + 5x_2 - 9x_3 - 4x_4 = 16$
$x_1 + 5x_2 - 7x_3 - 7x_4 = 13$

46. $2x_1 + 4x_2 + 5x_3 + 4x_4 = 8$
$x_1 + 2x_2 + 2x_3 + x_4 = 3$

47. $x_1 - x_2 + 3x_3 - 2x_4 = 1$
$-2x_1 + 4x_2 - 3x_3 + x_4 = 0.5$
$3x_1 - x_2 + 10x_3 - 4x_4 = 2.9$
$4x_1 - 3x_2 + 8x_3 - 2x_4 = 0.6$

48. $x_1 + x_2 + 4x_3 + x_4 = 1.3$
$-x_1 + x_2 - x_3 = 1.1$
$2x_1 + x_3 + 3x_4 = -4.4$
$2x_1 + 5x_2 + 11x_3 + 3x_4 = 5.6$

49. $x_1 - 2x_2 + x_3 + x_4 + 2x_5 = 2$
$-2x_1 + 4x_2 + 2x_3 + 2x_4 - 2x_5 = 0$
$3x_1 - 6x_2 + x_3 + x_4 + 5x_5 = 4$
$-x_1 + 2x_2 + 3x_3 + x_4 + x_5 = 3$

50. $x_1 - 3x_2 + x_3 + x_4 + 2x_5 = 2$
$-x_1 + 5x_2 + 2x_3 + 2x_4 - 2x_5 = 0$
$2x_1 - 6x_2 + 2x_3 + 2x_4 + 4x_5 = 4$
$-x_1 + 3x_2 - x_3 - x_5 = -3$

APPLICATIONS

Solve Problems 51–72 using Gauss–Jordan elimination.

★ **51. Puzzle.** A friend of yours came out of the post office after spending $14.00 on 15¢, 20¢, and 35¢ stamps. If she bought 45 stamps in all, how many of each type did she buy?

★ **52. Puzzle.** A parking meter accepts only nickels, dimes, and quarters. If the meter contains 32 coins with a total value of $6.80, how many of each type are there?

★★ **53. Chemistry.** A chemist can purchase a 10% saline solution in 500 cubic centimeter containers, a 20% saline solution in 500 cubic centimeter containers, and a 50% saline solution in 1,000 cubic centimeter containers. He needs 12,000 cubic centimeters of 30% saline solution. How many containers of each type of solution should he purchase in order to form this solution?

★★ 54. Chemistry. Repeat Problem 53 if the 50% saline solution is available only in 1,500 cubic centimeter containers.

55. Geometry. Find a, b, and c so that the graph of the parabola with equation $y = a + bx + cx^2$ passes through the points $(-2, 3)$, $(-1, 2)$, and $(1, 6)$.

56. Geometry. Find a, b, and c so that the graph of the parabola with equation $y = a + bx + cx^2$ passes through the points $(1, 3)$, $(2, 2)$, and $(3, 5)$.

57. Geometry. Find a, b, and c so that the graph of the circle with equation $x^2 + y^2 + ax + by + c = 0$ passes through the points $(6, 2)$, $(4, 6)$, and $(-3, -1)$.

58. Geometry. Find a, b, and c so that the graph of the circle with equation $x^2 + y^2 + ax + by + c = 0$ passes through the points $(-4, 1)$, $(-1, 2)$, and $(3, -6)$.

59. Production Scheduling. A small manufacturing plant makes three types of inflatable boats: one-person, two-person, and four-person models. Each boat requires the services of three departments, as listed in the table. The cutting, assembly, and packaging departments have available a maximum of 380, 330, and 120 labor-hours per week, respectively. How many boats of each type must be produced each week for the plant to operate at full capacity?

	One-person boat	Two-person boat	Four-person boat
Cutting department	0.5 h	1.0 h	1.5 h
Assembly department	0.6 h	0.9 h	1.2 h
Packaging department	0.2 h	0.3 h	0.5 h

60. Production Scheduling. Repeat Problem 59 assuming the cutting, assembly, and packaging departments have available a maximum of 350, 330, and 115 labor-hours per week, respectively.

★ 61. Production Scheduling. Rework Problem 59 assuming the packaging department is no longer used.

★ 62. Production Scheduling. Rework Problem 60 assuming the packaging department is no longer used.

★ 63. Production Scheduling. Rework Problem 59 assuming the four-person boat is no longer produced.

★ 64. Production Scheduling. Rework Problem 60 assuming the four-person boat is no longer produced.

65. Nutrition. A dietitian in a hospital is to arrange a special diet using three basic foods. The diet is to include exactly 340 units of calcium, 180 units of iron, and 220 units of vitamin A. The number of units per ounce of each special ingredient for each of the foods is indicated in the table. How many ounces of each food must be used to meet the diet requirements?

	Units per ounce		
	Food A	**Food B**	**Food C**
Calcium	30	10	20
Iron	10	10	20
Vitamin A	10	30	20

66. Nutrition. Repeat Problem 65 if the diet is to include exactly 400 units of calcium, 160 units of iron, and 240 units of vitamin A.

★ 67. Nutrition. Solve Problem 65 with the assumption that food C is no longer available.

★ 68. Nutrition. Solve Problem 66 with the assumption that food C is no longer available.

★ 69. Nutrition. Solve Problem 65 assuming the vitamin A requirement is deleted.

★ 70. Nutrition. Solve Problem 66 assuming the vitamin A requirement is deleted.

71. Sociology. Two sociologists have grant money to study school busing in a particular city. They wish to conduct an opinion survey using 600 telephone contacts and 400 house contacts. Survey company A has personnel to do 30 telephone and 10 house contacts per hour; survey company B can handle 20 telephone and 20 house contacts per hour. How many hours should be scheduled for each firm to produce exactly the number of contacts needed?

72. Sociology. Repeat Problem 71 if 650 telephone contacts and 350 house contacts are needed.

SECTION 6-3   Systems Involving Second-Degree Equations

- Solution by Substitution
- Other Solution Methods

If a system of equations contains any equations that are not linear, then the system is called a **nonlinear system.** In this section we investigate nonlinear systems involving second-degree terms such as

$$x^2 + y^2 = 5 \qquad x^2 - 2y^2 = 2 \qquad x^2 + 3xy + y^2 = 20$$
$$3x + y = 1 \qquad\qquad xy = 2 \qquad\qquad xy - y^2 = 0$$

It can be shown that such systems have at most four solutions, some of which may be imaginary. Since we are interested in finding both real and imaginary solutions to the systems we consider, we now assume that the replacement set for each variable is the set of complex numbers, rather than the set of real numbers.

• Solution by Substitution   The substitution method used to solve linear systems of two equations in two variables is also an effective method for solving nonlinear systems. This process is best illustrated by examples.

EXAMPLE 1   **Solving a Nonlinear System by Substitution**

Solve the system: $x^2 + y^2 = 5$
$3x + y = 1$

Solution   Solve the second equation for y in terms of x; then substitute for y in the first equation to obtain an equation that involves x alone.

$$3x + y = 1$$
$$y = 1 - 3x \qquad \text{Substitute this expression for } y \text{ in the first equation.}$$
$$x^2 + y^2 = 5$$
$$x^2 + (1 - 3x)^2 = 5$$
$$10x^2 - 6x - 4 = 0 \qquad \text{Simplify and write in standard quadratic form.}$$
$$5x^2 - 3x - 2 = 0 \qquad \text{Divide through by 2 to simplify further.}$$
$$(x - 1)(5x + 2) = 0$$
$$x = 1, -\tfrac{2}{5}$$

If we substitute these values back into the equation $y = 1 - 3x$, we obtain two solutions to the system:

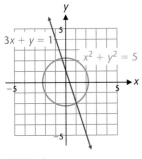

$$x = 1 \qquad\qquad\qquad x = -\tfrac{2}{5}$$
$$y = 1 - 3(1) = -2 \qquad y = 1 - 3(-\tfrac{2}{5}) = \tfrac{11}{5}$$

A check, which you should provide, verifies that $(1, -2)$ and $(-\tfrac{2}{5}, \tfrac{11}{5})$ are both solutions to the system. These solutions are illustrated in Figure 1. However, if we substitute the values of x back into the equation $x^2 + y^2 = 5$, we obtain

$$x = 1 \qquad\qquad\qquad x = -\tfrac{2}{5}$$
$$1^2 + y^2 = 5 \qquad\quad (-\tfrac{2}{5})^2 + y^2 = 5$$
$$y^2 = 4 \qquad\qquad\qquad y^2 = \tfrac{121}{25}$$
$$y = \pm 2 \qquad\qquad\qquad y = \pm \tfrac{11}{5}$$

FIGURE 1

It appears that we have found two additional solutions, $(1, 2)$ and $(-\tfrac{2}{5}, -\tfrac{11}{5})$. But neither of these solutions satisfies the equation $3x + y = 1$, which you should verify. So, neither is a solution of the original system. We have produced two **extraneous roots,** apparent solutions that do not actually satisfy both equations in the system. This is a common occurrence when solving nonlinear systems.

It is always very important to check the solutions of any nonlinear system to ensure that extraneous roots have not been introduced.

Matched Problem 1   Solve the system: $x^2 + y^2 = 10$
                                              $2x + y = 1$

EXPLORE-DISCUSS 1   In Example 1, we saw that the line $3x + y = 1$ intersected the circle $x^2 + y^2 = 5$ in two points.

(A) Consider the system

$$x^2 + y^2 = 5$$
$$3x + y = 10$$

Graph both equations in the same coordinate system. Are there any real solutions to this system? Are there any complex solutions? Find any real or complex solutions.

(B) Consider the family of lines given by

$$3x + y = b \qquad b \text{ any real number}$$

What do all these lines have in common? Illustrate graphically the lines in this family that intersect the circle $x^2 + y^2 = 5$ in exactly one point. How many such lines are there? What are the corresponding value(s) of b? What are the intersection points? How are these lines related to the circle?

EXAMPLE 2   **Solving a Nonlinear System by Substitution**

Solve: $x^2 - 2y^2 = 2$
$xy = 2$

Solution   Solve the second equation for y, substitute in the first equation, and proceed as before.

$$xy = 2$$

$$y = \frac{2}{x}$$

$$x^2 - 2\left(\frac{2}{x}\right)^2 = 2$$

$$x^2 - \frac{8}{x^2} = 2$$

$$x^4 - 2x^2 - 8 = 0 \qquad \text{Multiply both sides by } x^2 \text{ and simplify.}$$

$$u^2 - 2u - 8 = 0 \qquad \text{Substitute } u = x^2 \text{ to transform to quadratic form and solve.}$$

$$(u - 4)(u + 2) = 0$$

$$u = 4, -2$$

Thus,

$$x^2 = 4 \qquad \text{or} \qquad x^2 = -2$$

$$x = \pm 2 \qquad\qquad x = \pm\sqrt{-2} = \pm i\sqrt{2}$$

$$\text{For } x = 2, y = \frac{2}{2} = 1. \qquad \text{For } x = i\sqrt{2}, y = \frac{2}{i\sqrt{2}} = -i\sqrt{2}.$$

$$\text{For } x = -2, y = \frac{2}{-2} = -1. \qquad \text{For } x = -i\sqrt{2}, y = \frac{2}{-i\sqrt{2}} = i\sqrt{2}.$$

FIGURE 2

Thus, the four solutions to this system are $(2, 1)$, $(-2, -1)$, $(i\sqrt{2}, -i\sqrt{2})$, and $(-i\sqrt{2}, i\sqrt{2})$. Notice that two of the solutions involve imaginary numbers. These imaginary solutions cannot be illustrated graphically (see Fig. 2); however, they do satisfy both equations in the system (verify this).

Matched Problem 2   Solve: $3x^2 - y^2 = 6$
$xy = 3$

EXPLORE-DISCUSS 2   (A) Refer to the system in Example 2. Could a graphing utility be used to find the real solutions of this system? The imaginary solutions?

(B) In general, explain why graphic approximation techniques can be used to approximate the real solutions of a system, but not the complex solutions.

EXAMPLE 3   **Design**

An engineer is to design a rectangular computer screen with a 19-inch diagonal and a 175-square-inch area. Find the dimensions of the screen to the nearest tenth of an inch.

Solution   Sketch a rectangle letting x be the width and y the height (Fig. 3). We obtain the following system using the Pythagorean theorem and the formula for the area of a rectangle:

$$x^2 + y^2 = 19^2$$

$$xy = 175$$

FIGURE 3

This system is solved using the procedures outlined in Example 2. However, in this case, we are only interested in real solutions. We start by solving the second equation for y in terms of x and substituting the result into the first equation.

$$y = \frac{175}{x}$$

$$x^2 + \frac{175^2}{x^2} = 19^2$$

$$x^4 + 30{,}625 = 361x^2 \quad \text{Multiply both sides by } x^2 \text{ and simplify}$$

$$x^4 - 361x^2 + 30{,}625 = 0 \qquad \text{Quadratic in } x^2$$

Solve the last equation for x^2 using the quadratic formula, then solve for x:

$$x = \sqrt{\frac{361 \pm \sqrt{361^2 - 4(1)(30{,}625)}}{2}}$$

$$= 15.0 \text{ inches or } 11.7 \text{ inches}$$

Substitute each choice of x into $y = 175/x$ to find the corresponding y values:

For $x = 15.0$ inches,   For $x = 11.7$ inches,

$$y = \frac{175}{15} = 11.7 \text{ inches}$$ $$y = \frac{175}{11.7} = 15.0 \text{ inches}$$

Assuming the screen is wider than it is high, the dimensions are 15.0 by 11.7 inches.

Matched Problem 3   An engineer is to design a rectangular television screen with a 21-inch diagonal and a 209-square-inch area. Find the dimensions of the screen to the nearest tenth of an inch.

Since Example 3 is only concerned with real solutions, graphic techniques can also be used to approximate the solutions (see Fig. 4). As we saw in Section 3-1, graphing a circle on a graphing utility requires two functions, one for the upper half of the circle and another for the lower half. [Note: since x and y must be nonnegative real numbers, we ignore the intersection points in the third quadrant—see Fig 4(a).]

FIGURE 4 Graphic solution of $x^2 + y^2 = 19^2$, $xy = 175$.

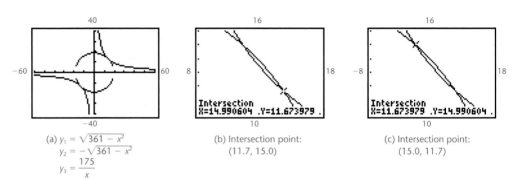

(a) $y_1 = \sqrt{361 - x^2}$
$y_2 = -\sqrt{361 - x^2}$
$y_3 = \dfrac{175}{x}$

(b) Intersection point:
(11.7, 15.0)

(c) Intersection point:
(15.0, 11.7)

● **Other Solution Methods**   We now look at some other techniques for solving nonlinear systems of equations.

EXAMPLE 4   **Solving a Nonlinear System by Elimination**

Solve: $x^2 - y^2 = 5$
         $x^2 + 2y^2 = 17$

Solution   This type of system can be solved using elimination by addition. Multiply the second equation by -1 and add:

$$
\begin{array}{r}
x^2 - y^2 = 5 \\
-x^2 - 2y^2 = -17 \\
\hline
-3y^2 = -12 \\
y^2 = 4 \\
y = \pm 2
\end{array}
$$

Now substitute $y = 2$ and $y = -2$ back into either original equation to find x.

For $y = 2$,   For $y = -2$,

$$x^2 - (2)^2 = 5 \qquad\qquad\qquad x^2 - (-2)^2 = 5$$
$$x = \pm 3 \qquad\qquad\qquad\qquad x = \pm 3$$

at the same time and follow the same route on the 75-mile trip across the English Channel to Cherbourg, France. The average speed of boat A is 5 miles per hour greater than the average speed of boat B. Consequently, boat A arrives at Cherbourg 30 minutes before boat B. Find the average speed of each boat.

★★ **58. Transportation.** Bus A leaves Milwaukee at noon and travels west on Interstate 94. Bus B leaves Milwaukee 30 minutes later, travels the same route, and overtakes bus A at a point 210 miles west of Milwaukee. If the average speed of bus B is 10 miles per hour greater than the average speed of bus A, at what time did bus B overtake bus A?

SECTION 6-4 Systems of Linear Inequalities in Two Variables

- Graphing Linear Inequalities in Two Variables
- Solving Systems of Linear Inequalities Graphically
- Application

Many applications of mathematics involve systems of inequalities rather than systems of equations. A graph is often the most convenient way to represent the solutions of a system of inequalities in two variables. In this section, we discuss techniques for graphing both a single linear inequality in two variables and a system of linear inequalities in two variables.

• Graphing Linear Inequalities in Two Variables

We know how to graph first-degree equations such as

$$y = 2x - 3 \quad \text{and} \quad 2x - 3y = 5$$

but how do we graph first-degree inequalities such as

$$y \leq 2x - 3 \quad \text{and} \quad 2x - 3y > 5$$

Actually, graphing these inequalities is almost as easy as graphing the equations. But before we begin, we must discuss some important subsets of a plane in a rectangular coordinate system.

A line divides a plane into two halves called **half-planes.** A vertical line divides a plane into **left** and **right half-planes** [Fig. 1(a)]; a nonvertical line divides a plane into **upper** and **lower half-planes** [Fig. 1(b)].

FIGURE 1 Half-planes.

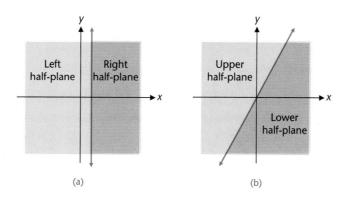

(a)

(b)

EXPLORE-DISCUSS 1 Consider the following linear equation and related linear inequalities:

(1) $2x - 3y = 12$ (2) $2x - 3y < 12$ (3) $2x - 3y > 12$

(A) Graph the line with equation (1).

(B) Find the point on this line with x coordinate 3 and draw a vertical line through this point. Discuss the relationship between the y coordinates of the points on this line and statements (1), (2), and (3).

(C) Repeat part B for $x = -3$. For $x = 9$.

(D) Based on your observations in parts B and C, write a verbal description of all the points in the plane that satisfy equation (1), those that satisfy inequality (2), and those that satisfy inequality (3).

Now let's investigate the half-planes determined by the linear equation $y = 2x - 3$. We start by graphing $y = 2x - 3$ (Fig. 2). For any given value of x, there is exactly one value for y such that (x, y) lies on the line. For the same x, if the point (x, y) is below the line, then $y < 2x - 3$. Thus, the lower half-plane corresponds to the solution of the inequality $y < 2x - 3$. Similarly, the upper half-plane corresponds to the solution of the inequality $y > 2x - 3$, as shown in Figure 2.

FIGURE 2

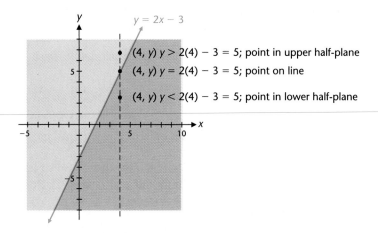

The four inequalities formed from $y = 2x - 3$ by replacing the $=$ sign by $\geq$, $>$, $\leq$, and $<$, respectively, are

$$y \geq 2x - 3 \qquad y > 2x - 3 \qquad y \leq 2x - 3 \qquad y < 2x - 3$$

The graph of each is a half-plane. The line $y = 2x - 3$, called the **boundary line** for the half-plane, is included for $\geq$ and $\leq$ and excluded for $>$ and $<$. In Figure 3, the half-planes are indicated with small arrows on the graph of $y = 2x - 3$ and then graphed as shaded regions. Included boundary lines are shown as solid lines, and excluded boundary lines are shown as dashed lines.

FIGURE 3

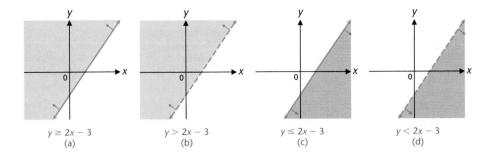

$y \geq 2x - 3$
(a)

$y > 2x - 3$
(b)

$y \leq 2x - 3$
(c)

$y < 2x - 3$
(d)

Theorem 1   **Graphs of Linear Inequalities in Two Variables**

The graph of a linear inequality

$$Ax + By < C \qquad \text{or} \qquad Ax + By > C$$

with $B \neq 0$, is either the upper half-plane or the lower half-plane (but not both) determined by the line $Ax + By = C$.

If $B = 0$, then the graph of

$$Ax < C \qquad \text{or} \qquad Ax > C$$

is either the left half-plane or the right half-plane (but not both) determined by the line $Ax = C$.

As a consequence of Theorem 1, we state a simple and fast mechanical procedure for graphing linear inequalities.

Procedure for Graphing Linear Inequalities in Two Variables

Step 1. Graph $Ax + By = C$ as a dashed line if equality is not included in the original statement or as a solid line if equality is included.

Step 2. Choose a test point anywhere in the plane not on the line and substitute the coordinates into the inequality. The origin $(0, 0)$ often requires the least computation.

Step 3. The graph of the original inequality includes the half-plane containing the test point if the inequality is satisfied by that point, or the half-plane not containing that point if the inequality is not satisfied by that point.

EXAMPLE 1   **Graphing a Linear Inequality**

Graph: $3x - 4y \leq 12$

Solution

Step 1. Graph $3x - 4y = 12$ as a solid line, since equality is included in the original statement (Fig. 4).

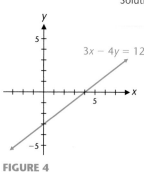

$3x - 4y = 12$

FIGURE 4

Step 2. Pick a convenient test point above or below the line. The origin $(0, 0)$ requires the least computation. Substituting $(0, 0)$ into the inequality

$$3x - 4y \leq 12$$

$$3(0) - 4(0) = 0 \leq 12$$

produces a true statement; therefore, $(0, 0)$ is in the solution set.

Step 3. The line $3x - 4y = 12$ and the half-plane containing the origin form the graph of $3x - 4y \leq 12$ (Fig. 5).

FIGURE 5

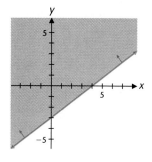

Matched Problem 1   Graph: $2x + 3y < 6$

EXAMPLE 2   **Graphing a Linear Inequality**

Graph: (A) $y > -3$ (B) $2x \leq 5$

Solutions   (A) The graph of $y > -3$ is shown in Figure 6.

(B) The graph of $2x \leq 5$ is shown in Figure 7.

FIGURE 6

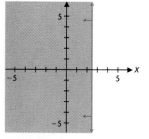

FIGURE 7

Matched Problem 2   Graph: (A) $y \leq 2$ (B) $3x > -8$

• **Solving Systems of Linear Inequalities Graphically**

We now consider systems of linear inequalities such as

$$x + y \geq 6 \qquad \text{and} \qquad 2x + y \leq 22$$
$$2x - y \geq 0 \qquad\qquad\qquad\quad x + y \leq 13$$
$$\qquad\qquad\qquad\qquad\qquad\qquad 2x + 5y \leq 50$$
$$\qquad\qquad\qquad\qquad\qquad\qquad x \geq 0$$
$$\qquad\qquad\qquad\qquad\qquad\qquad y \geq 0$$

We wish to **solve** such systems **graphically**—that is, to find the graph of all ordered pairs of real numbers (x, y) that simultaneously satisfy all the inequalities in the system. The graph is called the **solution region** for the system. To find the solution region, we graph each inequality in the system and then take the intersection of all the graphs. To simplify the discussion that follows, **we will consider only systems of linear inequalities where equality is included in each statement in the system.**

EXAMPLE 3   **Solving a System of Linear Inequalities Graphically**

Solve the following system of linear inequalities graphically:

$$x + y \geq 6$$
$$2x - y \geq 0$$

Solution   First, graph the line $x + y = 6$ and shade the region that satisfies the inequality $x + y \geq 6$. This region is shaded in blue in Figure 8(a). Next, graph the line $2x - y = 0$ and shade the region that satisfies the inequality $2x - y \geq 0$. This region is shaded in red in Figure 8(a). The solution region for the system of inequalities is the intersection of these two regions. This is the region shaded in both red and blue in Figure 8(a), which is redrawn in Figure 8(b) with only the solution region shaded for clarity. The coordinates of any point in the shaded region of Figure 8(b) specify a solution to the system. For example, the points $(2, 4)$, $(6, 3)$, and $(7.43, 8.56)$ are three of infinitely many solutions, as can be easily checked. The intersection point $(2, 4)$ can be obtained by solving the equations $x + y = 6$ and $2x - y = 0$ simultaneously.

FIGURE 8

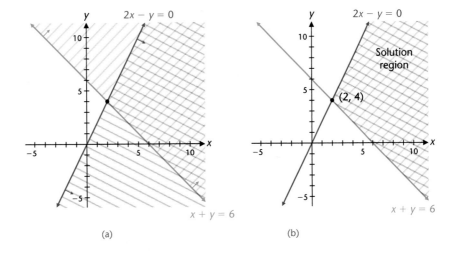

(a)   (b)

Matched Problem 3   Solve the following system of linear inequalities graphically: $\begin{aligned} 3x + y &\le 21 \\ x - 2y &\le 0 \end{aligned}$

EXPLORE-DISCUSS 2   Refer to Example 3. Graph each boundary line and shade the regions obtained by reversing each inequality. That is, shade the region of the plane that corresponds to the inequality $x + y < 6$ and then shade the region that corresponds to the inequality $2x - y < 0$. What portion of the plane is left unshaded? Compare this method with the one used in the solution to Example 3.

The points of intersection of the lines that form the boundary of a solution region play a fundamental role in the solution of linear programming problems, which are discussed in the next section.

Definition 1   **Corner Point**

A **corner point** of a solution region is a point in the solution region that is the intersection of two boundary lines.

The point $(2, 4)$ is the only corner point of the solution region in Example 3; see figure (b).

EXAMPLE 4   **Solving a System of Linear Inequalities Graphically**

Solve the following system of linear inequalities graphically, and find the corner points.

$$2x + y \le 22$$
$$x + y \le 13$$
$$2x + 5y \le 50$$
$$x \ge 0$$
$$y \ge 0$$

Solution   The inequalities $x \ge 0$ and $y \ge 0$, called **nonnegative restrictions,** occur frequently in applications involving systems of inequalities since x and y often represent quantities that can't be negative—number of units produced, number of hours worked, etc. The solution region lies in the first quadrant, and we can restrict our attention to that portion of the plane. First, we graph the lines

$$2x + y = 22 \qquad \text{Find the } x \text{ and } y \text{ intercepts of each line; then sketch}$$
the line through these points, as shown in Figure 9.

$$x + y = 13$$
$$2x + 5y = 50$$

Next, choosing (0, 0) as a test point, we see that the graph of each of the first three inequalities in the system consists of its corresponding line and the half-plane lying below the line, as indicated by the arrows in Figure 9. Thus, the solution region of the system consists of the points in the first quadrant that simultaneously lie on or below all three of these lines—see Figure 9.

FIGURE 9

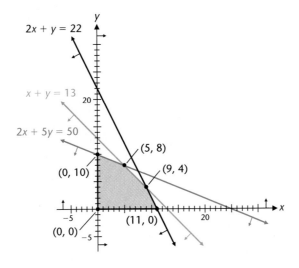

The corner points (0, 0), (0, 10), and (11, 0) can be determined from the graph. The other two corner points are determined as follows:

Solve the system   Solve the system

$$2x + 5y = 50$$ $$2x + y = 22$$
$$y + y = 13$$ $$x + y = 13$$

to obtain (5, 8).   to obtain (9, 4).

Note that the lines $2x + 5y = 50$ and $2x + y = 22$ also intersect, but the intersection point is not part of the solution region, and hence, is not a corner point.

Matched Problem 4   Solve the following system of linear inequalities graphically, and find the corner points:

$$5x + y \geq 20$$
$$x + y \geq 12$$
$$x + 3y \geq 18$$
$$x \geq 0$$
$$y \geq 0$$

As we saw in Section 6-1, a graphing utility is a useful tool for graphing lines and finding intersection points. Most graphing utilities also have some limited capabilities for shading regions that satisfy inequalities. Figure 10(a) shows a solution for Example 3 that could be produced on most graphing utilities. On newer models (for example, the TI-83 and TI-96), it is a simple operation to shade the region above or below a graph. Using this option to shade the points that do not satisfy a given inequality, as discussed in Explore–Discuss 2, clearly displays the solution to a system of inequalities. Figure 10(b) shows the result of applying this method to Example 4.

FIGURE 10

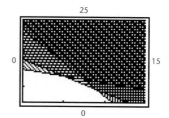

(a) Shaded region is the solution of the system in Example 3 [see Fig. 8(b)].

(b) Unshaded region is the solution of the system in Example 4 (see Fig. 9).

If we compare the solution regions of Examples 3 and 4, we see that there is a fundamental difference between these two regions. We can draw a circle around the solution region in Example 4. However, it is impossible to include all the points in the solution region in Example 3 in any circle, no matter how large we draw it. This leads to the following definition.

DEFINITION 2   Bounded and Unbounded Solution Regions

A solution region of a system of linear inequalities is **bounded** if it can be enclosed within a circle. If it cannot be enclosed within a circle, then it is **unbounded.**

Thus, the solution region for Example 4 is bounded and the solution region for Example 3 is unbounded. This definition will be important in the next section.

• Application

EXAMPLE 5   **Production Scheduling**

Sideview

Nose   Tail

Topview

A manufacturer of surfboards makes a standard model and a competition model. Each standard board requires 6 labor-hours for fabricating and 1 labor-hour for finishing. Each competition board requires 8 labor-hours for fabricating and 3 labor-hours for finishing. The maximum labor-hours available per week in the fabricating and finishing departments are 120 and 30, respectively. What combinations of boards can be produced each week so as not to exceed the number of labor-hours available in each department per week?

Solution   To clarify relationships, we summarize the information in the following table:

	Standard model (labor-hours per board)	Competition model (labor-hours per board)	Maximum labor-hours available per week
Fabricating	6	8	120
Finishing	1	3	30

Let

$$x = \text{Number of standard boards produced per week}$$
$$y = \text{Number of competition boards produced per week}$$

These variables are restricted as follows:

Fabricating department restriction:

$$\begin{pmatrix} \text{Weekly fabricating time} \\ \text{for } x \text{ standard boards} \end{pmatrix} + \begin{pmatrix} \text{Weekly fabricating time} \\ \text{for } y \text{ competition boards} \end{pmatrix} \leq \begin{pmatrix} \text{Maximum labor-hours} \\ \text{available per week} \end{pmatrix}$$
$$6x \qquad + \qquad 8y \qquad \leq \qquad 120$$

Finishing department restriction:

$$\begin{pmatrix} \text{Weekly finishing time} \\ \text{for } x \text{ standard boards} \end{pmatrix} + \begin{pmatrix} \text{Weekly finishing time} \\ \text{for } y \text{ competition boards} \end{pmatrix} \leq \begin{pmatrix} \text{Maximum labor-hours} \\ \text{available per week} \end{pmatrix}$$
$$1x \qquad + \qquad 3y \qquad \leq \qquad 30$$

Since it is not possible to manufacture a negative number of boards, x and y also must satisfy the nonnegative restrictions

$$x \geq 0$$
$$y \geq 0$$

Thus, x and y must satisfy the following system of linear inequalities:

$$6x + 8y \leq 120 \quad \text{Fabricating department restriction}$$
$$x + 3y \leq 30 \quad \text{Finishing department restriction}$$
$$x \geq 0 \quad \text{Nonnegative restriction}$$
$$y \geq 0 \quad \text{Nonnegative restriction}$$

Graphing this system of linear inequalities, we obtain the set of **feasible solutions,** or the **feasible region,** as shown in Figure 11. For problems of this type and for the linear programming problems we consider in the next section, solution regions are often referred to as feasible regions. Any point within the shaded area, including the boundary lines, represents a possible production schedule. Any point outside the shaded area represents an impossible schedule. For example, it would be possible to produce 12 standard boards and 5 competition boards per week, but it would not be possible to produce 12 standard boards and 7 competition boards per week (see the figure).

FIGURE 11

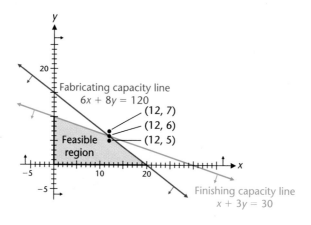

Matched Problem 5   Repeat Example 5 using 5 hours for fabricating a standard board and a maximum of 27 labor-hours for the finishing department.

Remark.   Refer to Example 5. How do we interpret a production schedule of 10.5 standard boards and 4.3 competition boards? It is not possible to manufacture a fraction of a board. But it is possible to *average* 10.5 standard and 4.3 competition boards per week. In general, we will assume that all points in the feasible region represent acceptable solutions, even though noninteger solutions might require special interpretation.

Answers to Matched Problems

1.

2. (A)

(B)

3.

4.

5.

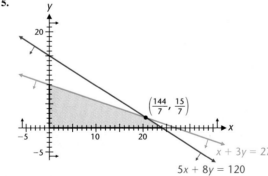

EXERCISE 6-4

A _____

Graph each inequality in Problems 1–10.

1. $2x - 3y < 6$ **2.** $3x + 4y < 12$ **3.** $3x + 2y \geq 18$

4. $3y - 2x \geq 24$ **5.** $y \leq \frac{2}{3}x + 5$ **6.** $y \geq \frac{1}{3}x - 2$

7. $y < 8$ **8.** $x > -5$ **9.** $-3 \leq y < 2$

10. $-1 < x \leq 3$

In Problems 11–14, match the solution region of each system of linear inequalities with one of the four regions shown in the figure below.

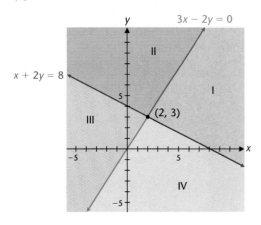

11. $x + 2y \leq 8$
   $3x - 2y \geq 0$

12. $x + 2y \geq 8$
   $3x - 2y \leq 0$

13. $x + 2y \geq 8$
   $3x - 2y \geq 0$

14. $x + 2y \leq 8$
   $3x - 2y \leq 0$

In Problems 15–20, solve each system of linear inequalities graphically.

15. $x \geq 5$
   $y \leq 6$

16. $x \leq 4$
   $y \geq 2$

17. $3x + y \geq 6$
   $x \leq 4$

18. $3x + 4y \leq 12$
   $y \geq -3$

19. $x - 2y \leq 12$
   $2x + y \geq 4$

20. $2x + 5y \leq 20$
   $x - 5y \leq -5$

B

In Problems 21–24, match the solution region of each system of linear inequalities with one of the four regions shown in the figure below. Identify the corner points of each solution region.

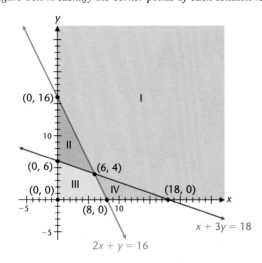

21. $x + 3y \leq 18$
   $2x + y \geq 16$
   $x \geq 0$
   $y \geq 0$

22. $x + 3y \leq 18$
   $2x + y \leq 16$
   $x \geq 0$
   $y \geq 0$

23. $x + 3y \geq 18$
   $2x + y \geq 16$
   $x \geq 0$
   $y \geq 0$

24. $x + 3y \geq 18$
   $2x + y \leq 16$
   $x \geq 0$
   $y \geq 0$

In Problems 25–36, solve the systems graphically, and indicate whether each solution region is bounded or unbounded. Find the coordinates of each corner point.

25. $2x + 3y \leq 6$
   $x \geq 0$
   $y \geq 0$

26. $4x + 3y \leq 12$
   $x \geq 0$
   $y \geq 0$

27. $4x + 5y \geq 20$
   $x \geq 0$
   $y \geq 0$

28. $5x + 6y \geq 30$
   $x \geq 0$
   $y \geq 0$

29. $2x + y \leq 8$
   $x + 3y \leq 12$
   $x \geq 0$
   $y \geq 0$

30. $x + 2y \leq 10$
   $3x + y \leq 15$
   $x \geq 0$
   $y \geq 0$

31. $4x + 3y \geq 24$
   $2x + 3y \geq 18$
   $x \geq 0$
   $y \geq 0$

32. $x + 2y \geq 8$
   $2x + y \geq 10$
   $x \geq 0$
   $y \geq 0$

33. $2x + y \leq 12$
   $x + y \leq 7$
   $x + 2y \leq 10$
   $x \geq 0$
   $y \geq 0$

34. $3x + y \leq 21$
   $x + y \leq 9$
   $x + 3y \leq 21$
   $x \geq 0$
   $y \geq 0$

35. $x + 2y \geq 16$
   $x + y \geq 12$
   $2x + y \geq 14$
   $x \geq 0$
   $y \geq 0$

36. $3x + y \geq 30$
   $x + y \geq 16$
   $x + 3y \geq 24$
   $x \geq 0$
   $y \geq 0$

C

In Problems 37–44, solve the systems graphically, and indicate whether each solution region is bounded or unbounded. Find the coordinates of each corner point.

37. $x + y \leq 11$
   $5x + y \geq 15$
   $x + 2y \geq 12$

38. $4x + y \leq 32$
   $x + 3y \leq 30$
   $5x + 4y \geq 51$

39. $3x + 2y \geq 24$
   $3x + y \leq 15$
   $x \geq 4$

40. $3x + 4y \leq 48$
   $x + 2y \geq 24$
   $y \leq 9$

41. $x + y \leq 10$
   $3x + 5y \geq 15$
   $3x - 2y \leq 15$
   $-5x + 2y \leq 6$

42. $3x - y \geq 1$
   $-x + 5y \geq 9$
   $x + y \leq 9$
   $y \leq 5$

43. $16x + 13y \leq 119$ **44.** $8x + 4y \leq 41$
$\quad\quad 12x + 16y \geq 101$ $\quad\quad\quad -15x + 5y \leq 19$
$\quad\quad -4x + 3y \leq 11$ $\quad\quad\quad2x + 6y \geq 37$

APPLICATIONS

45. Manufacturing—Resource Allocation. A manufacturing company makes two types of water skis: a trick ski and a slalom ski. The trick ski requires 6 labor-hours for fabricating and 1 labor-hour for finishing. The slalom ski requires 4 labor-hours for fabricating and 1 labor-hour for finishing. The maximum labor-hours available per day for fabricating and finishing are 108 and 24, respectively. If x is the number of trick skis and y is the number of slalom skis produced per day, write a system of inequalities that indicates appropriate restraints on x and y. Find the set of feasible solutions graphically for the number of each type of ski that can be produced.

46. Manufacturing—Resource Allocation. A furniture manufacturing company manufactures dining room tables and chairs. A table requires 8 labor-hours for assembling and 2 labor-hours for finishing. A chair requires 2 labor-hours for assembling and 1 labor-hour for finishing. The maximum labor-hours available per day for assembly and finishing are 400 and 120, respectively. If x is the number of tables and y is the number of chairs produced per day, write a system of inequalities that indicates appropriate restraints on x and y. Find the set of feasible solutions graphically for the number of tables and chairs that can be produced.

★ **47. Manufacturing—Resource Allocation.** Refer to Problem 45. The company makes a profit of $50 on each trick ski and a profit of $60 on each slalom ski.
   (A) If the company makes 10 trick and 10 slalom skis per day, the daily profit will be $1,100. Are there other feasible production schedules that will result in a daily profit of $1,100? How are these schedules related to the graph of the line $50x + 60y = 1,100$?
   (B) Find a feasible production schedule that will produce a daily profit greater than $1,100 and repeat part A for this schedule.
   (C) Discuss methods for using lines like those in parts A and B to find the largest possible daily profit.

★ **48. Manufacturing—Resource Allocation.** Refer to Problem 46. The company makes a profit of $50 on each table and a profit of $15 on each chair.
   (A) If the company makes 20 tables and 20 chairs per day, the daily profit will be $1,300. Are there other feasible production schedules that will result in a daily profit of $1,300? How are these schedules related to the graph of the line $50x + 15y = 1,300$?
   (B) Find a feasible production schedule that will produce a daily profit greater than $1,300 and repeat part A for this schedule.

   (C) Discuss methods for using lines like those in parts A and B to find the largest possible daily profit.

49. Nutrition—Plants. A farmer can buy two types of plant food, mix A and mix B. Each cubic yard of mix A contains 20 pounds of phosphoric acid, 30 pounds of nitrogen, and 5 pounds of potash. Each cubic yard of mix B contains 10 pounds of phosphoric acid, 30 pounds of nitrogen, and 10 pounds of potash. The minimum requirements are 460 pounds of phosphoric acid, 960 pounds of nitrogen, and 220 pounds of potash. If x is the number of cubic yards of mix A used and y is the number of cubic yards of mix B used, write a system of inequalities that indicates appropriate restraints on x and y. Find the set of feasible solutions graphically for the amount of mix A and mix B that can be used.

50. Nutrition. A dietitian in a hospital is to arrange a special diet using two foods. Each ounce of food M contains 30 units of calcium, 10 units of iron, and 10 units of vitamin A. Each ounce of food N contains 10 units of calcium, 10 units of iron, and 30 units of vitamin A. The minimum requirements in the diet are 360 units of calcium, 160 units of iron, and 240 units of vitamin A. If x is the number of ounces of food M used and y is the number of ounces of food N used, write a system of linear inequalities that reflects the conditions indicated. Find the set of feasible solutions graphically for the amount of each kind of food that can be used.

51. Sociology. A city council voted to conduct a study on inner-city community problems. A nearby university was contacted to provide sociologists and research assistants. Each sociologist will spend 10 hours per week collecting data in the field and 30 hours per week analyzing data in the research center. Each research assistant will spend 30 hours per week in the field and 10 hours per week in the research center. The minimum weekly labor-hour requirements are 280 hours in the field and 360 hours in the research center. If x is the number of sociologists hired for the study and y is the number of research assistants hired for the study, write a system of linear inequalities that indicates appropriate restrictions on x and y. Find the set of feasible solutions graphically.

52. Psychology. In an experiment on conditioning, a psychologist uses two types of Skinner (conditioning) boxes with mice and rats. Each mouse spends 10 minutes per day in box A and 20 minutes per day in box B. Each rat spends 20 minutes per day in box A and 10 minutes per day in box B. The total maximum time available per day is 800 minutes for box A and 640 minutes for box B. We are interested in the various numbers of mice and rats that can be used in the experiment under the conditions stated. If x is the number of mice used and y is the number of rats used, write a system of linear inequalities that indicates appropriate restrictions on x and y. Find the set of feasible solutions graphically.

SECTION 6-5 Linear Programming

- A Linear Programming Problem
- Linear Programming—A General Description
- Application

Several problems in Section 6-4 are related to the general type of problems called *linear programming problems.* Linear programming is a mathematical process that has been developed to help management in decision making, and it has become one of the most widely used and best-known tools of management science and industrial engineering. We will use an intuitive graphical approach based on the techniques discussed in Section 6-4 to illustrate this process for problems involving two variables.

The American mathematician George B. Dantzig (1914–   ) formulated the first linear programming problem in 1947 and introduced a solution technique, called the *simplex method,* that does not rely on graphing and is readily adaptable to computer solutions. Today, it is quite common to use a computer to solve applied linear programming problems involving thousands of variables and thousands of inequalities.

• A Linear Programming Problem

We begin our discussion with an example that will lead to a general procedure for solving linear programming problems in two variables.

EXAMPLE 1   Production Scheduling

A manufacturer of fiberglass camper tops for pickup trucks makes a compact model and a regular model. Each compact top requires 5 hours from the fabricating department and 2 hours from the finishing department. Each regular top requires 4 hours from the fabricating department and 3 hours from the finishing department. The maximum labor-hours available per week in the fabricating department and the finishing department are 200 and 108, respectively. If the company makes a profit of $40 on each compact top and $50 on each regular top, how many tops of each type should be manufactured each week to maximize the total weekly profit, assuming all tops can be sold? What is the maximum profit?

Solution   This is an example of a linear programming problem. To see relationships more clearly, we summarize the manufacturing requirements, objectives, and restrictions in the table:

	Compact model (labor-hours per top)	Regular model (labor-hours per top)	Maximum labor-hours available per week
Fabricating	5	4	200
Finishing	2	3	108
Profit per top	$40	$50	

Chapter 6 Review

6-1   SYSTEMS OF LINEAR EQUATIONS IN TWO VARIABLES; AUGMENTED MATRICES

A system of two linear equations with two variables is a system of the form

$$ax + by = h$$
$$cx + dy = k \tag{1}$$

where x and y are variables, a, b, c, and d are real numbers called the **coefficients** of x and y, and h and k are real numbers called the **constant terms** in the equations. The ordered pair of numbers (x_0, y_0) is a **solution** to system (1) if each equation is satisfied by the pair. The set of all such ordered pairs of numbers is called the **solution set** for the system. To **solve** a system is to find its solution set.

In general, a system of linear equations has exactly one solution, no solution, or infinitely many solutions. A system of linear equations is **consistent** if it has one or more solutions and **inconsistent** if no solutions exist. A consistent system is said to be **independent** if it has exactly one solution and **dependent** if it has more than one solution.

Two standard methods for solving system (1) were reviewed: **solution by graphing** and **solution using elimination by addition.**

Two systems of equations are **equivalent** if both have the same solution set. A system of linear equations is transformed into an equivalent system if:

1. Two equations are interchanged.

2. An equation is multiplied by a nonzero constant.

3. A constant multiple of another equation is added to a given equation.

These operations form the basis of solution using elimination by addition. The method of solution using elimination by addition can be transformed into a more efficient method for larger-scale systems by the introduction of an *augmented matrix.* A **matrix** is a rectangular array of numbers written within brackets. Each number in a matrix is called an **element** of the matrix. If a matrix has m rows and n columns, it is called an $m \times n$ **matrix** (read "m by n matrix"). The expression $m \times n$ is called the **size** of the matrix, and the numbers m and n are called the **dimensions** of the matrix. A matrix with n rows and n columns is called a **square matrix of order n.** A matrix with only one column is called a **column matrix,** and a matrix with only one row is called a **row matrix.** The **position** of an element in a matrix is the row and column containing the element. This is usually denoted using **double subscript notation** a_{ij}, where i is the row and j is the column containing the element a_{ij}.

For ease of generalization to larger systems, we change the notation for variables and constants in system (1) to a subscript form:

$$a_1 x_1 + b_1 x_2 = k_1$$
$$a_2 x_1 + b_2 x_2 = k_2 \tag{2}$$

Associated with each linear system of the form (2), where x_1 and x_2 are variables, is the **augmented matrix** of the system:

$$
\begin{array}{c}
\quad\quad\text{Column 1 } (C_1) \\
\quad\quad\quad\text{Column 2 } (C_2) \\
\quad\quad\quad\quad\text{Column 3 } (C_3) \\
\begin{bmatrix} a_1 & b_1 & k_1 \\ a_2 & b_2 & k_2 \end{bmatrix}
\begin{array}{l} \leftarrow \text{Row 1 } (R_1) \\ \leftarrow \text{Row 2 } (R_2) \end{array}
\end{array} \tag{3}
$$

Two augmented matrices are **row-equivalent,** denoted by the symbol $\sim$ between the two matrices, if they are augmented matrices of equivalent systems of equations. An augmented matrix is transformed into a row-equivalent matrix if any of the following **row operations** is performed:

1. Two rows are interchanged.

2. A row is multiplied by a nonzero constant.

3. A constant multiple of another row is added to a given row.

The following symbols are used to describe these row operations:

1. $R_i \leftrightarrow R_j$ means "interchange row i with row j."

2. $kR_i \rightarrow R_i$ means "multiply row i by the constant k."

3. $kR_j + R_i \rightarrow R_i$ means "multiply row j by the constant k and add to R_i."

In solving system (2) using row operations, the objective is to transform the augmented matrix (3) into the form

$$\begin{bmatrix} 1 & 0 & m \\ 0 & 1 & n \end{bmatrix}$$

If this can be done, then (m, n) is the unique solution of system (2). If (3) is transformed into the form

$$\begin{bmatrix} 1 & m & n \\ 0 & 0 & 0 \end{bmatrix}$$

then system (2) has infinitely many solutions. If (3) is transformed into the form

$$\begin{bmatrix} 1 & m & n \\ 0 & 0 & p \end{bmatrix} \quad p \neq 0$$

then system (2) does not have a solution.

6-2 GAUSS–JORDAN ELIMINATION

In the last part of Section 6-1 we were actually using *Gauss–Jordan elimination* to solve a system of two equations with two variables. The method generalizes completely for systems with more than two variables, and the number of variables does not have to be the same as the number of equations.

As before, our objective is to start with the augmented matrix of a linear system and transform it using row operations into a simple form where the solution can be read by inspection. The simple form, called the **reduced form,** is achieved if:

1. Each row consisting entirely of 0's is below any row having at least one nonzero element.

2. The leftmost nonzero element in each row is 1.

3. The column containing the leftmost 1 of a given row has 0's above and below the 1.

4. The leftmost 1 in any row is to the right of the leftmost 1 in the preceding row.

A **reduced system** is a system of linear equations that corresponds to a reduced augmented matrix. When a reduced system has more variables than equations and contains no contradictions, the system is dependent and has infinitely many solutions.

The **Gauss–Jordan elimination** procedure for solving a system of linear equations is given in step-by-step form as follows:

Step 1. Choose the leftmost nonzero column, and use appropriate row operations to get a 1 at the top.

Step 2. Use multiples of the row containing the 1 from step 1 to get zeros in all remaining places in the column containing this 1.

Step 3. Repeat step 1 with the **submatrix** formed by (mentally) deleting the row used in step 2 and all rows above this row.

Step 4. Repeat step 2 with the **entire matrix,** including the mentally deleted rows. Continue this process until it is impossible to go further.

If at any point in the above process we obtain a row with all 0's to the left of the vertical line and a nonzero number n to the right, we can stop, since we have a contradiction: $0 = n, n \neq 0$. We can then conclude that the system has no solution. If this does not happen and we obtain an augmented matrix in reduced form without any contradictions, the solution can be read by inspection.

6-3 SYSTEMS INVOLVING SECOND-DEGREE EQUATIONS

If a system of equations contains any equations that are not linear, then the system is called a **nonlinear system.** In this section we investigated nonlinear systems involving second-degree terms such as

$$x^2 + y^2 = 5 \qquad x^2 - 2y^2 = 2 \qquad x^2 + 3xy + y^2 = 20$$
$$3x + y = 1 \qquad xy = 2 \qquad xy - y^2 = 0$$

It can be shown that such systems have at most four solutions, some of which may be imaginary.

Several methods were used to solve nonlinear systems of the indicated form: **solution by substitution, solution using elimination by addition,** and **solution using factoring and substitution.** It is always important to check the solutions of any nonlinear system to ensure that extraneous roots have not been introduced.

6-4 SYSTEMS OF LINEAR INEQUALITIES IN TWO VARIABLES

A graph is often the most convenient way to represent the solution of a linear inequality in two variables or of a system of linear inequalities in two variables.

A vertical line divides a plane into **left** and **right half-planes.** A nonvertical line divides a plane into **upper** and **lower half-planes.** Let A, B, and C be real numbers with A and B not both zero, then the **graph of the linear inequality**

$$Ax + By < C \qquad \text{or} \qquad Ax + By > C$$

with $B \neq 0$, is either the upper half-plane or the lower half-plane (but not both) determined by the line $Ax + By = C$. If $B = 0$, then the graph of

$$Ax < C \qquad \text{or} \qquad Ax > C$$

is either the left half-plane or the right half-plane (but not both) determined by the line $Ax = C$. Out of these results follows an easy **step-by-step procedure for graphing a linear inequality in two variables:**

Step 1. Graph $Ax + By = C$ as a broken line if equality is not included in the original statement or as a solid line if equality is included.

Step 2. Choose a test point anywhere in the plane not on the line and substitute the coordinates into the inequality. The origin $(0, 0)$ often requires the least computation.

Step 3. The graph of the original inequality includes the half-plane containing the test point if the inequality is satisfied by that point, or the half-plane not containing that point if the inequality is not satisfied by that point.

We now turn to systems of linear inequalities in two variables. The **solution to a system of linear inequalities in two variables** is the set of all ordered pairs of real numbers that simultaneously satisfy all the inequalities in the system. The graph is called the **solution region.** In many applications the solution region is also referred to as the **feasible region. To find the solution region,** we graph each inequality in the system and then take the intersection of all the graphs. A **corner point** of a solution region is a point in the solution region that is the intersection of two boundary lines. A solution region is **bounded** if it can be enclosed within a circle. If it cannot be enclosed within a circle, then it is **unbounded.**

6-5 LINEAR PROGRAMMING

Linear programming is a mathematical process that has been developed to help management in decision making, and it has become one of the most widely used and best-known tools of management science and industrial engineering.

A **linear programming problem** is one that is concerned with finding the **optimal value** (maximum or minimum value) of a linear **objective function** of the form $z = ax + by$, where the **decision variables** x and y are subject to **problem constraints** in the form of linear inequalities and **nonnegative constraints** $x, y \geq 0$. The set of points satisfying both the problem constraints and the nonnegative constraints is called the **feasible region** for the problem. Any point in the feasible region that produces the optimal value of the objective function over the feasible region is called an **optimal solution. The fundamental theorem of linear programming** is basic to the solving of linear programming problems: Let S be the feasible region for a linear programming problem, and let $z = ax + by$ be the objective function. If S is bounded, then z has both a maximum and a minimum value on S and each of these occurs at a corner point of S. If S is unbounded, then a maximum or minimum value of z on S may not exist. However, if either does exist, then it must occur at a corner point of S.

Problems with unbounded feasible regions are not considered in this brief introduction. The theorem leads to a simple **step-by-step solution to linear programming problems with a bounded feasible region:**

Step 1. Form a mathematical model for the problem:
    (A) Introduce decision variables and write a linear objective function.
    (B) Write problem constraints in the form of linear inequalities.
    (C) Write nonnegative constraints.

Step 2. Graph the feasible region and find the corner points.

Step 3. Evaluate the objective function at each corner point to determine the optimal solution.

If two corner points are both optimal solutions of the same type (both produce the same maximum value or both produce the same minimum value) to a linear programming problem, then any point on the line segment joining the two corner points is also an optimal solution of that type.

Chapter 6 Review Exercise

Work through all the problems in this chapter review and check answers in the back of the book. Answers to all review problems are there, and following each answer is a number in italics indicating the section in which that type of problem is discussed. Where weaknesses show up, review appropriate sections in the text.

A

Solve Problems 1–6 using elimination by addition.

1. $2x + y = 7$
$3x - 2y = 0$

2. $3x - 6y = 5$
$-2x + 4y = 1$

3. $4x - 3y = -8$
$-2x + \frac{3}{2}y = 4$

4. $y = x^2 - 5x - 3$
$y = -x + 2$

5. $x^2 + y^2 = 2$
$2x - y = 3$

6. $3x^2 - y^2 = -6$
$2x^2 + 3y^2 = 29$

Solve Problems 7–9 by graphing.

7. $3x - 2y = 8$
$x + 3y = -1$

8. $3x - 4y \geq 24$

9. $2x + y \leq 2$
$x + 2y \geq -2$

Perform each of the row operations indicated in Problems 10–12 on the following augmented matrix:

$$\begin{bmatrix} 1 & -4 & | & 5 \\ 3 & -6 & | & 12 \end{bmatrix}$$

10. $R_1 \leftrightarrow R_2$

11. $\frac{1}{3}R_2 \rightarrow R_2$

12. $(-3)R_1 + R_2 \rightarrow R_2$

In Problems 13–15, write the linear system corresponding to each reduced augmented matrix and solve.

13. $\begin{bmatrix} 1 & 0 & | & 4 \\ 0 & 1 & | & -7 \end{bmatrix}$

14. $\begin{bmatrix} 1 & -1 & | & 4 \\ 0 & 0 & | & 1 \end{bmatrix}$

15. $\begin{bmatrix} 1 & -1 & | & 4 \\ 0 & 0 & | & 0 \end{bmatrix}$

16. Find the maximum and minimum values of $z = 5x + 3y$ over the feasible region S shown in the figure.

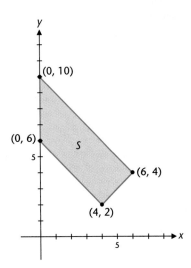

17. Use Gauss–Jordan elimination to solve the system

$$x_1 - x_2 = 4$$
$$2x_1 + x_2 = 2$$

Then write the linear system represented by each augmented matrix in your solution, and solve each of these systems graphically. Discuss the relationship between the solutions of these systems.

 18. Use an intersection routine on a graphing utility to approximate the solution of the following system to two decimal places:

$$x + 3y = 9$$
$$-2x + 7y = 10$$

B _____

Solve Problems 19–24 using Gauss–Jordan elimination.

19. $3x_1 + 2x_2 = 3$
$x_1 + 3x_2 = 8$

20. $x_1 + x_2 = 1$
$x_1 - x_3 = -2$
$x_2 + 2x_3 = 4$

21. $x_1 + 2x_2 + 3x_3 = 1$
$2x_1 + 3x_2 + 4x_3 = 3$
$x_1 + 2x_2 + x_3 = 3$

22. $x_1 + 2x_2 - x_3 = 2$
$2x_1 + 3x_2 + x_3 = -3$
$3x_1 + 5x_2 = -1$

23. $x_1 - 2x_2 = 1$
$2x_1 - x_2 = 0$
$x_1 - 3x_2 = -2$

24. $x_1 + 2x_2 - x_3 = 2$
$3x_1 - x_2 + 2x_3 = -3$

Solve Problems 25–27.

25. $x^2 - y^2 = 2$
$y^2 = x$

26. $x^2 + 2xy + y^2 = 1$
$xy = -2$

27. $2x^2 + xy + y^2 = 8$
$x^2 - y^2 = 0$

Solve the systems in Problems 28–30 graphically, and indicate whether each solution region is bounded or unbounded. Find the coordinates of each corner point.

28. $2x + y \le 8$
$2x + 3y \le 12$
$x, y \ge 0$

29. $2x + y \ge 8$
$x + 3y \ge 12$
$x, y \ge 0$

30. $x + y \le 20$
$x + 4y \ge 20$
$x - y \ge 0$

Solve the linear programming problems in Problems 31–33.

31. Maximize $z = 7x + 9y$
Subject to $x + 2y \le 8$
$2x + y \le 10$
$x, y \ge 0$

32. Minimize $z = 5x + 10y$
Subject to $x + y \le 20$
$3x + y \ge 15$
$x + 2y \ge 15$
$x, y \ge 0$

33. Minimize and maximize $z = 5x + 8y$
Subject to $x + 2y \le 20$
$3x + y \le 15$
$x + y \ge 7$
$x, y \ge 0$

C _____

34. Solve using Gauss–Jordan elimination:

$$x_1 + x_2 + x_3 = 7{,}000$$
$$0.04x_1 + 0.05x_2 + 0.06x_3 = 360$$
$$0.04x_1 + 0.05x_2 - 0.06x_3 = 120$$

35. Solve:

$$x^2 - xy + y^2 = 4$$
$$x^2 + xy - 2y^2 = 0$$

36. Maximize $z = 30x + 20y$
Subject to $1.2x + 0.6y \le 960$
$0.04x + 0.03y \le 36$
$0.2x + 0.3y \le 270$
$x, y \ge 0$

37. Approximate all real solutions to two decimal places:

$$x^2 + 4xy + y^2 = 8$$
$$5x^2 + 2xy + y^2 = 25$$

38. Discuss the number of solutions for the system corresponding to the reduced form shown below if

    (A) $m \neq 0$ (B) $m = 0$ and $n \neq 0$
    (C) $m = 0$ and $n = 0$

    $$\begin{bmatrix} 1 & 0 & -3 & | & 4 \\ 0 & 1 & 2 & | & 5 \\ 0 & 0 & m & | & n \end{bmatrix}$$

APPLICATIONS

39. **Business.** A container holds 120 packages. Some of the packages weigh $\frac{1}{2}$ pound each, and the rest weigh $\frac{1}{3}$ pound each. If the total contents of the container weigh 48 pounds, how many are there of each type of package? Solve using two-equation–two-variable methods.

★ 40. **Geometry.** Find the dimensions of a rectangle with an area of 48 square meters and a perimeter of 28 meters. Solve using two-equation–two-variable methods.

★ 41. **Diet.** A laboratory assistant wishes to obtain a food mix that contains, among other things, 27 grams of protein, 5.4 grams of fat, and 19 grams of moisture. He has available mixes A, B, and C with the compositions listed in the table. How many grams of each mix should be used to get the desired diet mix? Set up a system of equations and solve using Gauss–Jordan elimination.

Mix	Protein (%)	Fat (%)	Moisture (%)
A	30	3	10
B	20	5	20
C	10	4	10

★★ 42. **Puzzle.** A piggy bank contains 30 coins worth $1.90.
    (A) If the bank contains only nickels and dimes, how many coins of each type does it contain?
    (B) If the bank contains nickels, dimes, and quarters, how many coins of each type does it contain?

★ 43. **Resource Allocation.** North Star Sail Loft manufactures regular and competition sails. Each regular sail takes 1 labor-hour to cut and 3 labor-hours to sew. Each competition sail takes 2 labor-hours to cut and 4 labor-hours to sew. There are 140 labor-hours available in the cutting department and 360 labor-hours available in the sewing department.
    (A) If the loft makes a profit of $60 on each regular sail and $100 on each competition sail, how many sails of each type should the company manufacture to maximize its profit? What is the maximum profit?
    (B) An increase in the demand for competition sails causes the profit on a competition sail to rise to $125. Discuss the effect of this change on the number of sails manufactured and on the maximum profit.
    (C) A decrease in the demand for competition sails causes the profit on a competition sail to drop to $75. Discuss the effect of this change on the number of sails manufactured and on the maximum profit.

★ 44. **Nutrition—Animals.** A special diet for laboratory animals is to contain at least 800 units of vitamins, at least 800 units of minerals, and at most 1,300 calories. There are two feed mixes available, mix A and mix B. A gram of mix A contains 5 units of vitamins, 2 units of minerals, and 4 calories. A gram of mix B contains 2 units of vitamins, 4 units of minerals, and 4 calories.
    (A) If mix A costs $0.07 per gram and mix B costs $0.04 per gram, how many grams of each mix should be used to satisfy the requirements of the diet at minimal cost? What is the minimal cost?
    (B) If the price of mix B decreases to $0.02 per gram, discuss the effect of this change on the solution in part A.
    (C) If the price of mix B increases to $0.15 per gram, discuss the effect of this change on the solution in part A.

MATRICES AND DETERMINANTS

CHAPTER 7

In this chapter we discuss matrices in more detail. In the first three sections we define and study some algebraic operations on matrices, including addition, multiplication, and inversion. The next three sections deal with the determinant of a matrix.

In the last chapter we used row operations and Gauss–Jordan elimination to solve systems of linear equations. Row operations play a prominent role in the development of several topics in this chapter. One consequence of our discussion will be the development of two additional methods for solving systems of linear equations: one method involves inverse matrices and the other determinants.

Matrices are both a very ancient and a very current mathematical concept. References to matrices and systems of equations can be found in Chinese manuscripts dating back to around 200 B.C. Over the years, mathematicians and scientists have found many applications of matrices. More recently, the advent of personal and large-scale computers has increased the use of matrices in a wide variety of applications. In 1979 Dan Bricklin and Robert Frankston introduced VisiCalc, the first electronic spreadsheet program for personal computers. Simply put, a *spreadsheet* is a computer program that allows the user to enter and manipulate numbers, often using matrix notation and operations. Spreadsheets were initially used by businesses in areas such as budgeting, sales projections, and cost estimation. However, many other applications have begun to appear. For example, a scientist can use a spreadsheet to analyze the results of an experiment, or a teacher can use one to record and average grades. There are even spreadsheets that can be used to help compute an individual's income tax.

SECTION **7-1** ## Matrices: Basic Operations

- Addition and Subtraction
- Multiplication of a Matrix by a Number
- Matrix Product

In Section 6-1 we introduced basic matrix terminology and solved systems of equations by performing row operations on augmented coefficient matrices. Matrices have many other useful applications and possess an interesting mathematical structure in their own right. As we will see, matrix addition and multiplication are similar to real number addition and multiplication in many respects, but there are some important differences. To help you understand the similarities and the differences, you should review the basic properties of real number operations discussed in Section 1-1.

● **Addition and Subtraction**   Before we can discuss arithmetic operations for matrices, we have to define equality for matrices. Two matrices are **equal** if they have the same size and their corresponding elements are equal. For example,

$$\underset{2 \times 3}{\begin{bmatrix} a & b & c \\ d & e & f \end{bmatrix}} = \underset{2 \times 3}{\begin{bmatrix} u & v & w \\ x & y & z \end{bmatrix}} \quad \text{if and only if} \quad \begin{matrix} a = u & b = v & c = w \\ d = x & e = y & f = z \end{matrix}$$

The **sum of two matrices** of the same size is a matrix with elements that are the sums of the corresponding elements of the two given matrices.

Addition is not defined for matrices of different sizes.

EXAMPLE 1 **Matrix Addition**

(A) $\begin{bmatrix} a & b \\ c & d \end{bmatrix} + \begin{bmatrix} w & x \\ y & z \end{bmatrix} = \begin{bmatrix} (a+w) & (b+x) \\ (c+y) & (d+z) \end{bmatrix}$

(B) $\begin{bmatrix} 2 & -3 & 0 \\ 1 & 2 & -5 \end{bmatrix} + \begin{bmatrix} 3 & 1 & 2 \\ -3 & 2 & 5 \end{bmatrix} = \begin{bmatrix} 5 & -2 & 2 \\ -2 & 4 & 0 \end{bmatrix}$

Matched Problem 1 Add: $\begin{bmatrix} 3 & 2 \\ -1 & -1 \\ 0 & 3 \end{bmatrix} + \begin{bmatrix} -2 & 3 \\ 1 & -1 \\ 2 & -2 \end{bmatrix}$

```
[A]
    [[2 -3 0 ]
     [1 2  -5]]
[B]
    [[3  1 2]
     [-3 2 5]]
[A]+[B]
    [[5  -2 2]
     [-2 4  0]]
```

FIGURE 1 Addition on a graphing calculator.

Graphing utilities can also be used to solve problems involving matrix operations. Figure 1 illustrates the solution to Example 1B on a graphing calculator.

Because we add two matrices by adding their corresponding elements, it follows from the properties of real numbers that matrices of the same size are commutative and associative relative to addition. That is, if A, B, and C are matrices of the same size, then

$$A + B = B + A \qquad \text{Commutative}$$

$$(A + B) + C = A + (B + C) \qquad \text{Associative}$$

A matrix with elements that are all 0's is called a **zero matrix.** For example, the following are zero matrices of different sizes:

$$\begin{bmatrix} 0 & 0 & 0 \end{bmatrix} \qquad \begin{bmatrix} 0 & 0 \\ 0 & 0 \end{bmatrix} \qquad \begin{bmatrix} 0 \\ 0 \\ 0 \\ 0 \end{bmatrix} \qquad \begin{bmatrix} 0 & 0 & 0 & 0 \\ 0 & 0 & 0 & 0 \\ 0 & 0 & 0 & 0 \end{bmatrix}$$

[*Note:* "0" may be used to denote the zero matrix of any size.]

The **negative of a matrix M,** denoted by $-M$, is a matrix with elements that are the negatives of the elements in M. Thus, if

$$M = \begin{bmatrix} a & b \\ c & d \end{bmatrix}$$

then

$$-M = \begin{bmatrix} -a & -b \\ -c & -d \end{bmatrix}$$

Note that $M + (-M) = 0$ (a zero matrix).

If A and B are matrices of the same size, then we define **subtraction** as follows:

$$A - B = A + (-B)$$

Thus, to subtract matrix B from matrix A, we simply subtract corresponding elements.

EXAMPLE 2   **Matrix Subtraction**

$$\begin{bmatrix} 3 & -2 \\ 5 & 0 \end{bmatrix} - \begin{bmatrix} -2 & 2 \\ 3 & 4 \end{bmatrix} = \begin{bmatrix} 3 & -2 \\ 5 & 0 \end{bmatrix} + \begin{bmatrix} 2 & -2 \\ -3 & -4 \end{bmatrix} = \begin{bmatrix} 5 & -4 \\ 2 & -4 \end{bmatrix}$$

Matched Problem 2   Subtract: $[2 \quad -3 \quad 5] - [3 \quad -2 \quad 1]$

● **Multiplication of a Matrix by a Number**   The **product of a number k and a matrix M,** denoted by kM, is a matrix formed by multiplying each element of M by k.

EXAMPLE 3   **Multiplication of a Matrix by a Number**

$$-2 \begin{bmatrix} 3 & -1 & 0 \\ -2 & 1 & 3 \\ 0 & -1 & -2 \end{bmatrix} = \begin{bmatrix} -6 & 2 & 0 \\ 4 & -2 & -6 \\ 0 & 2 & 4 \end{bmatrix}$$

Matched Problem 3   Find: $10 \begin{bmatrix} 1.3 \\ 0.2 \\ 3.5 \end{bmatrix}$

EXPLORE-DISCUSS 1   Multiplication of two numbers can be interpreted as repeated addition if one of the numbers is a positive integer. That is,

$$2a = a + a \qquad 3a = a + a + a \qquad 4a = a + a + a + a$$

and so on. Discuss this interpretation for the product of an integer k and a matrix M. Use specific examples to illustrate your remarks.

We now consider an application that uses various matrix operations.

EXAMPLE 4 **Sales and Commissions**

Ms. Fong and Mr. Petris are salespeople for a new car agency that sells only two models. August was the last month for this year's models, and next year's models were introduced in September. Gross dollar sales for each month are given in the following matrices:

| | AUGUST SALES | | | SEPTEMBER SALES | |
	Compact	Luxury		Compact	Luxury
Fong	$36,000	$72,000	= A	$144,000	$288,000	= B
Petris	$72,000	$0		$180,000	$216,000	

For example, Ms. Fong had $36,000 in compact sales in August and Mr. Petris had $216,000 in luxury car sales in September.

(A) What were the combined dollar sales in August and September for each salesperson and each model?
(B) What was the increase in dollar sales from August to September?
(C) If both salespeople receive a 3% commission on gross dollar sales, compute the commission for each salesperson for each model sold in September.

Solutions We use matrix addition for part A, matrix subtraction for part B, and multiplication of a matrix by a number for part C.

(A) $A + B = \begin{bmatrix} \$180,000 & \$360,000 \\ \$252,000 & \$216,000 \end{bmatrix} \begin{matrix} \text{Fong} \\ \text{Petris} \end{matrix}$
   (columns: Compact, Luxury)

(B) $B - A = \begin{bmatrix} \$108,000 & \$216,000 \\ \$108,000 & \$216,000 \end{bmatrix} \begin{matrix} \text{Fong} \\ \text{Petris} \end{matrix}$

(C) $0.03B = \begin{bmatrix} (0.03)(\$144,000) & (0.03)(\$288,000) \\ (0.03)(\$180,000) & (0.03)(\$216,000) \end{bmatrix}$
   (columns: Compact, Luxury)

$\qquad = \begin{bmatrix} \$4,320 & \$8,640 \\ \$5,400 & \$6,480 \end{bmatrix} \begin{matrix} \text{Fong} \\ \text{Petris} \end{matrix}$

Matched Problem 4   Repeat Example 4 with

$$A = \begin{bmatrix} \$72{,}000 & \$72{,}000 \\ \$36{,}000 & \$72{,}000 \end{bmatrix} \quad \text{and} \quad B = \begin{bmatrix} \$180{,}000 & \$216{,}000 \\ \$144{,}000 & \$216{,}000 \end{bmatrix}$$

Example 4 involved an agency with only two salespeople and two models. A more realistic problem might involve 20 salespeople and 15 models. Problems of this size are often solved with the aid of a spreadsheet on a personal computer. Figure 2 illustrates a computer spreadsheet solution for Example 4.

	A	B	C	D	E	F	G
1		Compact	Luxury	Compact	Luxury	Compact	Luxury
2		August Sales		September Sales		September Commissions	
3	Fong	$36,000	$72,000	$144,000	$288,000	$4,320	$8,640
4	Petris	$72,000	$0	$180,000	$216,000	$5,400	$6,480
5		Combined Sales		Sales Increases			
6	Fong	$180,000	$360,000	$108,000	$216,000		
7	Petris	$252,000	$216,000	$108,000	$216,000		

FIGURE 2

● **Matrix Product**   Now we are going to introduce a matrix multiplication that may at first seem rather strange. In spite of its apparent strangeness, this operation is well-founded in the general theory of matrices and, as we will see, is extremely useful in many practical problems.

Historically, matrix multiplication was introduced by the English mathematician Arthur Cayley (1821–1895) in studies of linear equations and linear transformations. In Section 7-3, you will see how matrix multiplication is central to the process of expressing systems of equations as matrix equations and to the process of solving matrix equations. Matrix equations and their solutions provide us with an alternate method of solving linear systems with the same number of variables as equations.

We start by defining the product of two special matrices, a row matrix and a column matrix.

DEFINITION 1   **Product of a Row Matrix and a Column Matrix**

The **product** of a $1 \times n$ row matrix and an $n \times 1$ column matrix is a 1×1 matrix given by

$$\underset{1 \times n}{[a_1 \quad a_2 \quad \cdots \quad a_n]} \; \underset{n \times 1}{\begin{bmatrix} b_1 \\ b_2 \\ \vdots \\ b_n \end{bmatrix}} = [a_1 b_1 + a_2 b_2 + \cdots + a_n b_n]$$

Note that the number of elements in the row matrix and in the column matrix must be the same for the product to be defined.

EXAMPLE 5   **Product of a Row Matrix and a Column Matrix**

$$[2 \quad -3 \quad 0]\begin{bmatrix} -5 \\ 2 \\ -2 \end{bmatrix} = [(2)(-5) + (-3)(2) + (0)(-2)]$$

$$= [-10 - 6 + 0] = [-16]$$

Matched Problem 5 $[-1 \quad 0 \quad 3 \quad 2]\begin{bmatrix} 2 \\ 3 \\ 4 \\ -1 \end{bmatrix} = ?$

Refer to Example 5. The distinction between the real number -16 and the 1×1 matrix $[-16]$ is a technical one, and it is common to see 1×1 matrices written as real numbers without brackets. In the work that follows, we will frequently refer to 1×1 matrices as real numbers and omit the brackets whenever it is convenient to do so.

EXAMPLE 6   **Production Scheduling**

A factory produces a slalom water ski that requires 4 labor-hours in the fabricating department and 1 labor-hour in the finishing department. Fabricating personnel receive $10 per hour, and finishing personnel receive $8 per hour. Total labor cost per ski is given by the product

$$[4 \quad 1]\begin{bmatrix} 10 \\ 8 \end{bmatrix} = [(4)(10) + (1)(8)] = [40 + 8] = [48] \text{ or } \$48 \text{ per ski}$$

Matched Problem 6   If the factory in Example 6 also produces a trick water ski that requires 6 labor-hours in the fabricating department and 1.5 labor-hours in the finishing department, write a product between appropriate row and column matrices that gives the total labor cost for this ski. Compute the cost.

We now use the product of a $1 \times n$ row matrix and an $n \times 1$ column matrix to extend the definition of matrix product to more general matrices.

DEFINITION 2

Matrix Product

If A is an $m \times p$ matrix and B is a $p \times n$ matrix, then the **matrix product** of A and B, denoted AB, is an $m \times n$ matrix whose element in the ith row and jth column is the real number obtained from the product of the ith row of A and the jth column of B. If the number of columns in A does not equal the number of rows in B, then the matrix product AB is **not defined.**

It is important to check sizes before starting the multiplication process. If A is an $a \times b$ matrix and B is a $c \times d$ matrix, then if $b = c$, the product AB will exist and will be an $a \times d$ matrix (see Fig. 3). If $b \neq c$, then the product AB does not exist.

FIGURE 3

Must be the same $(b = c)$

$$a \times b \qquad c \times d$$

Size of product $(a \times d)$

The definition is not as complicated as it might first seem. An example should help clarify the process. For

$$A = \begin{bmatrix} 2 & 3 & -1 \\ -2 & 1 & 2 \end{bmatrix} \quad \text{and} \quad B = \begin{bmatrix} 1 & 3 \\ 2 & 0 \\ -1 & 2 \end{bmatrix}$$

A is 2×3, B is 3×2, and so AB is 2×2. To find the first row of AB, we take the product of the first row of A with every column of B and write each result as a real number, not a 1×1 matrix. The second row of AB is computed in the same manner. The four products of row and column matrices used to produce the four elements in AB are shown in the dashed box below. These products are usually calculated mentally, or with the aid of a calculator, and need not be written out. The shaded portions highlight the steps involved in computing the element in the first row and second column of AB.

$$\underset{2 \times 3}{\begin{bmatrix} 2 & 3 & -1 \\ -2 & 1 & 2 \end{bmatrix}} \underset{3 \times 2}{\begin{bmatrix} 1 & 3 \\ 2 & 0 \\ -1 & 2 \end{bmatrix}} = \begin{bmatrix} \begin{bmatrix} 2 & 3 & -1 \end{bmatrix}\begin{bmatrix} 1 \\ 2 \\ -1 \end{bmatrix} & \begin{bmatrix} 2 & 3 & -1 \end{bmatrix}\begin{bmatrix} 3 \\ 0 \\ 2 \end{bmatrix} \\ \begin{bmatrix} -2 & 1 & 2 \end{bmatrix}\begin{bmatrix} 1 \\ 2 \\ -1 \end{bmatrix} & \begin{bmatrix} -2 & 1 & 2 \end{bmatrix}\begin{bmatrix} 3 \\ 0 \\ 2 \end{bmatrix} \end{bmatrix}$$

$$\underset{2 \times 2}{} = \begin{bmatrix} 9 & 4 \\ -2 & -2 \end{bmatrix}$$

EXAMPLE 7 **Matrix Product**

(A) $\overset{3 \times 2}{\begin{bmatrix} 2 & 1 \\ 1 & 0 \\ -1 & 2 \end{bmatrix}} \overset{2 \times 4}{\begin{bmatrix} 1 & -1 & 0 & 1 \\ 2 & 1 & 2 & 0 \end{bmatrix}} = \overset{3 \times 4}{\begin{bmatrix} 4 & -1 & 2 & 2 \\ 1 & -1 & 0 & 1 \\ 3 & 3 & 4 & -1 \end{bmatrix}}$

(B) $\overset{2 \times 4}{\begin{bmatrix} 1 & -1 & 0 & 1 \\ 2 & 1 & 2 & 0 \end{bmatrix}} \overset{3 \times 2}{\begin{bmatrix} 2 & 1 \\ 1 & 0 \\ -1 & 2 \end{bmatrix}}$ (C) $\begin{bmatrix} 2 & 6 \\ -1 & -3 \end{bmatrix} \begin{bmatrix} 1 & 2 \\ 3 & 6 \end{bmatrix} = \begin{bmatrix} 20 & 40 \\ -10 & -20 \end{bmatrix}$

Product is not defined

(D) $\begin{bmatrix} 1 & 2 \\ 3 & 6 \end{bmatrix} \begin{bmatrix} 2 & 6 \\ -1 & -3 \end{bmatrix} = \begin{bmatrix} 0 & 0 \\ 0 & 0 \end{bmatrix}$ (E) $\begin{bmatrix} 2 & -3 & 0 \end{bmatrix} \begin{bmatrix} -5 \\ 2 \\ -2 \end{bmatrix} = \begin{bmatrix} -16 \end{bmatrix}$

(F) $\begin{bmatrix} -5 \\ 2 \\ -2 \end{bmatrix} \begin{bmatrix} 2 & -3 & 0 \end{bmatrix} = \begin{bmatrix} -10 & 15 & 0 \\ 4 & -6 & 0 \\ -4 & 6 & 0 \end{bmatrix}$

Matched Problem 7 Find each product, if it is defined:

(A) $\begin{bmatrix} -1 & 0 & 3 & -2 \\ 1 & 2 & 2 & 0 \end{bmatrix} \begin{bmatrix} -1 & 1 \\ 2 & 3 \\ 1 & 0 \end{bmatrix}$ (B) $\begin{bmatrix} -1 & 1 \\ 2 & 3 \\ 1 & 0 \end{bmatrix} \begin{bmatrix} -1 & 0 & 3 & -2 \\ 1 & 2 & 2 & 0 \end{bmatrix}$

(C) $\begin{bmatrix} 1 & 2 \\ -1 & -2 \end{bmatrix} \begin{bmatrix} -2 & 4 \\ 1 & -2 \end{bmatrix}$ (D) $\begin{bmatrix} -2 & 4 \\ 1 & -2 \end{bmatrix} \begin{bmatrix} 1 & 2 \\ -1 & -2 \end{bmatrix}$

(E) $\begin{bmatrix} 3 & -2 & 1 \end{bmatrix} \begin{bmatrix} 4 \\ 2 \\ 3 \end{bmatrix}$ (F) $\begin{bmatrix} 4 \\ 2 \\ 3 \end{bmatrix} \begin{bmatrix} 3 & -2 & 1 \end{bmatrix}$

FIGURE 4 Multiplication on a graphing calculator.

Figure 4 illustrates a graphing calculator solution to Example 7A. What would you expect to happen if you tried to solve Example 7B on a graphing calculator?

In the arithmetic of real numbers it does not matter in which order we multiply; for example, $5 \times 7 = 7 \times 5$. In matrix multiplication, however, it does make a difference. That is, AB does not always equal BA, even if both multiplications are defined and both products are the same size (see Examples 7C and 7D). Thus,

Matrix multiplication is not commutative.

Also, AB may be zero with neither A nor B equal to zero (see Example 7D). Thus,

The zero property does not hold for matrix multiplication.

(See Section 1-1 for a discussion of the zero property for real numbers.)

Just as we used the familiar algebraic notation AB to represent the product of matrices A and B, we use the notation A^2 for AA, the product of A with itself, A^3 for AAA, and so on.

EXPLORE-DISCUSS 2   In addition to the commutative and zero properties, there are other significant differences between real number multiplication and matrix multiplication.

(A) In real number multiplication, the only real number whose square is 0 is the real number 0 ($0^2 = 0$). Find at least one 2×2 matrix A with all elements nonzero such that $A^2 = 0$, where 0 is the 2×2 zero matrix.

(B) In real number multiplication, the only nonzero real number that is equal to its square is the real number 1 ($1^2 = 1$). Find at least one 2×2 matrix A with all elements nonzero such that $A^2 = A$.

We will continue our discussion of properties of matrix multiplication later in this chapter. Now we consider an application of matrix multiplication.

EXAMPLE 8   **Labor Costs**

Let us combine the time requirements for slalom and trick water skis discussed in Example 6 and Matched Problem 6 into one matrix:

$$
\begin{array}{c}
\\
\text{Trick ski} \\
\text{Slalom ski}
\end{array}
\begin{array}{c}
\text{Labor-hours per ski} \\
\begin{array}{cc}
\text{Assembly} & \text{Finishing} \\
\text{department} & \text{department}
\end{array} \\
\begin{bmatrix} 6\text{ h} & 1.5\text{ h} \\ 4\text{ h} & 1\text{ h} \end{bmatrix} = L
\end{array}
$$

Now suppose that the company has two manufacturing plants, X and Y, in different parts of the country and that the hourly rates for each department are given in the following matrix:

$$
\begin{array}{c}
\\
\text{Assembly department} \\
\text{Finishing department}
\end{array}
\begin{array}{c}
\text{Hourly wages} \\
\begin{array}{cc}
\text{Plant} & \text{Plant} \\
X & Y
\end{array} \\
\begin{bmatrix} \$10 & \$12 \\ \$\ 8 & \$10 \end{bmatrix} = H
\end{array}
$$

Since H and L are both 2×2 matrices, we can take the product of H and L in either order and the result will be a 2×2 matrix:

$$
HL = \begin{bmatrix} 10 & 12 \\ 8 & 10 \end{bmatrix} \begin{bmatrix} 6 & 1.5 \\ 4 & 1 \end{bmatrix} = \begin{bmatrix} 108 & 27 \\ 88 & 22 \end{bmatrix}
$$

$$
LH = \begin{bmatrix} 6 & 1.5 \\ 4 & 1 \end{bmatrix} \begin{bmatrix} 10 & 12 \\ 8 & 10 \end{bmatrix} = \begin{bmatrix} 72 & 87 \\ 48 & 58 \end{bmatrix}
$$

How can we interpret the elements in these products? Let's begin with the product *HL*. The element 108 in the first row and first column of *HL* is the product of the first row matrix of *H* and the first column matrix of *L*:

$$
\begin{array}{cc} \text{Plant} & \text{Plant} \\ X & Y \end{array}
$$

$$
\begin{bmatrix} 10 & 12 \end{bmatrix} \begin{bmatrix} 6 \\ 4 \end{bmatrix} \begin{array}{l} \text{Trick} \\ \text{Slalom} \end{array} = 10(6) + 12(4) = 60 + 48 = 108
$$

Notice that $60 is the labor cost for assembling a trick ski at the California plant and $48 is the labor cost for assembling a slalom ski at the Wisconsin plant. Although both numbers represent labor costs, it makes no sense to add them together. They do not pertain to the same type of ski or to the same plant. Thus, even though the product *HL* happens to be defined mathematically, it has no useful interpretation in this problem.

Now let's consider the product *LH*. The element 72 in the first row and first column of *LH* is given by the following product:

$$
\begin{array}{cc} \text{Assembly} & \text{Finishing} \end{array}
$$

$$
\begin{bmatrix} 6 & 1.5 \end{bmatrix} \begin{bmatrix} 10 \\ 8 \end{bmatrix} \begin{array}{l} \text{Assembly} \\ \text{Finishing} \end{array} = 6(10) + 1.5(8)
$$

$$
= 60 + 12 = 72
$$

where $60 is the labor cost for assembling a trick ski at plant *X* and $12 is the labor cost for finishing a trick ski at plant *X*. Thus, the sum is the total labor cost for producing a trick ski at plant *X*. The other elements in *LH* also represent total labor costs, as indicated by the row and column labels shown below:

$$
\begin{array}{c} \text{Labor costs per ski} \\ \begin{array}{cc} \text{Plant} & \text{Plant} \\ X & Y \end{array} \end{array}
$$

$$
LH = \begin{bmatrix} \$72 & \$87 \\ \$48 & \$58 \end{bmatrix} \begin{array}{l} \text{Trick ski} \\ \text{Slalom ski} \end{array}
$$

Matched Problem 8   Refer to Example 8. The company wants to know how many hours to schedule in each department in order to produce 1,000 trick skis and 2,000 slalom skis. These production requirements can be represented by either of the following matrices:

$$
\begin{array}{cc} \text{Trick} & \text{Slalom} \\ \text{skis} & \text{skis} \end{array}
$$

$$
P = \begin{bmatrix} 1{,}000 & 2{,}000 \end{bmatrix} \qquad Q = \begin{bmatrix} 1{,}000 \\ 2{,}000 \end{bmatrix} \begin{array}{l} \text{Trick skis} \\ \text{Slalom skis} \end{array}
$$

Using the labor-hour matrix *L* from Example 8, find *PL* or *LQ*, whichever has a meaningful interpretation for this problem, and label the rows and columns accordingly.

CAUTION   Example 8 and Problem 8 illustrate an important point about matrix multiplication. Even if you are using a graphing utility to perform the calculations in a matrix product, it is still necessary for you to know the definition of matrix multiplication so that you can interpret the results correctly.

Answers to Matched Problems

1. $\begin{bmatrix} 1 & 5 \\ 0 & -2 \\ 2 & 1 \end{bmatrix}$ 2. $\begin{bmatrix} -1 & -1 & 4 \end{bmatrix}$ 3. $\begin{bmatrix} 13 \\ 2 \\ 35 \end{bmatrix}$

4. (A) $\begin{bmatrix} \$252,000 & \$288,000 \\ \$180,000 & \$288,000 \end{bmatrix}$ (B) $\begin{bmatrix} \$108,000 & \$144,000 \\ \$108,000 & \$144,000 \end{bmatrix}$ (C) $\begin{bmatrix} \$5,400 & \$6,480 \\ \$4,320 & \$6,480 \end{bmatrix}$

5. $\begin{bmatrix} 8 \end{bmatrix}$ 6. $\begin{bmatrix} 6 & 1.5 \end{bmatrix} \begin{bmatrix} 10 \\ 8 \end{bmatrix} = \begin{bmatrix} 72 \end{bmatrix}$ or $72

7. (A) Not defined   (B) $\begin{bmatrix} 2 & 2 & -1 & 2 \\ 1 & 6 & 12 & -4 \\ -1 & 0 & 3 & -2 \end{bmatrix}$ (C) $\begin{bmatrix} 0 & 0 \\ 0 & 0 \end{bmatrix}$ (D) $\begin{bmatrix} -6 & -12 \\ 3 & 6 \end{bmatrix}$

   (E) $\begin{bmatrix} 11 \end{bmatrix}$ (F) $\begin{bmatrix} 12 & -8 & 4 \\ 6 & -4 & 2 \\ 9 & -6 & 3 \end{bmatrix}$

8. $\quad\quad\quad$ Assembly Finishing
   $PL = \begin{bmatrix} 14,000 & 3,500 \end{bmatrix}$ Labor hours

EXERCISE 7-1

A

Perform the indicated operations in Problems 1–18, if possible.

1. $\begin{bmatrix} 5 & -2 \\ 3 & 0 \end{bmatrix} + \begin{bmatrix} -3 & 7 \\ 1 & -6 \end{bmatrix}$ 2. $\begin{bmatrix} 0 & 8 \\ 2 & -1 \end{bmatrix} + \begin{bmatrix} 9 & -4 \\ 7 & 5 \end{bmatrix}$

3. $\begin{bmatrix} 4 & 0 \\ -2 & 3 \\ 8 & 1 \end{bmatrix} + \begin{bmatrix} -1 & 2 \\ 0 & 5 \\ 4 & -6 \end{bmatrix}$

4. $\begin{bmatrix} 6 & -2 & 3 \\ 4 & -8 & -7 \end{bmatrix} + \begin{bmatrix} 3 & 9 & -1 \\ 6 & -2 & 4 \end{bmatrix}$

5. $\begin{bmatrix} 4 & 0 \\ -2 & 3 \\ 8 & 1 \end{bmatrix} + \begin{bmatrix} -1 & 0 & 4 \\ 2 & 5 & -6 \end{bmatrix}$

6. $\begin{bmatrix} 6 & -2 & 3 \\ 4 & -8 & -7 \end{bmatrix} + \begin{bmatrix} 3 & 6 \\ 9 & -2 \\ -1 & 4 \end{bmatrix}$

7. $\begin{bmatrix} 5 & -1 & 0 \\ 4 & 6 & 3 \end{bmatrix} - \begin{bmatrix} 2 & 4 & -6 \\ 3 & 5 & -5 \end{bmatrix}$

8. $\begin{bmatrix} 6 & 2 \\ -4 & 1 \\ 3 & 0 \end{bmatrix} - \begin{bmatrix} 0 & 5 \\ -7 & 2 \\ -1 & 0 \end{bmatrix}$

9. $4 \begin{bmatrix} 3 & -4 & 7 \\ -2 & 9 & 5 \end{bmatrix}$ 10. $5 \begin{bmatrix} -7 & 3 & 0 & 9 \\ 4 & -5 & 6 & 2 \end{bmatrix}$

11. $\begin{bmatrix} 5 & 3 \end{bmatrix} \begin{bmatrix} 4 \\ 7 \end{bmatrix}$ 12. $\begin{bmatrix} -2 & 4 \end{bmatrix} \begin{bmatrix} 3 \\ -8 \end{bmatrix}$

13. $\begin{bmatrix} -6 & 3 \\ 2 & -5 \end{bmatrix} \begin{bmatrix} 1 \\ 3 \end{bmatrix}$ 14. $\begin{bmatrix} 3 & 7 \\ -1 & -9 \end{bmatrix} \begin{bmatrix} 4 \\ -1 \end{bmatrix}$

15. $\begin{bmatrix} 5 & 1 \\ 4 & 6 \end{bmatrix} \begin{bmatrix} 2 & 0 \\ 3 & 8 \end{bmatrix}$ 16. $\begin{bmatrix} -2 & 7 \\ 3 & -1 \end{bmatrix} \begin{bmatrix} 4 & -1 \\ 0 & 5 \end{bmatrix}$

17. $\begin{bmatrix} 8 & -3 \\ -5 & 3 \end{bmatrix} \begin{bmatrix} 2 & 0 \\ 0 & 6 \end{bmatrix}$ 18. $\begin{bmatrix} 7 & 0 \\ 0 & 3 \end{bmatrix} \begin{bmatrix} 9 & -2 \\ -4 & -1 \end{bmatrix}$

FIGURE 1 Identity matrices.

Most graphing calculators have a built-in command for generating the identity matrix of a given order (see Fig. 1).

EXAMPLE 1 **Identity Matrix Multiplication**

(A) $\begin{bmatrix} 1 & 0 & 0 \\ 0 & 1 & 0 \\ 0 & 0 & 1 \end{bmatrix} \begin{bmatrix} a & b & c \\ d & e & f \\ g & h & i \end{bmatrix} = \begin{bmatrix} a & b & c \\ d & e & f \\ g & h & i \end{bmatrix}$

(B) $\begin{bmatrix} a & b & c \\ d & e & f \\ g & h & i \end{bmatrix} \begin{bmatrix} 1 & 0 & 0 \\ 0 & 1 & 0 \\ 0 & 0 & 1 \end{bmatrix} = \begin{bmatrix} a & b & c \\ d & e & f \\ g & h & i \end{bmatrix}$

(C) $\begin{bmatrix} 1 & 0 \\ 0 & 1 \end{bmatrix} \begin{bmatrix} a & b & c \\ d & e & f \end{bmatrix} = \begin{bmatrix} a & b & c \\ d & e & f \end{bmatrix}$

(D) $\begin{bmatrix} a & b & c \\ d & e & f \end{bmatrix} \begin{bmatrix} 1 & 0 & 0 \\ 0 & 1 & 0 \\ 0 & 0 & 1 \end{bmatrix} = \begin{bmatrix} a & b & c \\ d & e & f \end{bmatrix}$

Matched Problem 1 Multiply:

(A) $\begin{bmatrix} 1 & 0 \\ 0 & 1 \end{bmatrix} \begin{bmatrix} 3 & -5 \\ 4 & 6 \end{bmatrix}$ and $\begin{bmatrix} 3 & -5 \\ 4 & 6 \end{bmatrix} \begin{bmatrix} 1 & 0 \\ 0 & 1 \end{bmatrix}$

(B) $\begin{bmatrix} 1 & 0 & 0 \\ 0 & 1 & 0 \\ 0 & 0 & 1 \end{bmatrix} \begin{bmatrix} 5 & -7 \\ 2 & 4 \\ 6 & -8 \end{bmatrix}$ and $\begin{bmatrix} 5 & -7 \\ 2 & 4 \\ 6 & -8 \end{bmatrix} \begin{bmatrix} 1 & 0 \\ 0 & 1 \end{bmatrix}$

In general, we can show that if M is a square matrix of order n and I is the identity matrix of order n, then

$$IM = MI = M$$

If M is an $m \times n$ matrix that is not square ($m \neq n$), then it is still possible to multiply M on the left and on the right by an identity matrix, but not with the same-size identity matrix (see Examples 1C and 1D). In order to avoid the complications involved with associating two different identity matrices with each nonsquare matrix, we restrict our attention in this section to square matrices.

The only real number solutions to the equation $x^2 = 1$ are $x = 1$ and $x = -1$.

(A) Show that $A = \begin{bmatrix} 0 & 1 \\ 1 & 0 \end{bmatrix}$ satisfies $A^2 = I$, where I is the 2×2 identity.

(B) Show that $B = \begin{bmatrix} 0 & -1 \\ -1 & 0 \end{bmatrix}$ satisfies $B^2 = I$.

(C) Find a 2×2 matrix with all elements nonzero whose square is the 2×2 identity matrix.

• **Inverse of a Square Matrix**   In the set of real numbers, we know that for each real number a, except 0, there exists a real number a^{-1} such that

$$a^{-1}a = 1$$

The number a^{-1} is called the *inverse* of the number a relative to multiplication, or the *multiplicative inverse* of a. For example, 2^{-1} is the multiplicative inverse of 2, since $2^{-1}(2) = 1$. We use this idea to define the *inverse of a square matrix*.

DEFINITION 2

Inverse of a Square Matrix

If M is a square matrix of order n and if there exists a matrix M^{-1} (read "M inverse") such that

$$M^{-1}M = MM^{-1} = I$$

then M^{-1} is called the **multiplicative inverse of M** or, more simply, the **inverse of M.**

The multiplicative inverse of a nonzero real number a also can be written as $1/a$. This notation is not used for matrix inverses.

Let's use Definition 2 to find M^{-1}, if it exists, for

$$M = \begin{bmatrix} 2 & 3 \\ 1 & 2 \end{bmatrix}$$

We are looking for

$$M^{-1} = \begin{bmatrix} a & c \\ b & d \end{bmatrix}$$

such that

$$MM^{-1} = M^{-1}M = I$$

Thus, we write

$$
\underset{M}{\begin{bmatrix} 2 & 3 \\ 1 & 2 \end{bmatrix}} \underset{M^{-1}}{\begin{bmatrix} a & c \\ b & d \end{bmatrix}} = \underset{I}{\begin{bmatrix} 1 & 0 \\ 0 & 1 \end{bmatrix}}
$$

and try to find a, b, c, and d so that the product of M and M^{-1} is the identity matrix I. Multiplying M and M^{-1} on the left side, we obtain

$$
\begin{bmatrix} (2a + 3b) & (2c + 3d) \\ (a + 2b) & (c + 2d) \end{bmatrix} = \begin{bmatrix} 1 & 0 \\ 0 & 1 \end{bmatrix}
$$

which is true only if

$$
2a + 3b = 1 \qquad 2c + 3d = 0
$$
$$
a + 2b = 0 \qquad c + 2d = 1
$$

Solving these two systems, we find that $a = 2$, $b = -1$, $c = -3$, and $d = 2$. Thus,

$$
M^{-1} = \begin{bmatrix} 2 & -3 \\ -1 & 2 \end{bmatrix}
$$

as is easily checked:

$$
\underset{M}{\begin{bmatrix} 2 & 3 \\ 1 & 2 \end{bmatrix}} \underset{M^{-1}}{\begin{bmatrix} 2 & -3 \\ -1 & 2 \end{bmatrix}} = \underset{I}{\begin{bmatrix} 1 & 0 \\ 0 & 1 \end{bmatrix}} = \underset{M^{-1}}{\begin{bmatrix} 2 & -3 \\ -1 & 2 \end{bmatrix}} \underset{M}{\begin{bmatrix} 2 & 3 \\ 1 & 2 \end{bmatrix}}
$$

Unlike nonzero real numbers, inverses do not always exist for nonzero square matrices. For example, if

$$
N = \begin{bmatrix} 2 & 1 \\ 4 & 2 \end{bmatrix}
$$

then, proceeding as before, we are led to the systems

$$
2a + b = 1 \qquad 2c + d = 0
$$
$$
4a + 2b = 0 \qquad 4c + 2d = 1
$$

These systems are both inconsistent and have no solution. Hence, N^{-1} does not exist.

Being able to find inverses, when they exist, leads to direct and simple solutions to many practical problems. In the next section, for example, we will show how inverses can be used to solve systems of linear equations.

The method outlined above for finding the inverse, if it exists, gets very involved for matrices of order larger than 2. Now that we know what we are looking for, we can use augmented matrices as in Section 6-2 to make the process more efficient. Details are illustrated in Example 2.

EXAMPLE 2 **Finding an Inverse**

Find the inverse, if it exists, of

$$M = \begin{bmatrix} 1 & -1 & 1 \\ 0 & 2 & -1 \\ 2 & 3 & 0 \end{bmatrix}$$

Solution We start as before and write

$$\overset{M}{\begin{bmatrix} 1 & -1 & 1 \\ 0 & 2 & -1 \\ 2 & 3 & 0 \end{bmatrix}} \overset{M^{-1}}{\begin{bmatrix} a & d & g \\ b & e & h \\ c & f & i \end{bmatrix}} = \overset{I}{\begin{bmatrix} 1 & 0 & 0 \\ 0 & 1 & 0 \\ 0 & 0 & 1 \end{bmatrix}}$$

This is true only if

$$
\begin{array}{lll}
a - b + c = 1 & d - e + f = 0 & g - h + i = 0 \\
\quad\;\; 2b - c = 0 & \quad\;\; 2e - f = 1 & \quad\;\; 2h - i = 0 \\
2a + 3b \quad\;\;\; = 0 & 2d + 3e \quad\;\;\; = 0 & 2g + 3h \quad\;\;\; = 1
\end{array}
$$

Now we write augmented matrices for each of the three systems:

$$
\overset{\text{First}}{\left[\begin{array}{ccc|c} 1 & -1 & 1 & 1 \\ 0 & 2 & -1 & 0 \\ 2 & 3 & 0 & 0 \end{array}\right]}
\quad
\overset{\text{Second}}{\left[\begin{array}{ccc|c} 1 & -1 & 1 & 0 \\ 0 & 2 & -1 & 1 \\ 2 & 3 & 0 & 0 \end{array}\right]}
\quad
\overset{\text{Third}}{\left[\begin{array}{ccc|c} 1 & -1 & 1 & 0 \\ 0 & 2 & -1 & 0 \\ 2 & 3 & 0 & 1 \end{array}\right]}
$$

Since each matrix to the left of the vertical bar is the same, exactly the same row operations can be used on each augmented matrix to transform it into a reduced form. We can speed up the process substantially by combining all three augmented matrices into the single augmented matrix form

$$\left[\begin{array}{ccc|ccc} 1 & -1 & 1 & 1 & 0 & 0 \\ 0 & 2 & -1 & 0 & 1 & 0 \\ 2 & 3 & 0 & 0 & 0 & 1 \end{array}\right] = [M \mid I] \tag{1}$$

We now try to perform row operations on matrix (1) until we obtain a row-equivalent matrix that looks like matrix (2):

$$\left[\begin{array}{ccc|ccc} 1 & 0 & 0 & a & d & g \\ 0 & 1 & 0 & b & e & h \\ 0 & 0 & 1 & c & f & i \end{array}\right] = [I \mid B] \tag{2}$$

If this can be done, then the new matrix to the right of the vertical bar is M^{-1}! Now let's try to transform (1) into a form like (2). We follow the same sequence of steps as in the solution of linear systems by Gauss–Jordan elimination (see Section 6-2):

$$
\begin{array}{c}
\quad\quad M \quad\quad\quad\quad\quad I \\
\begin{bmatrix}
1 & -1 & 1 & | & 1 & 0 & 0 \\
0 & 2 & -1 & | & 0 & 1 & 0 \\
2 & 3 & 0 & | & 0 & 0 & 1
\end{bmatrix} \quad (-2)R_1 + R_3 \rightarrow R_3
\end{array}
$$

$$
\sim \begin{bmatrix}
1 & -1 & 1 & | & 1 & 0 & 0 \\
0 & 2 & -1 & | & 0 & 1 & 0 \\
0 & 5 & -2 & | & -2 & 0 & 1
\end{bmatrix} \quad \tfrac{1}{2}R_2 \rightarrow R_2
$$

$$
\sim \begin{bmatrix}
1 & -1 & 1 & | & 1 & 0 & 0 \\
0 & 1 & -\tfrac{1}{2} & | & 0 & \tfrac{1}{2} & 0 \\
0 & 5 & -2 & | & -2 & 0 & 1
\end{bmatrix} \quad \begin{array}{l} R_2 + R_1 \rightarrow R_1 \\ (-5)R_2 + R_3 \rightarrow R_3 \end{array}
$$

$$
\sim \begin{bmatrix}
1 & 0 & \tfrac{1}{2} & | & 1 & \tfrac{1}{2} & 0 \\
0 & 1 & -\tfrac{1}{2} & | & 0 & \tfrac{1}{2} & 0 \\
0 & 0 & \tfrac{1}{2} & | & -2 & -\tfrac{5}{2} & 1
\end{bmatrix} \quad 2R_3 \rightarrow R_3
$$

$$
\sim \begin{bmatrix}
1 & 0 & \tfrac{1}{2} & | & 1 & \tfrac{1}{2} & 0 \\
0 & 1 & -\tfrac{1}{2} & | & 0 & \tfrac{1}{2} & 0 \\
0 & 0 & 1 & | & -4 & -5 & 2
\end{bmatrix} \quad \begin{array}{l} (-\tfrac{1}{2})R_3 + R_1 \rightarrow R_1 \\ \tfrac{1}{2}R_3 + R_2 \rightarrow R_2 \end{array}
$$

$$
\sim \begin{bmatrix}
1 & 0 & 0 & | & 3 & 3 & -1 \\
0 & 1 & 0 & | & -2 & -2 & 1 \\
0 & 0 & 1 & | & -4 & -5 & 2
\end{bmatrix} = [I \,|\, B]
$$

Converting back to systems of equations equivalent to our three original systems (we won't have to do this step in practice), we have

$$
\begin{array}{lll}
a = 3 & d = 3 & g = -1 \\
b = -2 & e = -2 & h = 1 \\
c = -4 & f = -5 & i = 2
\end{array}
$$

And these are just the elements of M^{-1} that we are looking for! Hence,

$$
M^{-1} = \begin{bmatrix}
3 & 3 & -1 \\
-2 & -2 & 1 \\
-4 & -5 & 2
\end{bmatrix}
$$

Note that this is the matrix to the right of the vertical line in the last augmented matrix.

Check   Since the definition of matrix inverse requires that

$$
M^{-1}M = I \quad \text{and} \quad MM^{-1} = I \tag{3}
$$

it appears that we must compute both $M^{-1}M$ and MM^{-1} to check our work. However, it can be shown that if one of the equations in (3) is satisfied, then the other is also satisfied. Thus, for checking purposes it is sufficient to compute either $M^{-1}M$ or MM^{-1}—we don't need to do both.

$$M^{-1}M = \begin{bmatrix} 3 & 3 & -1 \\ -2 & -2 & 1 \\ -4 & -5 & 2 \end{bmatrix} \begin{bmatrix} 1 & -1 & 1 \\ 0 & 2 & -1 \\ 2 & 3 & 0 \end{bmatrix} = \begin{bmatrix} 1 & 0 & 0 \\ 0 & 1 & 0 \\ 0 & 0 & 1 \end{bmatrix} = I$$

Matched Problem 2   Let: $M = \begin{bmatrix} 3 & -1 & 1 \\ -1 & 1 & 0 \\ 1 & 0 & 1 \end{bmatrix}$

(A) Form the augmented matrix $[M \mid I]$.
(B) Use row operations to transform $[M \mid I]$ into $[I \mid B]$.
(C) Verify by multiplication that $B = M^{-1}$.

The procedure used in Example 2 can be used to find the inverse of any square matrix, if the inverse exists, and will also indicate when the inverse does not exist. These ideas are summarized in Theorem 1.

Theorem 1   **Inverse of a Square Matrix M**

If $[M \mid I]$ is transformed by row operations into $[I \mid B]$, then the resulting matrix B is M^{-1}. If, however, we obtain all 0's in one or more rows to the left of the vertical line, then M^{-1} does not exist.

EXPLORE-DISCUSS 2   (A) Suppose that the square matrix M has a row of all zeros. Explain why M has no inverse.
(B) Suppose that the square matrix M has a column of all zeros. Explain why M has no inverse.

EXAMPLE 3   **Finding a Matrix Inverse**

Find M^{-1}, given: $M = \begin{bmatrix} 4 & -1 \\ -6 & 2 \end{bmatrix}$

Solution

$$\begin{bmatrix} 4 & -1 & | & 1 & 0 \\ -6 & 2 & | & 0 & 1 \end{bmatrix} \quad \tfrac{1}{4}R_1 \to R_1$$

$$\sim \begin{bmatrix} 1 & -\tfrac{1}{4} & | & \tfrac{1}{4} & 0 \\ -6 & 2 & | & 0 & 1 \end{bmatrix} \quad 6R_1 + R_2 \to R_2$$

$$\sim \begin{bmatrix} 1 & -\tfrac{1}{4} & | & \tfrac{1}{4} & 0 \\ 0 & \tfrac{1}{2} & | & \tfrac{3}{2} & 1 \end{bmatrix} \quad 2R_2 \to R_2$$

$$\sim \begin{bmatrix} 1 & -\frac{1}{4} & \Big| & \frac{1}{4} & 0 \\ 0 & 1 & \Big| & 3 & 2 \end{bmatrix} \quad \frac{1}{4}R_2 + R_1 \rightarrow R_1$$

$$\sim \begin{bmatrix} 1 & 0 & \Big| & 1 & \frac{1}{2} \\ 0 & 1 & \Big| & 3 & 2 \end{bmatrix}$$

Thus,

$$M^{-1} = \begin{bmatrix} 1 & \frac{1}{2} \\ 3 & 2 \end{bmatrix} \qquad \text{Check by showing } M^{-1}M = I.$$

Matched Problem 3   Find M^{-1}, given: $M = \begin{bmatrix} 2 & -6 \\ 1 & -2 \end{bmatrix}$

EXAMPLE 4   Finding an Inverse

Find M^{-1}, if it exists, given: $M = \begin{bmatrix} 10 & -2 \\ -5 & 1 \end{bmatrix}$

Solution

$$\begin{bmatrix} 10 & -2 & \Big| & 1 & 0 \\ -5 & 1 & \Big| & 0 & 1 \end{bmatrix} \sim \begin{bmatrix} 1 & -\frac{1}{5} & \Big| & \frac{1}{10} & 0 \\ -5 & 1 & \Big| & 0 & 1 \end{bmatrix}$$

$$\sim \begin{bmatrix} 1 & -\frac{1}{5} & \Big| & \frac{1}{10} & 0 \\ 0 & 0 & \Big| & \frac{1}{2} & 1 \end{bmatrix}$$

We have all 0's in the second row to the left of the vertical line. Therefore, M^{-1} does not exist.

Matched Problem 4   Find M^{-1}, if it exists, given: $M = \begin{bmatrix} 6 & -3 \\ -2 & 1 \end{bmatrix}$

Most graphing utilities and computers can compute matrix inverses and can identify those matrices that do not have inverses (see Fig. 2). A matrix that does not have
an inverse is often referred to as a **singular matrix.**

FIGURE 2 Finding matrix inverses
on a graphing calculator.

(a) Example 3

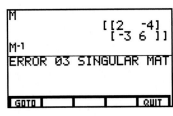

(b) Example 4

● **Application:**
Cryptography

Matrix inverses can be used to provide a simple and effective procedure for encoding and decoding messages. To begin, we assign the numbers 1 to 26 to the letters in the alphabet, as shown below. We also assign the number 27 to a blank to provide for space between words. (A more sophisticated code could include both uppercase and lowercase letters and punctuation symbols.)

A	B	C	D	E	F	G	H	I	J	K	L	M	N
1	2	3	4	5	6	7	8	9	10	11	12	13	14

O	P	Q	R	S	T	U	V	W	X	Y	Z	Blank
15	16	17	18	19	20	21	22	23	24	25	26	27

Thus, the message I LOVE MATH corresponds to the sequence

$$9 \quad 27 \quad 12 \quad 15 \quad 22 \quad 5 \quad 27 \quad 13 \quad 1 \quad 20 \quad 8$$

Any matrix whose elements are positive integers and whose inverse exists can be used as an **encoding matrix.** For example, to use the 2×2 matrix

$$A = \begin{bmatrix} 4 & 3 \\ 5 & 4 \end{bmatrix}$$

to encode the above message, first we divide the numbers in the sequence into groups of 2 and use these groups as the columns of a matrix with 2 rows. (Notice that we added an extra blank at the end of the message to make the columns come out even.) Then we multiply this matrix on the left by A:

$$\begin{bmatrix} 4 & 3 \\ 5 & 4 \end{bmatrix}\begin{bmatrix} 9 & 12 & 22 & 27 & 1 & 8 \\ 27 & 15 & 5 & 13 & 20 & 27 \end{bmatrix} = \begin{bmatrix} 117 & 93 & 103 & 147 & 64 & 113 \\ 153 & 120 & 130 & 187 & 85 & 148 \end{bmatrix}$$

The coded message is

$$117 \quad 153 \quad 93 \quad 120 \quad 103 \quad 130 \quad 147 \quad 187 \quad 64 \quad 85 \quad 113 \quad 148$$

This message can be decoded simply by putting it back into matrix form and multiplying on the left by the **decoding matrix** A^{-1}. Since A^{-1} is easily determined if A is known, the encoding matrix A is the only key needed to decode messages encoded in this manner. Although simple in concept, codes of this type can be very difficult to crack.

Answers to Matched Problems

1. (A) $\begin{bmatrix} 3 & -5 \\ 4 & 6 \end{bmatrix}$ (B) $\begin{bmatrix} 5 & -7 \\ 2 & 4 \\ 6 & -8 \end{bmatrix}$

2. (A) $\begin{bmatrix} 3 & -1 & 1 & | & 1 & 0 & 0 \\ -1 & 1 & 0 & | & 0 & 1 & 0 \\ 1 & 0 & 1 & | & 0 & 0 & 1 \end{bmatrix}$ (B) $\begin{bmatrix} 1 & 0 & 0 & | & 1 & 1 & -1 \\ 0 & 1 & 0 & | & 1 & 2 & -1 \\ 0 & 0 & 1 & | & -1 & -1 & 2 \end{bmatrix}$

(C) $\begin{bmatrix} 1 & 1 & -1 \\ 1 & 2 & -1 \\ -1 & -1 & 2 \end{bmatrix}\begin{bmatrix} 3 & -1 & 1 \\ -1 & 1 & 0 \\ 1 & 0 & 1 \end{bmatrix} = \begin{bmatrix} 1 & 0 & 0 \\ 0 & 1 & 0 \\ 0 & 0 & 1 \end{bmatrix}$

3. $\begin{bmatrix} -1 & 3 \\ -\frac{1}{2} & 1 \end{bmatrix}$ **4.** Does not exist

EXERCISE 7-2

A

Perform the indicated operations in Problems 1–8.

1. $\begin{bmatrix} 6 & -3 \\ 5 & 8 \end{bmatrix}\begin{bmatrix} 1 & 0 \\ 0 & 1 \end{bmatrix}$
    2. $\begin{bmatrix} 1 & 0 \\ 0 & 1 \end{bmatrix}\begin{bmatrix} -4 & 9 \\ 7 & -2 \end{bmatrix}$

3. $\begin{bmatrix} 1 & 0 \\ 0 & 1 \end{bmatrix}\begin{bmatrix} 2 & -5 \\ 8 & -4 \end{bmatrix}$
    4. $\begin{bmatrix} 2 & 0 \\ -7 & -6 \end{bmatrix}\begin{bmatrix} 1 & 0 \\ 0 & 1 \end{bmatrix}$

5. $\begin{bmatrix} 4 & -2 & 5 \\ 6 & 0 & -3 \\ -1 & 2 & 7 \end{bmatrix}\begin{bmatrix} 1 & 0 & 0 \\ 0 & 1 & 0 \\ 0 & 0 & 1 \end{bmatrix}$

6. $\begin{bmatrix} 1 & 0 & 0 \\ 0 & 1 & 0 \\ 0 & 0 & 1 \end{bmatrix}\begin{bmatrix} 8 & 0 & 3 \\ -2 & -1 & -3 \\ 4 & -6 & 1 \end{bmatrix}$

7. $\begin{bmatrix} 1 & 0 & 0 \\ 0 & 1 & 0 \\ 0 & 0 & 1 \end{bmatrix}\begin{bmatrix} -3 & 2 & -5 \\ 4 & -1 & 6 \\ -3 & 4 & -2 \end{bmatrix}$

8. $\begin{bmatrix} 5 & 1 & 0 \\ 1 & -4 & 1 \\ 0 & 1 & 9 \end{bmatrix}\begin{bmatrix} 1 & 0 & 0 \\ 0 & 1 & 0 \\ 0 & 0 & 1 \end{bmatrix}$

For Problems 9–14, show that the two matrices are inverses of each other by showing that their product is the identity matrix I.

9. $\begin{bmatrix} 1 & 6 \\ 0 & 1 \end{bmatrix}\begin{bmatrix} 1 & -6 \\ 0 & 1 \end{bmatrix}$
    10. $\begin{bmatrix} 1 & 0 \\ -3 & 1 \end{bmatrix}\begin{bmatrix} 1 & 0 \\ 3 & 1 \end{bmatrix}$

11. $\begin{bmatrix} 3 & 4 \\ 5 & 7 \end{bmatrix}\begin{bmatrix} 7 & -4 \\ -5 & 3 \end{bmatrix}$
    12. $\begin{bmatrix} -6 & 5 \\ -5 & 4 \end{bmatrix}\begin{bmatrix} 4 & -5 \\ 5 & -6 \end{bmatrix}$

13. $\begin{bmatrix} 1 & -3 & 5 \\ 0 & 1 & 4 \\ 0 & 0 & 1 \end{bmatrix}\begin{bmatrix} 1 & 3 & -17 \\ 0 & 1 & -4 \\ 0 & 0 & 1 \end{bmatrix}$

14. $\begin{bmatrix} 1 & 0 & 0 \\ 2 & 1 & 0 \\ -6 & -3 & 1 \end{bmatrix}\begin{bmatrix} 1 & 0 & 0 \\ -2 & 1 & 0 \\ 0 & 3 & 1 \end{bmatrix}$

B

Given M in Problems 15–24, find M^{-1}, and show that $M^{-1}M = I$.

15. $\begin{bmatrix} 1 & 9 \\ 0 & 1 \end{bmatrix}$
  16. $\begin{bmatrix} 0 & -1 \\ -1 & 3 \end{bmatrix}$
  17. $\begin{bmatrix} -1 & -2 \\ 2 & 5 \end{bmatrix}$

18. $\begin{bmatrix} 3 & -4 \\ -2 & 3 \end{bmatrix}$
  19. $\begin{bmatrix} -5 & 7 \\ 2 & -3 \end{bmatrix}$
  20. $\begin{bmatrix} 11 & 4 \\ 3 & 1 \end{bmatrix}$

21. $\begin{bmatrix} 1 & -1 & 0 \\ -1 & 1 & -1 \\ 0 & -1 & 1 \end{bmatrix}$
  22. $\begin{bmatrix} 2 & -1 & 0 \\ 0 & 1 & 1 \\ 1 & 0 & 1 \end{bmatrix}$

23. $\begin{bmatrix} 1 & 2 & 5 \\ 3 & 5 & 9 \\ 1 & 1 & -2 \end{bmatrix}$
  24. $\begin{bmatrix} 1 & -1 & 1 \\ -2 & 3 & 2 \\ 3 & -3 & 2 \end{bmatrix}$

Find the inverse of each matrix in Problems 25–28, if it exists.

25. $\begin{bmatrix} 3 & 9 \\ 2 & 6 \end{bmatrix}$
    26. $\begin{bmatrix} 2 & -4 \\ -3 & 6 \end{bmatrix}$

27. $\begin{bmatrix} 2 & 3 \\ 3 & 5 \end{bmatrix}$
    28. $\begin{bmatrix} -5 & 4 \\ 4 & -3 \end{bmatrix}$

C

Find the inverse of each matrix in Problems 29–34, if it exists.

29. $\begin{bmatrix} 2 & 2 & -1 \\ 0 & 4 & -1 \\ -1 & -2 & 1 \end{bmatrix}$
  30. $\begin{bmatrix} 4 & 2 & -1 \\ 1 & 1 & -1 \\ -3 & -1 & 1 \end{bmatrix}$

31. $\begin{bmatrix} 2 & 1 & 1 \\ 1 & 1 & 0 \\ -1 & -1 & 0 \end{bmatrix}$
  32. $\begin{bmatrix} 1 & -1 & 0 \\ 2 & -1 & 1 \\ 0 & 1 & 1 \end{bmatrix}$

33. $\begin{bmatrix} 1 & 5 & 10 \\ 0 & 1 & 4 \\ 1 & 6 & 15 \end{bmatrix}$
  34. $\begin{bmatrix} 1 & -5 & -10 \\ 0 & 1 & 6 \\ 1 & -4 & -3 \end{bmatrix}$

35. Show that $(A^{-1})^{-1} = A$ for

$$A = \begin{bmatrix} 3 & 4 \\ 2 & 3 \end{bmatrix}$$

36. Show that $(AB)^{-1} = B^{-1}A^{-1}$ for

$$A = \begin{bmatrix} 3 & 4 \\ 2 & 3 \end{bmatrix} \quad \text{and} \quad B = \begin{bmatrix} 3 & 7 \\ 2 & 5 \end{bmatrix}$$

37. Discuss the existence of M^{-1} for 2×2 diagonal matrices of the form

$$M = \begin{bmatrix} a & 0 \\ 0 & d \end{bmatrix}$$

38. Discuss the existence of M^{-1} for 2×2 upper triangular matrices of the form

$$M = \begin{bmatrix} a & b \\ 0 & d \end{bmatrix}$$

In Problems 39–42, use a graphing utility to find the inverse of each matrix, if it exists.

39. $\begin{bmatrix} 1 & -5 & -2 & -2 \\ 2 & 7 & 0 & -1 \\ 3 & -5 & -2 & 0 \\ 4 & -1 & -2 & 0 \end{bmatrix}$ **40.** $\begin{bmatrix} 1 & -2 & -1 & 0 \\ 2 & 1 & 3 & 5 \\ 1 & 0 & -3 & 2 \\ 0 & 4 & 2 & 6 \end{bmatrix}$

41. $\begin{bmatrix} 1 & 3 & 1 & 6 & 0 \\ 4 & 4 & 2 & 6 & 1 \\ 3 & 4 & 1 & 2 & -2 \\ 3 & -5 & 1 & -4 & 3 \\ -1 & 2 & -1 & -2 & -4 \end{bmatrix}$

42. $\begin{bmatrix} 1 & -1 & -6 & -1 & 3 \\ -2 & 2 & 0 & 2 & 1 \\ 4 & -2 & 0 & -3 & 1 \\ 1 & 3 & 1 & 4 & 2 \\ -1 & 3 & 1 & 1 & 4 \end{bmatrix}$

APPLICATIONS

Problems 43–46 refer to the encoding matrix $A = \begin{bmatrix} 3 & 5 \\ 1 & 2 \end{bmatrix}$

43. Cryptography. Encode the message CAT IN THE HAT with the matrix *A* given above.

44. Cryptography. Encode the message FOX IN SOCKS with the matrix *A* given above.

45. Cryptography. The following message was encoded with the matrix *A* given above. Decode this message.

111 43 40 15 177 68 50 19 116 45 86
29 62 22 121 43 68 27

46. Cryptography. The following message was encoded with the matrix *A* given above. Decode this message.

99 38 154 58 115 43 121 43 20 7 149
56 86 29 196 73 99 38

Problems 47–50 require the use of a graphing utility. To use the 5 × 5 encoding matrix B given below, form a matrix with 5 rows and as many columns as necessary to accommodate each message.

$$B = \begin{bmatrix} 1 & 0 & 1 & 0 & 1 \\ 0 & 1 & 1 & 0 & 3 \\ 2 & 1 & 1 & 1 & 1 \\ 0 & 0 & 1 & 0 & 2 \\ 1 & 1 & 1 & 2 & 1 \end{bmatrix}$$

47. Cryptography. Encode the message DWIGHT DAVID EISENHOWER with the matrix *B* given above.

48. Cryptography. Encode the message JOHN FITZGER-ALD KENNEDY with the matrix *B* given above.

49. Cryptography. The following message was encoded with the matrix *B* given above. Decode this message.

41 84 82 44 74 25 56 67 20 54 43 54
89 39 102 44 67 86 44 90 68 135 136
81 149

50. Cryptography. The following message was encoded with the matrix *B* given above. Decode this message.

22 15 57 5 47 54 58 89 45 84 46 80
87 53 96 51 68 116 39 113 68 135 136
81 149

SECTION 7-3 Matrix Equations and Systems of Linear Equations

- Matrix Equations
- Matrix Equations and Systems of Linear Equations
- Application

The identity matrix and inverse matrix discussed in the last section can be put to immediate use in the solving of certain simple matrix equations. Being able to solve a matrix equation gives us another important method of solving a system of equations having the same number of variables as equations. If the system either has fewer variables than equations or more variables than equations, then we must return to the Gauss–Jordan method of elimination.

• **Matrix Equations**   Before we discuss the solution of matrix equations, you will probably find it helpful to briefly review the basic properties of real numbers and linear equations discussed in Sections 1-1 and 2-1.

EXPLORE-DISCUSS 1   Let a, b, and c be real numbers, with $a \neq 0$. Solve each equation for x.

(A) $ax = b$ (B) $ax + b = c$

Solving simple matrix equations follows very much the same procedures used in solving real number equations. We have, however, less freedom with matrix equations, because matrix multiplication is not commutative. In solving matrix equations, we will be guided by the properties of matrices summarized in Theorem 1.

Theorem 1   **Basic Properties of Matrices**

Assuming all products and sums are defined for the indicated matrices A, B, C, I, and 0, then

Addition Properties
Associative: $(A + B) + C = A + (B + C)$
Commutative: $A + B = B + A$
Additive Identity: $A + 0 = 0 + A = A$
Additive Inverse: $A + (-A) = (-A) + A = 0$

Multiplication Properties
Associative Property: $A(BC) = (AB)C$
Multiplicative Identity: $AI = IA = A$
Multiplicative Inverse: If A is a square matrix and A^{-1} exists,
then $AA^{-1} = A^{-1}A = I$.

Combined Properties
Left Distributive: $A(B + C) = AB + AC$
Right Distributive: $(B + C)A = BA + CA$

Equality
Addition: If $A = B$, then $A + C = B + C$.
Left Multiplication: If $A = B$, then $CA = CB$.
Right Multiplication: If $A = B$, then $AC = BC$.

The process of solving certain types of simple matrix equations is best illustrated by an example.

EXAMPLE 1   **Solving a Matrix Equation**

Given an $n \times n$ matrix A and $n \times 1$ column matrices B and X, solve $AX = B$ for X. Assume all necessary inverses exist.

Solution   We are interested in finding a column matrix X that satisfies the matrix equation $AX = B$. To solve this equation, we multiply both sides, on the left, by A^{-1}, assuming it exists, to isolate X on the left side.

$$AX = B$$

$$A^{-1}(AX) = A^{-1}B \quad \text{Use the left multiplication property.}$$

$$(A^{-1}A)X = A^{-1}B \quad \text{Associative property}$$

$$IX = A^{-1}B \quad A^{-1}A = I$$

$$X = A^{-1}B \quad IX = X$$

CAUTION   Do not mix the left multiplication property and the right multiplication property. If $AX = B$, then

$$A^{-1}(AX) \neq BA^{-1}$$

Matched Problem 1   Given an $n \times n$ matrix A and $n \times 1$ column matrices B, C, and X, solve $AX + C = B$ for X. Assume all necessary inverses exist.

● Matrix Equations and Systems of Linear Equations

We now show how independent systems of linear equations with the same number of variables as equations can be solved by first converting the system into a matrix equation of the form $AX = B$ and using $X = A^{-1}B$ as obtained in Example 1.

EXAMPLE 2   **Using Inverses to Solve Systems of Equations**

Use matrix inverse methods to solve the system:

$$\begin{aligned}
x_1 - x_2 + x_3 &= 1 \\
2x_2 - x_3 &= 1 \\
2x_1 + 3x_2 &= 1
\end{aligned} \qquad (1)$$

Solution   The inverse of the coefficient matrix

$$A = \begin{bmatrix} 1 & -1 & 1 \\ 0 & 2 & -1 \\ 2 & 3 & 0 \end{bmatrix}$$

provides an efficient method for solving this system. To see how, we convert system (1) into a matrix equation:

EXAMPLE 1   **Taking Out a Common Factor of a Column**

$$\begin{vmatrix} 6 & 1 & 3 \\ -2 & 7 & -2 \\ 4 & 5 & 0 \end{vmatrix} = 2 \begin{vmatrix} 3 & 1 & 3 \\ -1 & 7 & -2 \\ 2 & 5 & 0 \end{vmatrix}$$

where 2 is a common factor of the first column.

Matched Problem 1   Take out factors common to any row or any column:

$$\begin{vmatrix} 3 & 2 & 1 \\ 6 & 3 & -9 \\ 1 & 0 & -5 \end{vmatrix}$$

EXPLORE-DISCUSS 1   (A) How are $\begin{vmatrix} a & b \\ c & d \end{vmatrix}$ and $\begin{vmatrix} ka & kb \\ kc & kd \end{vmatrix}$ related?

(B) How are $\begin{vmatrix} a & b & c \\ d & e & f \\ g & h & i \end{vmatrix}$ and $\begin{vmatrix} ka & kb & kc \\ kd & ke & kf \\ kg & kh & ki \end{vmatrix}$ related?

Theorem 2   **Row or Column of Zeros**

If every element in a row (or column) is 0, the value of the determinant is 0.

Theorem 2 is an immediate consequence of Theorem 1, and its proof is left as an exercise. It is illustrated in the following example:

$$\begin{vmatrix} 3 & -2 & 5 \\ 0 & 0 & 0 \\ -1 & 4 & 9 \end{vmatrix} = 0$$

Theorem 3   **Interchanging Rows or Columns**

If two rows (or two columns) of a determinant are interchanged, the new determinant is the negative of the original.

A proof of Theorem 3 even for a determinant of order 3 is notationally involved. We suggest that you partially prove the theorem by direct expansion of the determinants before and after the interchange of two rows (or columns). The theorem is illustrated by the following example, where the second and third columns are interchanged:

$$\begin{vmatrix} 1 & 0 & 9 \\ -2 & 1 & 5 \\ 3 & 0 & 7 \end{vmatrix} = -\begin{vmatrix} 1 & 9 & 0 \\ -2 & 5 & 1 \\ 3 & 7 & 0 \end{vmatrix}$$

EXPLORE-DISCUSS 2   (A) What are the cofactors of each element in the first row of the following determinant? What is the value of the determinant?

$$\begin{vmatrix} a & b & c \\ d & e & f \\ d & e & f \end{vmatrix}$$

(B) What are the cofactors of each element in the second column of the following determinant? What is the value of the determinant?

$$\begin{vmatrix} a & b & a \\ d & e & d \\ g & h & g \end{vmatrix}$$

Theorem 4   **Equal Rows or Columns**

If the corresponding elements are equal in two rows (or columns), the value of the determinant is 0.

Proof   The general proof of Theorem 4 follows directly from Theorem 3. If we start with a determinant D that has two rows (or columns) equal and we interchange the equal rows (or columns), the new determinant will be the same as the original. But by Theorem 3,

$$D = -D$$

hence,

$$2D = 0$$
$$D = 0$$

Theorem 5   **Addition of Rows or Columns**

If a multiple of any row (or column) of a determinant is added to any other row (or column), the value of the determinant is not changed.

The distributor has just received the order shown in Table 4 from a retail toy store.

TABLE 4 Customer Order	
Set	**Quantity**
Limited	48
Empire	24
Comet	12

The distributor wants to store the information in each table in a matrix and use matrix operations to find the following information:

1. The inventory (parts and labor) required to fill the order
2. The cost (parts and labor) of filling the order
3. The revenue (sales) received from the customer
4. The profit realized on the order

(A) Use a single letter to designate the matrix representing each table, and write matrix expressions in terms of these letters that will provide the required information. Discuss the size of the matrix you must use to represent each table so that all the pertinent matrix operations are defined.
(B) Evaluate the matrix expressions in part A.

Shortly after filling the order in Table 4, a supplier informs the distributor that the cars and locomotives used in these train sets are no longer available. The distributor currently has 30 locomotives and 134 cars in stock.

(C) How many train sets of each type can the distributor produce using all the available locomotives and cars? Assume that the distributor has unlimited quantities of the other components used in these sets.
(D) How much profit will the distributor make if all these sets are sold? If there is more than one way to use all the available locomotives and cars, which one will produce the largest profit?

Chapter 7 Review

7-1   MATRICES: BASIC OPERATIONS

Two matrices are **equal** if they are the same size and their corresponding elements are equal. The **sum of two matrices** of the same size is a matrix with elements that are the sums of the corresponding elements of the two given matrices. Matrix addition is **commutative** and **associative.** A matrix with all zero elements is called the **zero matrix.** The **negative of a matrix M,** denoted $-M$, is a matrix with elements that are the negatives of the elements in M. If A and B are matrices of the same size, then we define **subtraction** as follows: $A - B = A + (-B)$. The **product of a number k and a matrix M,** denoted by kM, is a matrix formed by multiplying each element of M by k. The **product** of a $1 \times n$ row matrix and an $n \times 1$ column matrix is a 1×1 matrix given by

$$
\underset{1 \times n}{[a_1 \quad a_2 \quad \cdots \quad a_n]} \underset{n \times 1}{\begin{bmatrix} b_1 \\ b_2 \\ \vdots \\ b_n \end{bmatrix}} = \underset{1 \times 1}{[a_1 b_1 + a_2 b_2 + \cdots + a_n b_n]}
$$

If A is an $m \times p$ matrix and B is a $p \times n$ matrix, then the **matrix**

product of A and B, denoted AB, is an $m \times n$ matrix whose element in the ith row and jth column is the real number obtained from the product of the ith row of A and the jth column of B. If the number of columns in A does not equal the number of rows in B, then the matrix product AB is **not defined. Matrix multiplication is not commutative,** and the **zero property does not hold for matrix multiplication.** That is, for matrices A and B, the matrix product AB can be zero without either A or B being the zero matrix.

7-2   INVERSE OF A SQUARE MATRIX

The **identity matrix for multiplication** for the set of all square matrices of order n is the square matrix of order n, denoted by I, with 1's along the **principal diagonal** (from upper left corner to lower right corner) and 0's elsewhere. If M is a square matrix of order n and I is the identity matrix of order n, then

$$IM = MI = M$$

If M is a square matrix of order n and if there exists a matrix M^{-1} (read "M inverse") such that

$$M^{-1}M = MM^{-1} = I$$

then M^{-1} is called the **multiplicative inverse of M** or, more simply, the **inverse of M.** If the augmented matrix $[M \mid I]$ is transformed by row operations into $[I \mid B]$, then the resulting matrix B is M^{-1}. If, however, we obtain all 0's in one or more rows to the left of the vertical line, then M^{-1} does not exist and M is called a **singular matrix.**

7-3   MATRIX EQUATIONS AND SYSTEMS OF LINEAR EQUATIONS

The following properties of matrices are fundamental to the process of solving matrix equations. Assuming all products and sums are defined for the indicated matrices A, B, C, I, and 0, then:

Addition Properties
Associative: $(A + B) + C = A + (B + C)$
Commutative: $A + B = B + A$
Additive Identity: $A + 0 = 0 + A = A$
Additive Inverse: $A + (-A) = (-A) + A = 0$

Multiplication Properties
Associative Property: $A(BC) = (AB)C$
Multiplicative Identity: $AI = IA = A$
Multiplicative Inverse: If A is a square matrix and A^{-1} exists, then $AA^{-1} = A^{-1}A = I$.

Combined Properties
Left Distributive: $A(B + C) = AB + AC$
Right Distributive: $(B + C)A = BA + CA$

Equality
Addition: If $A = B$, then $A + C = B + C$.
Left Multiplication: If $A = B$, then $CA = CB$.
Right Multiplication: If $A = B$, then $AC = BC$.

A system of linear equations with the same number of variables as equations such as

$$a_{11}x_1 + a_{12}x_2 + a_{13}x_3 = k_1$$
$$a_{21}x_1 + a_{22}x_2 + a_{23}x_3 = k_2$$
$$a_{31}x_1 + a_{32}x_2 + a_{33}x_3 = k_3$$

can be written as the matrix equation

$$\begin{array}{ccc} A & X & B \end{array}$$
$$\begin{bmatrix} a_{11} & a_{12} & a_{13} \\ a_{21} & a_{22} & a_{23} \\ a_{31} & a_{32} & a_{33} \end{bmatrix} \begin{bmatrix} x_1 \\ x_2 \\ x_3 \end{bmatrix} = \begin{bmatrix} k_1 \\ k_2 \\ k_3 \end{bmatrix}$$

If the inverse of A exists, then the matrix equation has a unique solution given by

$$X = A^{-1}B$$

After multiplying B by A^{-1} from the left, it is easy to read the solution to the original system of equations.

7-4   DETERMINANTS

Associated with each square matrix A is a real number called the **determinant** of the matrix. The determinant of A is denoted by **det A,** or simply by writing the array of elements in A using vertical lines in place of square brackets. For example,

$$\det \begin{bmatrix} a_{11} & a_{12} \\ a_{21} & a_{22} \end{bmatrix} = \begin{vmatrix} a_{11} & a_{12} \\ a_{21} & a_{22} \end{vmatrix}$$

A determinant of **order n** is a determinant with n rows and n columns.

The **value of a second-order determinant** is the real number given by

$$\begin{vmatrix} a_{11} & a_{12} \\ a_{21} & a_{22} \end{vmatrix} = a_{11}a_{22} - a_{21}a_{12}$$

The **value of a third-order determinant** is the sum of three products obtained by multiplying each element of any one row (or each element of any one column) by its cofactor. The **cofactor of an element** a_{ij} (from the ith row and jth column) is the product of the minor of a_{ij} and $(-1)^{i+j}$. The **minor of an element** a_{ij} is the determinant remaining after deleting the ith row and jth column. A similar process can be used to evaluate determinants of order higher than 3.

7-5   PROPERTIES OF DETERMINANTS

The use of the following five determinant properties can greatly reduce the effort in evaluating determinants of order 3 or greater:

1. If each element of any row (or column) of a determinant is multiplied by a constant k, the new determinant is k times the original.

$$\begin{vmatrix} 2a & 2b \\ c & d \end{vmatrix} = 2\begin{vmatrix} a & b \\ c & d \end{vmatrix}$$

$$3\begin{vmatrix} a & b \\ c & d \end{vmatrix} = \begin{vmatrix} 3a & b \\ 3c & d \end{vmatrix}$$

2. If every element in a row (or column) is 0, the value of the determinant is 0.

$$\begin{vmatrix} a & b \\ 0 & 0 \end{vmatrix} = 0$$

$$\begin{vmatrix} 0 & b \\ 0 & d \end{vmatrix} = 0$$

3. If two rows (or two columns) of a determinant are interchanged, the new determinant is the negative of the original.

$$\begin{vmatrix} a & b \\ c & d \end{vmatrix} = -\begin{vmatrix} c & d \\ a & b \end{vmatrix}$$

$$\begin{vmatrix} a & b \\ c & d \end{vmatrix} = -\begin{vmatrix} b & a \\ d & c \end{vmatrix}$$

4. If the corresponding elements are equal in two rows (or columns), the value of the determinant is 0.

$$\begin{vmatrix} a & b \\ a & b \end{vmatrix} = 0$$

$$\begin{vmatrix} a & a \\ c & c \end{vmatrix} = 0$$

5. If a multiple of any row (or column) of a determinant is added to any other row (or column), the value of the determinant is not changed.

$$\begin{vmatrix} a & b \\ c & d \end{vmatrix} = \begin{vmatrix} a & b \\ c + ka & d + kb \end{vmatrix}$$

$$\begin{vmatrix} a & b \\ c & d \end{vmatrix} = \begin{vmatrix} a + kb & b \\ c + kd & d \end{vmatrix}$$

7-6 CRAMER'S RULE

Systems of equations having the same number of variables as equations can also be solved using determinants and Cramer's rule. **Cramer's rule for three equations and three variables** is as follows: Given the system

$$\begin{aligned} a_{11}x + a_{12}y + a_{13}z &= k_1 \\ a_{21}x + a_{22}y + a_{23}z &= k_2 \\ a_{31}x + a_{32}y + a_{33}z &= k_3 \end{aligned} \quad \text{with} \quad D = \begin{vmatrix} a_{11} & a_{12} & a_{13} \\ a_{21} & a_{22} & a_{23} \\ a_{31} & a_{32} & a_{33} \end{vmatrix} \neq 0$$

then

$$x = \frac{\begin{vmatrix} k_1 & a_{12} & a_{13} \\ k_2 & a_{22} & a_{23} \\ k_3 & a_{32} & a_{33} \end{vmatrix}}{D} \quad y = \frac{\begin{vmatrix} a_{11} & k_1 & a_{13} \\ a_{21} & k_2 & a_{23} \\ a_{31} & k_3 & a_{33} \end{vmatrix}}{D} \quad z = \frac{\begin{vmatrix} a_{11} & a_{12} & k_1 \\ a_{21} & a_{22} & k_2 \\ a_{31} & a_{32} & k_3 \end{vmatrix}}{D}$$

Cramer's rule can be generalized completely for any size linear system that has the same number of variables as equations. The formulas are easily remembered if you observe the following:

1. Determinant D is formed from the coefficients of x, y, and z, keeping the same relative position in the determinant as found in the system of equations.

2. Determinant D appears in the denominators for x, y, and z.

3. The numerator for x can be obtained from D by replacing the coefficients of x (a_{11}, a_{21}, a_{31}) with the constants k_1, k_2, and k_3, respectively. Similar statements can be made for the numerators for y and z.

Cramer's rule is rarely used to solve systems of order higher than 3 by hand, since more efficient methods are available. Cramer's rule, however, is a valuable tool in more advanced theoretical and applied mathematics.

Chapter 7 Review Exercise

Work through all the problems in this chapter review and check answers in the back of the book. Answers to all review problems are there, and following each answer is a number in italics indicating the section in which that type of problem is discussed. Where weaknesses show up, review appropriate sections in the text.

A

In Problems 1–9, perform the operations that are defined, given the following matrices:

$$A = \begin{bmatrix} 4 & -2 \\ 0 & 3 \end{bmatrix} \quad B = \begin{bmatrix} -1 & 5 \\ -4 & 6 \end{bmatrix} \quad C = \begin{bmatrix} -1 & 4 \end{bmatrix} \quad D = \begin{bmatrix} 3 \\ -2 \end{bmatrix}$$

1. AB **2.** CD **3.** CB

4. AD **5.** $A + B$ **6.** $C + D$

7. $A + C$ **8.** $2A - 5B$ **9.** $CA + C$

10. Find the inverse of

$$A = \begin{bmatrix} 4 & 7 \\ -1 & -2 \end{bmatrix}$$

Show that $A^{-1}A = I$.

11. Write the system

$$\begin{aligned} 3x_1 + 2x_2 &= k_1 \\ 4x_1 + 3x_2 &= k_2 \end{aligned}$$

as a matrix equation, and solve using matrix inverse methods for:

(A) $k_1 = 3, k_2 = 5$ (B) $k_1 = 7, k_2 = 10$
(C) $k_1 = 4, k_2 = 2$

Evaluate the determinants in Problems 12 and 13.

12. $\begin{vmatrix} 2 & -3 \\ -5 & -1 \end{vmatrix}$

13. $\begin{vmatrix} 2 & 3 & -4 \\ 0 & 5 & 0 \\ 1 & -4 & -2 \end{vmatrix}$

14. Solve the system using Cramer's rule:

$$3x - 2y = 8$$
$$x + 3y = -1$$

15. Use properties of determinants to find each of the following, given that

$$\begin{vmatrix} a & b & c \\ d & e & f \\ g & h & i \end{vmatrix} = 2$$

(A) $\begin{vmatrix} g & h & i \\ d & e & f \\ a & b & c \end{vmatrix}$ (B) $\begin{vmatrix} a & 3b & c \\ d & 3e & f \\ g & 3h & i \end{vmatrix}$

(C) $\begin{vmatrix} a & b & a+b+c \\ d & e & d+e+f \\ g & h & g+h+i \end{vmatrix}$

B _____

In Problems 16–21, perform the operations that are defined, given the following matrices:

$$A = \begin{bmatrix} 1 & 2 \\ 4 & 5 \\ -3 & -1 \end{bmatrix} \quad B = \begin{bmatrix} 6 \\ 0 \\ -4 \end{bmatrix} \quad C = \begin{bmatrix} 2 & 4 & -1 \end{bmatrix}$$

$$D = \begin{bmatrix} 7 & 0 & -5 \\ 0 & 8 & -2 \end{bmatrix} \quad E = \begin{bmatrix} 9 & -3 \\ -6 & 2 \end{bmatrix}$$

16. AD **17.** DA **18.** BC

19. CB **20.** DE **21.** ED

22. Find the inverse of

$$A = \begin{bmatrix} 1 & 0 & 4 \\ -2 & 1 & 0 \\ 4 & -1 & 4 \end{bmatrix}$$

Show that $AA^{-1} = I$.

23. Write the system

$$x_1 + 2x_2 + 3x_3 = k_1$$
$$2x_1 + 3x_2 + 4x_3 = k_2$$
$$x_1 + 2x_2 + x_3 = k_3$$

as a matrix equation, and solve using matrix inverse methods for:
(A) $k_1 = 1, k_2 = 3, k_3 = 3$
(B) $k_1 = 0, k_2 = 0, k_3 = -2$
(C) $k_1 = -3, k_2 = -4, k_3 = 1$

Evaluate the determinants in Problems 24 and 25.

24. $\begin{vmatrix} -\frac{1}{4} & \frac{3}{2} \\ \frac{1}{2} & \frac{2}{3} \end{vmatrix}$

25. $\begin{vmatrix} 2 & -1 & 1 \\ -3 & 5 & 2 \\ 1 & -2 & 4 \end{vmatrix}$

26. Solve for y only using Cramer's rule:

$$x - 2y + z = -6$$
$$y - z = 4$$
$$2x + 2y + z = 2$$

(Find the numerator and denominator first; then reduce.)

27. Discuss the number of solutions for a system of n equations in n variables if the coefficient matrix:
(A) Has an inverse.   (B) Does not have an inverse.

28. If A is a nonzero square matrix of order n satisfying $A^2 = 0$, can A^{-1} exist? Explain.

C _____

29. For $n \times n$ matrices A and C and $n \times 1$ column matrices B and X, solve for X assuming all necessary inverses exist:

$$AX - B = CX.$$

30. Find the inverse of

$$A = \begin{bmatrix} 4 & 5 & 6 \\ 4 & 5 & -6 \\ 1 & 1 & 1 \end{bmatrix}$$

Show that $A^{-1}A = I$.

31. Clear the decimals in the system

$$0.04x_1 + 0.05x_2 + 0.06x_3 = 360$$
$$0.04x_1 + 0.05x_2 - 0.06x_3 = 120$$
$$x_1 + x_2 + x_3 = 7{,}000$$

by multiplying the first two equations by 100. Then write the resulting system as a matrix equation and solve using the inverse found in Problem 30.

32.
$$\begin{vmatrix} -1 & 4 & 1 & 1 \\ 5 & -1 & 2 & -1 \\ 2 & -1 & 0 & 3 \\ -3 & 3 & 0 & 3 \end{vmatrix} = ?$$

33. Show that

$$\begin{vmatrix} u & v \\ w & x \end{vmatrix} = \begin{vmatrix} u + kv & v \\ w + kx & x \end{vmatrix}$$

34. Explain why the points $(1, 2)$ and $(-1, 5)$ must satisfy the equation

$$\begin{vmatrix} x & y & 1 \\ 1 & 2 & 1 \\ -1 & 5 & 1 \end{vmatrix} = 0$$

Describe the set of all points that satisfy this equation.

APPLICATIONS

35. Resource Allocation. A Colorado mining company operates mines at Big Bend and Saw Pit. The Big Bend mine produces ore that is 5% nickel and 7% copper. The Saw Pit mine produces ore that is 3% nickel and 4% copper. How many tons of ore should be produced at each mine to obtain the amounts of nickel and copper listed in the table? Set up a matrix equation and solve using matrix inverses.

	Nickel	Copper
(A)	3.6 tons	5 tons
(B)	3 tons	4.1 tons
(C)	3.2 tons	4.4 tons

36. Labor Costs. A company with manufacturing plants in North and South Carolina has labor-hour and wage requirements for the manufacturing of computer desks and printer stands as given in matrices L and H:

Labor-hour requirements

	Fabricating department	Assembly department	Packaging department	
$L =$	1.7 h	2.4 h	0.8 h	Desk
	0.9 h	1.8 h	0.6 h	Stand

Hourly wages

	North Carolina plant	South Carolina plant	
$H =$	$11.50	$10.00	Fabricating department
	$9.50	$8.50	Assembly department
	$5.00	$4.50	Packaging department

(A) Find the labor cost for producing one printer stand at the South Carolina plant.

(B) Discuss possible interpretations of the elements in the matrix products HL and LH.

(C) If either of the products HL or LH has a meaningful interpretation, find the product and label its rows and columns.

37. Labor Costs. The monthly production of computer desks and printer stands for the company in Problem 36 for the months of January and February are given in matrices J and F:

January production

	North Carolina plant	South Carolina plant	
$J =$	1,500	1,650	Desks
	850	700	Stands

February production

	North Carolina plant	South Carolina plant	
$F =$	1,700	1,810	Desks
	930	740	Stands

(A) Find the average monthly production for the months of January and February.

(B) Find the increase in production from January to February.

(C) Find $J \begin{bmatrix} 1 \\ 1 \end{bmatrix}$ and interpret.

38. Cryptography. The following message was encoded with the matrix B shown below. Decode the message:

25 8 26 24 25 33 21 41 48 41 30 50
21 32 41 52 52 79

$$B = \begin{bmatrix} 1 & 1 & 0 \\ 1 & 0 & 1 \\ 1 & 1 & 1 \end{bmatrix}$$

Cumulative Review Exercise Chapters 6 and 7

Work through all the problems in this cumulative review and check answers in the back of the book. Answers to all review problems are there, and following each answer is a number in italics indicating the section in which that type of problem is discussed. Where weaknesses show up, review appropriate sections in the text.

A

1. Solve using substitution or elimination by addition:

$$3x - 5y = 11$$
$$2x + 3y = 1$$

2. Solve by graphing: $2x - y = -4$
$$3x + y = -1$$

3. Solve by substitution or elimination by addition:

$$x^2 + y^2 = 2$$
$$2x - y = 1$$

4. Solve by graphing: $3x + 5y \le 15$
$$x, y \ge 0$$

5. Find the maximum and minimum value of $z = 2x + 3y$ over the feasible region S:

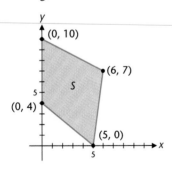

6. Perform the operations that are defined, given the following matrices:

$$M = \begin{bmatrix} 2 & 1 \\ 1 & -3 \end{bmatrix} \quad N = \begin{bmatrix} 1 & 2 \\ -1 & 3 \end{bmatrix}$$

$$P = \begin{bmatrix} 1 & 2 \end{bmatrix} \quad Q = \begin{bmatrix} -1 \\ 2 \end{bmatrix}$$

(A) $M - 2N$ (B) $P + Q$ (C) PQ
(D) MN (E) PN (F) QM

7. Evaluate: $\begin{vmatrix} 0 & 2 & 0 \\ 1 & 3 & 2 \\ -1 & 4 & 3 \end{vmatrix}$

8. Write the linear system corresponding to each augmented matrix and solve:

(A) $\begin{bmatrix} 1 & 0 & | & 3 \\ 0 & 1 & | & -4 \end{bmatrix}$ (B) $\begin{bmatrix} 1 & -2 & | & 3 \\ 0 & 0 & | & 0 \end{bmatrix}$

(C) $\begin{bmatrix} 1 & -2 & | & 3 \\ 0 & 0 & | & 1 \end{bmatrix}$

9. Given the system: $x_1 + x_2 = 3$
$$-x_1 + x_2 = 5$$
(A) Write the augmented matrix for the system.
(B) Transform the augmented matrix into reduced form.
(C) Write the solution to the system.

10. Given the system: $x_1 - 3x_2 = k_1$
$$2x_1 - 5x_2 = k_2$$
(A) Write the system as a matrix equation of the form $AX = B$.
(B) Find the inverse of the coefficient matrix A.
(C) Use A^{-1} to find the solution for $k_1 = -2$ and $k_2 = 1$.
(D) Use A^{-1} to find the solution for $k_1 = 1$ and $k_2 = -2$.

11. Given the system: $2x - 3y = 1$
$$4x - 5y = 2$$
(A) Find the determinant of the coefficient matrix.
(B) Solve the system using Cramer's rule.

12. Use Gauss–Jordan elimination to solve the system

$$x_1 + 3x_2 = 10$$
$$2x_1 - x_2 = -1$$

Then write the linear system represented by each augmented matrix in your solution, and solve each of these systems graphically. Discuss the relationship between the solutions of these systems.

13. Use an intersection routine on a graphing utility to approximate the solution of the following system to two decimal places:

$$-2x + 3y = 7$$
$$3x + 4y = 18$$

B

Solve Problems 14–16 using Gauss–Jordan elimination.

14.
$$x_1 + 2x_2 - x_3 = 3$$
$$x_2 + x_3 = -2$$
$$2x_1 + 3x_2 + x_3 = 0$$

15.
$$x_1 + x_2 - x_3 = 2$$
$$4x_2 + 6x_3 = -1$$
$$6x_2 + 9x_3 = 0$$

16.
$$x_1 - 2x_2 + x_3 = 1$$
$$3x_1 - 2x_2 - x_3 = -5$$

In Problems 17 and 18, solve each system.

17.
$$x^2 - 3xy + 3y^2 = 1$$
$$xy = 1$$

18.
$$x^2 - 3xy + y^2 = -1$$
$$x^2 - xy = 0$$

19. Given $M = \begin{bmatrix} 1 & 2 & -1 \end{bmatrix}$ and $N = \begin{bmatrix} 1 \\ -1 \\ 2 \end{bmatrix}$. Find:

(A) MN (B) NM

20. Given

$$L = \begin{bmatrix} 2 & -1 & 0 \\ 1 & 2 & 1 \end{bmatrix} \quad M = \begin{bmatrix} 1 & 2 \\ -1 & 0 \\ 1 & 1 \end{bmatrix} \quad N = \begin{bmatrix} 2 & 1 \\ -1 & 0 \end{bmatrix}$$

Find, if defined: (A) $LM - 2N$ (B) $ML + N$

21. Solve graphically and indicate whether the solution region is bounded or unbounded. Find the coordinates of each corner point.

$$3x + 2y \geq 12$$
$$x + 2y \geq 8$$
$$x, y \geq 0$$

22. Solve the linear programming problem:

$$\text{Maximize} \quad z = 4x + 9y$$
$$\text{Subject to} \quad x + 2y \leq 14$$
$$2x + y \leq 16$$
$$x, y \geq 0$$

23. Given the system:
$$x_1 + 4x_2 + 2x_3 = k_1$$
$$2x_1 + 6x_2 + 3x_3 = k_2$$
$$2x_1 + 5x_2 + 2x_3 = k_3$$

(A) Write the system as a matrix equation in the form $AX = B$.
(B) Find the inverse of the coefficient matrix A.
(C) Use A^{-1} to solve the system when $k_1 = -1, k_2 = 2$, and $k_3 = 1$.
(D) Use A^{-1} to solve the system when $k_1 = 2, k_2 = 0$, and $k_3 = -1$.

24. Given the system:
$$x + 2y - z = 1$$
$$2x + 8y + z = -2$$
$$-x + 3y + 5z = 2$$

(A) Evaluate the coefficient determinant D.
(B) Solve for z using Cramer's rule.

C

25. Use a graphing utility to approximate all real solutions to two decimal places:

$$x^2 + 2xy - y^2 = 1$$
$$9x^2 + 4xy + y^2 = 15$$

26. Discuss the number of solutions for the system corresponding to the reduced form shown below if
(A) $m = 0$ and $n = 0$ (B) $m = 0$ and $n \neq 0$
(C) $m \neq 0$

$$\begin{bmatrix} 1 & 0 & -5 & | & 2 \\ 0 & 1 & 3 & | & 6 \\ 0 & 0 & m & | & n \end{bmatrix}$$

27. If a square matrix A satisfies the equation $A^2 = A$, find A. Assume that A^{-1} exists.

28. Which of the following augmented matrices are in reduced form?

$$L = \begin{bmatrix} 1 & 0 & 0 & | & 2 \\ 0 & 1 & 0 & | & 0 \\ 0 & 0 & 1 & | & -1 \end{bmatrix} \quad M = \begin{bmatrix} 1 & 0 & 3 & | & 3 \\ 0 & 1 & -2 & | & 2 \\ 0 & 0 & 0 & | & 0 \end{bmatrix}$$

$$N = \begin{bmatrix} 0 & 0 & | & 0 \\ 1 & 0 & | & 2 \\ 0 & 1 & | & -3 \end{bmatrix} \quad P = \begin{bmatrix} 1 & 2 & 0 & 2 & | & -2 \\ 0 & 0 & 1 & 3 & | & 1 \end{bmatrix}$$

29. Show that

$$k\begin{vmatrix} a & b \\ c & d \end{vmatrix} = \begin{vmatrix} ka & b \\ kc & d \end{vmatrix}$$

30. Show that

$$\begin{vmatrix} a & b \\ c & d \end{vmatrix} = \begin{vmatrix} a & b \\ c + ka & d + kb \end{vmatrix}$$

31. If $M = \begin{vmatrix} a & b \\ c & d \end{vmatrix}$ and $\det M \neq 0$, show that

$$M^{-1} = \frac{1}{\det M}\begin{bmatrix} d & -b \\ -c & a \end{bmatrix}$$

*Recall that a square matrix is called **upper triangular** if all elements below the principal diagonal are zero, and it is called **diagonal** if all elements not on the principal diagonal are zero. A square matrix is called **lower triangular** if all elements above the principal diagonal are zero. In Problems 32–39, determine whether the statement is true or false. If true, explain why. If false, give a counterexample.*

32. The sum of two upper triangular matrices is upper triangular.

33. The product of two lower triangular matrices is lower triangular.

34. The sum of an upper triangular matrix and a lower triangular matrix is a diagonal matrix.

35. The product of an upper triangular matrix and a lower triangular matrix is a diagonal matrix.

36. A matrix that is both upper triangular and lower triangular is a diagonal matrix.

37. If a diagonal matrix has no zero elements on the principal diagonal, then it has an inverse.

38. The determinant of a diagonal matrix is the product of the elements on the principal diagonal.

39. The determinant of a lower triangular matrix is the product of the elements on the principal diagonal.

APPLICATIONS

40. **Finance.** An investor has $12,000 to invest. If part is invested at 8% and the rest in a higher-risk investment at 14%, how much should be invested at each rate to produce the same yield as if all had been invested at 10%?

41. **Diet.** In an experiment involving mice, a zoologist needs a food mix that contains, among other things, 23 grams of protein, 6.2 grams of fat, and 16 grams of moisture. She has on hand mixes of the following compositions: Mix A contains 20% protein, 2% fat, and 15% moisture; mix B contains 10% protein, 6% fat, and 10% moisture; and mix C contains 15% protein, 5% fat, and 5% moisture. How many grams of each mix should be used to get the desired diet mix?

42 **Purchasing.** A soft-drink distributor has budgeted $300,000 for the purchase of 12 new delivery trucks. If a model A truck costs $18,000, a model B truck costs $22,000, and a model C truck costs $30,000, how many trucks of each model should the distributor purchase to use exactly all the budgeted funds?

43. **Geometry.** Find the dimensions of a rectangle with perimeter 24 meters and area 32 square meters.

44. **Manufacturing.** A manufacturer makes two types of day packs, a standard model and a deluxe model. Each standard model requires 0.5 labor-hour from the fabricating department and 0.3 labor-hour from the sewing department. Each deluxe model requires 0.5 labor-hour from the fabricating department and 0.6 labor-hour from the sewing department. The maximum number of labor-hours available per week in the fabricating department and the sewing department are 300 and 240, respectively.
    (A) If the profit on a standard day pack is $8 and the profit on a deluxe day pack is $12, how many of each type of pack should be manufactured each day to realize a maximum profit? What is the maximum profit?
    (B) Discuss the effect on the production schedule and the maximum profit if the profit on a standard day pack decreases by $3 and the profit on a deluxe day pack increases by $3.
    (C) Discuss the effect on the production schedule and the maximum profit if the profit on a standard day pack increases by $3 and the profit on a deluxe day pack decreases by $3.

45. **Averaging Tests.** A teacher has given four tests to a class of five students and stored the results in the following matrix:

$$
\begin{array}{c}
 \\
\text{Ann} \\
\text{Bob} \\
\text{Carol} \\
\text{Dan} \\
\text{Eric}
\end{array}
\begin{array}{cccc}
 & \text{Tests} & & \\
1 & 2 & 3 & 4
\end{array}
\left[
\begin{array}{cccc}
78 & 84 & 81 & 86 \\
91 & 65 & 84 & 92 \\
95 & 90 & 92 & 91 \\
75 & 82 & 87 & 91 \\
83 & 88 & 81 & 76
\end{array}
\right] = M
$$

Discuss methods of matrix multiplication that the teacher can use to obtain the indicated information in parts A–C below. In each case, state the matrices to be used and then perform the necessary multiplications.
    (A) The average on all four tests for each student, assuming that all four tests are given equal weight
    (B) The average on all four tests for each student, assuming that the first three tests are given equal weight and the fourth is given twice this weight
    (C) The class average on each of the four tests

SEQUENCES AND SERIES

CHAPTER 8

If someone asked you to list all natural numbers that are perfect squares, you might begin by writing

$$1, \quad 4, \quad 9, \quad 16, \quad 25, \quad 36$$

But you would soon realize that it is impossible to actually list all the perfect squares, since there are an infinite number of them. However, this collection of numbers can be represented in several different ways. One common method is to write

$$1, 4, 9, \dots, n^2, \dots \quad\quad n \in N$$

where N is the set of natural numbers. A list of numbers such as this is generally called a *sequence*. Sequences and related topics form the subject matter of this chapter. One of the related topics involves a method of proof we have referred to several times earlier in this book—*mathematical induction*. This method enables us to prove conjectures involving infinite sets of successive integers.

SECTION 8-1 Sequences and Series

- Sequences
- Series

In this section we introduce special notation and formulas for representing and generating sequences and sums of sequences.

● Sequences

Consider the function f given by

$$f(n) = 2n - 1 \tag{1}$$

where the domain of f is the set of natural numbers N. Note that

$$f(1) = 1, f(2) = 3, f(3) = 5, \dots$$

The function f is an example of a sequence. A **sequence** is a function with domain a set of successive integers. However, a sequence is hardly ever represented in the form of equation (1). A special notation for sequences has evolved, which we describe here.

To start, the range value $f(n)$ is usually symbolized more compactly with a symbol such as a_n. Thus, in place of equation (1) we write

$$a_n = 2n - 1$$

The domain is understood to be the set of natural numbers N unless stated to the contrary or the context indicates otherwise. The elements in the range are called **terms** of the sequence: a_1 is the first term, a_2 the second term, and a_n the nth term, or the **general term:**

$$a_1 = 2(1) - 1 = 1 \quad \text{First term}$$
$$a_2 = 2(2) - 1 = 3 \quad \text{Second term}$$
$$a_3 = 2(3) - 1 = 5 \quad \text{Third term}$$
$$\vdots \qquad\qquad \vdots$$

The ordered list of elements

$$1, 3, 5, \ldots, 2n - 1, \ldots$$

in which the terms of a sequence are written in their natural order with respect to the domain values, is often informally referred to as a sequence. A sequence is also represented in the abbreviated form $\{a_n\}$, where a symbol for the nth term is placed between braces. For example, we can refer to the sequence

$$1, 3, 5, \ldots, 2n - 1, \ldots$$

as the sequence $\{2n - 1\}$.

If the domain of a function is a finite set of successive integers, then the sequence is called a **finite sequence.** If the domain is an infinite set of successive integers, then the sequence is called an **infinite sequence.** The sequence $\{2n - 1\}$ above is an example of an infinite sequence.

EXPLORE-DISCUSS 1

The sequence $\{2n - 1\}$ is a function whose domain is the set of natural numbers, and so it may be graphed in the same way as any function whose domain and range are sets of real numbers (see Fig. 1).

FIGURE 1 Graph of $\{2n - 1\}$.

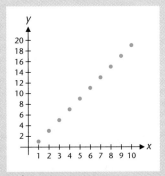

(A) Explain why the graph of the sequence $\{2n - 1\}$ is not continuous.

(B) Explain why the points on the graph of $\{2n - 1\}$ lie on a line. Find an equation for that line.

(C) Graph the sequence $\left\{\dfrac{2n^2 - n + 1}{n}\right\}$. How are the graphs of $\{2n - 1\}$ and $\left\{\dfrac{2n^2 - n + 1}{n}\right\}$ related?

Some sequences are specified by a **recursion formula**—that is, a formula that defines each term in terms of one or more preceding terms. The sequence we have chosen to illustrate a recursion formula is a very famous sequence in the history of mathematics called the **Fibonacci sequence.** It is named after the most celebrated mathematician of the thirteenth century, Leonardo Fibonacci from Italy (1180?–1250?).

EXAMPLE 1   Fibonacci Sequence

List the first six terms of the sequence specified by

$$a_1 = 1$$
$$a_2 = 1$$
$$a_n = a_{n-1} + a_{n-2} \qquad n \geq 3$$

Solution

$$a_1 \quad = 1$$
$$a_2 \quad = 1$$
$$a_3 \quad = a_2 + a_1 = 1 + 1 \quad = 2$$
$$a_4 \quad = a_3 + a_2 = 2 + 1 \quad = 3$$
$$a_5 \quad = a_4 + a_3 = 3 + 2 \quad = 5$$
$$a_6 \quad = a_5 + a_4 = 5 + 3 \quad = 8$$

The formula $a_n = a_{n-1} + a_{n-2}$ is a recursion formula that can be used to generate the terms of a sequence in terms of preceding terms. Of course, starting terms a_1 and a_2 must be provided in order to use the formula. Recursion formulas are particularly suitable for use with calculators and computers (see Problems 57 and 58 in Exercise 8-1).

Matched Problem 1   List the first five terms of the sequence specified by

$$a_1 = 4$$
$$a_n = \tfrac{1}{2}a_{n-1} \qquad n \geq 2$$

Now we consider the reverse problem. That is, can a sequence be defined just by listing the first three or four terms of the sequence? And can we then use these initial terms to find a formula for the nth term? In general, without other information, the answer to the first question is no. Many different sequences may start off with the same terms. For example, each of the following sequences starts off with the same three terms:

$$1, 3, 9, \ldots, 3^{n-1}, \ldots$$
$$1, 3, 9, \ldots, 1 + 2(n - 1)^2, \ldots$$
$$1, 3, 9, \ldots, 8n + \frac{12}{n} - 19, \ldots$$

However, these are certainly different sequences. You should verify that these sequences agree for the first three terms and differ in the fourth term by evaluating the general term for each sequence at $n = 1, 2, 3, 4$. Thus, simply listing the first three terms, or any other finite number of terms, does not specify a particular sequence. In fact, it can be shown that given any list of m numbers, there are an infinite number of sequences whose first m terms agree with these given numbers. What about the second question? That is, given a few terms, can we find the general formula for at least one sequence whose first few terms agree with the given terms? The answer to this question is a qualified yes. If we can observe a simple pattern in the given terms, then we may be able to construct a general term that will produce the pattern. The next example illustrates this approach.

EXAMPLE 2 **Finding the General Term of a Sequence**

Find the general term of a sequence whose first four terms are:

(A) $5, 6, 7, 8, \ldots$ (B) $2, -4, 8, -16, \ldots$

Solutions   (A) Since these terms are consecutive integers, one solution is $a_n = n$, $n \geq 5$. If we want the domain of the sequence to be all natural numbers, then another solution is $b_n = n + 4$.

(B) Each of these terms can be written as the product of a power of 2 and a power of -1:

$$2 = (-1)^0 2^1$$
$$-4 = (-1)^1 2^2$$
$$8 = (-1)^2 2^3$$
$$-16 = (-1)^3 2^4$$

If we choose the domain to be all natural numbers, then a solution is

$$a_n = (-1)^{n-1} 2^n$$

Matched Problem 2   Find the general term of a sequence whose first four terms are:

(A) $2, 4, 6, 8, \ldots$ (B) $1, -\frac{1}{2}, \frac{1}{4}, -\frac{1}{8}, \ldots$

In general, there is usually more than one way of representing the nth term of a given sequence. This was seen in the solution of Example 2A. However, unless stated to the contrary, we assume the domain of the sequence is the set of natural numbers N.

EXPLORE-DISCUSS 2   The sequence with general term $b_n = \dfrac{\sqrt{5}}{5}\left(\dfrac{1+\sqrt{5}}{2}\right)^n$ is closely related to the Fibonacci sequence. Compute the first 20 terms of both sequences and discuss the relationship. (The values of b_1, b_2, and b_3 are shown in Fig. 2.)

FIGURE 2

```
(1+√(5))/2→A
        1.618033989
√(5)/5*A
         .7236067977
Ans*A
        1.170820393
        1.894427191
```

• **Series**   If $a_1, a_2, a_3, \ldots, a_n, \ldots$ is a sequence, then the expression $a_1 + a_2 + a_3 + \ldots + a_n + \ldots$ is called a **series.** If the sequence is finite, the corresponding series is a **finite series.** If the sequence is infinite, the corresponding series is an **infinite series.** For example,

$$1, 2, 4, 8, 16 \qquad \text{Finite sequence}$$

$$1 + 2 + 4 + 8 + 16 \qquad \text{Finite series}$$

We restrict our discussion to finite series in this section.

Series are often represented in a compact form called **summation notation** using the symbol $\sum$, which is a stylized version of the Greek letter sigma. Consider the following examples:

$$\sum_{k=1}^{4} a_k = a_1 + a_2 + a_3 + a_4$$

$$\sum_{k=3}^{7} b_k = b_3 + b_4 + b_5 + b_6 + b_7$$

$$\sum_{k=0}^{n} c_k = c_0 + c_1 + c_2 + \cdots + c_n \qquad \begin{array}{l}\text{Domain is the set of integers}\\ \text{greater than or equal to 0.}\end{array}$$

The terms on the right are obtained from the expression on the left by successively replacing the **summing index** k with integers, starting with the first number indicated below $\sum$ and ending with the number that appears above $\sum$. Thus, for example, if we are given the sequence

$$\frac{1}{2}, \frac{1}{4}, \frac{1}{8}, \ldots, \frac{1}{2^n}$$

the corresponding series is

$$\frac{1}{2} + \frac{1}{4} + \frac{1}{8} + \cdots + \frac{1}{2^n}$$

or, more compactly,

$$\sum_{k=1}^{n} \frac{1}{2^k}$$

EXAMPLE 3 **Writing the Terms of a Series**

Write without summation notation: $\displaystyle\sum_{k=1}^{5} \frac{k-1}{k}$

Solution
$$\sum_{k=1}^{5} \frac{k-1}{k} = \frac{1-1}{1} + \frac{2-1}{2} + \frac{3-1}{3} + \frac{4-1}{4} + \frac{5-1}{5}$$

$$= 0 + \frac{1}{2} + \frac{2}{3} + \frac{3}{4} + \frac{4}{5}$$

Matched Problem 3 Write without summation notation: $\displaystyle\sum_{k=0}^{5} \frac{(-1)^k}{2k+1}$

If the terms of a series are alternately positive and negative, it is called an **alternating series**. Example 4 deals with the representation of such a series.

EXAMPLE 4 **Writing a Series in Summation Notation**

Write the following series using summation notation:

$$1 - \frac{1}{2} + \frac{1}{3} - \frac{1}{4} + \frac{1}{5} - \frac{1}{6}$$

(A) Start the summing index at $k = 1$.
(B) Start the summing index at $k = 0$.

Solutions (A) $(-1)^{k-1}$ provides the alternation of sign, and $1/k$ provides the other part of each term. Thus, we can write

$$\sum_{k=1}^{6} \frac{(-1)^{k-1}}{k}$$

as can be easily checked.

(B) $(-1)^k$ provides the alternation of sign, and $1/(k+1)$ provides the other part of each term. Thus, we write

$$\sum_{k=0}^{5} \frac{(-1)^k}{k+1}$$

as can be checked.

Matched Problem 4   Write the following series using summation notation:

$$1 - \frac{2}{3} + \frac{4}{9} - \frac{8}{27} + \frac{16}{81}$$

(A) Start with $k = 1$.   (B) Start with $k = 0$.

EXPLORE-DISCUSS 3   (A) Find the smallest number of terms of the infinite series

$$1 + \frac{1}{2} + \frac{1}{3} + \cdots + \frac{1}{n} + \cdots$$

that, when added together, give a number greater than 3.

(B) Find the smallest number of terms of the infinite series

$$\frac{1}{2} + \frac{1}{4} + \cdots + \frac{1}{2^n} + \cdots$$

that, when added together, give a number greater than 0.99. Greater than 0.999. Can the sum ever exceed 1? Explain.

Answers to Matched Problems

1. $4, 2, 1, \frac{1}{2}, \frac{1}{4}$ **2.** (A) $a_n = 2n$ (B) $a_n = (-1)^{n-1}(\frac{1}{2})^{n-1}$ **3.** $1 - \frac{1}{3} + \frac{1}{5} - \frac{1}{7} + \frac{1}{9} - \frac{1}{11}$

4. (A) $\sum_{k=1}^{5} (-1)^{k-1} \left(\frac{2}{3}\right)^{k-1}$ (B) $\sum_{k=0}^{4} (-1)^k \left(\frac{2}{3}\right)^k$

EXERCISE **8-1**

A

Write the first four terms for each sequence in Problems 1–6.

1. $a_n = 2n + 5$

2. $a_n = n^2 + 1$

3. $a_n = \dfrac{3n+1}{n+1}$

4. $a_n = (n+1)^n$

5. $a_n = 3^n + (-2)^n$

6. $a_n = \dfrac{(-1)^n}{n^3}$

7. Write the one-hundredth term of the sequence in Problem 1.

8. Write the fiftieth term of the sequence in Problem 2.

9. Write the two-hundredth term of the sequence in Problem 3.

10. Write the ninety-ninth term of the sequence in Problem 4.

In Problems 11–16, write each series in expanded form without summation notation.

11. $\sum_{k=1}^{5} k$

12. $\sum_{k=1}^{4} k^2$

13. $\sum_{k=1}^{3} \dfrac{1}{10^k}$

14. $\sum_{k=1}^{5} \left(\dfrac{1}{3}\right)^k$

15. $\sum_{k=1}^{4} (-1)^k$

16. $\sum_{k=1}^{6} (-1)^{k+1} k$

B

Write the first five terms of each sequence in Problems 17–26.

17. $a_n = (-1)^{n+1}n^2$

18. $a_n = (-1)^{n+1}\left(\dfrac{1}{2^n}\right)$

19. $a_n = \dfrac{1}{3}\left(1 - \dfrac{1}{10^n}\right)$

20. $a_n = n[1 - (-1)^n]$

21. $a_n = (-\tfrac{1}{2})^{n-1}$

22. $a_n = (-\tfrac{3}{2})^{n-1}$

23. $a_1 = 7; a_n = a_{n-1} - 4, n \geq 2$

24. $a_1 = a_2 = 1; a_n = a_{n-1} + a_{n-2}, n \geq 3$

25. $a_1 = 4; a_n = \tfrac{1}{4}a_{n-1}, n \geq 2$

26. $a_1 = 2; a_n = 2a_{n-1}, n \geq 2$

In Problems 27–38, find the general term of a sequence whose first four terms are given.

27. $3, 5, 7, 9, \ldots$

28. $2, -1, -4, -7, \ldots$

29. $4, 9, 16, 25, \ldots$

30. $24, 35, 48, 63, \ldots$

31. $4, 8, 16, 32, \ldots$

32. $1, 4, 27, 256, \ldots$

33. $1, \tfrac{3}{2}, \tfrac{5}{3}, \tfrac{7}{4}, \ldots$

34. $\tfrac{1}{2}, \tfrac{3}{4}, \tfrac{7}{8}, \tfrac{15}{16}, \ldots$

35. $7, -7, 7, -7, \ldots$

36. $1, -\tfrac{1}{5}, \tfrac{1}{25}, -\tfrac{1}{125}, \ldots$

37. $x, \dfrac{x^2}{2}, \dfrac{x^3}{4}, \dfrac{x^4}{8}, \ldots$

38. $-x, x^4, -x^7, x^{10}, \ldots$

Some graphing utilities use special routines to graph sequences (consult your manual). In Problems 39–42, use such routines to graph the first 20 terms of each sequence.

39. $a_n = 1/n$

40. $a_n = 2 + \pi n$

41. $a_n = (-0.9)^n$

42. $a_1 = -1, a_n = \tfrac{2}{3}a_{n-1} + \tfrac{1}{2}$

In Problems 43–48, write each series in expanded form without summation notation.

43. $\displaystyle\sum_{k=1}^{4} \dfrac{(-2)^{k+1}}{k}$

44. $\displaystyle\sum_{k=1}^{5} (-1)^{k+1}(2k-1)^2$

45. $\displaystyle\sum_{k=1}^{3} \dfrac{1}{k}x^{k+1}$

46. $\displaystyle\sum_{k=1}^{5} x^{k-1}$

47. $\displaystyle\sum_{k=1}^{5} \dfrac{(-1)^{k+1}}{k}x^k$

48. $\displaystyle\sum_{k=0}^{4} \dfrac{(-1)^k x^{2k+1}}{2k+1}$

In Problems 49–56, write each series using summation notation with the summing index k starting at k = 1.

49. $1^2 + 2^2 + 3^2 + 4^2$

50. $2 + 3 + 4 + 5 + 6$

51. $\dfrac{1}{2} + \dfrac{1}{2^2} + \dfrac{1}{2^3} + \dfrac{1}{2^4} + \dfrac{1}{2^5}$

52. $1 - \dfrac{1}{2} + \dfrac{1}{3} - \dfrac{1}{4}$

53. $1 + \dfrac{1}{2^2} + \dfrac{1}{3^2} + \cdots + \dfrac{1}{n^2}$

54. $2 + \dfrac{3}{2} + \dfrac{4}{3} + \cdots + \dfrac{n+1}{n}$

55. $1 - 4 + 9 - \cdots + (-1)^{n+1}n^2$

56. $\dfrac{1}{2} - \dfrac{1}{4} + \dfrac{1}{8} - \cdots + \dfrac{(-1)^{n+1}}{2^n}$

C

The sequence

$$a_n = \dfrac{a_{n-1}^2 + M}{2a_{n-1}} \qquad n \geq 2, M \text{ a positive real number}$$

can be used to find $\sqrt{M}$ to any decimal-place accuracy desired. To start the sequence, choose a_1 arbitrarily from the positive real numbers. Problems 57 and 58 are related to this sequence.

57. (A) Find the first four terms of the sequence

$$a_1 = 3 \qquad a_n = \dfrac{a_{n-1}^2 + 2}{2a_{n-1}} \qquad n \geq 2$$

(B) Compare the terms with $\sqrt{2}$ from a calculator.

(C) Repeat parts A and B letting a_1 be any other positive number, say 1.

58. (A) Find the first four terms of the sequence

$$a_1 = 2 \qquad a_n = \dfrac{a_{n-1}^2 + 5}{2a_{n-1}} \qquad n \geq 2$$

(B) Find $\sqrt{5}$ with a calculator, and compare with the results of part A.

(C) Repeat parts A and B letting a_1 be any other positive number, say 3.

59. Let $\{a_n\}$ denote the Fibonacci sequence and let $\{b_n\}$ denote the sequence defined by $b_1 = 1$, $b_2 = 3$, $b_n = b_{n-1} + b_{n-2}$ for $n \geq 3$. Compute 10 terms of the sequence $\{c_n\}$, where $c_n = b_n/a_n$. Describe the terms of $\{c_n\}$ for large values of n.

60. Define sequences $\{u_n\}$ and $\{v_n\}$ by $u_1 = 1$, $v_1 = 0$, $u_n = u_{n-1} + v_{n-1}$ and $v_n = u_{n-1}$ for $n \geq 2$. Find the first 10 terms of each sequence, and explain their relationship to the Fibonacci sequence.

∫ *In calculus, it can be shown that*

$$e^x = \sum_{k=0}^{\infty} \dfrac{x^k}{k!} \approx 1 + \dfrac{x}{1!} + \dfrac{x^2}{2!} + \dfrac{x^3}{3!} + \cdots + \dfrac{x^n}{n!}$$

where the larger n is, the better the approximation. Problems 61 and 62 refer to this series. Note that n!, read "n factorial," is defined by $0! = 1$ and $n! = 1 \cdot 2 \cdot 3 \cdots \cdot n$ for $n \in N$.

61. Approximate $e^{0.2}$ using the first five terms of the series. Compare this approximation with your calculator evaluation of $e^{0.2}$.

62. Approximate $e^{-0.5}$ using the first five terms of the series. Compare this approximation with your calculator evaluation of $e^{-0.5}$.

63. Show that: $\displaystyle\sum_{k=1}^{n} ca_k = c \sum_{k=1}^{n} a_k$

64. Show that: $\displaystyle\sum_{k=1}^{n} (a_k + b_k) = \sum_{k=1}^{n} a_k + \sum_{k=1}^{n} b_k$

SECTION 8-2 Mathematical Induction

- Introduction
- Mathematical Induction
- Additional Examples of Mathematical Induction
- Three Famous Problems

● **Introduction**

In common usage, the word "induction" means the generalization from particular cases or facts. The ability to formulate general hypotheses from a limited number of facts is a distinguishing characteristic of a creative mathematician. The creative process does not stop here, however. These hypotheses must then be proved or disproved. In mathematics, a special method of proof called **mathematical induction** ranks among the most important basic tools in a mathematician's toolbox. In this section mathematical induction will be used to prove a variety of mathematical statements, some new and some that up to now we have just assumed to be true.

We illustrate the process of formulating hypotheses by an example. Suppose we are interested in the sum of the first n consecutive odd integers, where n is a positive integer. We begin by writing the sums for the first few values of n to see if we can observe a pattern:

$$1 = 1 \qquad n = 1$$
$$1 + 3 = 4 \qquad n = 2$$
$$1 + 3 + 5 = 9 \qquad n = 3$$
$$1 + 3 + 5 + 7 = 16 \qquad n = 4$$
$$1 + 3 + 5 + 7 + 9 = 25 \qquad n = 5$$

Is there any pattern to the sums 1, 4, 9, 16, and 25? You no doubt observed that each is a perfect square and, in fact, each is the square of the number of terms in the sum. Thus, the following conjecture seems reasonable:

Conjecture P_n: For each positive integer n,

$$1 + 3 + 5 + \cdots + (2n - 1) = n^2$$

That is, the sum of the first n odd integers is n^2 for each positive integer n.

So far ordinary induction has been used to generalize the pattern observed in the first few cases listed above. But at this point conjecture P_n is simply that—a conjecture. How do we prove that P_n is a true statement? Continuing to list specific cases will never provide a general proof—not in your lifetime or all your descendants' lifetimes! Mathematical induction is the tool we will use to establish the validity of conjecture P_n.

Before discussing this method of proof, let's consider another conjecture:

Conjecture Q_n: For each positive integer n, the number $n^2 - n + 41$ is a prime number.

TABLE 1

n	$n^2 - n + 41$	Prime?
1	41	Yes
2	43	Yes
3	47	Yes
4	53	Yes
5	61	Yes

It is important to recognize that a conjecture can be proved false if it fails for only one case. A single case or example for which a conjecture fails is called a **counterexample.** We check the conjecture for a few particular cases in Table 1. From the table, it certainly appears that conjecture Q_n has a good chance of being true. You may want to check a few more cases. If you persist, you will find that conjecture Q_n is true for n up to 41. What happens at $n = 41$?

$$41^2 - 41 + 41 = 41^2$$

which is not prime. Thus, since $n = 41$ provides a counterexample, conjecture Q_n is false. Here we see the danger of generalizing without proof from a few special cases. This example was discovered by Euler (1707–1783).

EXPLORE-DISCUSS 1   Prove that the following statement is false by finding a counterexample: If $n \geq 2$, then at least one-third of the positive integers less than or equal to n are prime.

• **Mathematical Induction**   We begin by stating the *principle of mathematical induction,* which forms the basis for all our work in this section.

Theorem 1   **Principle of Mathematical Induction**

Let P_n be a statement associated with each positive integer n, and suppose the following conditions are satisfied:

1. P_1 is true.

2. For any positive integer k, if P_k is true, then P_{k+1} is also true.

Then the statement P_n is true for all positive integers n.

Theorem 1 must be read very carefully. At first glance, it seems to say that if we assume a statement is true, then it is true. But that is not the case at all. If the two conditions in Theorem 1 are satisfied, then we can reason as follows:

P_1 is true.	Condition 1
P_2 is true, because P_1 is true.	Condition 2
P_3 is true, because P_2 is true.	Condition 2
P_4 is true, because P_3 is true.	Condition 2
$\vdots$	$\vdots$

Condition 1: The first domino
can be pushed over.
(a)

Condition 2: If the kth domino
falls, then so does the $(k + 1)$st.
(b)

Conclusion: All the dominoes
will fall.
(c)

FIGURE 1 Interpreting mathematical induction.

Since this chain of implications never ends, we will eventually reach P_n for any positive integer n.

To help visualize this process, picture a row of dominoes that goes on forever (see Fig. 1) and interpret the conditions in Theorem 1 as follows: Condition 1 says that the first domino can be pushed over. Condition 2 says that if the kth domino falls, then so does the $(k + 1)$st domino. Together, these two conditions imply that all the dominoes must fall.

Now, to illustrate the process of proof by mathematical induction, we return to the conjecture P_n discussed earlier, which we restate below:

$$P_n: \quad 1 + 3 + 5 + \cdots + (2n - 1) = n^2 \qquad n \text{ any positive integer}$$

We already know that P_1 is a true statement. In fact, we demonstrated that P_1 through P_5 are all true by direct calculation. Thus, condition 1 in Theorem 1 is satisfied. To show that condition 2 is satisfied, we assume that P_k is a true statement:

$$P_k: 1 + 3 + 5 + \cdots + (2k - 1) = k^2$$

Now we must show that this assumption implies that P_{k+1} is also a true statement:

$$P_{k+1}: \quad 1 + 3 + 5 + \cdots + (2k - 1) + (2k + 1) = (k + 1)^2$$

Since we have assumed that P_k is true, we can perform operations on this equation. Note that the left side of P_{k+1} is the left side of P_k plus $(2k + 1)$. So we start by adding $(2k + 1)$ to both sides of P_k:

$$1 + 3 + 5 + \cdots + (2k - 1) = k^2 \qquad P_k$$

$$1 + 3 + 5 + \cdots + (2k - 1) + (2k + 1) = k^2 + (2k + 1) \qquad \text{Add } 2k + 1 \text{ to both sides.}$$

Factoring the right side of this equation, we have

$$1 + 3 + 5 + \cdots + (2k - 1) + (2k + 1) = (k + 1)^2 \qquad P_{k+1}$$

But this last equation is P_{k+1}. Thus, we have started with P_k, the statement we assumed true, and performed valid operations to produce P_{k+1}, the statement we want to be true. In other words, we have shown that if P_k is true, then P_{k+1} is also true. Since both conditions in Theorem 1 are satisfied, P_n is true for all positive integers n.

● **Additional Examples of Mathematical Induction**

Now we will consider some additional examples of proof by induction. The first is another summation formula. Mathematical induction is the primary tool for proving that formulas of this type are true.

EXAMPLE 1 **Proving a Summation Formula**

Prove that for all positive integers n

$$\frac{1}{2} + \frac{1}{4} + \frac{1}{8} + \cdots + \frac{1}{2^n} = \frac{2^n - 1}{2^n}$$

Proof   State the conjecture:

$$P_n\colon\quad \frac{1}{2} + \frac{1}{4} + \frac{1}{8} + \cdots + \frac{1}{2^n} = \frac{2^n - 1}{2^n}$$

Part 1   Show that P_1 is true.

$$P_1\colon\quad \frac{1}{2} = \frac{2^1 - 1}{2^1}$$

$$= \frac{1}{2}$$

Thus, P_1 is true.

Part 2   Show that if P_k is true, then P_{k+1} is true. It is a good practice to always write out both P_k and P_{k+1} at the beginning of any induction proof to see what is assumed and what must be proved:

$$P_k\colon\quad \frac{1}{2} + \frac{1}{4} + \frac{1}{8} + \cdots + \frac{1}{2^k} = \frac{2^k - 1}{2^k} \qquad \text{We assume } P_k \text{ is true.}$$

$$P_{k+1}\colon\quad \frac{1}{2} + \frac{1}{4} + \frac{1}{8} + \cdots + \frac{1}{2^k} + \frac{1}{2^{k+1}} = \frac{2^{k+1} - 1}{2^{k+1}} \qquad \text{We must show that } P_{k+1} \text{ follows from } P_k.$$

We start with the true statement P_k, add $1/2^{k+1}$ to both sides, and simplify the right side:

$$\frac{1}{2} + \frac{1}{4} + \frac{1}{8} + \cdots + \frac{1}{2^k} = \frac{2^k - 1}{2^k} \qquad\qquad P_k$$

$$\frac{1}{2} + \frac{1}{4} + \frac{1}{8} + \cdots + \frac{1}{2^k} + \frac{1}{2^{k+1}} = \frac{2^k - 1}{2^k} + \frac{1}{2^{k+1}}$$

$$= \frac{2^k - 1}{2^k} \cdot \frac{2}{2} + \frac{1}{2^{k+1}}$$

$$= \frac{2^{k+1} - 2 + 1}{2^{k+1}}$$

$$= \frac{2^{k+1} - 1}{2^{k+1}}$$

Thus,

$$\frac{1}{2} + \frac{1}{4} + \frac{1}{8} + \cdots + \frac{1}{2^k} + \frac{1}{2^{k+1}} = \frac{2^{k+1} - 1}{2^{k+1}} \qquad P_{k+1}$$

and we have shown that if P_k is true, then P_{k+1} is true.

Conclusion   Both conditions in Theorem 1 are satisfied. Thus, P_n is true for all positive integers n.

Matched Problem 1   Prove that for all positive integers n

$$1 + 2 + 3 + \cdots + n = \frac{n(n + 1)}{2}$$

The next example provides a proof of a law of exponents that previously we had to assume was true. First we redefine a^n for n a positive integer, using a recursion formula:

DEFINITION 1   **Recursive Definition of a^n**

For n a positive integer

$$a^1 = a$$
$$a^{n+1} = a^n a \quad\quad n \geq 1$$

EXAMPLE 2   **Proving a Law of Exponents**

Prove that $(xy)^n = x^n y^n$ for all positive integers n.

Proof   State the conjecture:

$$P_n: \quad (xy)^n = x^n y^n$$

Part 1   Show that P_1 is true.

$$(xy)^1 = xy \quad\quad \text{Definition 1}$$
$$= x^1 y^1 \quad\quad \text{Definition 1}$$

Thus, P_1 is true.

Part 2   Show that if P_k is true, then P_{k+1} is true.

$$P_k: \quad (xy)^k = x^k y^k \quad\quad \text{Assume } P_k \text{ is true.}$$
$$P_{k+1}: \quad (xy)^{k+1} = x^{k+1} y^{k+1} \quad\quad \text{Show that } P_{k+1} \text{ follows from } P_k.$$

Here we start with the left side of P_{k+1} and use P_k to find the right side of P_{k+1}:

$$(xy)^{k+1} = (xy)^k (xy)^1 \quad\quad \text{Definition 1}$$
$$= x^k y^k xy \quad\quad \text{Use } P_k: \; (xy)^k = x^k y^k$$
$$= (x^k x)(y^k y) \quad\quad \text{Property of real numbers}$$
$$= x^{k+1} y^{k+1} \quad\quad \text{Definition 1}$$

Thus, $(xy)^{k+1} = x^{k+1}y^{k+1}$, and we have shown that if P_k is true, then P_{k+1} is true.

Conclusion   Both conditions in Theorem 1 are satisfied. Thus, P_n is true for all positive integers n.

Matched Problem 2   Prove that $(x/y)^n = x^n/y^n$ for all positive integers n.

Our last example deals with factors of integers. Before we start, recall that an integer p is *divisible* by an integer q if $p = qr$ for some integer r.

EXAMPLE 3   **Proving a Divisibility Property**

Prove that $4^{2n} - 1$ is divisible by 5 for all positive integers n.

Proof   Use the definition of divisibility to state the conjecture as follows:

$$P_n: \quad 4^{2n} - 1 = 5r \qquad \text{for some integer } r$$

Part 1   Show that P_1 is true.

$$P_1: \quad 4^2 - 1 = 15 = 5 \cdot 3$$

Thus, P_1 is true.

Part 2   Show that if P_k is true, then P_{k+1} is true.

$$P_k: \quad 4^{2k} - 1 = 5r \qquad \text{for some integer } r \qquad \text{Assume } P_k \text{ is true.}$$
$$P_{k+1}: \quad 4^{2(k+1)} - 1 = 5s \qquad \text{for some integer } s \qquad \text{Show that } P_{k+1} \text{ must follow.}$$

As before, we start with the true statement P_k:

$$
\begin{aligned}
4^{2k} - 1 &= 5r & & P_k \\
4^2(4^{2k} - 1) &= 4^2(5r) & & \text{Multiply both sides by } 4^2. \\
4^{2k+2} - 16 &= 80r & & \text{Simplify.} \\
4^{2(k+1)} - 1 &= 80r + 15 & & \text{Add 15 to both sides.} \\
&= 5(16r + 3) & & \text{Factor out 5.}
\end{aligned}
$$

Thus,

$$4^{2(k+1)} - 1 = 5s \qquad\qquad P_{k+1}$$

where $s = 16r + 3$ is an integer, and we have shown that if P_k is true, then P_{k+1} is true.

Conclusion   Both conditions in Theorem 1 are satisfied. Thus, P_n is true for all positive integers n.

Matched Problem 3   Prove that $8^n - 1$ is divisible by 7 for all positive integers n.

In some cases, a conjecture may be true only for $n \geq m$, where m is a positive integer, rather than for all $n \geq 0$. For example, see Problems 49 and 50 in Exercise 8-2. The principle of mathematical induction can be extended to cover cases like this as follows:

Theorem 2   **Extended Principle of Mathematical Induction**

Let m be a positive integer, let P_n be a statement associated with each integer $n \geq m$, and suppose the following conditions are satisfied:

1. P_m is true.
2. For any integer $k \geq m$, if P_k is true, then P_{k+1} is also true.

Then the statement P_n is true for all integers $n \geq m$.

• **Three Famous Problems**

The problem of determining whether a certain statement about the positive integers is true may be extremely difficult. Proofs may require remarkable insight and ingenuity and the development of techniques far more advanced than mathematical induction. Consider, for example, the famous problems of proving the following statements:

1. **Lagrange's Four Square Theorem, 1772:** Each positive integer can be expressed as the sum of four or fewer squares of positive integers.

2. **Fermat's Last Theorem, 1637:** For $n > 2$, $x^n + y^n = z^n$ does not have solutions in the natural numbers.

3. **Goldbach's Conjecture, 1742:** Every positive even integer greater than 2 is the sum of two prime numbers.

The first statement was considered by the early Greeks and finally proved in 1772 by Lagrange. Fermat's last theorem, defying the best mathematical minds for over 350 years, finally succumbed to a 200-page proof by Prof. Andrew Wiles of Princeton University in 1993. To this date no one has been able to prove or disprove Goldbach's conjecture.

EXPLORE-DISCUSS 2   (A) Explain the difference between a theorem and a conjecture.

(B) Why is "Fermat's last theorem" a misnomer? Suggest more accurate names for the result.

Answers to Matched Problems

1. Sketch of proof. State the conjecture: P_n: $\quad 1 + 2 + 3 + \cdots + n = \dfrac{n(n+1)}{2}$

$Part\ 1.$ $1 = \dfrac{1(1+1)}{2}$. P_1 is true.

$Part\ 2.$ Show that if P_k is true, then P_{k+1} is true.

$$1 + 2 + 3 + \cdots + k = \frac{k(k+1)}{2} \qquad\qquad P_k$$

$$1 + 2 + 3 + \ldots + k + (k+1) = \frac{k(k+1)}{2} + (k+1)$$

$$= \frac{(k+1)(k+2)}{2} \qquad P_{k+1}$$

Conclusion: P_n is true.

2. Sketch of proof. State the conjecture: P_n: $\quad \left(\dfrac{x}{y}\right)^n = \dfrac{x^n}{y^n}$

$Part\ 1.$ $\left(\dfrac{x}{y}\right)^1 = \dfrac{x}{y} = \dfrac{x^1}{y^1}$. P_1 is true.

$Part\ 2.$ Show that if P_k is true, then P_{k+1} is true.

$$\left(\frac{x}{y}\right)^{k+1} = \left(\frac{x}{y}\right)^k\left(\frac{x}{y}\right) = \frac{x^k}{y^k}\left(\frac{x}{y}\right) = \frac{x^k x}{y^k y} = \frac{x^{k+1}}{y^{k+1}}$$

Conclusion: P_n is true.

3. Sketch of proof. State the conjecture: P_n: $\quad 8^n - 1 = 7r \qquad$ for some integer r

$Part\ 1.$ $8^1 - 1 = 7 = 7 \cdot 1$. P_1 is true.

$Part\ 2.$ Show that if P_k is true, then P_{k+1} is true.

$$8^k - 1 = 7r \qquad\qquad P_k$$

$$8(8^k - 1) = 8(7r)$$

$$8^{k+1} - 1 = 56r + 7 = 7(8r+1) = 7s \qquad P_{k+1}$$

Conclusion: P_n is true.

EXERCISE 8-2

A

In Problems 1–4, find the first positive integer n that causes the statement to fail.

1. $3^n + 4^n \geq 5^n$ **2.** $n^2 - 3n < 100$

3. $17^n - 1$ is divisible by 2^n **4.** $n^2 = 5n - 6$

Verify each statement P_n in Problems 5–10 for $n = 1, 2,$ and 3.

5. P_n: $2 + 6 + 10 + \ldots + (4n - 2) = 2n^2$

6. P_n: $4 + 8 + 12 + \ldots + 4n = 2n(n+1)$

7. P_n: $a^5 a^n = a^{5+n}$ **8.** P_n: $(a^5)^n = a^{5n}$

9. P_n: $9^n - 1$ is divisible by 4

10. P_n: $4^n - 1$ is divisible by 3

Write P_k and P_{k+1} for P_n as indicated in Problems 11–16.

11. P_n in Problem 5   **12.** P_n in Problem 6

13. P_n in Problem 7   **14.** P_n in Problem 8

15. P_n in Problem 9   **16.** P_n in Problem 10

In Problems 17–22, use mathematical induction to prove that each P_n holds for all positive integers n.

17. P_n in Problem 5   **18.** P_n in Problem 6

19. P_n in Problem 7   **20.** P_n in Problem 8

21. P_n in Problem 9   **22.** P_n in Problem 10

B

In Problems 23–26, prove the statement is false by finding a counterexample.

23. If $n > 2$, then any polynomial of degree n has at least one real zero.

24. Any positive integer $n > 7$ can be written as the sum of three or fewer squares of positive integers.

25. If n is a positive integer, then there is at least one prime number p such that $n < p < n + 6$.

26. If a, b, c, d are positive integers such that $a^2 + b^2 = c^2 + d^2$, then $a = c$ or $a = d$.

In Problems 27–42, use mathematical induction to prove each proposition for all positive integers n, unless restricted otherwise.

27. $2 + 2^2 + 2^3 + \cdots + 2^n = 2^{n+1} - 2$

28. $\dfrac{1}{2} + \dfrac{1}{4} + \dfrac{1}{8} + \cdots + \dfrac{1}{2^n} = 1 - \left(\dfrac{1}{2}\right)^n$

29. $1^2 + 3^2 + 5^2 + \cdots + (2n - 1)^2 = \frac{1}{3}(4n^3 - n)$

30. $1 + 8 + 16 + \cdots + 8(n - 1) = (2n - 1)^2; n > 1$

31. $1^2 + 2^2 + 3^2 + \cdots + n^2 = \dfrac{n(n + 1)(2n + 1)}{6}$

32. $1 \cdot 2 + 2 \cdot 3 + 3 \cdot 4 + \cdots$
$$+ n(n + 1) = \dfrac{n(n + 1)(n + 2)}{3}$$

33. $\dfrac{a^n}{a^3} = a^{n-3}; n > 3$ **34.** $\dfrac{a^5}{a^n} = \dfrac{1}{a^{n-5}}; n > 5$

35. $a^m a^n = a^{m+n}; m, n \in N$
[Hint: Choose m as an arbitrary element of N, and then use induction on n.]

36. $(a^n)^m = a^{mn}; m, n \in N$

37. $x^n - 1$ is divisible by $x - 1$; $x \neq 1$
[Hint: Divisible means that $x^n - 1 = (x - 1)Q(x)$ for some polynomial $Q(x)$.]

38. $x^n - y^n$ is divisible by $x - y$; $x \neq y$

39. $x^{2n} - 1$ is divisible by $x - 1$; $x \neq 1$

40. $x^{2n} - 1$ is divisible by $x + 1$; $x \neq -1$

41. $1^3 + 2^3 + 3^3 + \cdots + n^3 = (1 + 2 + 3 + \cdots + n)^2$
[Hint: See Matched Problem 1 following Example 1.]

42. $\dfrac{1}{1 \cdot 2 \cdot 3} + \dfrac{1}{2 \cdot 3 \cdot 4} + \dfrac{1}{3 \cdot 4 \cdot 5} + \cdots$
$$+ \dfrac{1}{n(n + 1)(n + 2)} = \dfrac{n(n + 3)}{4(n + 1)(n + 2)}$$

C

In Problems 43–46, suggest a formula for each expression, and prove your hypothesis using mathematical induction, $n \in N$.

43. $2 + 4 + 6 + \cdots + 2n$

44. $\dfrac{1}{1 \cdot 2} + \dfrac{1}{2 \cdot 3} + \dfrac{1}{3 \cdot 4} + \cdots + \dfrac{1}{n(n + 1)}$

45. The number of lines determined by n points in a plane, no three of which are collinear

46. The number of diagonals in a polygon with n sides

In Problems 47–50, prove the statement is true for all integers n as specified.

47. $a > 1 \Rightarrow a^n > 1; n \in N$

48. $0 < a < 1 \Rightarrow 0 < a^n < 1; n \in N$

49. $n^2 > 2n; n \geq 3$ **50.** $2^n > n^2; n \geq 5$

51. Prove or disprove the generalization of the following two facts:
$$3^2 + 4^2 = 5^2$$
$$3^3 + 4^3 + 5^3 = 6^3$$

52. Prove or disprove: $n^2 + 21n + 1$ is a prime number for all natural numbers n.

If $\{a_n\}$ and $\{b_n\}$ are two sequences, we write $\{a_n\} = \{b_n\}$ if and only if $a_n = b_n$, $n \in N$. In Problems 53–56, use mathematical induction to show that $\{a_n\} = \{b_n\}$.

53. $a_1 = 1, a_n = a_{n-1} + 2; b_n = 2n - 1$

54. $a_1 = 2, a_n = a_{n-1} + 2; b_n = 2n$

55. $a_1 = 2, a_n = 2^2 a_{n-1}; b_n = 2^{2n-1}$

56. $a_1 = 2, a_n = 3a_{n-1}; b_n = 2 \cdot 3^{n-1}$

SECTION 8-3 Arithmetic and Geometric Sequences

- Arithmetic and Geometric Sequences
- nth-Term Formulas
- Sum Formulas for Finite Arithmetic Series
- Sum Formulas for Finite Geometric Series
- Sum Formula for Infinite Geometric Series

For most sequences it is difficult to sum an arbitrary number of terms of the sequence without adding term by term. But particular types of sequences, *arithmetic sequences* and *geometric sequences,* have certain properties that lead to convenient and useful formulas for the sums of the corresponding *arithmetic series* and *geometric series.*

• Arithmetic and Geometric Sequences

The sequence 5, 7, 9, 11, 13, ..., $5 + 2(n-1), \ldots$, where each term after the first is obtained by adding 2 to the preceding term, is an example of an arithmetic sequence. The sequence 5, 10, 20, 40, 80, ..., $5(2)^{n-1}, \ldots$, where each term after the first is obtained by multiplying the preceding term by 2, is an example of a geometric sequence.

DEFINITION 1

Arithmetic Sequence

A sequence

$$a_1, a_2, a_3, \ldots, a_n, \ldots$$

is called an **arithmetic sequence,** or **arithmetic progression,** if there exists a constant d, called the **common difference,** such that

$$a_n - a_{n-1} = d$$

That is,

$$a_n = a_{n-1} + d \qquad \text{for every } n > 1$$

DEFINITION 2

Geometric Sequence

A sequence

$$a_1, a_2, a_3, \ldots, a_n, \ldots$$

is called a **geometric sequence,** or **geometric progression,** if there exists a nonzero constant r, called the **common ratio,** such that

$$\frac{a_n}{a_{n-1}} = r$$

That is,

$$a_n = r a_{n-1} \qquad \text{for every } n > 1$$

EXPLORE-DISCUSS 1   (A) Graph the arithmetic sequence 5, 7, 9, ...
Describe the graphs of all arithmetic sequences with common difference 2.

(B) Graph the geometric sequence 5, 10, 20, ...
Describe the graphs of all geometric sequences with common ratio 2.

EXAMPLE 1   **Recognizing Arithmetic and Geometric Sequences**

Which of the following can be the first four terms of an arithmetic sequence? Of a geometric sequence?

(A) 1, 2, 3, 5, . . .   (B) −1, 3, −9, 27, . . .
(C) 3, 3, 3, 3, . . .   (D) 10, 8.5, 7, 5.5, . . .

Solution   (A) Since $2 - 1 \neq 5 - 3$, there is no common difference, so the sequence is not an arithmetic sequence. Since $2/1 \neq 3/2$, there is no common ratio, so the sequence is not geometric either.

(B) The sequence is geometric with common ratio −3, but it is not arithmetic.

(C) The sequence is arithmetic with common difference 0 and it is also geometric with common ratio 1.

(D) The sequence is arithmetic with common difference −1.5, but it is not geometric.

Matched Problem 1   Which of the following can be the first four terms of an arithmetic sequence? Of a geometric sequence?

(A) 8, 2, 0.5, 0.125, . . .
(B) −7, −2, 3, 8, . . .
(C) 1, 5, 25, 100, . . .

• **nth-Term Formulas**   If $\{a_n\}$ is an arithmetic sequence with common difference d, then

$$a_2 = a_1 + d$$
$$a_3 = a_2 + d = a_1 + 2d$$
$$a_4 = a_3 + d = a_1 + 3d$$

This suggests Theorem 1, which can be proved by mathematical induction (see Problem 63 in Exercise 8-3).

49. $a_n = 1,000 (0.99)^n$, $b_n = 2n + 1$

50. $a_n = 500 - n$, $b_n = 1.05^n$

In Problems 51–56, find the sum of each infinite geometric series that has a sum.

51. $3 + 1 + \frac{1}{3} + \cdots$

52. $16 + 4 + 1 + \cdots$

53. $2 + 4 + 8 + \cdots$

54. $4 + 6 + 9 + \cdots$

55. $2 - \frac{1}{2} + \frac{1}{8} - \cdots$

56. $21 - 3 + \frac{3}{7} - \cdots$

In Problems 57–62, represent each repeating decimal as the quotient of two integers.

57. $0.\overline{7} = 0.7777 \cdots$

58. $0.\overline{5} = 0.5555 \cdots$

59. $0.\overline{54} = 0.545\,454 \cdots$

60. $0.\overline{27} = 0.272\,727 \cdots$

61. $3.\overline{216} = 3.216\,216\,216 \cdots$

62. $5.\overline{63} = 5.636\,363 \cdots$

C

63. Prove, using mathematical induction, that if $\{a_n\}$ is an arithmetic sequence, then

$$a_n = a_1 + (n - 1)d \qquad \text{for every } n > 1$$

64. Prove, using mathematical induction, that if $\{a_n\}$ is an arithmetic sequence, then

$$S_n = \frac{n}{2} [2a_1 + (n - 1)d]$$

65. If in a given sequence, $a_1 = -2$ and $a_n = -3a_{n-1}$, $n > 1$, find a_n in terms of n.

66. For the sequence in Problem 65, find $S_n = \sum_{k=1}^{n} a_k$ in terms of n.

67. Show that $(x^2 + xy + y^2)$, $(z^2 + xz + x^2)$, and $(y^2 + yz + z^2)$ are consecutive terms of an arithmetic progression if x, y, and z form an arithmetic progression. (From U.S.S.R. Mathematical Olympiads, 1955–1956, Grade 9.)

68. Take 121 terms of each arithmetic progression 2, 7, 12, . . . and 2, 5, 8, ... How many numbers will there be in common? (From U.S.S.R. Mathematical Olympiads, 1955–1956, Grade 9.)

69. Prove, using mathematical induction, that if $\{a_n\}$ is a geometric sequence, then

$$a_n = a_1 r^{n-1} \qquad n \in N$$

70. Prove, using mathematical induction, that if $\{a_n\}$ is a geometric sequence, then

$$S_n = \frac{a_1 - a_1 r^n}{1 - r} \qquad n \in N, r \neq 1$$

71. Given the system of equations

$$ax + by = c$$
$$dx + ey = f$$

where a, b, c, d, e, f is any arithmetic progression with a nonzero constant difference, show that the system has a unique solution.

72. The sum of the first and fourth terms of an arithmetic sequence is 2, and the sum of their squares is 20. Find the sum of the first eight terms of the sequence.

***

APPLICATIONS

73. **Business.** In investigating different job opportunities, you find that firm A will start you at $25,000 per year and guarantee you a raise of $1,200 each year while firm B will start you at $28,000 per year but will guarantee you a raise of only $800 each year. Over a period of 15 years, how much would you receive from each firm?

74. **Business.** In Problem 73, what would be your annual salary at each firm for the tenth year?

75. **Economics.** The government, through a subsidy program, distributes $1,000,000. If we assume that each individual or agency spends 0.8 of what is received, and 0.8 of this is spent, and so on, how much total increase in spending results from this government action?

76. **Economics.** Due to reduced taxes, an individual has an extra $600 in spendable income. If we assume that the individual spends 70% of this on consumer goods, that the producers of these goods in turn spend 70% of what they receive on consumer goods, and that this process continues indefinitely, what is the total amount spent on consumer goods?

★ **77.** **Business.** If $P is invested at $100r\%$ compounded annually, the amount A present after n years forms a geometric progression with a common ratio $1 + r$. Write a formula for the amount present after n years. How long will it take a sum of money P to double if invested at 6% interest compounded annually?

★ **78.** **Population Growth.** If a population of A_0 people grows at the constant rate of $100r\%$ per year, the population after t years forms a geometric progression with a common ratio $1 + r$. Write a formula for the total population after t years. If the world's population is increasing at the rate of 2% per year, how long will it take to double?

79. **Finance.** Eleven years ago an investment earned $7,000 for the year. Last year the investment earned $14,000. If the earnings from the investment have increased the same amount each year, what is the yearly increase and how much income has accrued from the investment over the past 11 years?

80. Air Temperature. As dry air moves upward, it expands. In so doing, it cools at the rate of about 5°F for each 1,000-foot rise. This is known as the **adiabatic process.**
(A) Temperatures at altitudes that are multiples of 1,000 feet form what kind of a sequence?
(B) If the ground temperature is 80°F, write a formula for the temperature T_n in terms of n, if n is in thousands of feet.

81. Engineering. A rotating flywheel coming to rest rotates 300 revolutions the first minute (see figure). If in each subsequent minute it rotates two-thirds as many times as in the preceding minute, how many revolutions will the wheel make before coming to rest?

82. Physics. The first swing of a bob on a pendulum is 10 inches. If on each subsequent swing it travels 0.9 as far as on the preceding swing, how far will the bob travel before coming to rest?

83. Food Chain. A plant is eaten by an insect, an insect by a trout, a trout by a salmon, a salmon by a bear, and the bear is eaten by you. If only 20% of the energy is transformed from one stage to the next, how many calories must be supplied by plant food to provide you with 2,000 calories from the bear meat?

★ **84. Genealogy.** If there are 30 years in a generation, how many direct ancestors did each of us have 600 years ago? By *direct* ancestors we mean parents, grandparents, great-grandparents, and so on.

★ **85. Physics.** An object falling from rest in a vacuum near the surface of the Earth falls 16 feet during the first second, 48 feet during the second second, 80 feet during the third second, and so on.
(A) How far will the object fall during the eleventh second?
(B) How far will the object fall in 11 seconds?
(C) How far will the object fall in t seconds?

★ **86. Physics.** In Problem 85, how far will the object fall during:
(A) The twentieth second?   (B) The tth second?

★ **87. Bacteria Growth.** A single cholera bacterium divides every $\frac{1}{2}$ hour to produce two complete cholera bacteria. If we start with a colony of A_0 bacteria, how many bacteria will we have in t hours, assuming adequate food supply?

★ **88. Cell Division.** One leukemic cell injected into a healthy mouse will divide into two cells in about $\frac{1}{2}$ day. At the end of the day these two cells will divide again, with the doubling process continuing each $\frac{1}{2}$ day until there are 1 billion cells, at which time the mouse dies. On which day after the experiment is started does this happen?

★★ **89. Astronomy.** Ever since the time of the Greek astronomer Hipparchus, second century B.C., the brightness of stars has been measured in terms of magnitude. The brightest stars, excluding the sun, are classed as magnitude 1, and the dimmest visible to the eye are classed as magnitude 6. In 1856, the English astronomer N. R. Pogson showed that first-magnitude stars are 100 times brighter than sixth-magnitude stars. If the ratio of brightness between consecutive magnitudes is constant, find this ratio. [*Hint:* If b_n is the brightness of an nth-magnitude star, find r for the geometric progression $b_1, b_2, b_3, \ldots$, given $b_1 = 100b_6$.]

★ **90. Music.** The notes on a piano, as measured in cycles per second, form a geometric progression.
(A) If A is 400 cycles per second and A′, 12 notes higher, is 800 cycles per second, find the constant ratio r.
(B) Find the cycles per second for C, three notes higher than A.

91. Puzzle. If you place 1¢ on the first square of a chessboard, 2¢ on the second square, 4¢ on the third, and so on, continuing to double the amount until all 64 squares are covered, how much money will be on the sixty-fourth square? How much money will there be on the whole board?

point of view it appears that he will never finish the race. This famous paradox is attributed to the Greek philosopher Zeno, 495–435 B.C. If we assume the runner runs at 440 yards per minute, the times between tape breakings form an infinite geometric progression. What is the sum of this progression?

95. Geometry. If the midpoints of the sides of an equilateral triangle are joined by straight lines, the new figure will be an equilateral triangle with a perimeter equal to half the original. If we start with an equilateral triangle with perimeter 1 and form a sequence of "nested" equilateral triangles proceeding as described, what will be the total perimeter of all the triangles that can be formed in this way?

96. Photography. The shutter speeds and f-stops on a camera are given as follows:

Shutter speeds: $1, \frac{1}{2}, \frac{1}{4}, \frac{1}{8}, \frac{1}{15}, \frac{1}{30}, \frac{1}{60}, \frac{1}{125}, \frac{1}{250}, \frac{1}{500}$

f-stops: 1.4, 2, 2.8, 4, 5.6, 8, 11, 16, 22

These are very close to being geometric progressions. Estimate their common ratios.

★★ **97. Geometry.** We know that the sum of the interior angles of a triangle is 180°. Show that the sums of the interior angles of polygons with 3, 4, 5, 6, . . . sides form an arithmetic sequence. Find the sum of the interior angles for a 21-sided polygon.

★ **92. Puzzle.** If a sheet of very thin paper 0.001 inch thick is torn in half, and each half is again torn in half, and this process is repeated for a total of 32 times, how high will the stack of paper be if the pieces are placed one on top of the other? Give the answer to the nearest mile.

★ **93. Atmospheric Pressure.** If atmospheric pressure decreases roughly by a factor of 10 for each 10-mile increase in altitude up to 60 miles, and if the pressure is 15 pounds per square inch at sea level, what will the pressure be 40 miles up?

94. Zeno's Paradox. Visualize a hypothetical 440-yard oval racetrack that has tapes stretched across the track at the halfway point and at each point that marks the halfway point of each remaining distance thereafter. A runner running around the track has to break the first tape before the second, the second before the third, and so on. From this

---

SECTION 8-4 Binomial Formula

- Factorial
- Binomial Formula

The binomial form

$$(a + b)^n$$

where n is a natural number, appears more frequently than you might expect. The coefficients in the expansion play an important role in probability studies. The *binomial formula*, which we derive below, enables us to expand $(a + b)^n$ directly for n any natural number. Since the formula involves *factorials*, we digress for a moment to introduce this important concept.

● **Factorial**   For n a natural number, *n factorial*—denoted by $n!$—is the product of the first n natural numbers. **Zero factorial** is defined to be 1.

DEFINITION 1   *n* **Factorial**

For *n* a natural number

$$n! = n(n - 1) \cdot \cdots \cdot 2 \cdot 1$$

$$1! = 1$$

$$0! = 1$$

It is also useful to note that:

Theorem 1   **Recursion Formula for *n* Factorial**

$$n! = n \cdot (n - 1)!$$

EXAMPLE 1   **Evaluating Factorials**

(A) $4! = 4 \cdot 3! = 4 \cdot 3 \cdot 2! = 4 \cdot 3 \cdot 2 \cdot 1! = 4 \cdot 3 \cdot 2 \cdot 1 = 24$

(B) $5! = 5 \cdot 4 \cdot 3 \cdot 2 \cdot 1 = 120$

(C) $\dfrac{7!}{6!} = \dfrac{7 \cdot 6!}{6!} = 7$

(D) $\dfrac{8!}{5!} = \dfrac{8 \cdot 7 \cdot 6 \cdot 5!}{5!} = 336$

Matched Problem 1   Find: (A) $6!$ (B) $\dfrac{6!}{5!}$ (C) $\dfrac{9!}{6!}$

CAUTION   When reducing fractions involving factorials, don't confuse the single integer *n* with the symbol *n*!, which represents the product of *n* consecutive integers.

$$\frac{6!}{3!} \neq 2! \qquad \frac{6!}{3!} = \frac{6 \cdot 5 \cdot 4 \cdot 3!}{3!} = 6 \cdot 5 \cdot 4 = 120$$

Factorials are used in the definition of the important symbol $\dbinom{n}{r}$. This symbol is frequently used in probability studies. It is called the **combinatorial symbol** and is defined for nonnegative *r* and *n*, as follows:

DEFINITION 2   **Combinatorial Symbol**

For nonnegative integers r and n, $0 \leq r \leq n$.

$$\binom{n}{r} = \frac{n!}{r!(n-r)!}$$

$$= \frac{n(n-1)(n-2) \cdot \cdots \cdot (n-r+1)}{r(r-1) \cdot \cdots \cdot 2 \cdot 1}$$

The combinatorial symbol $\binom{n}{r}$ also can be denoted by $C_{n,r}$, ${}_nC_r$, or $C(n, r)$ and read as "n choose r." Many calculators use ${}_nC_r$ to denote the function that evaluates the combinatorial symbol.

EXAMPLE 2   **Evaluating the Combinatorial Symbol**

(A) $\binom{8}{3} = \dfrac{8!}{3!(8-3)!} = \dfrac{8!}{3!5!} = \dfrac{8 \cdot 7 \cdot 6 \cdot 5!}{3 \cdot 2 \cdot 1 \cdot 5!} = 56$

(B) $\binom{7}{0} = \dfrac{7!}{0!(7-0)!} = \dfrac{7!}{7!} = 1$ Remember, 0! = 1.

Matched Problem 2   Find: (A) $\binom{9}{2}$ (B) $\binom{5}{5}$

EXPLORE-DISCUSS 1   (A) Compute the terms of the finite sequence $\binom{8}{0}, \binom{8}{1}, \binom{8}{2}, \ldots, \binom{8}{8}$ (the sequence is graphed in Fig. 1). Is the sequence arithmetic? Geometric? Which term is the largest? The smallest? Find the sum of the corresponding series.

FIGURE 1

(B) Answer the same questions for the finite sequence $\binom{9}{0}, \binom{9}{1}, \binom{9}{2}, \ldots, \binom{9}{9}$.

● Binomial Formula

We are now ready to try to discover a formula for the expansion of $(a + b)^n$ using ordinary induction; that is, we will look at a few special cases and postulate a general formula from them. We will then try to prove that the formula holds for all natural numbers, using mathematical induction. To start, we calculate directly the first five natural number powers of $(a + b)^n$, arranging the terms in decreasing powers of a:

$$(a + b)^1 = a + b$$

$$(a + b)^2 = a^2 + 2ab + b^2$$

$$(a + b)^3 = a^3 + 3a^2b + 3ab^2 + b^3$$

$$(a + b)^4 = a^4 + 4a^3b + 6a^2b^2 + 4ab^3 + b^4$$

$$(a + b)^5 = a^5 + 5a^4b + 10a^3b^2 + 10a^2b^3 + 5ab^4 + b^5$$

Observations

1. The expansion of $(a + b)^n$ has $n + 1$ terms.

2. The power of a decreases by 1 for each term as we move from left to right.

3. The power of b increases by 1 for each term as we move from left to right.

4. In each term, the sum of the powers of a and b always adds up to n.

5. Starting with a given term, we can get the coefficient of the next term by multiplying the coefficient of the given term by the exponent of a and dividing by the number that represents the position of the term in the series of terms. For example, in the expansion of $(a + b)^4$, the coefficient of the third term is found from the second term by multiplying 4 and 3 and then dividing by 2. Thus, the coefficient of the third term is $(4 \cdot 3)/2 = 6$.

We now postulate the properties for the general case:

$$(a + b)^n = a^n + \frac{n}{1} a^{n-1}b + \frac{n(n - 1)}{1 \cdot 2} a^{n-2}b^2$$

$$+ \frac{n(n - 1)(n - 2)}{1 \cdot 2 \cdot 3} a^{n-3}b^3 + \cdots + b^n$$

$$= \frac{n!}{0!(n - 0)!} a^n + \frac{n!}{1!(n - 1)!} a^{n-1}b + \frac{n!}{2!(n - 2)!} a^{n-2}b^2$$

$$+ \frac{n!}{3!(n - 3)!} a^{n-3}b^3 + \cdots + \frac{n!}{n!(n - n)!} b^n$$

$$= \binom{n}{0} a_n + \binom{n}{1} a^{n-1}b + \binom{n}{2} a^{n-2}b^2 + \binom{n}{3} a^{n-3}b^3 + \cdots + \binom{n}{n} b^n$$

Thus, we have arrived at the **binomial formula** using ordinary induction:

Theorem 2   **Binomial Formula**

For n a positive integer

$$(a + b)^n = \sum_{k=0}^{n} \binom{n}{k} a^{n-k} b^k$$

We now proceed to prove that the binomial formula holds for all natural numbers n using mathematical induction.

Proof   State the conjecture.

$$P_n: \quad (a + b)^n = \sum_{j=0}^{n} \binom{n}{j} a^{n-j} b^j$$

Part 1   Show that P_1 is true.

$$\sum_{j=0}^{1} \binom{1}{j} a^{1-j} b^j = \binom{1}{0} a + \binom{1}{1} b = a + b = (a + b)^1$$

Thus, P_1 is true.

Part 2   Show that if P_k is true, then P_{k+1} is true.

$$P_k: \quad (a + b)^k = \sum_{j=0}^{k} \binom{k}{j} a^{k-j} b^j \qquad \text{Assume } P_k \text{ is true.}$$

$$P_{k+1}: \quad (a + b)^{k+1} = \sum_{j=0}^{k+1} \binom{k+1}{j} a^{k+1-j} b^j \qquad \text{Show } P_{k+1} \text{ is true.}$$

We begin by multiplying both sides of P_k by $(a + b)$:

$$(a + b)^k (a + b) = \left[\sum_{j=0}^{k} \binom{k}{j} a^{k-j} b^j \right] (a + b)$$

The left side of this equation is the left side of P_{k+1}. Now we multiply out the right side of the equation and try to obtain the right side of P_{k+1}:

$$(a + b)^{k+1} = \left[\binom{k}{0} a^k + \binom{k}{1} a^{k-1} b + \binom{k}{2} a^{k-2} b^2 + \cdots + \binom{k}{k} b^k \right] (a + b)$$

$$= \left[\binom{k}{0} a^{k+1} + \binom{k}{1} a^k b + \binom{k}{2} a^{k-1} b^2 + \cdots + \binom{k}{k} ab^k \right]$$

$$+ \left[\binom{k}{0} a^k b + \binom{k}{1} a^{k-1} b^2 + \cdots + \binom{k}{k-1} ab^k + \binom{k}{k} b^{k+1} \right]$$

$$= \binom{k}{0} a^{k+1} + \left[\binom{k}{0} + \binom{k}{1} \right] a^k b + \left[\binom{k}{1} + \binom{k}{2} \right] a^{k-1} b^2 + \cdots$$

$$+ \left[\binom{k}{k-1} + \binom{k}{k} \right] ab^k + \binom{k}{k} b^{k+1}$$

We now use the following facts (the proofs are left as exercises; see Problems 49–51, Exercise 11-4)

$$\binom{k}{r-1} + \binom{k}{r} = \binom{k+1}{r} \qquad \binom{k}{0} = \binom{k+1}{0} \qquad \binom{k}{k} = \binom{k+1}{k+1}$$

to rewrite the right side as

$$\binom{k+1}{0}a^{k+1} + \binom{k+1}{1}a^k b + \binom{k+1}{2}a^{k-1}b^2 + \cdots$$

$$+ \binom{k+1}{k}ab^k + \binom{k+1}{k+1}b^{k+1} = \sum_{j=0}^{k+1}\binom{k+1}{j}a^{k+1-j}b^j$$

Since the right side of the last equation is the right side of P_{k+1}, we have shown that P_{k+1} follows from P_k.

Conclusion P_n is true. That is, the binomial formula holds for all positive integers n.

EXAMPLE 3   **Using the Binomial Formula**

Use the binomial formula to expand $(x + y)^6$.

Solution $(x + y)^6 = \displaystyle\sum_{k=0}^{6}\binom{6}{k}x^{6-k}y^k$

$$= \binom{6}{0}x^6 + \binom{6}{1}x^5 y + \binom{6}{2}x^4 y^2 + \binom{6}{3}x^3 y^3 + \binom{6}{4}x^2 y^4 + \binom{6}{5}xy^5 + \binom{6}{6}y^6$$

$$= x^6 + 6x^5 y + 15x^4 y^2 + 20x^3 y^3 + 15x^2 y^4 + 6xy^5 + y^6$$

Matched Problem 3   Use the binomial formula to expand $(x + 1)^5$.

EXAMPLE 4   **Using the Binomial Formula**

Use the binomial formula to expand $(3p - 2q)^4$.

Solution $(3p - 2q)^4 = [(3p) + (-2q)]^4$ $a = 3p, \ b = -2q$

$$= \sum_{k=0}^{4}\binom{4}{k}(3p)^{4-k}(-2q)^k$$

$$= \binom{4}{0}(3p)^4 + \binom{4}{1}(3p)^3(-2q) + \binom{4}{2}(3p)^2(-2q)^2$$

$$+ \binom{4}{3}(3p)(-2q)^3 + \binom{4}{4}(-2q)^4$$

$$= 81p^4 - 216p^3 q + 216p^2 q^2 - 96pq^3 + 16q^4$$

Matched Problem 4   Use the binomial formula to expand $(2m - 5n)^3$.

EXPLORE-DISCUSS 2   (A) Compute each term and also the sum of the alternating series
$$\binom{6}{0} - \binom{6}{1} + \binom{6}{2} - \cdots + \binom{6}{6}.$$

(B) What result about an alternating series can be deduced by letting $a = 1$ and $b = -1$ in the binomial formula?

EXAMPLE 5   **Using the Binomial Formula**

Use the binomial formula to find the fourth and sixteenth terms in the expansion of $(x - 2)^{20}$.

Solution   In the expansion of $(a + b)^n$, the exponent of b in the rth term is $r - 1$ and the exponent of a is $n - (r - 1)$. Thus,

Fourth term:

$$\binom{20}{3}x^{17}(-2)^3$$

$$= \frac{20 \cdot 19 \cdot 18}{3 \cdot 2 \cdot 1}x^{17}(-8)$$

$$= -9{,}120x^{17}$$

Sixteenth term:

$$\binom{20}{15}x^5(-2)^{15}$$

$$= \frac{20 \cdot 19 \cdot 18 \cdot 17 \cdot 16}{5 \cdot 4 \cdot 3 \cdot 2 \cdot 1}x^5(-32{,}768)$$

$$= -508{,}035{,}072x^5$$

Matched Problem 5   Use the binomial formula to find the fifth and twelfth terms in the expansion of $(u - 1)^{18}$.

Answers to Matched Problems

1. (A) 720   (B) 6   (C) 504   **2.** (A) 36   (B) 1
3. $x^5 + 5x^4 + 10x^3 + 10x^2 + 5x + 1$ **4.** $8m^3 - 60m^2n + 150mn^2 - 125n^3$
5. $3{,}060u^{14}$; $-31{,}824u^7$

EXERCISE 8-4

A

Evaluate each expression in Problems 1–12.

1. 7!

2. 5!

3. $\dfrac{15!}{13!}$

4. $\dfrac{20!}{17!}$

5. $4! + 5!$

6. $(4 + 5)!$

7. $\dfrac{8!}{3!5!}$

8. $\dfrac{10!}{2!8!}$

9. $\dfrac{7!}{0!(7 - 0)!}$

10. $\dfrac{12!}{12!(12-12)!}$ **11.** $\dfrac{10!}{7!}$ **12.** $\dfrac{10!}{3!}$

Write each expression in Problems 13–16 as the quotient of two factorials.

13. 9   **14.** 12

15. $6 \cdot 7 \cdot 8$ **16.** $9 \cdot 10 \cdot 11 \cdot 12$

B
_____

Evaluate each expression in Problems 17–22.

17. $\dbinom{13}{9}$ **18.** $\dbinom{12}{5}$ **19.** $\dbinom{14}{7}$

20. $\dbinom{16}{8}$ **21.** $\dbinom{100}{97}$ **22.** $\dbinom{100}{3}$

23. Find the smallest positive integer n such that $n!$ produces an overflow error on your calculator.

24. Find the smallest positive integer n such that $\dbinom{2n}{n}$ produces an overflow error on your calculator.

Expand Problems 25–30 using the binomial formula.

25. $(2x-3y)^3$ **26.** $(3u+2v)^3$ **27.** $(x-2)^4$

28. $(3p-q)^4$ **29.** $(2x-y)^5$ **30.** $(2x-y)^6$

In Problems 31–38, find the indicated term in each expansion.

31. $(u+v)^{15}$; seventh term   **32.** $(a+b)^{12}$; fifth term

33. $(2m+n)^{10}$; ninth term   **34.** $(x+3y)^{13}$; third term

35. $(w-3)^{20}$; fifth term   **36.** $(2x-5y)^8$; sixth term

37. $(3x-2y)^8$; sixth term   **38.** $(2p-3q)^7$; fourth term

In Problems 39–42, use a graphing utility to graph each sequence and to display it in table form.

39. Find the number of terms of the sequence

$$\dbinom{20}{0}, \dbinom{20}{1}, \dbinom{20}{2}, \dots, \dbinom{20}{20}$$

that are greater than one-half of the largest term.

40. Find the number of terms of the sequence

$$\dbinom{40}{0}, \dbinom{40}{1}, \dbinom{40}{2}, \dots, \dbinom{40}{40}$$

that are greater than one-half of the largest term.

41. (A) Find the largest term of the sequence a_0, a_1, a_2, $\dots$, a_{10} to three decimal places, where

$$a_k = \dbinom{10}{k} (0.6)^{10-k}(0.4)^k.$$

(B) According to the binomial formula, what is the sum of the series $a_0 + a_1 + a_2 + \cdots + a_{10}$?

42. (A) Find the largest term of the sequence a_0, a_1, a_2, $\dots$, a_{10} to three decimal places, where

$$a_k = \dbinom{10}{k} (0.3)^{10-k}(0.7)^k.$$

(B) According to the binomial formula, what is the sum of the series $a_0 + a_1 + a_2 + \cdots + a_{10}$?

C
_____

43. Evaluate $(1.01)^{10}$ to four decimal places, using the binomial formula. [*Hint:* Let $1.01 = 1 + 0.01.$]

44. Evaluate $(0.99)^6$ to four decimal places, using the binomial formula.

In Problems 45–48, determine whether the statement is true or false. If true, explain why. If false, give a counterexample.

45. $\dbinom{n}{r} = \dbinom{n}{n-r}$

46. If the positive integer n is divisible by a prime p, then $\dbinom{n}{r}$ is divisible by p for $1 \le r \le n-1$.

47. If p is a prime, then $\dbinom{p}{r}$ is divisible by p for $1 \le r \le p-1$.

48. If p is an odd prime, then $\dbinom{2p}{p}$ is not divisible by p.

49. Show that: $\dbinom{k}{r-1} + \dbinom{k}{r} = \dbinom{k+1}{r}$

50. Show that: $\dbinom{k}{0} = \dbinom{k+1}{0}$

51. Show that: $\dbinom{k}{k} = \dbinom{k+1}{k+1}$

52. Show that $\dbinom{n}{r}$ is given by the recursion formula

$$\dbinom{n}{r} = \frac{n-r+1}{r}\dbinom{n}{r-1}$$

where $\dbinom{n}{0} = 1.$

53. Write $2^n = (1+1)^n$ and expand, using the binomial formula to obtain

$$2^n = \dbinom{n}{0} + \dbinom{n}{1} + \dbinom{n}{2} + \cdots + \dbinom{n}{n}$$

54. Can you guess what the next two rows in **Pascal's triangle,** shown at right, are? Compare the numbers in the triangle with the binomial coefficients obtained with the binomial formula.

$$1$$
$$1 \; 1$$
$$1 \; 2 \; 1$$
$$1 \; 3 \; 3 \; 1$$
$$1 \; 4 \; 6 \; 4 \; 1$$

SECTION 8-5 Multiplication Principle, Permutations, and Combinations

- Multiplication Principle
- Permutations
- Combinations

We may expand the binomial form $(a + b)^n$ in two steps: first, expand into a sum of 2^n terms, each with coefficient 1; second, group together those terms in which b appears to the same power, obtaining the sum of the $n + 1$ terms of the binomial formula. For example,

$$(a + b)^3 = (a + b)(a + b)^2 = (a + b)(aa + ab + ba + bb)$$

$$= aaa + aab + aba + abb + baa + bab + bba + bbb \quad \text{Step 1}$$

$$= a^3 + 3a^2b + 3ab^2 + b^3 \quad \text{Step 2}$$

Consider the term aba of step 1: The first a comes from the first factor of $a + b$, the b comes from the second factor of $a + b$, and the final a from the third factor. Therefore, $\binom{3}{1} = 3$, the coefficient of a^2b in step 2, is the number of ways of choosing b from exactly one of the three factors of $a + b$ in $(a + b)^3$.

In the same way, $\binom{52}{5} = 2{,}598{,}960$ is the number of ways of choosing b from exactly five of the 52 factors of $a + b$ in $(a + b)^{52}$. Analogously, 2,598,960 is the number of 5-card hands which can be chosen from a standard 52-card deck. In this section we study such counting techniques that are related to the sequence $\binom{n}{0}, \binom{n}{1}, \binom{n}{2}, \ldots, \binom{n}{n}$, and we develop important counting tools that form the foundation of probability theory.

• Multiplication Principle

We start with an example.

EXAMPLE 1 **Combined Outcomes**

Suppose we flip a coin and then throw a single die (see Fig. 1). What are the possible combined outcomes?

Solution   To solve this problem, we use a **tree diagram:**

Coin outcomes

Die outcomes

FIGURE 1 Coin and die outcomes.

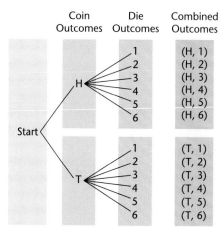

Thus, there are 12 possible combined outcomes—two ways in which the coin can come up followed by six ways in which the die can come up.

Matched Problem 1   Use a tree diagram to determine the number of possible outcomes of throwing a single die followed by flipping a coin.

Now suppose you are asked, "From the 26 letters in the alphabet, how many ways can 3 letters appear in a row on a license plate if no letter is repeated?" To try to count the possibilities using a tree diagram would be extremely tedious, to say the least. The following **multiplication principle,** also called the **fundamental counting principle,** enables us to solve this problem easily. In addition, it forms the basis for several other counting techniques developed later in this section.

Multiplication Principle

1. If two operations O_1 and O_2 are performed in order, with N_1 possible outcomes for the first operation and N_2 possible outcomes for the second operation, then there are

$$N_1 \cdot N_2$$

possible combined outcomes of the first operation followed by the second.

2. In general, if n operations $O_1, O_2, \ldots, O_n$ are performed in order, with possible number of outcomes $N_1, N_2, \ldots, N_n$, respectively, then there are

$$N_1 \cdot N_2 \cdot \cdots \cdot N_n$$

possible combined outcomes of the operations performed in the given order.

In Example 1, we see that there are two possible outcomes from the first operation of flipping a coin and six possible outcomes from the second operation of throwing a die. Hence, by the multiplication principle, there are $2 \cdot 6 = 12$ possible combined outcomes of flipping a coin followed by throwing a die. Use the multiplication principle to solve Matched Problem 1.

To answer the license plate question, we reason as follows: There are 26 ways the first letter can be chosen. After a first letter is chosen, 25 letters remain; hence there are 25 ways a second letter can be chosen. And after 2 letters are chosen, there are 24 ways a third letter can be chosen. Hence, using the multiplication principle, there are $26 \cdot 25 \cdot 24 = 15,600$ possible ways 3 letters can be chosen from the alphabet without allowing any letter to repeat. By not allowing any letter to repeat, earlier selections affect the choice of subsequent selections. If we allow letters to repeat, then earlier selections do not affect the choice in subsequent selections, and there are 26 possible choices for each of the 3 letters. Thus, if we allow letters to repeat, there are $26 \cdot 26 \cdot 26 = 26^3 = 17,576$ possible ways the 3 letters can be chosen from the alphabet.

EXAMPLE 2   **Computer-Generated Tests**

Many universities and colleges are now using computer-assisted testing procedures. Suppose a screening test is to consist of 5 questions, and a computer stores 5 equivalent questions for the first test question, 8 equivalent questions for the second, 6 for the third, 5 for the fourth, and 10 for the fifth. How many different 5-question tests can the computer select? Two tests are considered different if they differ in one or more questions.

Solution

O_1:	Select the first question	N_1:	5 ways
O_2:	Select the second question	N_2:	8 ways
O_3:	Select the third question	N_3:	6 ways
O_4:	Select the fourth question	N_4:	5 ways
O_5:	Select the fifth question	N_5:	10 ways

Thus, the computer can generate

$$5 \cdot 8 \cdot 6 \cdot 5 \cdot 10 = 12,000 \text{ different tests}$$

Matched Problem 2   Each question on a multiple-choice test has 5 choices. If there are 5 such questions on a test, how many different response sheets are possible if only 1 choice is marked for each question?

EXAMPLE 3   **Counting Code Words**

How many 3-letter code words are possible using the first 8 letters of the alphabet if:

(A) No letter can be repeated?   (B) Letters can be repeated?
(C) Adjacent letters cannot be alike?

Solutions   (A) No letter can be repeated.

> O_1:   Select first letter N_1:   8 ways
>
> O_2:   Select second letter N_2:   7 ways   Because 1 letter has been used
>
> O_3:   Select third letter N_3:   6 ways   Because 2 letters have been used

Thus, there are

$$8 \cdot 7 \cdot 6 = 336 \text{ possible code words}$$

(B) Letters can be repeated.

> O_1:   Select first letter N_1:   8 ways
>
> O_2:   Select second letter N_2:   8 ways   Repeats are allowed.
>
> O_3:   Select third letter N_3:   8 ways   Repeats are allowed.

Thus, there are

$$8 \cdot 8 \cdot 8 = 8^3 = 512 \text{ possible code words}$$

(C) Adjacent letters cannot be alike.

> O_1:   Select first letter N_1:   8 ways
>
> O_2:   Select second letter N_2:   7 ways   Cannot be the same as the first
>
> O_3:   Select third letter N_3:   7 ways   Cannot be the same as the second, but can be the same as the first

Thus, there are

$$8 \cdot 7 \cdot 7 = 392 \text{ possible code words}$$

Matched Problem 3   How many 4-letter code words are possible using the first 10 letters of the alphabet under the three conditions stated in Example 3?

EXPLORE-DISCUSS 1   The postal service of a developing country is choosing a five-character postal code consisting of letters (of the English alphabet) and digits. At least half a million postal codes must be accommodated. Which format would you recommend to make the codes easy to remember?

The multiplication principle can be used to develop two additional methods for counting that are extremely useful in more complicated counting problems. Both of these methods use the factorial function, which was introduced in Section 8-4.

● Permutations

Suppose 4 pictures are to be arranged from left to right on one wall of an art gallery. How many arrangements are possible? Using the multiplication principle, there are 4 ways of selecting the first picture. After the first picture is selected, there are 3 ways of selecting the second picture. After the first 2 pictures are selected, there are 2 ways of selecting the third picture. And after the first 3 pictures are selected, there is only 1 way to select the fourth. Thus, the number of arrangements possible for the 4 pictures is

$$4 \cdot 3 \cdot 2 \cdot 1 = 4! \quad \text{or} \quad 24$$

In general, we refer to a particular arrangement, or **ordering,** of n objects without repetition as a **permutation** of the n objects. How many permutations of n objects are there? From the reasoning above, there are n ways in which the first object can be chosen, there are $n - 1$ ways in which the second object can be chosen, and so on. Applying the multiplication principle, we have Theorem 1:

Theorem 1

Permutations of n Objects

The number of permutations of n objects, denoted by $P_{n,n}$, is given by

$$P_{n,n} = n \cdot (n - 1) \cdots \cdots 1 = n!$$

Now suppose the director of the art gallery decides to use only 2 of the 4 available pictures on the wall, arranged from left to right. How many arrangements of 2 pictures can be formed from the 4? There are 4 ways the first picture can be selected. After selecting the first picture, there are 3 ways the second picture can be selected. Thus, the number of arrangements of 2 pictures from 4 pictures, denoted by $P_{4,2}$, is given by

$$P_{4,2} = 4 \cdot 3 = 12$$

Or, in terms of factorials, multiplying $4 \cdot 3$ by 1 in the form $2!/2!$, we have

$$P_{4,2} = 4 \cdot 3 = \frac{4 \cdot 3 \cdot 2!}{2!} = \frac{4!}{2!}$$

This last form gives $P_{4,2}$ in terms of factorials, which is useful in some cases.

A **permutation of a set of n objects taken r at a time** is an arrangement of the r objects in a specific order. Thus, reasoning in the same way as in the example above, we find that the number of permutations of n objects taken r at a time, $0 \le r \le n$, denoted by $P_{n,r}$, is given by

$$P_{n,r} = n(n - 1)(n - 2) \cdots \cdots (n - r + 1)$$

Multiplying the right side of this equation by 1 in the form $(n - r)!/(n - r)!$, we obtain a factorial form for $P_{n,r}$:

$$P_{n,r} = n(n - 1)(n - 2) \cdots \cdots (n - r + 1) \frac{(n - r)!}{(n - r)!}$$

But

$$n(n - 1)(n - 2) \cdot \cdots \cdot (n - r + 1)(n - r)! = n!$$

Hence, we have Theorem 2:

Theorem 2   **Permutation of n Objects Taken r at a Time**

The number of permutations of n objects taken r at a time is given by

$$P_{n,r} = \underbrace{n(n - 1)(n - 2) \cdot \cdots \cdot (n - r + 1)}_{r \text{ factors}}$$

or

$$P_{n,r} = \frac{n!}{(n - r)!} \qquad 0 \le r \le n$$

Note that if $r = n$, then the number of permutations of n objects taken n at a time is

$$P_{n,n} = \frac{n!}{(n - n)!} = \frac{n!}{0!} = n! \qquad \text{Recall, } 0! = 1.$$

which agrees with Theorem 1, as it should.

The permutation symbol $P_{n,r}$ also can be denoted by P_r^n, $_nP_r$, or $P(n, r)$. Many calculators use $_nP_r$ to denote the function that evaluates the permutation symbol.

EXAMPLE 4   **Selecting Officers**

From a committee of 8 people, in how many ways can we choose a chair and a vice-chair, assuming one person cannot hold more than one position?

Solution   We are actually asking for the number of permutations of 8 objects taken 2 at a time—that is, $P_{8,2}$:

$$P_{8,2} = \frac{8!}{(8 - 2)!} = \frac{8!}{6!} = \frac{8 \cdot 7 \cdot 6!}{6!} = 56$$

Matched Problem 4   From a committee of 10 people, in how many ways can we choose a chair, vice-chair, and secretary, assuming one person cannot hold more than one position?

CAUTION

Remember to use the definition of factorial when simplifying fractions involving factorials.

$$\frac{6!}{3!} \neq 2! \qquad \frac{6!}{3!} = \frac{6 \cdot 5 \cdot 4 \cdot 3!}{3!} = 120$$

EXAMPLE 5 **Evaluating $P_{n,r}$**

Find the number of permutations of 25 objects taken 8 at a time. Compute the answer to 4 significant digits using a calculator.

Solution

$$P_{25,8} = \frac{25!}{(25-8)!} = \frac{25!}{17!} = 4.361 \times 10^{10} \qquad \text{A very large number}$$

Matched Problem 5

Find the number of permutations of 30 objects taken 4 at a time. Compute the answer exactly using a calculator.

• **Combinations**

Now suppose that an art museum owns 8 paintings by a given artist and another art museum wishes to borrow 3 of these paintings for a special show. How many ways can 3 paintings be selected for shipment out of the 8 available? Here, the order of the items selected doesn't matter. What we are actually interested in is how many subsets of 3 objects can be formed from a set of 8 objects. We call such a subset a **combination** of 8 objects taken 3 at a time. The total number of combinations is denoted by the symbol

$$C_{8,3} \qquad \text{or} \qquad \binom{8}{3}$$

To find the number of combinations of 8 objects taken 3 at a time, $C_{8,3}$, we make use of the formula for $P_{n,r}$ and the multiplication principle. We know that the number of permutations of 8 objects taken 3 at a time is given by $P_{8,3}$, and we have a formula for computing this quantity. Now suppose we think of $P_{8,3}$ in terms of two operations:

O_1: Select a subset of 3 objects (paintings)

N_1: $C_{8,3}$ ways

O_2: Arrange the subset in a given order

N_2: 3! ways

The combined operation, O_1 followed by O_2, produces a permutation of 8 objects taken 3 at a time. Thus,

$$P_{8,3} = C_{8,3} \cdot 3!$$

To find $C_{8,3}$, we replace $P_{8,3}$ in the above equation with $8!/(8-3)!$ and solve for $C_{8,3}$:

$$\frac{8!}{(8-3)!} = C_{8,3} \cdot 3!$$

$$C_{8,3} = \frac{8!}{3!(8-3)!} = \frac{8 \cdot 7 \cdot 6 \cdot 5!}{3 \cdot 2 \cdot 1 \cdot 5!} = 56$$

Thus, the museum can make 56 different selections of 3 paintings from the 8 available.

A **combination of a set of n objects taken r at a time** is an r-element subset of the n objects. Reasoning in the same way as in the example, the number of combinations of n objects taken r at a time, $0 \le r \le n$, denoted by $C_{n,r}$, can be obtained by solving for $C_{n,r}$ in the relationship

$$P_{n,r} = C_{n,r} \cdot r!$$

$$C_{n,r} = \frac{P_{n,r}}{r!}$$

$$= \frac{n!}{r!(n-r)!} \qquad P_{n,r} = \frac{n!}{(n-r)!}$$

Theorem 3

Combination of n Objects Taken r at a Time

The number of combinations of n objects taken r at a time is given by

$$C_{n,r} = \binom{n}{r} = \frac{P_{n,r}}{r!} = \frac{n!}{r!(n-r)!} \qquad 0 \le r \le n$$

Note that we used the combination formula in Section 8-4 to represent binomial coefficients.

The combination symbols $C_{n,r}$ and $\binom{n}{r}$ also can be denoted by C_r^n, $_nC_r$, or $C(n, r)$.

EXAMPLE 6 **Selecting Subcommittees**

From a committee of 8 people, in how many ways can we choose a subcommittee of 2 people?

Solution Notice how this example differs from Example 4, where we wanted to know how many ways a chair and a vice-chair can be chosen from a committee of 8 people. In Example 4, ordering matters. In choosing a subcommittee of 2 people, the ordering does not matter. Thus, we are actually asking for the number of combinations of 8 objects taken 2 at a time. The number is given by

$$C_{8,2} = \binom{8}{2} = \frac{8!}{2!(8-2)!} = \frac{8 \cdot 7 \cdot 6!}{2 \cdot 1 \cdot 6!} = 28$$

Matched Problem 6   How many subcommittees of 3 people can be chosen from a committee of 8 people?

EXAMPLE 7   **Evaluating $C_{n,r}$**

Find the number of combinations of 25 objects taken 8 at a time. Compute the answer to 4 significant digits using a calculator.

Solution

$$C_{25,8} = \binom{25}{8} = \frac{25!}{8!(25-8)!} = \frac{25!}{8!17!} = 1.082 \times 10^6$$

Compare this result with that obtained in Example 5.

Matched Problem 7   Find the number of combinations of 30 objects taken 4 at a time. Compute the answer exactly using a calculator.

Remember: In a permutation, order counts. In a combination, order does not count.

To determine whether a permutation or combination is needed, decide whether rearranging the collection or listing makes a difference. If so, use permutations. If not, use combinations.

EXPLORE-DISCUSS 2   Each of the following is a selection without repetition. Would you consider the selection to be a combination? A permutation? Discuss your reasoning.

(A) A student checks out three books from the library.

(B) A baseball manager names his starting lineup.

(C) The newly elected President names his Cabinet members.

(D) The President selects a delegation of three Cabinet members to attend the funeral of a head of state.

(E) An orchestra conductor chooses three pieces of music for a symphony program.

A standard deck of 52 cards involves four suits, hearts, spades, diamonds, and clubs, as shown in Figure 2. Example 8, as well as other examples and exercises in this chapter, refer to this standard deck.

FIGURE 2 A standard deck of cards.

EXAMPLE 8   **Counting Card Hands**

Out of a standard 52-card deck, how many 5-card hands will have 3 aces and 2 kings?

Solution

O_1:   Choose 3 aces out of 4 possible   Order is not important.

N_1: $C_{4,3}$

O_2:   Choose 2 kings out of 4 possible   Order is not important.

N_2: $C_{4,2}$

Using the multiplication principle, we have

$$\text{Number of hands} = C_{4,3} \cdot C_{4,2} = 4 \cdot 6 = 24$$

Matched Problem 8   From a standard 52-card deck, how many 5-card hands will have 3 hearts and 2 spades?

EXAMPLE 9   **Counting Serial Numbers**

Serial numbers for a product are to be made using 2 letters followed by 3 numbers. If the letters are to be taken from the first 8 letters of the alphabet with no repeats and the numbers from the 10 digits 0 through 9 with no repeats, how many serial numbers are possible?

Solution

O_1:   Choose 2 letters out of 8 available   Order is important.

N_1: $P_{8,2}$

O_2:   Choose 3 numbers out of 10 available   Order is important.

N_2: $P_{10,3}$

Using the multiplication principle, we have

$$\text{Number of serial numbers} = P_{8,2} \cdot P_{10,3} = 40{,}320$$

Matched Problem 9   Repeat Example 9 under the same conditions, except the serial numbers are now to have 3 letters followed by 2 digits with no repeats.

Answers to Matched Problems

1.

2. 5^5, or 3,125

3. (A) $10 \cdot 9 \cdot 8 \cdot 7 = 5{,}040$ (B) $10 \cdot 10 \cdot 10 \cdot 10 = 10{,}000$ (C) $10 \cdot 9 \cdot 9 \cdot 9 = 7{,}290$

4. $P_{10,3} = \dfrac{10!}{(10-3)!} = 720$ **5.** $P_{30,4} = \dfrac{30!}{(30-4)!} = 657{,}720$ **6.** $C_{8,3} = \dfrac{8!}{3!(8-3)!} = 56$

7. $C_{30,4} = \dfrac{30!}{4!(30-4)!} = 27{,}405$ **8.** $C_{13,3} \cdot C_{13,2} = 22{,}308$ **9.** $P_{8,3} \cdot P_{10,2} = 30{,}240$

EXERCISE 8-5

A

Evaluate Problems 1–16.

1. $\dfrac{15!}{12!}$ **2.** $\dfrac{20!}{18!}$ **3.** $\dfrac{32!}{0!32!}$

4. $\dfrac{25!}{24!1!}$ **5.** $\dfrac{9!}{6!3!}$ **6.** $\dfrac{7!}{5!2!}$

7. $\dfrac{16!}{4!(16-4)!}$ **8.** $\dfrac{18!}{3!(18-3)!}$ **9.** $P_{8,5}$

10. $C_{8,5}$ **11.** $P_{52,3}$ **12.** $P_{13,5}$

13. $C_{13,5}$ **14.** $C_{13,4}$ **15.** $C_{52,5}$

16. $P_{20,4}$

17. A particular new car model is available with 5 choices of color, 3 choices of transmission, 4 types of interior, and 2 types of engine. How many different variations of this model car are possible?

18. A deli serves sandwiches with the following options: 3 kinds of bread, 5 kinds of meat, and lettuce or sprouts. How many different sandwiches are possible, assuming one item is used out of each category?

19. In a horse race, how many different finishes among the first 3 places are possible for a 10-horse race? Exclude ties.

20. In a long-distance foot race, how many different finishes among the first 5 places are possible for a 50-person race? Exclude ties.

21. How many ways can a subcommittee of 3 people be selected from a committee of 7 people? How many ways can a president, vice president, and secretary be chosen from a committee of 7 people?

22. Suppose 9 cards are numbered with the 9 digits from 1 to 9. A 3-card hand is dealt, 1 card at a time. How many hands are possible where:
(A) Order is taken into consideration?
(B) Order is not taken into consideration?

23. There are 10 teams in a league. If each team is to play every other team exactly once, how many games must be scheduled?

24. Given 7 points, no 3 of which are on a straight line, how many lines can be drawn joining 2 points at a time?

B

25. How many 4-letter code words are possible from the first 6 letters of the alphabet, with no letter repeated? Allowing letters to repeat?

26. A small combination lock on a suitcase has 3 wheels, each labeled with digits from 0 to 9. How many opening combinations of 3 numbers are possible, assuming no digit is repeated? Assuming digits can be repeated?

27. From a standard 52-card deck, how many 5-card hands will have all hearts?

28. From a standard 52-card deck, how many 5-card hands will have all face cards? All face cards, but no kings? Consider only jacks, queens, and kings to be face cards.

29. How many different license plates are possible if each contains 3 letters followed by 3 digits? How many of these license plates contain no repeated letters and no repeated digits?

30. How many 5-digit zip codes are possible? How many of these codes contain no repeated digits?

31. From a standard 52-card deck, how many 7-card hands have exactly 5 spades and 2 hearts?

32. From a standard 52-card deck, how many 5-card hands will have 2 clubs and 3 hearts?

33. A catering service offers 8 appetizers, 10 main courses, and 7 desserts. A banquet chairperson is to select 3 appetizers, 4 main courses, and 2 desserts for a banquet. How many ways can this be done?

34. Three research departments have 12, 15, and 18 members, respectively. If each department is to select a delegate and an alternate to represent the department at a conference, how many ways can this be done?

35. (A) Use a graphing utility to display the sequences $P_{10,0}$, $P_{10,1}, \ldots, P_{10,10}$ and $0!, 1!, \ldots, 10!$ in table form, and show that $P_{10,r} \geq r!$ for $r = 0, 1, \ldots, 10$.
    (B) Find all values of r such that $P_{10,r} = r!$
    (C) Explain why $P_{n,r} \geq r!$ whenever $0 \leq r \leq n$.

36. (A) How are the sequences $\dfrac{P_{10,0}}{0!}, \dfrac{P_{10,1}}{1!}, \ldots, \dfrac{P_{10,10}}{10!}$ and $C_{10,0}$, $C_{10,1}, \ldots, C_{10,10}$ related?

    (B) Use a graphing utility to graph each sequence and confirm the relationship of part A.

C

37. A sporting goods store has 12 pairs of ski gloves of 12 different brands thrown loosely in a bin. The gloves are all the same size. In how many ways can a left-hand glove and a right-hand glove be selected that do not match relative to brand?

38. A sporting goods store has 6 pairs of running shoes of 6 different styles thrown loosely in a basket. The shoes are all the same size. In how many ways can a left shoe and a right shoe be selected that do not match?

39. Eight distinct points are selected on the circumference of a circle.
    (A) How many chords can be drawn by joining the points in all possible ways?
    (B) How many triangles can be drawn using these 8 points as vertices?
    (C) How many quadrilaterals can be drawn using these 8 points as vertices?

40. Five distinct points are selected on the circumference of a circle.
    (A) How many chords can be drawn by joining the points in all possible ways?
    (B) How many triangles can be drawn using these 5 points as vertices?

41. How many ways can 2 people be seated in a row of 5 chairs? 3 people? 4 people? 5 people?

42. Each of 2 countries sends 5 delegates to a negotiating conference. A rectangular table is used with 5 chairs on each long side. If each country is assigned a long side of the table, how many seating arrangements are possible? [*Hint:* Operation 1 is assigning a long side of the table to each country.]

43. A basketball team has 5 distinct positions. Out of 8 players, how many starting teams are possible if:
    (A) The distinct positions are taken into consideration?
    (B) The distinct positions are not taken into consideration?
    (C) The distinct positions are not taken into consideration, but either Mike or Ken, but not both, must start?

44. How many committees of 4 people are possible from a group of 9 people if:
    (A) There are no restrictions?
    (B) Both Juan and Mary must be on the committee?
    (C) Either Juan or Mary, but not both, must be on the committee?

45. A 5-card hand is dealt from a standard 52-card deck. Which is more likely: the hand contains exactly 1 king or the hand contains no hearts?

46. A 10-card hand is dealt from a standard 52-card deck. Which is more likely: all cards in the hand are red or the hand contains all four aces?

47. A parent is placing an order for five single-dip ice cream cones. If today's flavors are vanilla, chocolate, and strawberry, how many orders are possible? Explain. (*Note:* This type of selection, in which repetition is allowed but order is irrelevant, is neither a combination nor a permutation.)

48. One dozen identical doughnuts are to be distributed among nine students. If each student must receive at least one doughnut, how many distributions are possible? Explain.

CHAPTER 8 GROUP ACTIVITY   Sequences Specified by Recursion Formulas

The recursion formula $a_n = 5a_{n-1} - 6a_{n-2}$, together with the initial values $a_1 = 4$, $a_2 = 14$, specifies the sequence $\{a_n\}$ whose first several terms are 4, 14, 46, 146, 454, 1394, ... The sequence $\{a_n\}$ is neither arithmetic nor geometric. Nevertheless, because it satisfies a simple recursion formula, it is possible to obtain an nth-term formula for $\{a_n\}$ that is analogous to the nth-term formulas for arithmetic and geometric sequences. Such an nth-term formula is valuable because it allows us to estimate a term of a sequence without computing all the preceding terms.

If the geometric sequence $\{r^n\}$ satisfies the recursion formula above, then $r^n = 5r^{n-1} - 6r^{n-2}$. Dividing by r^{n-2} leads to the quadratic equation $r^2 - 5r + 6 = 0$, whose solutions are $r = 2$ and $r = 3$. Now it is easy to check that the geometric sequences $\{2^n\} = 2, 4, 8, 16, \ldots$ and $\{3^n\} = 3, 9, 27, 81, \ldots$ satisfy the recursion formula. Therefore, any sequence of the form $\{u2^n + v3^n\}$, where u and v are constants, will satisfy the same recursion formula.

We now find u and v so that the first two terms of $\{u2^n + v3^n\}$ are $a_1 = 4$, $a_2 = 14$. Letting $n = 1$ and $n = 2$ we see that u and v must satisfy the following linear system:

$$2u + 3v = 4$$

$$4u + 9v = 14$$

Solving the system gives $u = -1$, $v = 2$. Therefore, an nth-term formula for the original sequence is $a_n = (-1)2^n + (2)3^n$.

Note that the nth-term formula was obtained by solving a quadratic equation and a system of two linear equations in two variables.

(A) Compute $(-1)2^n + (2)3^n$ for $n = 1, 2, \ldots, 6$, and compare with the terms of $\{a_n\}$.

(B) Estimate the one-hundredth term of $\{a_n\}$.

(C) Show that any sequence of the form $\{u2^n + v3^n\}$, where u and v are constants, satisfies the recursion formula $a_n = 5a_{n-1} - 6a_{n-2}$.

(D) Find an nth-term formula for the sequence $\{b_n\}$ that is specified by $b_1 = 5$, $b_2 = 55$, $b_n = 3b_{n-1} + 4b_{n-2}$.

(E) Find an nth-term formula for the Fibonacci sequence.

(F) Find an nth-term formula for the sequence $\{c_n\}$ that is specified by $c_1 = -3$, $c_2 = 15$, $c_3 = 99$, $c_n = 6c_{n-1} - 3c_{n-2} - 10c_{n-3}$. (Since the recursion formula involves the three terms which precede c_n, our method will involve the solution of a cubic equation and a system of three linear equations in three variables.)

Chapter 8 Review

8-1   SEQUENCES AND SERIES

A **sequence** is a function with the domain a set of successive integers. The symbol a_n, called the **nth term,** or **general term,** represents the range value associated with the domain value n. Unless specified otherwise, the domain is understood to be the set of natural numbers. A **finite sequence** has a finite domain, and an **infinite sequence** has an infinite domain. A **recursion formula** defines each term of a sequence in terms of one or more of the preceding terms. For example, the **Fibonacci sequence** is defined by $a_n = a_{n-1} + a_{n-2}$ for $n \geq 3$, where $a_1 = a_2 = 1$. If $a_1, a_2, \ldots, a_n, \ldots$ is a sequence, then the expression $a_1 + a_2 + \cdots + a_n + \cdots$ is called a **series.** A finite sequence produces a **finite series,** and an infinite sequence produces an **infinite series.** Series can be represented using summation notation:

$$\sum_{k=m}^{n} a_k = a_m + a_{m+1} + \cdots + a_n$$

where k is called the **summing index.** If the terms in the series are alternately positive and negative, the series is called an **alternating series.**

8-2 MATHEMATICAL INDUCTION

A wide variety of statements can be proven using the **principle of mathematical induction:** Let P_n be a statement associated with each positive integer n and suppose the following conditions are satisfied:

1. P_1 is true.

2. For any positive integer k, if P_k is true, then P_{k+1} is also true.

Then the statement P_n is true for all positive integers n.

To use mathematical induction to prove statements involving laws of exponents, it is convenient to state a **recursive definition of a^n:**

$$a^1 = a \quad \text{and} \quad a^{n+1} = a^n a \quad \text{for any integer } n \geq 1$$

To deal with conjectures that may be true only for $n \geq m$, where m is a positive integer, we use the **extended principle of mathematical induction:** Let m be a positive integer, let P_n be a statement associated with each integer $n \geq m$, and suppose the following conditions are satisfied:

1. P_m is true.

2. For any integer $k \geq m$, if P_k is true, then P_{k+1} is also true.

Then the statement P_n is true for all integers $n \geq m$.

8-3 ARITHMETIC AND GEOMETRIC SEQUENCES

A sequence is called an **arithmetic sequence,** or **arithmetic progression,** if there exists a constant d, called the **common difference,** such that

$$a_n - a_{n-1} = d \quad \text{or} \quad a_n = a_{n-1} + d$$

$$\text{for every } n > 1$$

The following formulas are useful when working with arithmetic sequences and their corresponding series:

$$a_n = a_1 + (n-1)d \qquad \textbf{nth-Term Formula}$$

$$S_n = \frac{n}{2}[2a_1 + (n-1)d] \qquad \textbf{Sum Formula—First Form}$$

$$S_n = \frac{n}{2}(a_1 + a_n) \qquad \textbf{Sum Formula—Second Form}$$

A sequence is called a **geometric sequence,** or a **geometric progression,** if there exists a nonzero constant r, called the **common ratio,** such that

$$\frac{a_n}{a_{n-1}} = r \quad \text{or} \quad a_n = ra_{n-1} \quad \text{for every } n > 1$$

The following formulas are useful when working with geometric sequences and their corresponding series:

$$a_n = a_1 r^{n-1} \qquad \textbf{nth-Term Formula}$$

$$S_n = \frac{a_1 - a_1 r^n}{1 - r} \quad r \neq 1 \qquad \textbf{Sum Formula—First Form}$$

$$S_n = \frac{a_1 - ra_n}{1 - r} \quad r \neq 1 \qquad \textbf{Sum Formula—Second Form}$$

$$S_\infty = \frac{a_1}{1 - r} \quad |r| < 1 \qquad \textbf{Sum of an Infinite Geometric Series}$$

8-4 BINOMIAL FORMULA

For n a natural number, **n factorial**—denoted $n!$—is defined by

$$n! = n(n-1) \cdot \cdots \cdot 2 \cdot 1 \qquad 1! = 1 \qquad 0! = 1$$

Also, n factorial is given by the **recursion formula**

$$n! = n \cdot (n-1)!$$

For nonnegative integers r and n, $0 \leq r \leq n$, the **combinatorial symbol** $\binom{n}{r}$ is defined by

$$\binom{n}{r} = \frac{n!}{r!(n-r)!} = \frac{n(n-1) \cdot \cdots \cdot (n-r+1)}{r(r-1) \cdot \cdots \cdot 2 \cdot 1}$$

For n a positive integer, the **binomial formula** is

$$(a+b)^n = \sum_{k=0}^{n} \binom{n}{k} a^{n-k} b^k$$

8-5 MULTIPLICATION PRINCIPLE, PERMUTATIONS, AND COMBINATIONS

Given a sequence of operations, **tree diagrams** are often used to list all the possible combined outcomes. To count the number of combined outcomes without actually listing them, we use the **multiplication principle:**

1. If operations O_1 and O_2 are performed in order with N_1 possible outcomes for the first operation and N_2 possible outcomes for the second operation, then there are

$$N_1 \cdot N_2$$

possible outcomes of the first operation followed by the second.

2. In general, if n operations $O_1, O_2, \ldots, O_n$ are performed in order, with possible number of outcomes $N_1, N_2, \ldots, N_n$, respectively, then there are

$$N_1 \cdot N_2 \cdot \cdots \cdot N_n$$

possible combined outcomes of the operations performed in the given order.

A particular arrangement or ordering of n objects without repetition is called a **permutation.** The number of permutations of n objects is given by

$$P_{n,n} = n \cdot (n-1) \cdot \cdots \cdot 1 = n!$$

and the number of permutations of n objects taken r at a time is given by

$$P_{n,r} = \frac{n!}{(n-r)!} \qquad 0 \le r \le n$$

A **combination of a set of n elements taken r at a time** is an r-element subset of the n objects. The number of combinations of n objects taken r at a time is given by

$$C_{n,r} = \binom{n}{r} = \frac{P_{n,r}}{r!} = \frac{n!}{r!(n-r)!} \qquad 0 \le r \le n$$

In a permutation, order is important. In a combination, order is not important.

Chapter 8 Review Exercise

Work through all the problems in this chapter review and check answers in the back of the book. Answers to all review problems are there, and following each answer is a number in italics indicating the section in which that type of problem is discussed. Where weaknesses show up, review appropriate sections in the text.

A

1. Determine whether each of the following can be the first three terms of a geometric sequence, an arithmetic sequence, or neither.
   (A) $16, -8, 4, \ldots$ (B) $5, 7, 9, \ldots$
   (C) $-8, -5, -2, \ldots$ (D) $2, 3, 5, \ldots$
   (E) $-1, 2, -4, \ldots$

In Problems 2–5:
(A) Write the first four terms of each sequence.
(B) Find a_{10}.   (C) Find S_{10}.

2. $a_n = 2n + 3$ 3. $a_n = 32(\tfrac{1}{2})^n$

4. $a_1 = -8; a_n = a_{n-1} + 3, n \ge 2$

5. $a_1 = -1; a_n = (-2)a_{n-1}, n \ge 2$

6. Find S_∞ in Problem 3.

Evaluate Problems 7–10.

7. $10!$ 8. $\dfrac{30!}{25!}$

9. $\dfrac{13!}{5!(13-5)!}$ 10. $P_{8,4}$ and $C_{8,4}$

11. A single die is rolled and a coin is flipped. How many combined outcomes are possible? Solve:
    (A) By using a tree diagram
    (B) By using the multiplication principle

12. How many seating arrangements are possible with 6 people

and 6 chairs in a row? Solve by using the multiplication principle.

13. Solve Problem 12 using permutations or combinations, whichever is applicable.

Verify Problems 14–16 for $n = 1, 2,$ and 3.

14. P_n: $5 + 7 + 9 + \cdots + (2n + 3) = n^2 + 4n$

15. P_n: $2 + 4 + 8 + \cdots + 2^n = 2^{n+1} - 2$

16. P_n: $49^n - 1$ is divisible by 6

In Problems 17–19, write P_k and P_{k+1}.

17. For P_n in Problem 14   18. For P_n in Problem 15

19. For P_n in Problem 16

20. Either prove the statement is true or prove it is false by finding a counterexample: If n is a positive integer, then the sum of the series $1 + \dfrac{1}{2} + \dfrac{1}{3} + \cdots + \dfrac{1}{n}$ is less than 4.

B

Write Problems 21 and 22 without summation notation, and find the sum.

21. $S_{10} = \displaystyle\sum_{k=1}^{10} (2k - 8)$

22. $S_7 = \displaystyle\sum_{k=1}^{7} \frac{16}{2^k}$

23. $S_\infty = 27 - 18 + 12 + \cdots = ?$

24. Write

$$S_n = \frac{1}{3} - \frac{1}{9} + \frac{1}{27} + \cdots + \frac{(-1)^{n+1}}{3^n}$$

using summation notation, and find S_∞.

25. Six distinct points are selected on the circumference of a circle. How many triangles can be formed using these points as vertices?

26. In an arithmetic sequence, $a_1 = 13$ and $a_7 = 31$. Find the common difference d and the fifth term a_5.

27. The sum of the first ten terms of an arithmetic series is 81. If the sixth term is 10, find the first term a_1 and the common difference d.

28. How many 3-letter code words are possible using the first 8 letters of the alphabet if no letter can be repeated? If letters can be repeated? If adjacent letters cannot be alike?

29. Use the formula for the sum of an infinite geometric series to write $0.727\ 272 \cdots = 0.\overline{72}$ as the quotient of two integers.

30. Solve the following problems using $P_{n,r}$ or $C_{n,r}$, as appropriate:
(A) How many 3-digit opening combinations are possible on a combination lock with 6 digits if the digits cannot be repeated?
(B) Suppose 5 tennis players have made the finals. If each of the 5 players is to play every other player exactly once, how many games must be scheduled?

Evaluate Problems 31–33.

31. $\dfrac{20!}{18!(20-18)!}$ 32. $\dbinom{16}{12}$ 33. $\dbinom{11}{11}$

34. Which is larger, $\dfrac{987!}{493!}$ or $\dfrac{987!}{(987-493)!}$? Explain.

35. Which is smaller, $\dbinom{1000}{500}$ or $\dbinom{1000}{501}$? Explain.

36. Expand $(x-y)^5$ using the binomial formula.

37. Find the tenth term in the expansion of $(2x-y)^{12}$.

Establish each statement in Problems 38–40 for all natural numbers, using mathematical induction.

38. P_n in Problem 14

39. P_n in Problem 15

40. P_n in Problem 16

In Problems 41–42, find the smallest positive integer n such that $a_n < b_n$ by graphing the sequences $\{a_n\}$ and $\{b_n\}$ with a graphing utility. Check your answer by using a graphing utility to display both sequences in table form.

41. $a_n = C_{50,n},\ b_n = 3^n$

42. $a_1 = 100,\ a_n = 0.99a_{n-1} + 5,\ b_n = 9 + 7n$

C

43. How many different families with 5 children are possible, excluding multiple births, where the sex of each child in the order of their birth is taken into consideration? How many families are possible if the order pattern is not taken into account?

44. A free-falling body travels $g/2$ feet in the first second, $3g/2$ feet during the next second, $5g/2$ feet the next, and so on. Find the distance fallen during the twenty-fifth second and the total distance fallen from the start to the end of the twenty-fifth second.

45. How many ways can 2 people be seated in a row of 4 chairs?

46. Expand $(x+i)^6$, where i is the imaginary unit, using the binomial formula.

47. **Transportation.** A distribution center A wishes to distribute its products to 5 different retail stores, $B, C, D, E,$ and F, in a city. How many different route plans can be constructed so that a single truck can start from A, deliver to each store exactly once, and then return to the center?

Prove that each statement in Problems 48–52 holds for all positive integers, using mathematical induction.

48. $\displaystyle\sum_{k=1}^{n} k^3 = \left(\sum_{k=1}^{n} k\right)^2$

49. $x^{2n} - y^{2n}$ is divisible by $x - y,\ x \neq y$

50. $\dfrac{a^n}{a^m} = a^{n-m};\ n > m,\ n, m$ positive integers

51. $\{a_n\} = \{b_n\}$, where $a_n = a_{n-1} + 2,\ a_1 = -3,\ b_n = -5 + 2n$

52. $(1!)1 + (2!)2 + (3!)3 + \cdots + (n!)n = (n+1)! - 1$ (From U.S.S.R. Mathematical Olympiads, 1955–1956, Grade 10.)

Problems 53–56 refer to the sequences $\{a_n\}$ and $\{b_n\}$ where $a_n = \dbinom{2n}{n}$ and $b_n = 4^{n-1}$.

53. Find the smallest positive integer n such that $a_n \leq b_n$.

54. Are $\{a_n\}$ and $\{b_n\}$ arithmetic sequences? Geometric sequences?

55. Show that $\dfrac{a_{n+1}}{a_n} < 4$ for all positive integers n.

56. Use mathematical induction to show that $a_n < b_n$ for $n \geq 5$.

ADDITIONAL TOPICS IN ANALYTIC GEOMETRY

CHAPTER 9

Analytic geometry, a union of geometry and algebra, enables us to analyze certain geometric concepts algebraically and to interpret certain algebraic relationships geometrically. Our two main concerns center around graphing algebraic equations and finding equations of useful geometric figures. We have discussed a number of topics in analytic geometry, such as straight lines and circles, in earlier chapters. In this chapter we discuss additional analytic geometry topics: conic sections and translation of axes.

René Descartes (1596–1650), the French philosopher–mathematician, is generally recognized as the founder of analytic geometry.

SECTION 9-1   Conic Sections; Parabola

- Conic Sections
- Definition of a Parabola
- Drawing a Parabola
- Standard Equations and Their Graphs
- Applications

In this section we introduce the general concept of a conic section and then discuss the particular conic section called a *parabola*. In the next two sections we will discuss two other conic sections called *ellipses* and *hyperbolas*.

• Conic Sections   In Section 3-2 we found that the graph of a first-degree equation in two variables,

$$Ax + By = C \tag{1}$$

where A and B are not both 0, is a straight line, and every straight line in a rectangular coordinate system has an equation of this form. What kind of graph will a second-degree equation in two variables,

$$Ax^2 + Bxy + Cy^2 + Dx + Ey + F = 0 \tag{2}$$

where A, B, and C are not all 0, yield for different sets of values of the coefficients? The graphs of equation (2) for various choices of the coefficients are plane curves obtainable by intersecting a cone* with a plane, as shown in Figure 1. These curves are called **conic sections.**

If a plane cuts clear through one nappe, then the intersection curve is called a **circle** if the plane is perpendicular to the axis and an **ellipse** if the plane is not perpendicular to the axis. If a plane cuts only one nappe, but does not cut clear through,

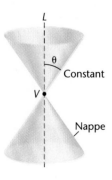

---

*Starting with a fixed line L and a fixed point V on L, the surface formed by all straight lines through V making a constant angle θ with L is called a **right circular cone**. The fixed line L is called the axis of the cone, and V is its **vertex**. The two parts of the cone separated by the vertex are called **nappes.**

FIGURE 1 Conic sections.

Circle   Ellipse   Parabola   Hyperbola

then the intersection curve is called a **parabola.** Finally, if a plane cuts through both nappes, but not through the vertex, the resulting intersection curve is called a **hyperbola.** A plane passing through the vertex of the cone produces a **degenerate conic**— a point, a line, or a pair of lines.

Conic sections are very useful and are readily observed in your immediate surroundings: wheels (circle), the path of water from a garden hose (parabola), some serving platters (ellipses), and the shadow on a wall from a light surrounded by a cylindrical or conical lamp shade (hyperbola) are some examples (see Fig. 2). We will discuss many applications of conics throughout the remainder of this chapter.

FIGURE 2 Examples of conics.

Wheel (circle) (a)   Water from garden hose (parabola) (b)   Serving platter (ellipse) (c)   Lamp light shadow (hyperbola) (d)

A definition of a conic section that does not depend on the coordinates of points in any coordinate system is called a **coordinate-free definition.** In Section 3-1 we gave a coordinate-free definition of a circle and developed its standard equation in a rectangular coordinate system. In this and the next two sections we will give coordinate-free definitions of a parabola, ellipse, and hyperbola, and we will develop standard equations for each of these conics in a rectangular coordinate system.

• Definition of a Parabola

The following definition of a parabola does not depend on the coordinates of points in any coordinate system:

DEFINITION 1   **Parabola**

A **parabola** is the set of all points in a
plane equidistant from a fixed point F and
a fixed line L in the plane. The fixed point
F is called the **focus,** and the fixed line L
is called the **directrix.** A line through the
focus perpendicular to the directrix is
called the **axis,** and the point on the axis
halfway between the directrix and focus
is called the **vertex.**

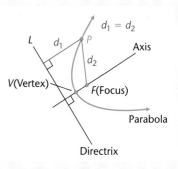

● **Drawing a**
Parabola

Using the definition, we can draw a parabola with fairly simple equipment—a
straightedge, a right-angle drawing triangle, a piece of string, a thumbtack, and a
pencil. Referring to Figure 3, tape the straightedge along the line AB and place the
thumbtack above the line AB. Place one leg of the triangle along the straightedge as
indicated, then take a piece of string the same length as the other leg, tie one end to
the thumbtack, and fasten the other end with tape at C on the triangle. Now press the
string to the edge of the triangle, and keeping the string taut, slide the triangle along
the straightedge. Since DE will always equal DF, the resulting curve will be part
of a parabola with directrix AB lying along the straightedge and focus F at the
thumbtack.

FIGURE 3 Drawing a parabola.

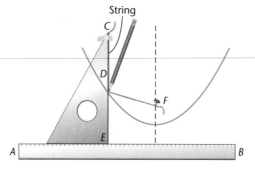

EXPLORE-DISCUSS 1   The line through the focus F that is perpendicular to the axis of a parabola intersects the parabola in two points G and H. Explain why the distance from G to H
is twice the distance from F to the directrix of the parabola.

● **Standard Equations**
and Their Graphs

Using the definition of a parabola and the distance-between-two-points formula

$$d = \sqrt{(x_2 - x_1)^2 + (y_2 - y_1)^2} \qquad (3)$$

FIGURE 4 Parabola with center at the origin and axis the x axis.

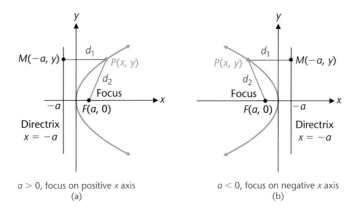

$a > 0$, focus on positive x axis
(a)

$a < 0$, focus on negative x axis
(b)

we can derive simple standard equations for a parabola located in a rectangular coordinate system with its vertex at the origin and its axis along a coordinate axis. We start with the axis of the parabola along the x axis and the focus at $F(a, 0)$. We locate the parabola in a coordinate system as in Figure 4 and label key lines and points. This is an important step in finding an equation of a geometric figure in a coordinate system. Note that the parabola opens to the right if $a > 0$ and to the left if $a < 0$. The vertex is at the origin, the directrix is $x = -a$, and the coordinates of M are $(-a, y)$.

The point $P(x, y)$ is a point on the parabola if and only if

$$d_1 = d_2$$

$$d(P, M) = d(P, F)$$

$$\sqrt{(x + a)^2 + (y - y)^2} = \sqrt{(x - a)^2 + (y - 0)^2} \quad \text{Use equation (3).}$$

$$(x + a)^2 = (x - a)^2 + y^2 \quad \text{Square both sides.}$$

$$x^2 + 2ax + a^2 = x^2 - 2ax + a^2 + y^2 \quad \text{Simplify.}$$

$$y^2 = 4ax \tag{4}$$

Equation (4) is the standard equation of a parabola with vertex at the origin, axis the x axis, and focus at $(a, 0)$.

Now we locate the vertex at the origin and focus on the y axis at $(0, a)$. Looking at Figure 5 on the following page, we note that the parabola opens upward if $a > 0$ and downward if $a < 0$. The directrix is $y = -a$, and the coordinates of N are $(x, -a)$. The point $P(x, y)$ is a point on the parabola if and only if

$$d_1 = d_2$$

$$d(P, N) = d(P, F)$$

$$\sqrt{(x - x)^2 + (y + a)^2} = \sqrt{(x - 0)^2 + (y - a)^2} \quad \text{Use equation (3).}$$

$$(y + a)^2 = x^2 + (y - a)^2 \quad \text{Square both sides.}$$

$$y^2 + 2ay + a^2 = x^2 + y^2 - 2ay + a^2 \quad \text{Simplify.}$$

$$x^2 = 4ay \tag{5}$$

Equation (5) is the standard equation of a parabola with vertex at the origin, axis the y axis, and focus at $(0, a)$.

FIGURE 5 Parabola with center at the origin and axis the y axis.

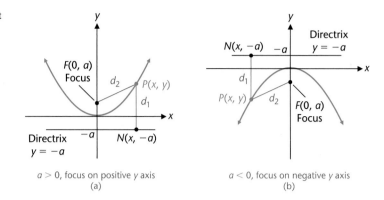

$a > 0$, focus on positive y axis
(a)

$a < 0$, focus on negative y axis
(b)

We summarize these results for easy reference in Theorem 1:

Theorem 1   **Standard Equations of a Parabola with Vertex at (0, 0)**

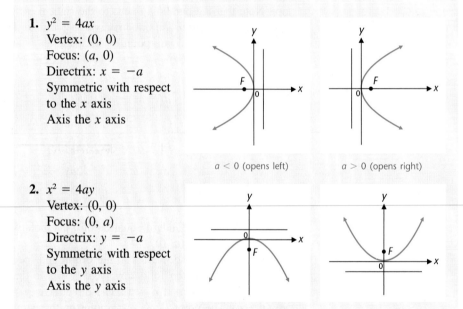

1. $y^2 = 4ax$
   Vertex: (0, 0)
   Focus: $(a, 0)$
   Directrix: $x = -a$
   Symmetric with respect to the x axis
   Axis the x axis

   $a < 0$ (opens left) $a > 0$ (opens right)

2. $x^2 = 4ay$
   Vertex: (0, 0)
   Focus: $(0, a)$
   Directrix: $y = -a$
   Symmetric with respect to the y axis
   Axis the y axis

   $a < 0$ (opens down) $a > 0$ (opens up)

EXAMPLE 1   **Graphing $x^2 = 4ay$**

Graph $x^2 = -16y$, and locate the focus and directrix.

Solution   To graph $x^2 = -16y$, it is convenient to assign y values that make the right side a perfect square, and solve for x. Note that y must be 0 or negative for x to be real. Since the coefficient of y is negative, a must be negative, and the parabola opens downward (Fig. 6).

x	0	± 4	± 8
y	0	-1	-4

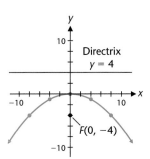

FIGURE 6 $x^2 = -16y$.

Focus: $\overset{a}{x^2 = -16y} = 4(-4)y$
          $F(0, a) = F(0, -4)$
Directrix: $y = -a$
          $= -(-4) = 4$

Matched Problem 1   Graph $y^2 = -8x$, and locate the focus and directrix.

 Remark.   To graph the equation $x^2 = -16y$ of Example 1 on a graphing utility, we first solve the equation for y and then graph the function $y = -\frac{1}{16}x^2$. If that same approach is used to graph the equation $y^2 = -8x$ of Matched Problem 1, then $y = \pm\sqrt{-8x}$, and there are two functions to graph. The graph of $y = \sqrt{-8x}$ is the upper half of the parabola, and the graph of $y = -\sqrt{-8x}$ is the lower half (see Fig. 7).

FIGURE 7

CAUTION   A common error in making a quick sketch of $y^2 = 4ax$ or $x^2 = 4ay$ is to sketch the first with the y axis as its axis and the second with the x axis as its axis. The graph of $y^2 = 4ax$ is symmetric with respect to the x axis, and the graph of $x^2 = 4ay$ is symmetric with respect to the y axis, as a quick symmetry check will reveal.

EXAMPLE 2   **Finding the Equation of a Parabola**

(A) Find the equation of a parabola having the origin as its vertex, the y axis as its axis, and $(-10, -5)$ on its graph.

(B) Find the coordinates of its focus and the equation of its directrix.

Solutions   (A) The parabola is opening down and has an equation of the form $x^2 = 4ay$. Since $(-10, -5)$ is on the graph, we have

$$x^2 = 4ay$$
$$(-10)^2 = 4a(-5)$$
$$100 = -20a$$
$$a = -5$$

Thus, the equation of the parabola is

$$x^2 = 4(-5)y$$
$$= -20y$$

(B) Focus: $x^2 = -20y$ $\overset{a}{= 4(-5)y}$
              $F(0, a) = F(0, -5)$
      Directrix: $y = -a$
              $= -(-5)$
              $= 5$

Matched Problem 2   (A) Find the equation of a parabola having the origin as its vertex, the x axis as its axis, and $(4, -8)$ on its graph.

(B) Find the coordinates of its focus and the equation of its directrix.

EXPLORE-DISCUSS 2   Consider the graph of an equation in the variables x and y. The equation of its magnification by a factor $k > 0$ is obtained by replacing x and y in the equation by x/k and y/k, respectively. (Of course, a magnification by a factor k between 0 and 1 means an actual reduction in size.)

(A) Show that the magnification by a factor 3 of the circle with equation $x^2 + y^2 = 1$ has equation $x^2 + y^2 = 9$.

(B) Explain why every circle with center at $(0, 0)$ is a magnification of the circle with equation $x^2 + y^2 = 1$.

(C) Find the equation of the magnification by a factor 3 of the parabola with equation $x^2 = y$. Graph both equations.

(D) Explain why every parabola with vertex $(0, 0)$ that opens upward is a magnification of the parabola with equation $x^2 = y$.

● **Applications** Parabolic forms are frequently encountered in the physical world. Suspension bridges, arch bridges, microphones, symphony shells, satellite antennas, radio and optical tele-scopes, radar equipment, solar furnaces, and searchlights are only a few of many items that utilize parabolic forms in their design.

Figure 8(a) illustrates a parabolic reflector used in all reflecting telescopes—from 3- to 6-inch home type to the 200-inch research instrument on Mount Palomar in California. Parallel light rays from distant celestial bodies are reflected to the focus off a parabolic mirror. If the light source is the sun, then the parallel rays are focused at *F* and we have a solar furnace. Temperatures of over 6,000°C have been achieved by such furnaces. If we locate a light source at *F*, then the rays in Figure 8(a) reverse, and we have a spotlight or a searchlight. Automobile headlights can use parabolic reflectors with special lenses over the light to diffuse the rays into useful patterns.

Figure 8(b) shows a suspension bridge, such as the Golden Gate Bridge in San Francisco. The suspension cable is a parabola. It is interesting to note that a free-hanging cable, such as a telephone line, does not form a parabola. It forms another curve called a *catenary*.

Figure 8(c) shows a concrete arch bridge. If all the loads on the arch are to be compression loads (concrete works very well under compression), then using physics and advanced mathematics, it can be shown that the arch must be parabolic.

FIGURE 8 Uses of parabolic forms.

Parabolic reflector

(a)

Suspension bridge

(b)

Arch bridge

(c)

EXAMPLE 3 **Parabolic Reflector**

A **paraboloid** is formed by revolving a parabola about its axis. A spotlight in the form of a paraboloid 5 inches deep has its focus 2 inches from the vertex. Find, to one decimal place, the radius *R* of the opening of the spotlight.

Solution *Step 1.* Locate a parabolic cross section containing the axis in a rectangular coordinate system, and label all known parts and parts to be found. This is a very important step and can be done in infinitely many ways. Since we are in charge, we can make things simpler for ourselves by locating the vertex at the origin and choosing a coordinate axis as the axis. We choose the *y* axis as the axis of the parabola with the parabola opening upward. See Figure 9 on the following page.

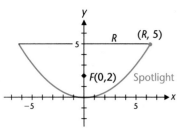

FIGURE 9

Step 2. Find the equation of the parabola in the figure. Since the parabola has the y axis as its axis and the vertex at the origin, the equation is of the form

$$x^2 = 4ay$$

We are given $F(0, a) = F(0, 2)$; thus, $a = 2$, and the equation of the parabola is

$$x^2 = 8y$$

Step 3. Use the equation found in step 2 to find the radius R of the opening. Since $(R, 5)$ is on the parabola, we have

$$R^2 = 8(5)$$
$$R = \sqrt{40} \approx 6.3 \text{ inches}$$

Matched Problem 3 Repeat Example 3 with a paraboloid 12 inches deep and a focus 9 inches from the vertex.

Answers to Matched Problems

1. Focus: $(-2, 0)$
Directrix: $x = 2$

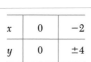

x	0	-2
y	0	±4

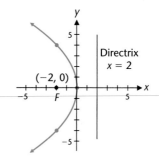

2. (A) $y^2 = 16x$ (B) Focus: $(4, 0)$; Directrix: $x = -4$
3. $R = 20.8$ in.

EXERCISE 9-1

A

In Problems 1–12, graph each equation, and locate the focus and directrix.

1. $y^2 = 4x$ **2.** $y^2 = 8x$ **3.** $x^2 = 8y$

4. $x^2 = 4y$ **5.** $y^2 = -12x$ **6.** $y^2 = -4x$

7. $x^2 = -4y$ **8.** $x^2 = -8y$ **9.** $y^2 = -20x$

10. $x^2 = -24y$ **11.** $x^2 = 10y$ **12.** $y^2 = 6x$

Find the coordinates to two decimal places of the focus for each parabola in Problems 13–18.

13. $y^2 = 39x$ **14.** $x^2 = 58y$ **15.** $x^2 = -105y$

16. $y^2 = -93x$ **17.** $y^2 = -77x$ **18.** $x^2 = -205y$

B

In Problems 19–24, find the equation of a parabola with vertex at the origin, axis the x or y axis, and:

19. Focus $(-6, 0)$ **20.** Directrix $y = 8$

21. Directrix $y = -5$ **22.** Focus $(3, 0)$

23. Focus $\left(0, -\dfrac{1}{3}\right)$ **24.** Directrix $y = -\dfrac{1}{2}$

In Problems 25–30, find the equation of the parabola having its vertex at the origin, its axis as indicated, and passing through the indicated point.

25. x axis; $(-4, -20)$ **26.** y axis; $(30, -15)$

27. y axis; $(9, -27)$ **28.** x axis; $(121, 11)$

29. x axis; $(-8, -2)$ **30.** y axis; $(-\sqrt{2}, 3)$

In Problems 31–34, find the first-quadrant points of intersection for each system of equations to three decimal places.

 Check Problems 31–34 with a graphing utility.

31. $x^2 = 4y$ **32.** $y^2 = 3x$
     $y^2 = 4x$ $x^2 = 3y$

33. $y^2 = 6x$ **34.** $x^2 = 7y$
     $x^2 = 5y$ $y^2 = 2x$

35. Consider the parabola with equation $x^2 = 4ay$.
   (A) How many lines through $(0, 0)$ intersect the parabola in exactly one point? Find their equations.
   (B) Find the coordinates of all points of intersection of the parabola with the line through $(0, 0)$ having slope $m \neq 0$.

36. Find the coordinates of all points of intersection of the parabola with equation $x^2 = 4ay$ and the parabola with equation $y^2 = 4bx$.

37. If a line through the focus contains two points A and B of a parabola, then the line segment AB is called a **focal chord**. Find the coordinates of A and B for the focal chord that is perpendicular to the axis of the parabola $x^2 = 4ay$.

Figure for 37 and 38

38. Find the length of the focal chord AB that is perpendicular to the axis of the parabola $x^2 = 4ay$.

In Problems 39–42, determine whether the statement is true or false. If true, explain why. If false, give a counterexample.

39. If a is real, then the graph of $y^2 = 4ax$ is a parabola.

40. If a is negative, then the graph of $y^2 = 4ax$ is a parabola.

41. Every vertical line intersects the graph of $x^2 = 4y$.

42. Every nonhorizontal line intersects the graph of $x^2 = 4y$.

In Problems 43–46, use the definition of a parabola and the distance formula to find the equation of a parabola with:

43. Directrix $y = -4$ and focus $(2, 2)$

44. Directrix $y = 2$ and focus $(-3, 6)$

45. Directrix $x = 2$ and focus $(6, -4)$

46. Directrix $x = -3$ and focus $(1, 4)$

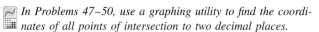 *In Problems 47–50, use a graphing utility to find the coordinates of all points of intersection to two decimal places.*

47. $x^2 = 8y,\ y = 5x + 4$ **48.** $x^2 = 3y,\ 7x + 4y = 11$

49. $x^2 = -8y,\ y^2 = -5x$ **50.** $y^2 = 6x,\ 2x - 9y = 13$

_____

APPLICATIONS

51. Engineering. The parabolic arch in the concrete bridge in the figure must have a clearance of 50 feet above the water and span a distance of 200 feet. Find the equation of the parabola after inserting a coordinate system with the origin at the vertex of the parabola and the vertical y axis (pointing upward) along the axis of the parabola.

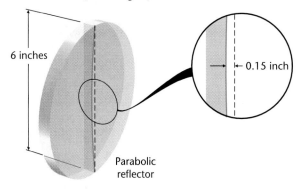

52. Astronomy. The cross section of a parabolic reflector with 6-inch diameter is ground so that its vertex is 0.15 inch below the rim (see the figure).

(A) Find the equation of the parabola after inserting an xy coordinate system with the vertex at the origin, the y axis (pointing upward) the axis of the parabola.

(B) How far is the focus from the vertex?

53. Space Science. A designer of a 200-foot-diameter parabolic electromagnetic antenna for tracking space probes wants to place the focus 100 feet above the vertex (see the figure).

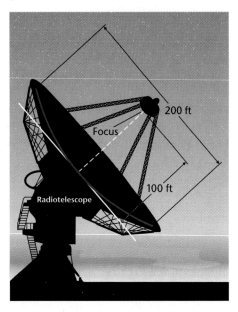

(A) Find the equation of the parabola using the axis of the parabola as the y axis (up positive) and vertex at the origin.

(B) Determine the depth of the parabolic reflector.

54. Signal Light. A signal light on a ship is a spotlight with parallel reflected light rays (see the figure). Suppose the parabolic reflector is 12 inches in diameter and the light source is located at the focus, which is 1.5 inches from the vertex.

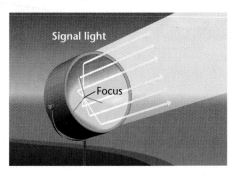

(A) Find the equation of the parabola using the axis of the parabola as the x axis (right positive) and vertex at the origin.

(B) Determine the depth of the parabolic reflector.

SECTION 9-2 Ellipse

- Definition of an Ellipse
- Drawing an Ellipse
- Standard Equations and Their Graphs
- Applications

We start our discussion of the ellipse with a coordinate-free definition. Using this definition, we show how an ellipse can be drawn and we derive standard equations for ellipses specially located in a rectangular coordinate system.

- **Definition of an Ellipse**

The following is a coordinate-free definition of an ellipse:

DEFINITION 1   Ellipse

An **ellipse** is the set of all points P in a plane such that the sum of the distances of P from two fixed points in the plane is constant. Each of the fixed points, F' and F, is called a **focus,** and together they are called **foci.** Referring to the figure, the line segment $V'V$ through the foci is the **major axis.** The perpendicular bisector $B'B$ of the major axis is the **minor axis.** Each end of the major axis,

V' and V, is called a **vertex.** The midpoint of the line segment $F'F$ is called the **center** of the ellipse.

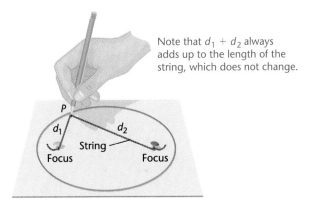

Drawing an Ellipse

An ellipse is easy to draw. All you need is a piece of string, two thumbtacks, and a pencil or pen (see Fig. 1). Place the two thumbtacks in a piece of cardboard. These form the foci of the ellipse. Take a piece of string longer than the distance between the two thumbtacks—this represents the constant in the definition—and tie each end to a thumbtack. Finally, catch the tip of a pencil under the string and move it while keeping the string taut. The resulting figure is by definition an ellipse. Ellipses of different shapes result, depending on the placement of thumbtacks and the length of the string joining them.

FIGURE 1 Drawing an ellipse.

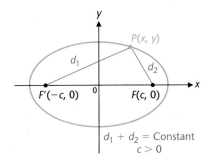

Note that $d_1 + d_2$ always adds up to the length of the string, which does not change.

Standard Equations and Their Graphs

Using the definition of an ellipse and the distance-between-two-points formula, we can derive standard equations for an ellipse located in a rectangular coordinate system. We start by placing an ellipse in the coordinate system with the foci on the x axis equidistant from the origin at $F'(-c, 0)$ and $F(c, 0)$, as in Figure 2.

FIGURE 2 Ellipse with foci on x axis.

For reasons that will become clear soon, it is convenient to represent the constant sum $d_1 + d_2$ by $2a$, $a > 0$. Also, the geometric fact that the sum of the lengths of any two sides of a triangle must be greater than the third side can be applied to Figure 2 to derive the following useful result:

$$d(F', P) + d(P, F) > d(F', F)$$

$$d_1 + d_2 > 2c$$

$$2a > 2c$$

$$a > c \tag{1}$$

We will use this result in the derivation of the equation of an ellipse, which we now begin.

Referring to Figure 2, the point $P(x, y)$ is on the ellipse if and only if

$$d_1 + d_2 = 2a$$

$$d(P, F') + d(P, F) = 2a$$

$$\sqrt{(x + c)^2 + (y - 0)^2} + \sqrt{(x - c)^2 + (y - 0)^2} = 2a$$

After eliminating radicals and simplifying, a good exercise for you, we obtain

$$(a^2 - c^2)x^2 + a^2 y^2 = a^2(a^2 - c^2) \tag{2}$$

$$\frac{x^2}{a^2} + \frac{y^2}{a^2 - c^2} = 1 \tag{3}$$

Dividing both sides of equation (2) by $a^2(a^2 - c^2)$ is permitted, since neither a^2 nor $a^2 - c^2$ is 0. From equation (1), $a > c$; thus $a^2 > c^2$ and $a^2 - c^2 > 0$. The constant a was chosen positive at the beginning.

To simplify equation (3) further, we let

$$b^2 = a^2 - c^2 \qquad b > 0 \tag{4}$$

to obtain

$$\frac{x^2}{a^2} + \frac{y^2}{b^2} = 1 \tag{5}$$

From equation (5) we see that the x intercepts are $x = \pm a$ and the y intercepts are $y = \pm b$. The x intercepts are also the vertices. Thus,

$$\text{Major axis length} = 2a$$

$$\text{Minor axis length} = 2b$$

To see that the major axis is longer than the minor axis, we show that $2a > 2b$. Returning to equation (4),

$$b^2 = a^2 - c^2 \qquad a, b, c > 0$$

$$b^2 + c^2 = a^2$$

$$b^2 < a^2 \qquad\qquad \text{Definition of } <$$

$$b^2 - a^2 < 0$$

$$(b - a)(b + a) < 0$$

$$b - a < 0 \qquad\qquad \text{Since } b + a \text{ is positive, } b - a \text{ must be negative.}$$

$$b < a$$

$$2b < 2a$$

$$2a > 2b$$

$$\begin{pmatrix} \text{Length of} \\ \text{major axis} \end{pmatrix} > \begin{pmatrix} \text{Length of} \\ \text{minor axis} \end{pmatrix}$$

If we start with the foci on the y axis at $F(0, c)$ and $F'(0, -c)$ as in Figure 3, instead of on the x axis as in Figure 2, then, following arguments similar to those used for the first derivation, we obtain

$$\frac{x^2}{b^2} + \frac{y^2}{a^2} = 1 \qquad a > b \qquad\qquad (6)$$

where the relationship among a, b, and c remains the same as before:

$$b^2 = a^2 - c^2 \qquad\qquad (7)$$

The center is still at the origin, but the major axis is now along the y axis and the minor axis is along the x axis.

FIGURE 3 Ellipse with foci on y axis.

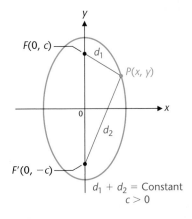

$$d_1 + d_2 = \text{Constant}$$
$$c > 0$$

To sketch graphs of equations of the form (5) or (6) is an easy matter. We find the x and y intercepts and sketch in an appropriate ellipse. Since replacing x with $-x$, or y with $-y$ produces an equivalent equation, we conclude that the graphs are symmetric with respect to the x axis, y axis, and origin. If further accuracy is required, additional points can be found with the aid of a calculator and the use of symmetry properties.

Given an equation of the form (5) or (6), how can we find the coordinates of the foci without memorizing or looking up the relation $b^2 = a^2 - c^2$? There is a simple geometric relationship in an ellipse that enables us to get the same result using the Pythagorean theorem. To see this relationship, refer to Figure 4(a). Then, using the definition of an ellipse and $2a$ for the constant sum, as we did in deriving the standard equations, we see that

$$d + d = 2a$$

$$2d = 2a$$

$$d = a$$

Thus:

> The length of the line segment from the end of a minor axis to a focus is the same as half the length of a major axis.

This geometric relationship is illustrated in Figure 4(b). Using the Pythagorean theorem for the triangle in Figure 4(b), we have

$$b^2 + c^2 = a^2$$

or

$$b^2 = a^2 - c^2 \qquad \text{Equations (4) and (7)}$$

or

$$c^2 = a^2 - b^2 \qquad \text{Useful for finding the foci, given } a \text{ and } b$$

Thus, we can find the foci of an ellipse given the intercepts a and b simply by using the triangle in Figure 4(b) and the Pythagorean theorem.

FIGURE 4 Geometric relationships.

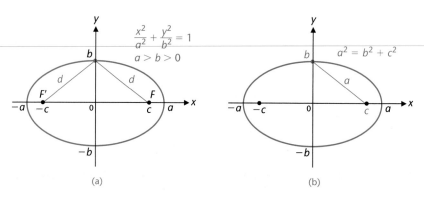

(a)   (b)

We summarize all of these results for convenient reference in Theorem 1.

Theorem 1   **Standard Equations of an Ellipse with Center at (0, 0)**

1. $\dfrac{x^2}{a^2} + \dfrac{y^2}{b^2} = 1 \qquad a > b > 0$

x intercepts: $\pm a$ (vertices)
y intercepts: $\pm b$
Foci: $F'(-c, 0)$, $F(c, 0)$

$$c^2 = a^2 - b^2$$

Major axis length $= 2a$
Minor axis length $= 2b$

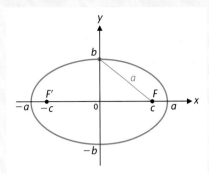

2. $\dfrac{x^2}{b^2} + \dfrac{y^2}{a^2} = 1 \qquad a > b > 0$

x intercepts: $\pm b$
y intercepts: $\pm a$ (vertices)
Foci: $F'(0, -c)$, $F(0, c)$

$$c^2 = a^2 - b^2$$

Major axis length $= 2a$
Minor axis length $= 2b$

[*Note:* Both graphs are symmetric with respect to the x axis, y axis, and origin. Also, the major axis is always longer than the minor axis.]

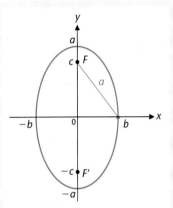

EXPLORE-DISCUSS 1   The line through a focus F of an ellipse that is perpendicular to the major axis intersects the ellipse in two points G and H. For each of the two standard equations of an ellipse with center $(0, 0)$, find an expression in terms of a and b for the distance from G to H.

EXAMPLE 1   **Graphing Ellipses**

Sketch the graph of each equation, find the coordinates of the foci, and find the lengths of the major and minor axes.

(A) $9x^2 + 16y^2 = 144$ (B) $2x^2 + y^2 = 10$

Solutions   (A) First, write the equation in standard form by dividing both sides by 144:

$$9x^2 + 16y^2 = 144$$

$$\frac{9x^2}{144} + \frac{16y^2}{144} = \frac{144}{144}$$

$$\frac{x^2}{16} + \frac{y^2}{9} = 1 \qquad a^2 = 16 \text{ and } b^2 = 9$$

Locate the intercepts:

$$x \text{ intercepts:} \quad \pm 4$$

$$y \text{ intercepts:} \quad \pm 3$$

and sketch in the ellipse, as shown in Figure 5.

$$\text{Foci:} \quad c^2 = a^2 - b^2$$
$$= 16 - 9$$
$$= 7$$
$$c = \sqrt{7} \qquad c \text{ is positive}$$

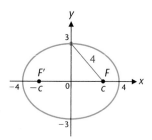

FIGURE 5 $9x^2 + 16y^2 = 144$.

Thus, the foci are $F'(-\sqrt{7}, 0)$ and $F(\sqrt{7}, 0)$.

$$\text{Major axis length} = 2(4) = 8$$
$$\text{Minor axis length} = 2(3) = 6$$

(B) Write the equation in standard form by dividing both sides by 10:

$$2x^2 + y^2 = 10$$

$$\frac{x^2}{5} + \frac{y^2}{10} = 1 \qquad a^2 = 10 \text{ and } b^2 = 5$$

Locate the intercepts:

$$x \text{ intercepts:} \quad \pm\sqrt{5} \approx \pm 2.24$$

$$y \text{ intercepts:} \quad \pm\sqrt{10} \approx \pm 3.16$$

and sketch in the ellipse, as shown in Figure 6.

$$\text{Foci:} \quad c^2 = a^2 - b^2$$
$$= 10 - 5$$
$$= 5$$
$$c = \sqrt{5}$$

FIGURE 6 $2x^2 + y^2 = 10$.

Thus, the foci are $F'(0, -\sqrt{5})$ and $F(0, \sqrt{5})$.

$$\text{Major axis length} = 2\sqrt{10} \approx 6.32$$
$$\text{Minor axis length} = 2\sqrt{5} \approx 4.47$$

Remark. To graph the equation $9x^2 + 16y^2 = 144$ of Example 1A on a graphing utility we first solve the equation for y, obtaining $y = \pm\sqrt{\dfrac{144 - 9x^2}{16}}$. We then graph each of the two functions. The graph of $y = \sqrt{\dfrac{144 - 9x^2}{16}}$ is the upper half of the ellipse, and the graph of $y = -\sqrt{\dfrac{144 - 9x^2}{16}}$ is the lower half.

Matched Problem 1 Sketch the graph of each equation, find the coordinates of the foci, and find the lengths of the major and minor axes.

(A) $x^2 + 4y^2 = 4$ (B) $3x^2 + y^2 = 18$

EXAMPLE 2 **Finding the Equation of an Ellipse**

Find an equation of an ellipse in the form

$$\frac{x^2}{M} + \frac{y^2}{N} = 1 \qquad M, N > 0$$

if the center is at the origin, the major axis is along the y axis, and:

(A) Length of major axis = 20   (B) Length of major axis = 10
     Length of minor axis = 12   Distance of foci from center = 4

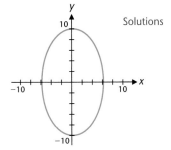

FIGURE 7 $\dfrac{x^2}{36} + \dfrac{y^2}{100} = 1$.

Solutions (A) Compute x and y intercepts and make a rough sketch of the ellipse, as shown in Figure 7.

$$\frac{x^2}{b^2} + \frac{y^2}{a^2} = 1$$

$$a = \frac{20}{2} = 10 \qquad b = \frac{12}{2} = 6$$

$$\frac{x^2}{36} + \frac{y^2}{100} = 1$$

(B) Make a rough sketch of the ellipse, as shown in Figure 8; locate the foci and y intercepts, then determine the x intercepts using the special triangle relationship discussed earlier.

$$\frac{x^2}{b^2} + \frac{y^2}{a^2} = 1$$

$$a = \frac{10}{2} = 5 \qquad b^2 = 5^2 - 4^2 = 25 - 16 = 9$$

$$b = 3$$

$$\frac{x^2}{9} + \frac{y^2}{25} = 1$$

FIGURE 8 $\dfrac{x^2}{9} + \dfrac{y^2}{25} = 1$.

Matched Problem 2   Find an equation of an ellipse in the form

$$\frac{x^2}{M} + \frac{y^2}{N} = 1 \qquad M, N > 0$$

if the center is at the origin, the major axis is along the x axis, and:

(A) Length of major axis = 50   (B) Length of minor axis = 16
        Length of minor axis = 30   Distance of foci from center = 6

EXPLORE-DISCUSS 2   Consider the graph of an equation in the variables x and y. The equation of its magnification by a factor $k > 0$ is obtained by replacing x and y in the equation by x/k and y/k, respectively.

(A) Find the equation of the magnification by a factor 3 of the ellipse with equation $(x^2/4) + y^2 = 1$. Graph both equations.

(B) Give an example of an ellipse with center $(0, 0)$ with $a > b$ that is not a magnification of $(x^2/4) + y^2 = 1$.

(C) Find the equations of all ellipses that are magnifications of $(x^2/4) + y^2 = 1$.

● Applications   You are no doubt aware of many occurrences and uses of elliptical forms: orbits of satellites, planets, and comets; shapes of galaxies; gears and cams; some airplane wings, boat keels, and rudders; tabletops; public fountains; and domes in buildings are a few examples (see Fig. 9). A fairly recent application in medicine is the use of elliptical reflectors and ultrasound to break up kidney stones.

Planetary motion
(a)

Elliptical gears
(b)

Elliptical dome
(c)

FIGURE 9 Uses of elliptical forms.

Johannes Kepler (1571–1630), a German astronomer, discovered that planets move in elliptical orbits, with the sun at a focus, and not in circular orbits as had been thought before [Fig. 9(a)]. Figure 9(b) shows a pair of elliptical gears with pivot points at foci. Such gears transfer constant rotational speed to variable rotational speed, and vice versa. Figure 9(c) shows an elliptical dome. An interesting property of such a dome is that a sound or light source at one focus will reflect off the dome and pass through the other focus. One of the chambers in the Capitol Building in Washington, D.C., has such a dome, and is referred to as a whispering room because a whispered sound at one focus can be easily heard at the other focus.

Answers to Matched Problems

1. (A)

Foci: $F'(-\sqrt{3}, 0)$, $F(\sqrt{3}, 0)$
Major axis length $= 4$
Minor axis length $= 2$

(B)

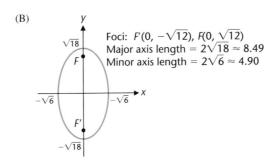

Foci: $F'(0, -\sqrt{12})$, $F(0, \sqrt{12})$
Major axis length $= 2\sqrt{18} \approx 8.49$
Minor axis length $= 2\sqrt{6} \approx 4.90$

2. (A) $\dfrac{x^2}{625} + \dfrac{y^2}{225} = 1$ (B) $\dfrac{x^2}{100} + \dfrac{y^2}{64} = 1$

EXERCISE 9-2

A

In Problems 1–6, sketch a graph of each equation, find the coordinates of the foci, and find the lengths of the major and minor axes.

1. $\dfrac{x^2}{25} + \dfrac{y^2}{4} = 1$

2. $\dfrac{x^2}{9} + \dfrac{y^2}{4} = 1$

3. $\dfrac{x^2}{4} + \dfrac{y^2}{25} = 1$

4. $\dfrac{x^2}{4} + \dfrac{y^2}{9} = 1$

5. $x^2 + 9y^2 = 9$

6. $4x^2 + y^2 = 4$

B

In Problems 7–12, sketch a graph of each equation, find the coordinates of the foci, and find the lengths of the major and minor axes.

7. $25x^2 + 9y^2 = 225$

8. $16x^2 + 25y^2 = 400$

9. $2x^2 + y^2 = 12$

10. $4x^2 + 3y^2 = 24$

11. $4x^2 + 7y^2 = 28$

12. $3x^2 + 2y^2 = 24$

In Problems 13–18, find an equation of an ellipse in the form

$$\frac{x^2}{M} + \frac{y^2}{N} = 1 \qquad M, N > 0$$

if the center is at the origin, and:

13. Major axis on y axis
Major axis length $= 6$
Minor axis length $= 2$

14. Major axis on x axis
Major axis length $= 32$
Minor axis length $= 30$

15. Major axis on x axis
Minor axis length $= 10$
Distance of foci from center $= 4$

16. Major axis on y axis
Major axis length $= 16$
Distance of foci from center $= 7$

17. Major axis on y axis
Major axis length = 24
Distance between foci = 2

18. Major axis on x axis
Major axis length = 4
Distance between foci = 50

19. Explain why an equation whose graph is an ellipse does not define a function.

20. Consider all ellipses having $(0, \pm 1)$ as the ends of the minor axis. Describe the connection between the elongation of the ellipse and the distance from a focus to the origin.

In Problems 21–24, graph each system of equations in the same rectangular coordinate system and find the coordinates of any points of intersection. Find noninteger coordinates to three decimal places.

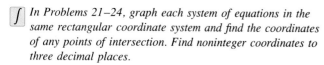 *Check Problems 21–24 with a graphing utility.**

21. $16x^2 + 25y^2 = 400$
$\quad\;\; 2x - 5y = 10$

22. $25x^2 + 16y^2 = 400$
$\quad\;\; 5x + 8y = 20$

23. $25x^2 + 16y^2 = 400$
$\quad\;\; 25x^2 - 36y = 0$

24. $16x^2 + 25y^2 = 400$
$\quad\;\; 3x^2 - 20y = 0$

In Problems 25–28, find the first-quadrant points of intersection for each system of equations to three decimal places.

 Check Problems 25–28 with a graphing utility.

25. $5x^2 + 2y^2 = 63$
$\quad\;\; 2x - y = 0$

26. $3x^2 + 4y^2 = 57$
$\quad\;\; x - 2y = 0$

27. $2x^2 + 3y^2 = 33$
$\quad\;\; x^2 - 8y = 0$

28. $3x^2 + 2y^2 = 43$
$\quad\;\; x^2 - 12y = 0$

 In Problems 29–32, determine whether the statement is true or false. If true, explain why. If false, give a counterexample.

29. The line segment joining the foci of an ellipse has greater length than the minor axis.

30. There is exactly one ellipse with center $(0, 0)$ and foci $(\pm 1, 0)$.

31. Every line through the center of $x^2 + 4y^2 = 16$ intersects the ellipse in exactly two points.

32. Every nonvertical line through a vertex of $x^2 + 4y^2 = 16$ intersects the ellipse in exactly two points.

C

33. Find an equation of the set of points in a plane, each of whose distance from $(2, 0)$ is one-half its distance from the line $x = 8$. Identify the geometric figure.

34. Find an equation of the set of points in a plane, each of whose distance from $(0, 9)$ is three-fourths its distance from the line $y = 16$. Identify the geometric figure.

*Please note that use of a graphing utility is not required to complete these exercises. Checking them with a g.u. is optional.

 In Problems 35–38, use a graphing utility to find the coordinates of all points of intersection to two decimal places.

35. $x^2 + 3y^2 = 20, \; 4x + 5y = 11$

36. $8x^2 + 35y^2 = 3,600, \; x^2 = -25y$

37. $50x^2 + 4y^2 = 1,025, \; 9x^2 + 2y^2 = 300$

38. $2x^2 + 7y^2 = 95, \; 13x^2 + 6y^2 = 63$

APPLICATIONS

39. Engineering. The semielliptical arch in the concrete bridge in the figure must have a clearance of 12 feet above the water and span a distance of 40 feet. Find the equation of the ellipse after inserting a coordinate system with the center of the ellipse at the origin and the major axis on the x axis. The y axis points up, and the x axis points to the right. How much clearance above the water is there 5 feet from the bank?

Elliptical bridge

40. Design. A 4×8 foot elliptical tabletop is to be cut out of a 4×8 foot rectangular sheet of teak plywood (see the figure). To draw the ellipse on the plywood, how far should the foci be located from each edge and how long a piece of string must be fastened to each focus to produce the ellipse (see Figure 1 in the text)? Compute the answer to two decimal places.

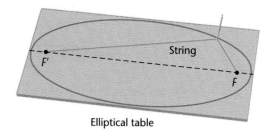

String
F'
F

Elliptical table

★ **41. Aeronautical Engineering.** Of all possible wing shapes, it has been determined that the one with the least drag along the trailing edge is an ellipse. The leading edge may be a straight line, as shown in the figure. One of the most famous planes with this design was the World War II British Spitfire. The plane in the figure has a wingspan of 48.0 feet.

Radiotelescope

(b)

For the receiving antenna shown in the figure, the common focus F is located 120 feet above the vertex of the parabola, and focus F' (for the hyperbola) is 20 feet above the vertex. The vertex of the reflecting hyperbola is 110 feet above the vertex for the parabola. Introduce a coordinate system by using the axis of the parabola as the y axis (up positive), and let the x axis pass through the center of the hyperbola (right positive). What is the equation of the reflecting hyperbola? Write y in terms of x.

SECTION 9-4 Translation of Axes

- Translation of Axes
- Standard Equations of Translated Conics
- Graphing Equations of the Form $Ax^2 + Cy^2 + Dx + Ey + F = 0$
- Finding Equations of Conics

In the last three sections we found standard equations for parabolas, ellipses, and hyperbolas located with their axes on the coordinate axes and centered relative to the origin. What happens if we move conics away from the origin while keeping their axes parallel to the coordinate axes? We will show that we can obtain new standard equations that are special cases of the equation $Ax^2 + Cy^2 + Dx + Ey + F = 0$, where A and C are not both zero. The basic mathematical tool used in this endeavor is *translation of axes*. The usefulness of translation of axes is not limited to graphing conics, however. Translation of axes can be put to good use in many other graphing situations.

• Translation of Axes

A **translation of coordinate axes** occurs when the new coordinate axes have the same direction as and are parallel to the original coordinate axes. To see how coordinates in the original system are changed when moving to the translated system, and vice versa, refer to Figure 1.

FIGURE 1 Translation of coordinates.

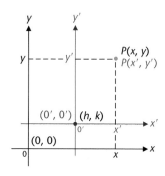

A point P in the plane has two sets of coordinates: (x, y) in the original system and (x', y') in the translated system. If the coordinates of the origin of the translated system are (h, k) relative to the original system, then the old and new coordinates are related as given in Theorem 1.

Theorem 1   **Translation Formulas**

1. $x = x' + h$ **2.** $x' = x - h$
       $y = y' + k$ $y' = y - k$

It can be shown that these formulas hold for (h, k) located anywhere in the original coordinate system.

EXAMPLE 1   **Equation of a Curve in a Translated System**

A curve has the equation

$$(x - 4)^2 + (y + 1)^2 = 36$$

If the origin is translated to $(4, -1)$, find the equation of the curve in the translated system and identify the curve.

Solution   Since $(h, k) = (4, -1)$, use translation formulas

$$x' = x - h = x - 4$$
$$y' = y - k = y + 1$$

to obtain, after substitution,

$$x'^2 + y'^2 = 36$$

This is the equation of a circle of radius 6 with center at the new origin. The coordinates of the new origin in the original coordinate system are $(4, -1)$ (Fig. 2). Note that this result agrees with our general treatment of the circle in Section 3-1.

FIGURE 2
$(x - 4)^2 + (y + 1)^2 = 36$.

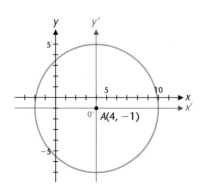

Matched Problem 1 A curve has the equation $(y + 2)^2 = 8(x - 3)$. If the origin is translated to $(3, -2)$, find an equation of the curve in the translated system and identify the curve.

• Standard Equations of Translated Conics

We now proceed to find standard equations of conics translated away from the origin. We do this by first writing the standard equations found in earlier sections in the $x'y'$ coordinate system with $0'$ at (h, k). We then use translation equations to find the standard forms relative to the original xy coordinate system. The equations of translation in all cases are

$$x' = x - h$$
$$y' = y - k$$

For parabolas we have

$$x'^2 = 4ay' \qquad (x - h)^2 = 4a(y - k)$$
$$y'^2 = 4ax' \qquad (y - k)^2 = 4a(x - h)$$

For circles we have

$$x'^2 + y'^2 = r^2 \qquad (x - h)^2 + (y - k)^2 = r^2$$

For ellipses we have for $a > b > 0$

$$\frac{x'^2}{a^2} + \frac{y'^2}{b^2} = 1 \qquad \frac{(x - h)^2}{a^2} + \frac{(y - k)^2}{b^2} = 1$$

$$\frac{x'^2}{b^2} + \frac{y'^2}{a^2} = 1 \qquad \frac{(x - h)^2}{b^2} + \frac{(y - k)^2}{a^2} = 1$$

For hyperbolas we have

$$\frac{x'^2}{a^2} - \frac{y'^2}{b^2} = 1 \qquad \frac{(x - h)^2}{a^2} - \frac{(y - k)^2}{b^2} = 1$$

$$\frac{y'^2}{a^2} - \frac{x'^2}{b^2} = 1 \qquad \frac{(y - k)^2}{a^2} - \frac{(x - h)^2}{b^2} = 1$$

Table 1 summarizes these results with appropriate figures and some properties discussed earlier.

TABLE 1   Standard Equations for Translated Conics

Parabolas

$$(x - h)^2 = 4a(y - k)$$ $$(y - k)^2 = 4a(x - h)$$

Vertex (h, k)
Focus $(h, k + a)$
$a > 0$ opens up
$a < 0$ opens down

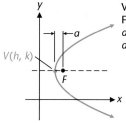

Vertex (h, k)
Focus $(h + a, k)$
$a < 0$ opens left
$a > 0$ opens right

Circles

$$(x - h)^2 + (y - k)^2 = r^2$$

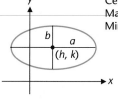

Center (h, k)
Radius r

Ellipses

$$\frac{(x - h)^2}{a^2} + \frac{(y - k)^2}{b^2} = 1 \qquad a > b > 0 \qquad \frac{(x - h)^2}{b^2} + \frac{(y - k)^2}{a^2} = 1$$

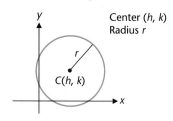

Center (h, k)
Major axis $2a$
Minor axis $2b$

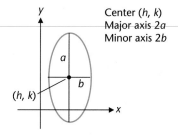

Center (h, k)
Major axis $2a$
Minor axis $2b$

Hyperbolas

$$\frac{(x - h)^2}{a^2} - \frac{(y - k)^2}{b^2} = 1 \qquad\qquad \frac{(y - k)^2}{a^2} - \frac{(x - h)^2}{b^2} = 1$$

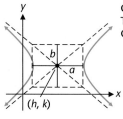

Center (h, k)
Transverse axis $2a$
Conjugate axis $2b$

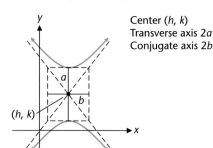

Center (h, k)
Transverse axis $2a$
Conjugate axis $2b$

The origin in the translated system is at $(h, k) = (-4, 3)$, and the translation formulas are

$$x' = x - h = x - (-4) = x + 4$$
$$y' = y - k = y - 3$$

Thus, the equation of the hyperbola in the original system is

$$\frac{(y - 3)^2}{4} - \frac{(x + 4)^2}{9} = 1$$

or, after simplifying and writing in the form of equation (1),

$$4x^2 - 9y^2 + 32x + 54y + 19 = 0$$

Matched Problem 4   Find the equation of an ellipse with foci on the line $x = 4$, minor axis on the line $y = -3$, length of the major axis $= 8$, and length of the minor axis $= 4$.

EXPLORE-DISCUSS 2   Use the strategy of completing the square to transform each equation to an equation in an $x'y'$ coordinate system. Note that the equation you obtain is not one of the standard forms in Table 1; instead, it is either the equation of a degenerate conic or the equation has no solution. If the solution set of the equation is not empty, graph it and identify the graph (a point, a line, two parallel lines, or two intersecting lines).

(A) $x^2 + 2y^2 - 2x + 16y + 33 = 0$

(B) $4x^2 - y^2 - 24x - 2y + 35 = 0$

(C) $y^2 - 2y - 15 = 0$

(D) $5x^2 + y^2 + 12y + 40 = 0$

(E) $x^2 - 18x + 81 = 0$

Answers to Matched Problems

1. $y'^2 = 8x'$; a parabola
2. $(x + 2)^2 = -4(y - 4)$; a parabola

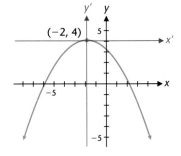

3. $\dfrac{(x + 2)^2}{16} + \dfrac{(y - 1)^2}{9} = 1$; ellipse   Foci: $F'(-\sqrt{7} - 2, 1)$, $F(\sqrt{7} - 2, 1)$

4. $\dfrac{(x - 4)^2}{4} + \dfrac{(y + 3)^2}{16} = 1$, or $4x^2 + y^2 - 32x + 6y + 57 = 0$

EXERCISE 9-4

A

In Problems 1–8:
(A) Find translation formulas that translate the origin to the indicated point (h, k).
(B) Write the equation of the curve for the translated system.
(C) Identify the curve.

1. $(x - 3)^2 + (y - 5)^2 = 81$; $(3, 5)$

2. $(x - 3)^2 = 8(y + 2)$; $(3, -2)$

3. $\dfrac{(x + 7)^2}{9} + \dfrac{(y - 4)^2}{16} = 1$; $(-7, 4)$

4. $(x + 2)^2 + (y + 6)^2 = 36$; $(-2, -6)$

5. $(y + 9)^2 = 16(x - 4)$; $(4, -9)$

6. $\dfrac{(y - 9)^2}{10} - \dfrac{(x + 5)^2}{6} = 1$; $(-5, 9)$

7. $\dfrac{(x + 8)^2}{12} + \dfrac{(y + 3)^2}{8} = 1$; $(-8, -3)$

8. $\dfrac{(x + 7)^2}{25} - \dfrac{(y - 8)^2}{50} = 1$; $(-7, 8)$

In Problems 9–14:
(A) Write each equation in one of the standard forms listed in Table 1.
(B) Identify the curve.

9. $16(x - 3)^2 - 9(y + 2)^2 = 144$

10. $(y + 2)^2 - 12(x - 3) = 0$

11. $6(x + 5)^2 + 5(y + 7)^2 = 30$

12. $12(y - 5)^2 - 8(x - 3)^2 = 24$

13. $(x + 6)^2 + 24(y - 4) = 0$

14. $4(x - 7)^2 + 7(y - 3)^2 = 28$

B

In Problems 15–22, transform each equation into one of the standard forms in Table 1. Identify the curve and graph it.

15. $4x^2 + 9y^2 - 16x - 36y + 16 = 0$

16. $16x^2 + 9y^2 + 64x + 54y + 1 = 0$

17. $x^2 + 8x + 8y = 0$

18. $y^2 + 12x + 4y - 32 = 0$

19. $x^2 + y^2 + 12x + 10y + 45 = 0$

20. $x^2 + y^2 - 8x - 6y = 0$

21. $-9x^2 + 16y^2 - 72x - 96y - 144 = 0$

22. $16x^2 - 25y^2 - 160x = 0$

23. If $A \neq 0$, $C = 0$, and $E \neq 0$, find h and k so that the translation of axes $x = x' + h$, $y = y' + k$ transforms the equation $Ax^2 + Cy^2 + Dx + Ey + F = 0$ into one of the standard forms of Table 1.

24. If $A = 0$, $C \neq 0$, and $D \neq 0$, find h and k so that the translation of axes $x = x' + h$, $y = y' + k$ transforms the equation $Ax^2 + Cy^2 + Dx + Ey + F = 0$ into one of the standard forms of Table 1.

In Problems 25–34, use the given information to find the equation of each conic. Express the answer in the form $Ax^2 + Cy^2 + Dx + Ey + F = 0$ with integer coefficients and $A \geq 0$.

25. A parabola with vertex at $(5, 3)$, and focus at $(5, 11)$.

26. A parabola with focus at $(2, 3)$, and directrix the y axis.

27. An ellipse with vertices $(-3, -2)$ and $(-3, 10)$ and length of minor axis $= 10$.

28. A hyperbola with vertices $(-2, -8)$ and $(4, -8)$ and length of conjugate axis $= 24$.

29. A hyperbola with foci $(2, 1)$ and $(6, 1)$ and vertices $(3, 1)$ and $(5, 1)$.

30. An ellipse with foci $(-3, 0)$ and $(-3, 6)$ and vertices $(-3, -2)$ and $(-3, 8)$.

31. A parabola with axis the y axis and passing through the points $(1, 0)$ and $(2, 4)$.

32. A parabola with vertex at $(-6, 2)$, axis the line $y = 2$, and passing through the point $(0, 7)$.

33. An ellipse with vertices $(1, -1)$, and $(-5, -1)$ that passes through the origin.

34. A hyperbola with vertices at $(2, 3)$, and $(2, 5)$ that passes through the point $(4, 0)$.

C

In Problems 35–40, find the coordinates of any foci relative to the original coordinate system:

35. Problem 15   **36.** Problem 16   **37.** Problem 17

38. Problem 18   **39.** Problem 21   **40.** Problem 22

In Problems 41–44, use a graphing utility to find the coordinates of all points of intersection to two decimal places.

41. $3x^2 - 5y^2 + 7x - 2y + 11 = 0, \ 6x + 4y = 15$

42. $8x^2 + 3y^2 - 14x + 17y - 39 = 0, \ 5x - 11y = 23$

43. $7x^2 - 8x + 5y - 25 = 0, \ x^2 + 4y^2 + 4x - y - 12 = 0$

44. $4x^2 - y^2 - 24x - 2y + 35 = 0, \ 2x^2 + 6y^2 - 3x - 34 = 0$

CHAPTER 9 GROUP ACTIVITY   Focal Chords

Many of the applications of the conic sections are based on their reflective or focal properties. One of the interesting algebraic properties of the conic sections concerns their focal chords.

If a line through a focus F contains two points G and H of a conic section, then the line segment GH is called a **focal chord.** Let $G(x_1, y_1)$ and $H(x_2, y_2)$ be points on the graph of $x^2 = 4ay$ such that GH is a focal chord. Let u denote the length of GF and v the length of FH (see Fig. 1).

FIGURE 1 Focal chord GH of the parabola $x^2 = 4ay$.

(A) Use the distance formula to show that $u = y_1 + a$.

(B) Show that G and H lie on the line $y - a = mx$, where $m = (y_2 - y_1)/(x_2 - x_1)$.

(C) Solve $y - a = mx$ for x and substitute in $x^2 = 4ay$, obtaining a quadratic equation in y. Explain why $y_1 y_2 = a^2$.

(D) Show that $\dfrac{1}{u} + \dfrac{1}{v} = \dfrac{1}{a}$.

(E) Show that $u + v - 4a = \dfrac{(u - 2a)^2}{u - a}$. Explain why this implies that $u + v \geq 4a$, with equality if and only if $u = v = 2a$.

(F) Which focal chord is the shortest? Is there a longest focal chord?

(G) Is $\dfrac{1}{u} + \dfrac{1}{v}$ a constant for focal chords of the ellipse? For focal chords of the hyperbola? Obtain evidence for your answers by considering specific examples.

Chapter 9 Review

9-1   CONIC SECTIONS; PARABOLA

The plane curves obtained by intersecting a right circular cone with a plane are called **conic sections.** If the plane cuts clear through one nappe, then the intersection curve is called a **circle** if the plane is perpendicular to the axis and an **ellipse** if the plane is not perpendicular to the axis. If a plane cuts only one nappe, but does not cut clear through, then the intersection curve is called a **parabola.** If a plane cuts through both nappes, but not through the vertex, the resulting intersection curve is called a **hyperbola.** A plane passing through the vertex of the cone produces a **degenerate conic**—a point, a line, or a pair of lines. The figure illustrates the four nondegenerate conics.

Circle   Ellipse   Parabola   Hyperbola

The graph of

$$Ax^2 + Bxy + Cy^2 + Dx + Ey + F = 0$$

where A, B, and C are not all 0, is a conic.

The following is a coordinate-free definition of a parabola:

Parabola

A **parabola** is the set of all points in a plane equidistant from a fixed point F and a fixed line L in the plane. The fixed point F is called the **focus,** and the fixed line L is called the **directrix.** A line through the focus perpendicular to the directrix is called the **axis,** and the point on the axis halfway between the directrix and focus is called the **vertex.**

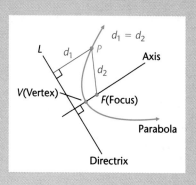

From the definition of a parabola, we can obtain the following standard equations:

Standard Equations of a Parabola with Vertex at (0, 0)

1. $y^2 = 4ax$
   Vertex: $(0, 0)$
   Focus: $(a, 0)$
   Directrix: $x = -a$
   Symmetric with
   respect to the x axis
   Axis the x axis

$a < 0$ (opens left) $a > 0$ (opens right)

2. $x^2 = 4ay$

Vertex: $(0, 0)$
Focus: $(0, a)$
Directrix: $y = -a$
Symmetric with respect to the y axis
Axis the y axis

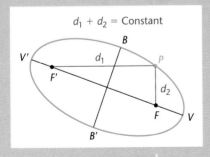

| $a < 0$ (opens down) | $a > 0$ (opens up) |

9-2 ELLIPSE

The following is a coordinate-free definition of an ellipse:

Ellipse

An **ellipse** is the set of all points P in a plane such that the sum of the distances of P from two fixed points in the plane is constant. Each of the fixed points, F' and F, is called a **focus,** and together they are called **foci.** Referring to the figure, the line segment $V'V$ through the foci is the **major axis.** The perpendicular bisector $B'B$ of the major axis is the **minor axis.** Each end of the major axis, V' and V, is called a **vertex.** The midpoint of the line segment $F'F$ is called the **center** of the ellipse.

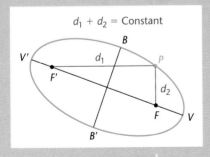

$d_1 + d_2$ = Constant

From the definition of an ellipse, we can obtain the following standard equations:

Standard Equations of an Ellipse with Center at (0, 0)

1. $\dfrac{x^2}{a^2} + \dfrac{y^2}{b^2} = 1 \qquad a > b > 0$

x intercepts: $\pm a$ (vertices)
y intercepts: $\pm b$
Foci: $F'(-c, 0), F(c, 0)$

$$c^2 = a^2 - b^2$$

Major axis length $= 2a$
Minor axis length $= 2b$

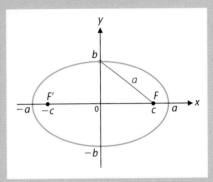

2. $\dfrac{x^2}{b^2} + \dfrac{y^2}{a^2} = 1 \qquad a > b > 0$

x intercepts: $\pm b$
y intercepts: $\pm a$ (vertices)
Foci: $F'(0, -c), F(0, c)$

$$c^2 = a^2 - b^2$$

Major axis length $= 2a$
Minor axis length $= 2b$

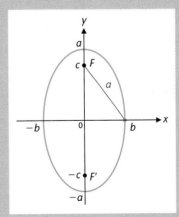

[*Note:* Both graphs are symmetric with respect to the x axis, y axis, and origin. Also, the major axis is always longer than the minor axis.]

9-3   HYPERBOLA

The following is a coordinate-free definition of a hyperbola:

Hyperbola

A **hyperbola** is the set of all points P in a plane such that the absolute value of the difference of the distances of P to two fixed points in the plane is a positive constant. Each of the fixed points, F' and F, is called a **focus.** The intersection points V' and V of the line through the foci and the two branches of the hyperbola are called **vertices,** and each is called a **vertex.** The line segment $V'V$ is called the **transverse axis.** The midpoint of the transverse axis is the **center** of the hyperbola.

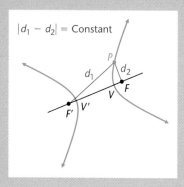

$|d_1 - d_2| = \text{Constant}$

From the definition of a hyperbola, we can obtain the following standard equations:

Standard Equations of a Hyperbola with Center at (0, 0)

1. $\dfrac{x^2}{a^2} - \dfrac{y^2}{b^2} = 1$

    x intercepts: $\pm a$ (vertices)
    y intercepts: none
    Foci: $F'(-c, 0)$, $F(c, 0)$

    $c^2 = a^2 + b^2$

    Transverse axis length $= 2a$
    Conjugate axis length $= 2b$

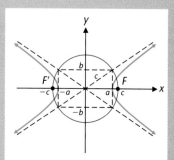

2. $\dfrac{y^2}{a^2} - \dfrac{x^2}{b^2} = 1$

    x intercepts: none
    y intercepts: $\pm a$ (vertices)
    Foci: $F'(0, -c)$, $F(0, c)$

    $c^2 = a^2 + b^2$

    Transverse axis length $= 2a$
    Conjugate axis length $= 2b$

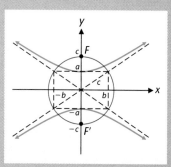

[*Note:* Both graphs are symmetric with respect to the x axis, y axis, and origin.]

9-4   TRANSLATION OF AXES

In the last three sections we found standard equations for parabolas, ellipses, and hyperbolas located with their axes on the coordinate axes and centered relative to the origin. We now move the conics away from the origin while keeping their axes parallel to the coordinate axes. In this process we obtain new standard equations that are special cases of the equation $Ax^2 + Cy^2 + Dx + Ey + F = 0$, where A and C are not both zero. The basic mathematical tool used is *translation of axes*.

A **translation of coordinate axes** occurs when the new co-ordinate axes have the same direction as and are parallel to the original coordinate axes. **Translation formulas** are as follows:

1. $x = x' + h$ **2.** $x' = x - h$
$\ y = y' + k$ $\ y' = y - k$

where (h, k) are the coordinates of the origin $0'$ relative to the original system.

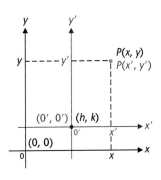

Table 1 lists the standard equations for translated conics.

TABLE 1   Standard Equations for Translated Conics

Parabolas

$$(x - h)^2 = 4a(y - k)$$

$$(y - k)^2 = 4a(x - h)$$

Vertex (h, k)
Focus $(h, k + a)$
$a > 0$ opens up
$a < 0$ opens down

Vertex (h, k)
Focus $(h + a, k)$
$a < 0$ opens left
$a > 0$ opens right

Circles

$$(x - h)^2 + (y - k)^2 = r^2$$

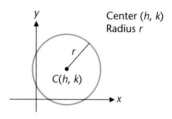

Center (h, k)
Radius r

Ellipses

$$\frac{(x - h)^2}{a^2} + \frac{(y - k)^2}{b^2} = 1 \qquad a > b > 0 \qquad \frac{(x - h)^2}{b^2} + \frac{(y - k)^2}{a^2} = 1$$

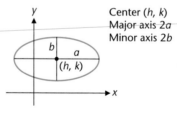

Center (h, k)
Major axis $2a$
Minor axis $2b$

Center (h, k)
Major axis $2a$
Minor axis $2b$

Hyperbolas

$$\frac{(x - h)^2}{a^2} - \frac{(y - k)^2}{b^2} = 1 \qquad \frac{(y - k)^2}{a^2} - \frac{(x - h)^2}{b^2} = 1$$

Center (h, k)
Transverse axis $2a$
Conjugate axis $2b$

Center (h, k)
Transverse axis $2a$
Conjugate axis $2b$

Chapter 9 Review Exercise

Work through all the problems in this chapter review and check answers in the back of the book. Answers to all review problems are there, and following each answer is a number in italics indicating the section in which that type of problem is discussed. Where weaknesses show up, review appropriate sections in the text.

A

In Problems 1–3, graph each equation and locate foci. Locate the directrix for any parabolas. Find the lengths of major, minor, transverse, and conjugate axes where applicable.

1. $9x^2 + 25y^2 = 225$

2. $x^2 = -12y$

3. $25y^2 - 9x^2 = 225$

In Problems 4–6:
(A) *Write each equation in one of the standard forms listed in Table 1 of the review.*
(B) *Identify the curve.*

4. $4(y + 2)^2 - 25(x - 4)^2 = 100$

5. $(x + 5)^2 + 12(y + 4) = 0$

6. $16(x - 6)^2 + 9(y - 4)^2 = 144$

B

7. Find the equation of the parabola having its vertex at the origin, its axis the x axis, and $(-4, 2\sqrt{7})$ on its graph.

8. Find an equation of an ellipse in the form

$$\frac{x^2}{M} + \frac{y^2}{N} = 1 \qquad M, N > 0$$

if the center is at the origin, the major axis is on the y axis, the minor axis has length 4, and the distance between the foci is 62.

9. Find an equation of a hyperbola in the form

$$\frac{x^2}{M} - \frac{y^2}{N} = 1, \qquad M, N > 0$$

if the center is at the origin, the transverse axis has length 10, and the foci are 6 units from the center.

 In Problems 10–12, graph each system of equations in the same coordinate system and find the coordinates of any points of intersection.

10. $x^2 + 4y^2 = 32$
$x + 2y = 0$

11. $16x^2 + 25y^2 = 400$
$16x^2 - 45y = 0$

12. $x^2 + y^2 = 10$
$16x^2 + y^2 = 25$

In Problems 13–15, transform each equation into one of the standard forms in Table 1 in the review. Identify the curve and graph it.

13. $16x^2 + 4y^2 + 96x - 16y + 96 = 0$

14. $x^2 - 4x - 8y - 20 = 0$

15. $4x^2 - 9y^2 + 24x - 36y - 36 = 0$

 16. Use a graphing utility to graph $x^2 = y$ and $x^2 = 50y$ in the viewing window $-10 \le x, y \le 10$. Find m so that the graph of $x^2 = y$ in the viewing window $-m \le x, y \le m$, has the same appearance as the graph of $x^2 = 50y$ in $-10 \le x, y \le 10$. Explain.

C

17. Use the definition of a parabola and the distance formula to find the equation of a parabola with directrix $x = 6$ and focus at $(2, 4)$.

18. Find an equation of the set of points in a plane each of whose distance from $(4, 0)$ is twice its distance from the line $x = 1$. Identify the geometric figure.

19. Find an equation of the set of points in a plane each of whose distance from $(4, 0)$ is two-thirds its distance from the line $x = 9$. Identify the geometric figure.

In Problems 20–22, find the coordinates of any foci relative to the original coordinate system.

20. Problem 14　　**21.** Problem 15　　**22.** Problem 16

 23. Use a graphing utility to find, to two decimal places, the coordinates of all points of intersection of $x^2 - 3y^2 + 9x + 7y - 22 = 0$ and $4x^2 + 5x + 10y - 53 = 0$.

APPLICATIONS 🌐

24. Communications. A parabolic satellite television antenna has a diameter of 8 feet and is 1 foot deep. How far is the focus from the vertex?

25. Engineering. An elliptical gear is to have foci 8 centimeters apart and a major axis 10 centimeters long. Letting the x axis lie along the major axis (right positive) and the y axis lie along the minor axis (up positive), write the equation of the ellipse in the standard form

$$\frac{x^2}{a^2} + \frac{y^2}{b^2} = 1$$

26. Space Science. A hyperbolic reflector for a radiotelescope (such as that illustrated in Problem 39, Exercise 9-3) has the equation

$$\frac{y^2}{40^2} - \frac{x^2}{30^2} = 1$$

If the reflector has a diameter of 30 feet, how deep is it? Compute the answer to 3 significant digits.

Cumulative Review Exercise
Chapters 8–9

Work through all the problems in this cumulative review and check answers in the back of the book. Answers to all review problems are there, and following each answer is a number in italics indicating the section in which that type of problem is discussed. Where weaknesses show up, review appropriate sections in the text.

A

1. Determine whether each of the following can be the first three terms of an arithmetic sequence, a geometric sequence, or neither.
(A) $5, 25, 100, \ldots$ (B) $15, 3, -9, \ldots$
(C) $1, 1, 1, \ldots$ (D) $-64, 16, -4, \ldots$
(E) $17, 119, 833, \ldots$ (F) $1, 3, 6, \ldots$

In Problems 2–4:
(A) Write the first four terms of each sequence.
(B) Find a_8.   (C) Find S_8.

2. $a_n = (-2)^n$ **3.** $a_n = 6n - 5$

4. $a_1 = -20; a_n = a_{n-1} + 4, n \geq 2$

5. Evaluate each of the following:
   (A) $7!$ (B) $\dfrac{25!}{22!}$ (C) $\dfrac{10!}{(10-4)! \, 4!}$

6. Evaluate each of the following:
   (A) $\begin{pmatrix} 12 \\ 6 \end{pmatrix}$ (B) $C_{9,4}$ (C) $P_{8,5}$

In Problems 7–9, graph each equation and locate foci. Locate the directrix for any parabolas. Find the lengths of major, minor, transverse, and conjugate axes where applicable.

7. $25x^2 - 36y^2 = 900$ **8.** $25x^2 + 36y^2 = 900$

9. $25x^2 - 36y = 0$

10. A coin is flipped three times. How many combined outcomes are possible? Solve:
   (A) By using a tree diagram
   (B) By using the multiplication principle

11. How many ways can 4 distinct books be arranged on a shelf? Solve:
   (A) By using the multiplication principle
   (B) By using permutations or combinations, whichever is applicable

Verify Problems 12 and 13 for n = 1, 2, and 3.

12. P_n: $1 + 5 + 9 + \cdots + (4n - 3) = n(2n - 1)$

13. P_n: $n^2 + n + 2$ is divisible by 2

In Problems 14 and 15, write P_k and P_{k+1}.

14. For P_n in Problem 12   **15.** For P_n in Problem 13

B

16. Find the equation of the parabola having its vertex at the origin, its axis the y axis, and $(2, -8)$ on its graph.

17. Find an equation of an ellipse in the form

$$\frac{x^2}{M} + \frac{y^2}{N} = 1 \qquad M, N > 0$$

if the center is at the origin, the major axis is the x axis, the major axis length is 10, and the distance of the foci from the center is 3.

18. Find an equation of a hyperbola in the form

$$\frac{x^2}{M} - \frac{y^2}{N} = 1 \qquad M, N > 0$$

if the center is at the origin, the transverse axis length is 16, and the distance of the foci from the center is $\sqrt{89}$.

19. Write $\displaystyle\sum_{k=1}^{5} k^k$ without summation notation and find the sum.

20. Write the series $\dfrac{2}{2!} - \dfrac{2^2}{3!} + \dfrac{2^3}{4!} - \dfrac{2^4}{5!} + \dfrac{2^5}{6!} - \dfrac{2^6}{7!}$ using summation notation with the summation index k starting at $k = 1$.

21. Find S_∞ for the geometric series $108 - 36 + 12 - 4 + \cdots$.

22. How many 4-letter code words are possible using the first 6 letters of the alphabet if no letter can be repeated? If letters can be repeated? If adjacent letters cannot be alike?

23. Let $a_n = 100(0.9)^n$ and $b_n = 10 + 0.03n$. Find the least positive integer n such that $a_n < b_n$ by graphing the sequences $\{a_n\}$ and $\{b_n\}$ with a graphing utility. Check your answer by using a graphing utility to display both sequences in table form.

24. Evaluate each of the following:

   (A) $P_{25,5}$ (B) $C(25, 5)$ (C) $\dbinom{25}{20}$

25. Expand $(a + \frac{1}{2}b)^6$ using the binomial formula.

26. Find the fifth and the eighth terms in the expansion of $(3x - y)^{10}$.

Establish each statement in Problems 27 and 28 for all positive integers using mathematical induction.

27. P_n in Problem 12   **28.** P_n in Problem 13

29. Find the sum of all the odd integers between 50 and 500.

30. Use the formula for the sum of an infinite geometric series to write $2.\overline{45} = 2.454\ 545 \cdots$ as the quotient of two integers.

31. Let $a_k = \dbinom{30}{k}(0.1)^{30-k}(0.9)^k$ for $k = 0, 1, \ldots, 30$. Use a graphing utility to find the largest term of the sequence $\{a_k\}$ and the number of terms that are greater than 0.01.

In Problems 32–34, use a translation of coordinates to transform each equation into a standard equation for a nondegenerate conic. Identify the curve and graph it.

32. $4x + 4y - y^2 + 8 = 0$

33. $x^2 + 2x - 4y^2 - 16y + 1 = 0$

34. $4x^2 - 16x + 9y^2 + 54y + 61 = 0$

35. How many 9-digit zip codes are possible? How many of these have no repeated digits?

36. Use a graphing utility to find, to two decimal places, the coordinates of all points of intersection of $5x^2 + 2y^2 - 7x + 8y - 48 = 0$ and $e^x - e^{-x} - 2y = 0$.

37. Use mathematical induction to prove that the following statement holds for all positive integers:

$$P_n: \quad \frac{1}{1 \cdot 3} + \frac{1}{3 \cdot 5} + \frac{1}{5 \cdot 7} + \cdots + $$
$$\frac{1}{(2n - 1)(2n + 1)} = \frac{n}{2n + 1}$$

C _____

38. Use the binomial formula to expand $(x - 2i)^6$, where i is the imaginary unit.

39. Use the definition of a parabola and the distance formula to find the equation of a parabola with directrix $y = 3$ and focus $(6, 1)$.

40. An ellipse has vertices $(\pm 4, 0)$ and foci $(\pm 2, 0)$. Find the y intercepts.

41. A hyperbola has vertices $(2, \pm 3)$ and foci $(2, \pm 5)$. Find the length of the conjugate axis.

In Problems 42–45, determine whether the sequence $\{a_n\}$ is arithmetic, geometric, both, or neither.

42. $a_n = \dfrac{2^n}{5^{n+1}}$

43. $a_1 = 3, a_{n+1} = a_n + \dfrac{1}{n}$ $n \geq 1$

44. $a_1 = 1, a_2 = 1, a_{n+2} = a_n a_{n+1}$ $n \geq 1$

45. $a_n = \dbinom{n}{2} - \dfrac{n^2}{2}$

In Problems 46–54, determine whether the statement is true or false. If true, explain why. If false, give a counterexample.

46. If n is a positive integer, then $n^{10} \geq n!$

47. If n is a positive integer, then $n^n \geq n!$

48. If the integer $n \geq 3$ is divisible by a prime p, then $\dbinom{n}{2}$ is divisible by p.

49. If p is a prime greater than 2, then $\dbinom{p}{2}$ is divisible by p.

50. If P is a point on an ellipse, then the sum of the distances from P to the foci is equal to the length of the major axis.

51. If P is a point on a parabola, then the distance from P to the directrix equals the distance from P to the vertex.

52. If P is a point on a hyperbola, then the absolute value of the differences of the distances from P to the foci is equal to the length of the conjugate axis.

53. The length of the major axis of an ellipse is greater than the length of the minor axis.

54. The length of the transverse axis of a hyperbola is greater than the length of the conjugate axis.

55. How many parabolas have the x axis as axis and pass through the points $(0, -3)$ and $(2, 5)$? Find the equation(s).

56. A hyperbola has its transverse axis on the line $y = -4$ and its conjugate axis on the line $x = 5$, and the transverse axis has twice the length of the conjugate axis. Find the equations of the asymptotes.

57. An ellipse passes through the point $P(8, -3)$, the distance between the foci is 12, and the length of the minor axis is 4. Find the sum of the distances from P to the foci.

58. Seven distinct points are selected on the circumference of a circle. How many triangles can be formed using these 7 points as vertices?

59. Use mathematical induction to prove that $2^n < n!$ for all integers $n > 3$.

60. Use mathematical induction to show that $\{a_n\} = \{b_n\}$, where $a_1 = 3$, $a_n = 2a_{n-1} - 1$ for $n > 1$, and $b_n = 2^n + 1$, $n \geq 1$.

61. Find an equation of the set of points in the plane each of whose distance from $(1, 4)$ is three times its distance from the x axis. Write the equation in the form $Ax^2 + Cy^2 + Dx + Ey + F = 0$, and identify the curve.

62. Economics. The government, through a subsidy program, distributes $2,000,000. If we assume that each individual or agency spends 75% of what it receives, and 75% of this is spent, and so on, how much total increase in spending results from this government action?

63. Engineering. An automobile headlight contains a parabolic reflector with a diameter of 8 inches. If the light source is located at the focus, which is 1 inch from the vertex, how deep is the reflector?

64. Architecture. A sound whispered at one focus of a whispering chamber can be easily heard at the other focus. Suppose that a cross section of this chamber is a semielliptical arch which is 80 feet wide and 24 feet high (see the figure). How far is each focus from the center of the arch? How high is the arch above each focus?

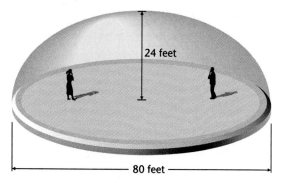

24 feet

80 feet

APPENDIX A
Significant Digits

Most calculations involving problems of the real world deal with figures that are only approximate. It therefore seems reasonable to assume that a final answer should not be any more accurate than the least accurate figure used in the calculation. This is an important point, since calculators tend to give the impression that greater accuracy is achieved than is warranted.

Suppose we wish to compute the length of the diagonal of a rectangular field from measurements of its sides of 237.8 meters and 61.3 meters. Using the Pythagorean theorem and a calculator, we find

$$d = \sqrt{237.8^2 + 61.3^2}$$
$$= 245.573\ 878 \cdots$$

61.3 meters

237.8 meters

The calculator answer suggests an accuracy that is not justified. What accuracy is justified? To answer this question, we introduce the idea of *significant digits*.

Whenever we write a measurement such as 61.3 meters, we assume that the measurement is accurate to the last digit written. Thus, the measurement 61.3 meters indicates that the measurement was made to the nearest tenth of a meter. That is, the actual width is between 61.25 meters and 61.35 meters. In general, the digits in a number that indicate the accuracy of the number are called **significant digits.** If all the digits in a number are nonzero, then they are all significant. Thus, the measurement 61.3 meters has 3 significant digits, and the measurement 237.8 meters has 4 significant digits.

What are the significant digits in the number 7,800? The accuracy of this number is not clear. It could represent a measurement with any of the following accuracies:

Between 7,750 and 7,850   Correct to the hundreds place

Between 7,795 and 7,805   Correct to the tens place

Between 7,799.5 and 7,800.5   Correct to the units place

In order to give a precise definition of significant digits that resolves this ambiguity, we use scientific notation.

A-1

DEFINITION 1　　**Significant Digits**

If a number x is written in scientific notation as

$$x = a \times 10^n \qquad 1 \le a < 10, \ n \text{ an integer}$$

then the number of significant digits in x is the number of digits in a.

Thus,

$$7.8 \times 10^3 \qquad \text{has 2 significant digits}$$
$$7.80 \times 10^3 \qquad \text{has 3 significant digits}$$
$$7.800 \times 10^3 \qquad \text{has 4 significant digits}$$

All three of these measurements have the same decimal representation (7,800), but each represents a different accuracy.

Definition 1 tells us how to write a number so that the number of significant digits is clear, but it does not tell us how to interpret the accuracy of a number that is not written in scientific notation. We will use the following convention for numbers that are written as decimal fractions:

Significant Digits in Decimal Fractions

The number of significant digits in a number with no decimal point is found by counting the digits from left to right, starting with the first digit and ending with the last *nonzero* digit.

The number of significant digits in a number containing a decimal point is found by counting the digits from left to right, starting with the first *nonzero* digit and ending with the last digit.

Applying this rule to the number 7,800, we conclude that this number has 2 significant digits. If we want to indicate that it has 3 or 4 significant digits, we must use scientific notation. The significant digits in the following numbers are underlined:

$$\underline{70,007} \qquad \underline{82},000 \qquad \underline{5.600} \qquad 0.000\underline{8} \qquad 0.000\ \underline{830}$$

In calculations involving multiplication, division, powers, and roots, we adopt the following convention:

Rounding Calculated Values

The result of a calculation is rounded to the same number of significant digits as the number used in the calculation that has the least number of significant digits.

Thus, in computing the length of the diagonal of the rectangular field shown earlier, we write the answer rounded to 3 significant digits because the width has 3 significant digits and the length has 4 significant digits:

$$d = 246 \text{ meters} \quad \text{3 significant digits}$$

One Final Note: In rounding a number that is exactly halfway between a larger and a smaller number, we use the convention of making the final result even.

EXAMPLE 1   **Rounding Numbers**

Round each number to 3 significant digits:

(A) 43.0690   (B) 48.05   (C) 48.15   (D) $8.017\ 632 \times 10^{-3}$

Solutions   (A) 43.1
(B) 48.0⎤   Use the convention of making the digit before the
(C) 48.2⎦   5 even if it is odd, or leaving it alone if it is even.
(D) 8.02×10^{-3}

Matched Problem 1   Round each number to 3 significant digits:

(A) 3.1495   (B) 0.004 135   (C) 32,450   (D) $4.314\ 764\ 09 \times 10^{12}$

Answers to Matched Problem
1. (A) 3.15   (B) 0.004 14   (C) 32,400   (D) 4.31×10^{12}

APPENDIX B
Geometric Formulas

- **Similar Triangles** (A) Two triangles are similar if two angles of one triangle have the same measure as two angles of the other.

  (B) If two triangles are similar, their corresponding sides are proportional:

  $$\frac{a}{a'} = \frac{b}{b'} = \frac{c}{c'}$$

  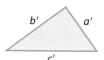

- **Pythagorean Theorem** $c^2 = a^2 + b^2$

  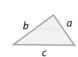

- **Rectangle** $A = ab$ Area

  $P = 2a + 2b$ Perimeter

- **Parallelogram** $h = $ Height

  $A = ah = ab \sin \theta$ Area

  $P = 2a + 2b$ Perimeter

- **Triangle** $h = $ Height

  $A = \frac{1}{2}hc$ Area

  $P = a + b + c$ Perimeter

  $s = \frac{1}{2}(a + b + c)$ Semiperimeter

  $A = \sqrt{s(s - a)(s - b)(s - c)}$ Area—Heron's formula

  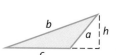

- **Trapezoid**   Base a is parallel to base b.

    h = Height

    $A = \frac{1}{2}(a + b)h$ Area

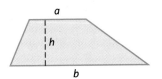

- **Circle** R = Radius

    D = Diameter

    $D = 2R$

    $A = \pi R^2 = \frac{1}{4}\pi D^2$ Area

    $C = 2\pi R = \pi D$ Circumference

    $\dfrac{C}{D} = \pi$ For all circles

    $\pi \approx 3.141\ 59$

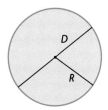

- **Rectangular Solid** $V = abc$ Volume

    $T = 2ab + 2ac + 2bc$ Total surface area

- **Right Circular Cylinder** R = Radius of base

    h = Height

    $V = \pi R^2 h$ Volume

    $S = 2\pi R h$ Lateral surface area

    $T = 2\pi R(R + h)$ Total surface area

- **Right Circular Cone** R = Radius of base

    h = Height

    s = Slant height

    $V = \frac{1}{3}\pi R^2 h$ Volume

    $S = \pi R s = \pi R\sqrt{R^2 + h^2}$ Lateral surface area

    $T = \pi R(R + s) = \pi R(R + \sqrt{R^2 + h^2})$ Total surface area

• **Sphere** $R = $ Radius

$D = $ Diameter

$D = 2R$

$V = \frac{4}{3}\pi R^3 = \frac{1}{6}\pi D^3$ Volume

$S = 4\pi R^2 = \pi D^2$ Surface area

ANSWERS

CHAPTER 1

Exercise 1-1

1. T **3.** F **5.** F **7.** T **9.** $7 + x$ **11.** $(xy)z$ **13.** $9m$ **15.** Commutative ($\cdot$) **17.** Distributive
19. Inverse ($\cdot$) **21.** Inverse ($+$) **23.** Identity ($+$) **25.** Negatives **27.** $\{-2, 0, 2, 4\}$ **29.** $\{a, s, t, u\}$ **31.** $\varnothing$
33. (A) 4 (B) 8 (C) 16 **35.** Negatives **37.** Associative ($\cdot$) **39.** Distributive **41.** Zero **43.** Yes
45. (A) T (B) F (C) T **47.** $\frac{3}{5}$ and -1.43 are two examples of infinitely many
49. (A) $S = \{1, 12\}$ (B) $S = \{-3, 0, 1, 12\}$ (C) $S = \{-3, -\frac{2}{3}, 0, 1, \frac{9}{5}, 12\}$
51. (A) $0.888\ 888\ldots$; repeating; repeated digit: 8 (B) $0.272\ 727\ldots$; repeating; repeated digits: 27
(C) $2.236\ 067\ 977\ldots$; nonrepeating and nonterminating (D) 1.375; terminating
53. (B) is false, since, for example, $5 - 3 \neq 3 - 5$ (D) is false, since, for example, $9 \div 3 \neq 3 \div 9$
55. (A) $\{1, 2, 3, 4, 6\}$ (B) $\{2, 4\}$ **57.** $\frac{1}{11}$ **59.**
$$\begin{array}{r} 23 \\ \times 12 \\ \hline 46 \\ 230 \\ \hline 276 \end{array} \qquad \begin{aligned} 23 \cdot 12 &= 23(2 + 10) \\ &= 23 \cdot 2 + 23 \cdot 10 \\ &= 46 + 230 \\ &= 276 \end{aligned}$$

Exercise 1-2

1. 3 **3.** $2x^3 - x^2 + 2x + 4$ **5.** $2x^3 - 5x^2 + 6$ **7.** $6x^4 - 13x^3 + 9x^2 + 13x - 10$ **9.** $4x - 6$ **11.** $6y^2 - 16y$
13. $m^2 - n^2$ **15.** $4t^2 - 11t + 6$ **17.** $3x^2 - 7xy - 6y^2$ **19.** $4m^2 - 49$ **21.** $30x^2 - 2xy - 12y^2$ **23.** $9x^2 - 4y^2$
25. $16x^2 - 8xy + y^2$ **27.** $a^3 + b^3$ **29.** $-x + 27$ **31.** $32a - 34$ **33.** $2x^4 - x^3 - 6x^2 + 7x - 2$ **35.** $x^4 - 8x^2y^2 + 16y^4$
37. $5u^2 - 12uv + 13v^2$ **39.** $z^3 - 1$ **41.** $8m^3 - 12m^2n + 6mn^2 - n^3$ **43.** $3h$ **45.** $4hx - 3h + 2h^2$ **47.** $4hx - 4h + 2h^2$
49. $3hx^2 - 4hx + 3h^2x - 2h^2 + h^3$ **51.** $m^3 - 3m^2 - 5$ **53.** $2x^3 - 13x^2 + 25x - 18$ **55.** $9x^3 - 9x^2 - 18x$
57. $(1 + 1)^2 \neq 1^2 + 1^2$; either a or b must be zero. **59.** m **61.** Now the degree is less than or equal to m.
63. Perimeter $= 2x + 2(x - 5) = 4x - 10$ **65.** Value $= 5x + 10(x - 5) + 25(x - 3) = 40x - 125$
67. Volume $= \frac{4}{3}\pi(x + 0.3)^3 - \frac{4}{3}\pi x^3 = 1.2\pi x^2 + 0.36\pi x + 0.036\pi$

Exercise 1-3

1. $2x^2(3x^2 - 4x - 1)$ **3.** $5xy(2x^2 + 4xy - 3y^2)$ **5.** $(5x - 3)(x + 1)$ **7.** $(2w - x)(y - 2z)$ **9.** $(x + 3)(x - 2)$
11. $(2m - 1)(3m + 5)$ **13.** $(2x - 3y)(x - 2y)$ **15.** $(4a - 3b)(2c - d)$ **17.** $(2x + 3)(x - 1)$ **19.** Prime
21. $(m - 2)(m + 3)$ **23.** $(2a + 3b)(2a - 3b)$ **25.** $(2x - 5)^2$ **27.** Prime **29.** $6(x + 2)(x + 6)$ **31.** $2y(y - 3)(y - 8)$
33. $y(4x - 1)^2$ **35.** Prime **37.** $xy(x - 3y)(x + 3y)$ **39.** $3m(m^2 - 2m + 5)$ **41.** $(m + n)(m^2 - mn + n^2)$
43. $2x(3x + 1)(x + 1)^3$ **45.** $2(3x - 5)(2x - 3)(12x - 19)$ **47.** $9x^4(9 - x)^3(5 - x)$ **49.** $2(x + 1)(x^2 - 5)(3x + 5)(x - 1)$
51. $[(a - b) - 2(c - d)][(a - b) + 2(c - d)]$ **53.** $(2m - 3n)(a + b)$ **55.** Prime **57.** $(x + 3)(x - 3)^2$
59. $(a - 2)(a + 1)(a - 1)$ **61.** Prime **63.** $(m - n)(m + n)(m^2 + n^2)$ **65.** Prime **67.** $(m + n)(m + n - 1)$
69. $2a[3a - 2(x + 4)][3a + 2(x + 4)]$ **71.** $(x^2 - x + 1)(x^2 + x + 1)$
73. (A) $4(10 - x)(10 + x) = 400 - 4x^2$ (B) $4x(10 - x)^2 = 400x - 80x^2 + 4x^3$

Exercise 1-4

1. $\dfrac{a^2}{2}$ **3.** $\dfrac{22y + 9}{252}$ **5.** $\dfrac{x^2 + 8}{8x^3}$ **7.** $\dfrac{1}{x + 2}$ **9.** $\dfrac{1}{2m + n}$ **11.** $\dfrac{2b^2}{(a + b)^2(a - b)}$ **13.** $\dfrac{m^2}{m - 1}$ **15.** $\dfrac{5}{x - 2}$
17. $\dfrac{1}{y + 2}$ **19.** $\dfrac{x - y}{y}$ **21.** $\dfrac{4(x + 1)(x - 1)(x^2 + 2)^2}{x^3}$ **23.** $\dfrac{x(2 + 3x)}{(1 - 3x)^4}$ **25.** $\dfrac{(x + 1)(x - 9)}{(x + 4)^4}$ **27.** $\dfrac{1}{y - 1}$ **29.** -1
31. $\dfrac{7y - 9x}{xy(a - b)}$ **33.** $\dfrac{x^2 - x + 1}{2(x - 9)}$ **35.** $\dfrac{(x - y)^2}{y^2(x + y)}$ **37.** $\dfrac{1}{x - 4}$ **39.** $\dfrac{x - 3}{x - 1}$ **41.** $\dfrac{-1}{x(x + h)}$ **43.** $\dfrac{x^2 + xh + 4x + 2h}{(x + h + 2)(x + 2)}$
45. (A) Incorrect (B) $x + 1$ **47.** (A) Incorrect (B) $2x + h$ **49.** (A) Incorrect (B) $\dfrac{x^2 - 2}{x + 1}$ **51.** (A) Correct
53. $\dfrac{-x(x + y)}{y}$ **55.** $\dfrac{a - 2}{a}$

Exercise 1-5

1. 1   **3.** $60x^{11}$ **5.** $4x^8/z^6$ **7.** $r^{10}s^{15}/(p^5q^{20})$ **9.** 10^7 **11.** $3a^4/b^3$ **13.** $1/y$ **15.** 4×10^{11} **17.** 4.532×10^7
19. 6.6×10^{-2} **21.** 8.4×10^{-8} **23.** 0.000 09   **25.** 3,480,000   **27.** 0.000 000 004 2   **29.** $3y^4/2$ **31.** x^{12}/y^8
33. $4x^8/y^6$ **35.** $w^{12}/(u^{20}v^4)$ **37.** $1/(x+y)^2$ **39.** $x/(x-1)$ **41.** $-1/(xy)$ **43.** $-9x^2/(x^3+3)^4$ **45.** 64
49. $2x - 6x^{-1}$ **51.** $\frac{5}{3}x - \frac{2}{3}x^{-2}$ **53.** $x - \frac{3}{2} + \frac{1}{2}x^{-1}$ **55.** 6.65×10^{-17} **57.** 1.54×10^{12} **59.** 1.0295×10^{11}
61. -4.3647×10^{-18} **63.** 9.4697×10^{29} **65.** $2(a+2b)^5$ **67.** $(x^2 + xy + y^2)/(xy)$ **69.** $(y-x)/y$ **71.** 1.3×10^{25} lb
73. 10^{10} or 10 billion; 6×10^{11} or 600 billion   **75.** 2.07×10^4 dollars per person; $20,700 per person

Exercise 1-6

1. 5   **3.** 27   **5.** -4 **7.** Not a real number   **9.** $\frac{9}{25}$ **11.** $\frac{1}{32}$ **13.** $a^{5/3}$ **15.** $c^{2/5}$ **17.** $1/u^{1/2}$ **19.** $2x^2/y$
21. $1/(a^{1/4}b^{1/3})$ **23.** $xy^2/2$ **25.** $2b^2/(3a^2)$ **27.** $2/(3x^{7/12})$ **29.** $a^{1/3}/b^2$ **31.** $6m - 2m^{19/3}$ **33.** $a - a^{1/2}b^{1/2} - 6b$
35. $4x - 9y$ **37.** $x + 4x^{1/2}y^{1/2} + 4y$ **39.** 29.52   **41.** 0.030 93   **43.** 5.421   **45.** 107.6
47. $x = y = 1$ is one of many choices.   **49.** $x = y = 1$ is one of many choices.   **51.** $3 - \frac{3}{4}x^{-1/2}$ **53.** $\frac{3}{5}x^{-1/3} + \frac{1}{5}x^{-1/2}$
55. $\frac{1}{2}x^{5/3} - 2x^{1/6}$ **57.** $a^{1/n}b^{1/m}$ **59.** $1/(x^{3m}y^{4n})$ **61.** (A) $x = -2$, for example   (B) $x = 2$, for example   (C) Not possible
63. No   **65.** $(x-3)/(2x-1)^{3/2}$ **67.** $(4x-3)/(3x-1)^{4/3}$ **69.** 1,920 units   **71.** 428 ft

Exercise 1-7

1. $\sqrt[3]{m^2}$ or $(\sqrt[3]{m})^2$ (first preferred)   **3.** $6\sqrt[5]{x^3}$ (not $\sqrt[5]{6x^3}$)   **5.** $\sqrt[5]{(4xy^3)^2}$ **7.** $\sqrt{x+y}$ **9.** $b^{1/5}$ **11.** $5x^{3/4}$ **13.** $(2x^2y)^{3/5}$
15. $x^{1/3} + y^{1/3}$ **17.** -2 **19.** $3x^4y^2$ **21.** $2mn^2$ **23.** $2ab^2\sqrt{2ab}$ **25.** $2xy^2\sqrt[3]{2xy}$ **27.** $\sqrt{m}$ **29.** $\sqrt[15]{xy}$
31. $3x\sqrt{3}$ **33.** $\sqrt{5}/5$ **35.** $2\sqrt{3x}$ **37.** $2\sqrt{2} + 2$ **39.** $\sqrt{3} - \sqrt{2}$ **41.** $3x^2y^2\sqrt[3]{3x^2y}$ **43.** $n\sqrt[4]{2m^3n}$
45. $\sqrt[3]{a^2(b-a)}$ **47.** $\sqrt[4]{a^3b}$ **49.** $x^2y\sqrt[3]{6xy^2}$ **51.** In simplified form   **53.** $\sqrt{2}/2$ or $\frac{1}{2}\sqrt{2}$ **55.** $2a^2b\sqrt[4]{4a^2b}$
57. $\sqrt[4]{12xy^3}/(2x)$ or $[1/(2x)]\sqrt[4]{12xy^3}$ **59.** $(6y + 9\sqrt{y})/(4y - 9)$ **61.** $(38 + 11\sqrt{10})/117$ **63.** $\sqrt{x^2+9} + 3$ **65.** $1/(\sqrt{t} + \sqrt{x})$
67. $1/(\sqrt{x+h} + \sqrt{x})$ **69.** 0.1816   **71.** 3.557   **73.** 0.092 75   **75.** 1.602   **77.** Both are 1.259   **79.** Both are 0.4807
81. $x \le 0$ **83.** All real numbers   **85.** A and E, B and F, C and D   **87.** $\dfrac{\sqrt[3]{a^2} + \sqrt[3]{ab} + \sqrt[3]{b^2}}{a - b}$

89. $\dfrac{(\sqrt{x} - \sqrt{y} - \sqrt{z})[(x+y-z) + 2\sqrt{xy}]}{(x+y-z)^2 - 4xy}$ **91.** $\dfrac{1}{\sqrt[3]{(x+h)^2} + \sqrt[3]{x(x+h)} + \sqrt[3]{x^2}}$

93. $\sqrt[kn]{x^{km}} = (x^{km})^{1/kn} = x^{km/kn} = x^{m/n} = \sqrt[n]{x^m}$

Chapter 1 Review Exercise*

1. (A) T   (B) T   (C) F   (D) T   (E) F   (F) F *(1-1)*   **2.** (A) $(y+z)x$ (B) $(2+x)+y$ (C) $2x + 3x$ *(1-1)*
3. $x^3 + 3x^2 + 5x - 2$ *(1-2)*   **4.** $x^3 - 3x^2 - 3x + 22$ *(1-2)*   **5.** $3x^5 + x^4 - 8x^3 + 24x^2 + 8x - 64$ *(1-2)*   **6.** 3 *(1-2)*
7. 1 *(1-2)*   **8.** $14x^2 - 30x$ *(1-2)*   **9.** $9m^2 - 25n^2$ *(1-2)*   **10.** $6x^2 - 5xy - 4y^2$ *(1-2)*   **11.** $4a^2 - 12ab + 9b^2$ *(1-2)*
12. $(3x-2)^2$ *(1-3)*   **13.** Prime *(1-3)*   **14.** $3n(2n-5)(n+1)$ *(1-3)*   **15.** $(12a^3b - 40b^2 - 5a)/(30a^3b^2)$ *(1-4)*
16. $(7x-4)/[6x(x-4)]$ *(1-4)*   **17.** $(y+2)/[y(y-2)]$ *(1-4)*   **18.** u *(1-4)*   **19.** $6x^5y^{15}$ *(1-5)*   **20.** $3u^4/v^2$ *(1-5)*
21. 6×10^2 *(1-5)*   **22.** x^6/y^4 *(1-5)*   **23.** $u^{7/3}$ *(1-6)*   **24.** $3a^2/b$ *(1-6)*   **25.** $3\sqrt[5]{x^2}$ *(1-7)*   **26.** $-3(xy)^{2/3}$ *(1-7)*
27. $3x^2y\sqrt[3]{x^2y}$ *(1-7)*   **28.** $6x^2y^3\sqrt{xy}$ *(1-7)*   **29.** $2b\sqrt{3a}$ *(1-7)*   **30.** $(3\sqrt{5} + 5)/4$ *(1-7)*   **31.** $\sqrt[4]{y^3}$ *(1-7)*
32. $\{-3, -1, 1\}$ *(1-1)*   **33.** Subtraction *(1-1)*   **34.** Commutative (+) *(1-1)*   **35.** Distributive *(1-1)*
36. Associative (·) *(1-1)*   **37.** Negatives *(1-1)*   **38.** Identity (+) *(1-1)*   **39.** (A) T   (B) F *(1-1)*
40. 0 and -3 are two examples of infinitely many. *(1-1)*   **41.** (A) (a) and (d)   (B) None *(1-2)*   **42.** $4xy - 2y^2$ *(1-2)*
43. $m^4 - 6m^2n^2 + n^4$ *(1-2)*   **44.** $10xh + 5h^2 - 7h$ *(1-2)*   **45.** $2x^3 - 4x^2 + 12x$ *(1-2)*   **46.** $x^3 - 6x^2y + 12xy^2 - 8y^3$ *(1-2)*
47. $(x-y)(7x-y)$ *(1-3)*   **48.** Prime *(1-3)*   **49.** $3xy(2x^2 + 4xy - 5y^2)$ *(1-3)*   **50.** $(y-b)(y-b-1)$ *(1-3)*
51. $3(x+2y)(x^2 - 2xy + 4y^2)$ *(1-3)*   **52.** $(y-2)(y+2)^2$ *(1-3)*   **53.** $x(x-4)^2(5x-8)$ *(1-3)*   **54.** $\dfrac{(x-4)(x+2)^2}{x^3}$ *(1-4)*
55. $2m/[(m+2)(m-2)^2]$ *(1-4)*   **56.** y^2/x *(1-4)*   **57.** $(x-y)/(x+y)$ *(1-4)*   **58.** $-ab/(a^2 + ab + b^2)$ *(1-4)*
59. Incorrect; correct final form is $(x^2 + 2x - 2)/(x-1)$ *(1-4)*   **60.** $\frac{1}{4}$ *(1-5)*   **61.** $\frac{5}{9}$ *(1-5)*   **62.** $3x^2/(2y^2)$ *(1-6)*
63. $27a^{1/6}/b^{1/2}$ *(1-6)*   **64.** $x + 2x^{1/2}y^{1/2} + y$ *(1-6)*   **65.** $6x + 7x^{1/2}y^{1/2} - 3y$ *(1-6)*   **66.** 2×10^{-7} *(1-5)*
67. 3.213×10^6 *(1-5)*   **68.** 4.434×10^{-5} *(1-5)*   **69.** -4.541×10^{-6} *(1-5)*   **70.** 128,800 *(1-6)*   **71.** 0.01507 *(1-6)*
72. 0.3664 *(1-7)*   **73.** 1.640 *(1-7)*   **74.** 0.08726 *(1-6)*   **75.** $-6x^2y^2\sqrt[3]{3x^2y}$ *(1-7)*   **76.** $x\sqrt[3]{2x^2}$ *(1-7)*
77. $\sqrt[4]{12x^3y^3}/(2x)$ *(1-7)*   **78.** $y\sqrt[3]{2x^2y}$ *(1-7)*   **79.** $\sqrt[3]{2x^2}$ *(1-7)*   **80.** $2x - 3\sqrt{xy} - 5y$ *(1-7)*
81. $(6x + 3\sqrt{xy})/(4x - y)$ *(1-7)*   **82.** $(4u - 12\sqrt{uv} + 9v)/(4u - 9v)$ *(1-7)*   **83.** $\sqrt{y^2 + 4} + 2$ *(1-7)*   **84.** $1/(\sqrt{t} + \sqrt{5})$ *(1-7)*
85. $2 - \frac{3}{2}x^{-1/2}$ *(1-7)*   **86.** $\frac{6}{11}$; rational *(1-1)*   **87.** (A) $\{-4, -3, 0, 2\}$ (B) $\{-3, 2\}$ *(1-1)*   **88.** 0 *(1-7)*
89. $x^3 + 8x^2 - 6x + 1$ *(1-2)*   **90.** $x(2a + 3x - 4)(2a - 3x - 4)$ *(1-3)*   **91.** All three have the same value. *(1-7)*
92. $\frac{2}{3}(x-2)(x+3)^4$ *(1-5)*   **93.** $a^2b^2/(a^3 + b^3)$ *(1-5)*   **94.** $x - y$ *(1-6)*   **95.** x^{m-1} *(1-6)*

*The number in parentheses after each answer to a chapter review problem refers to the section in which that type of problem is discussed.

96. $(1 + \sqrt[3]{x} + \sqrt[3]{x^2})/(1 - x)$ *(1-7)* **97.** $1/(\sqrt[3]{t^2} + \sqrt[3]{5t} + \sqrt[3]{25})$ *(1-7)* **98.** x^{n+1} *(1-7)*

99. Volume $= 3\pi(x + 2)^2 - 3\pi x^2 = 12\pi x + 12\pi$ ft^3 *(1-2)* **100.** 2.84×10^4 dollars per person; \$28,400 per person

101. (A) 24,000 units (B) Production doubles to 48,000 units
(C) At any production level, doubling the units of capital and labor doubles production. *(1-6)*

102. $R = \dfrac{R_1 R_2 R_3}{R_2 R_3 + R_1 R_3 + R_1 R_2}$ *(1-4)*

103. (A) $A = 480 - 6x^2 = 6(80 - x^2)$ (B) $V = x(16 - 2x)(15 - 1.5x) = 240x - 54x^2 + 3x^3$ *(1-3)*

CHAPTER 2

Exercise 2-1

1. $x = 16$ **3.** No solution **5.** $a = -\frac{23}{4}$ or -5.75 **7.** $x = \frac{16}{5}$ or 3.2 **9.** No solution **11.** $s = -5.93$
13. $y = \frac{16}{9}$ or $1.\overline{7}$ **15.** No solution **17.** $m = \frac{5}{2}$ or 2.5 **19.** No solution **21.** $x = -\frac{15}{2}$ or -7.5 **23.** $a = -\frac{9}{8}$ or -1.125
25. $x = 1.83$ **27.** $x = -8.55$ **29.** $d = (a_n - a_1)/(n - 1)$ **31.** $f = d_1 d_2/(d_1 + d_2)$ **33.** $a = (A - 2bc)/(2b + 2c)$
35. $x = (5y + 3)/(2 - 3y)$ **37.** Wrong. There is no solution. **39.** 4 **41.** All real numbers except 0 and 1
43. $x = (by + cy - ac)/(a - y)$ **45.** 24 **47.** 8, 10, 12, 14 **49.** 17 by 10 m **51.** 42 ft **53.** \$90 **55.** \$19,750
57. (A) $T = 30 + 25(x - 3)$ (B) 330°C (C) 13 km **59.** 90 mi **61.** 5,000 trout **63.** 10 gal **65.** 11.25 liters
67. 1.5 h **69.** (A) 216 mi (B) 225 mi **71.** 330 Hz; 396 Hz **73.** 150 cm **75.** 150 ft **77.** $5\frac{5}{11}$ min after 1 P.M.

Exercise 2-2

1. $x = -2, y = -1$ **3.** $x = 3, y = -4$ **5.** $x = -1, y = 2$ **7.** $s = -\frac{5}{2}$ or $-2.5, t = \frac{9}{2}$ or 4.5 **9.** $m = 0, n = 10/3$ or $3.\overline{3}$
11. $x = 10,500, y = 57,330$ **13.** $u = 0.8, v = 0.3$ **15.** $a = -7/4$ or $-1.75, b = -19/15$ or $-1.2\overline{6}$ **17.** The system has no solutions.
19. $q = x + y - 5, p = 3x + 2y - 12$ **21.** $x = \dfrac{dh - bk}{ad - bc}, y = \dfrac{ak - ch}{ad - bc}, ad - bc \ne 0$ **23.** Airspeed $= 330$ mph, wind rate $= 90$ mph
25. 2.475 km **27.** 40mL 50% solution, 60mL 80% solution **29.** 5,200 records **31.** \$7,200 at 10% and \$4,800 at 15%
33. Mexico plant: 75 h; Taiwan plant: 50 h **35.** Mix A: 80 g; Mix B: 60 g
37. (A) $p = 0.001q + 0.15$ (B) $p = -0.002q + 1.89$ (C) Equilibrium price $=$ \$0.73; Equilibrium quantity $= 580$ bushels
39. (A) $a = 196, b = -16$ (B) 196 ft (C) 3.5 s **41.** 40 s, 24 s, 120 mi

Exercise 2-3

1. $-8 \le x \le 7$ **3.** $-6 \le x < 6$ **5.** $x \ge -6$

7. $(-2, 6]$ **9.** $(-7, 8)$ **11.** $(-\infty, -2]$

13. $[-7, 2); -7 \le x < 2$ **15.** $(-\infty, 0]; x \le 0$ **17.** $x < 5$ or $(-\infty, 5)$ **19.** $x \ge 3$ or $[3, \infty)$

21. $N < -8$ or $(-\infty, -8)$ **23.** $t > 2$ or $(2, \infty)$ **25.** $m > 3$ or $(3, \infty)$

27. $B \ge -4$ or $[-4, \infty)$ **29.** $-2 < t \le 3$ or $(-2, 3]$ **31.** $(-5, 7]$

33. $(2, 4)$ **35.** $(-\infty, \infty)$ **37.** $(-\infty, -1) \cup [3, 7)$

39. $(1, 5)$ **41.** $(-\infty, 6]$ **43.** $q < -14$ or $(-\infty, -14)$

45. $x \ge 4.5$ or $[4.5, \infty)$ **47.** $-20 \le x \le 20$ or $[-20, 20]$

49. $-30 \le x < 18$ or $[-30, 18)$ **51.** $-8 \le x < -3$ or $[-8, -3)$

53. $-14 < x \le 11$ or $(-14, 11]$ **55.** $x \ge -0.60$ **57.** $-0.255 < x < 0.362$ **59.** $x \le 1$ **61.** $x \ge -\frac{5}{3}$

63. $x > -\frac{3}{2}$ **65.** (A) and (C) $a > 0$ and $b > 0$, or $a < 0$ and $b < 0$ (B) and (D) $a > 0$ and $b < 0$, or $a < 0$ and $b > 0$
67. (A) $>$ (B) $<$ **69.** Positive **71.** (A) F (B) T (C) T **77.** $9.8 \le x \le 13.8$ (from 9.8 to 13.8 km)
79. (A) $x > 40,625$ (B) $x = 40,625$ **81.** (B) $x > 52,000$ (C) Raise wholesale price \$3.50 to \$66.50
83. $2 \le I \le 25$ or $[2, 25]$ **85.** \$2,060 $\le$ Benefit reduction $\le$ \$3,560

Exercise 2-4

1. $\sqrt{5}$ **3.** 4   **5.** $5 - \sqrt{5}$ **7.** $5 - \sqrt{5}$ **9.** 12   **11.** 12   **13.** 4   **15.** 4   **17.** 9   **19.** $|x - 3| = 4$
21. $|m + 2| = 5$ **23.** $|x - 3| < 5$ **25.** $|p + 2| > 6$ **27.** $|q - 1| \geq 2$

29. x is no more than 7 units from the origin. $-7 \leq x \leq 7$ or $[-7, 7]$

31. x is at least 7 units from the origin. $x \leq -7$ or $x \geq 7$ or $(-\infty, -7] \cup [7, \infty)$

33. y is 3 units from 5. $y = 2, 8$ **35.** y is less than 3 units from 5. $2 < y < 8$ or $(2, 8)$

37. y is more than 3 units from 5. $y < 2$ or $y > 8$ or $(-\infty, 2) \cup (8, \infty)$

39. u is 3 units from -8. $u = -11, -5$

41. u is no more than 3 units from -8. $-11 \leq u \leq -5$ or $[-11, -5]$

43. u is at least 3 units from -8. $u \leq -11$ or $u \geq -5$ or $(-\infty, -11] \cup [-5, \infty)$

45. $1 \leq x \leq \frac{11}{3}$; $[1, \frac{11}{3}]$ **47.** $t < -1$ or $t > 5$; $(-\infty, -1) \cup (5, \infty)$ **49.** $m = -2, -\frac{8}{7}$ **51.** $-2.5 < w < 5.5$; $(-2.5, 5.5)$
53. $u \leq -11$ or $u \geq -6$; $(-\infty, -11] \cup [-6, \infty)$ **55.** $-35 < C < -\frac{5}{9}$ or $(-35, -\frac{5}{9})$ **57.** $-2 < x < 2$ or $(-2, 2)$
59. $-\frac{1}{3} \leq t \leq 1$ or $[-\frac{1}{3}, 1]$ **61.** $t < 0$ or $t > 3$ or $(-\infty, 0) \cup (3, \infty)$ **63.** $(2.9, 3) \cup (3, 3.1)$

65. $(c - d, c) \cup (c, c + d)$ **67.** $x \geq 2$ **69.** $x \leq 1.5$ **71.** $x = -2.2, 1$ **73.** $-3 \leq x \leq 0$

75. $x = 1.4, 5$ **77.** ± 1 **91.** $42.2 < x < 48.6$ **93.** $|P - 500| \leq 20$ **95.** $|A - 12.436| < 0.001$, (12.435, 12.437)
97. $|N - 2.37| \leq 0.005$

Exercise 2-5

1. $5 + 9i$ **3.** $3 + 3i$ **5.** $3 + 3i$ **7.** $8 - 10i$ **9.** $7 + 3i$ **11.** -8 **13.** $-12 - 8i$ **15.** $11 + 2i$
17. $13 - i$ **19.** 85   **21.** $0.1 - 0.2i$ **23.** $2 - i$ **25.** $3 - i$ **27.** $7 - 5i$ **29.** $-3 + 2i$ **31.** $8 + 25i$
33. $\frac{5}{7} - \frac{2}{7}i$ **35.** $\frac{2}{13} + \frac{3}{13}i$ **37.** $-\frac{2}{5}i$ or $0 - \frac{2}{5}i$ **39.** $\frac{3}{2} - \frac{1}{2}i$ **41.** $-6i$ or $0 - 6i$ **43.** 0 or $0 + 0i$
45. $i^{18} = -1$, $i^{32} = 1$, $i^{67} = -i$ **47.** $x = 3, y = -2$ **49.** $x > 3$ **51.** $x > \frac{2}{3}$ **53.** $33.89 - 20.38i$ **55.** $0.85 - 0.89i$
57. $(a + c) + (b + d)i$ **59.** $a^2 + b^2$ or $(a^2 + b^2) + 0i$ **61.** $(ac - bd) + (ad + bc)i$
63. $i^{4k} = (i^4)^k = (i^2 \cdot i^2)^k = [(-1)(-1)]^k = 1^k = 1$
65. (*1*) Definition of addition; (*2*) Commutative (+) property for R; (*3*) Definition of addition

Exercise 2-6

1. $x = 0, 4$ **3.** $t = \frac{3}{2}$ (double root)   **5.** $w = -5, \frac{2}{3}$ **7.** $m = \pm 5$ **9.** $c = \pm 3i$ **11.** $y = \pm \frac{3}{2}i$ **13.** $z = \pm 4\sqrt{2}/5$
15. $s = -1 \pm \sqrt{5}$ **17.** $n = 3 \pm 2i$ **19.** $x = 1 \pm \sqrt{2}$ **21.** $x = 1 \pm i\sqrt{2}$ **23.** $t = (3 \pm i\sqrt{7})/2$ **25.** $t = (3 \pm \sqrt{7})/2$
27. $x = 2 \pm \sqrt{5}$ **29.** $r = (-5 \pm \sqrt{3})/2$ **31.** $u = (-2 \pm i\sqrt{11})/2$ **33.** $w = (-2 \pm i\sqrt{5})/3$ **35.** $x = -\frac{5}{4}, \frac{2}{3}$
37. $y = (3 \pm \sqrt{5})/2$ **39.** $x = (3 \pm \sqrt{13})/2$ **41.** $n = -\frac{4}{7}, 0$ **43.** $x = 2 \pm 2i$ **45.** $m = -50, 2$
47. $x = (-5 \pm \sqrt{57})/2$ **49.** $x = (-3 \pm \sqrt{57})/4$ **51.** $u = 1, 2, (-3 \pm \sqrt{17})/2$ **53.** $t = \sqrt{2s/g}$
55. $I = (E + \sqrt{E^2 - 4RP})/2R$ **57.** $x = 1.35, 0.48$ **59.** $x = -1.05, 0.63$
61. If $c < 4$, there are two distinct real roots; if $c = 4$, there is one real double root; and if $c > 4$, there are two distinct imaginary roots.
63. Has real solutions, since discriminant is positive   **65.** Has no real solutions, since discriminant is negative
67. $x = \frac{4}{3}\sqrt{6} \pm \frac{2}{3}\sqrt{15}$ or $(4\sqrt{6} \pm 2\sqrt{15})/3$ **69.** $x = \sqrt{2} - i, -\sqrt{2} - i$ **71.** $x = 1, -\frac{1}{2} \pm \frac{1}{2}i\sqrt{3}$
75. $[(-b + \sqrt{b^2 - 4ac})/2a] \times [(-b - \sqrt{b^2 - 4ac})/2a] = [b^2 - (b^2 - 4ac)]/4a^2 = c/a$
77. The $\pm$ in front still yields the same two numbers even if a is negative.   **79.** 8, 13   **81.** 12, 14   **83.** 5.12 by 3.12 in.
85. 20%   **87.** 100 mph, 240 mph   **89.** 13.09 h and 8.09 h   **91.** 50 mph
93. 50 ft wide and 300 ft long or 150 ft wide and 100 ft long   **95.** 52 mi

Exercise 2-7

1. T   **3.** F   **5.** F   **7.** $x = 14$ **9.** $y = 6$ **11.** $w = -1, 2$ **13.** No solution   **15.** $m = \pm\sqrt{3}, \pm i\sqrt{5}$
17. $x = \frac{1}{2}i$ **19.** $y = -64, \frac{27}{8}$ **21.** $m = -1, 3, 1 \pm 2i$ **23.** No solution   **25.** $w = -2, 1$ **27.** $z = -1$

29. $x = -\frac{3}{2} + \frac{1}{2}i$ **31.** $y = \frac{1}{3} \pm \frac{i\sqrt{2}}{3}$ **33.** $t = \pm\frac{\sqrt{2}}{2}, \pm\sqrt{2}$ **35.** $z = \frac{3}{2} \pm \frac{3i\sqrt{3}}{2}$ **37.** $m = 0.25$ **39.** $w = 4$

41. $x = -1$ **43.** $x = \pm\sqrt{\dfrac{5 \pm \sqrt{13}}{6}}$ (four roots)   **45.** $x = -4, 39{,}596$ **47.** $x = \left(\dfrac{4}{5 \pm \sqrt{17}}\right)^5$ **49.** 13.1 in. by 9.1 in.

51. 1.65 ft or 3.65 ft   **53.** \$30; 1,600 telephones

Exercise 2-8

1. $-5 < x < 2$ **3.** $x < 3$ or $x > 7$ **5.** $0 \le x \le 8$ **7.** $-5 \le x \le 0$ **9.** $x < -2$ or $x > 2$
$(-5, 2)$ $(-\infty, 3) \cup (7, \infty)$ $[0, 8]$ $[-5, 0]$ $(-\infty, -2) \cup (2, \infty)$

11. $-4 < x \le 2$ **13.** $x \le -4$ or $x > 1$ **15.** $-5 \le x \le 0$ or $x > 3$ **17.** $-3 < x < 1$ **19.** $x < 0$ or $x > \frac{1}{4}$
$(-4, 2]$ $(-\infty, -4] \cup (1, \infty)$ $[-5, 0] \cup (3, \infty)$ $(-3, 1)$ $(-\infty, 0) \cup (\frac{1}{4}, \infty)$

21. $-4 < x \le \frac{3}{2}$ **23.** $-1 < x < 2$ or $x \ge 5$ **25.** $x \le -4$ or $0 \le x \le 2$ **27.** $x \le -3$ or $x \ge 3$ **29.** $x \le -2$ or $x \ge \frac{3}{2}$
$(-4, \frac{3}{2}]$ $(-1, 2) \cup [5, \infty)$ $(-\infty, -4] \cup [0, 2]$

31. $-7 \le x < 3$ **33.** If $a > 0$, the solution set is $(-\infty, r_1) \cup (r_2, \infty)$. If $a < 0$, the solution set is (r_1, r_2).
35. If $a > 0$, the solution set is R, the set of real numbers. If $a < 0$, the solution set is $\{r\}$.   **37.** $x^2 \ge 0$ **39.** No solution; $\varnothing$
41. No solution; $\varnothing$
43. $x \le 2 - \sqrt{5}$ or $x \ge 2 + \sqrt{5}$ **45.** $1 - \sqrt{2} < x < 0$ or $x > 1 + \sqrt{2}$ **47.** $-2 \le x \le -\frac{1}{2}$ or $\frac{1}{2} \le x \le 2$ **49.** $-2 \le x \le 2$
$(-\infty, 2 - \sqrt{5}] \cup [2 + \sqrt{5}, \infty)$ $(1 - \sqrt{2}, 0) \cup (1 + \sqrt{2}, \infty)$ $[-2, -\frac{1}{2}] \cup [\frac{1}{2}, 2]$ $[-2, 2]$

51. (A) Profit: $\$4 < p < \$7$ or $(\$4, \$7)$ (B) Loss: $\$0 \le p < \$4$ or $p > \$7$ or $[\$0, \$4) \cup (\$7, \infty)$ **53.** $2 \le t \le 5$
55. $v > 75$ mph   **57.** $5 \le t \le 20$

Chapter 2 Review Exercise

1. $x = 21$ *(2-1)*   **2.** $x = \frac{30}{11}$ *(2-1)*   **3.** $x = 3, y = 3$ *(2-2)*
4. $x \ge 1$ *(2-3)*   **5.** $-14 < y < -4$ *(2-4)*   **6.** $-1 \le x \le 4$ *(2-4)*   **7.** $-5 < x < 4$ *(2-8)*   **8.** $x \le -3$ or $x \ge 7$ *(2-8)*
$[1, \infty)$ $(-14, -4)$ $[-1, 4]$ $(-5, 4)$ $(-\infty, -3] \cup [7, \infty)$

9. (A) $3 - 6i$ (B) $15 + 3i$ (C) $2 + i$ *(2-5)*   **10.** $x = \pm\sqrt{\frac{7}{2}}$ or $\pm\frac{1}{2}\sqrt{14}$ *(2-6)*   **11.** $x = 0, 2$ *(2-6)*
12. $x = \frac{1}{2}, 3$ *(2-6)*   **13.** $m = -\frac{1}{2} \pm (\sqrt{3}/2)i$ *(2-6)*   **14.** $y = (3 \pm \sqrt{33})/4$ *(2-6)*   **15.** $x = 2, 3$ *(2-7)*
16. $x = 3, y = -2$ *(2-2)*   **17.** $x \le \frac{3}{5}$ *(2-3)*   **18.** $x = -15$ *(2-1)*   **19.** No solution   *(2-1)*   **20.** $m = 2, n = -\frac{4}{3}$ *(2-2)*
21. $x \ge -19$ *(2-3)*   **22.** $x < 2$ or $x > \frac{10}{3}$ *(2-4)*   **23.** $x < 0$ or $x > \frac{1}{2}$ *(2-8)*   **24.** $x \le 1$ or $3 < x < 4$ *(2-8)*
$[-19, \infty)$ $(-\infty, 2) \cup (\frac{10}{3}, \infty)$ $(-\infty, 0) \cup (\frac{1}{2}, \infty)$ $(-\infty, 1] \cup (3, 4)$

25. $-1 \le m \le 2$ *(2-4)*   **26.** $-4 \le x < 2$ or $[-4, 2)$ *(2-8)*   **27.** (A) 6   (B) 6   *(2-4)*   **28.** (A) $5 + 4i$ (B) $-i$ *(2-5)*
$[-1, 2]$

29. (A) $-1 + i$ (B) $\frac{4}{13} - \frac{7}{13}i$ (C) $\frac{5}{2} - 2i$ *(2-5)*   **30.** $u = (-5 \pm \sqrt{5})/2$ *(2-6)*   **31.** $u = 1 \pm i\sqrt{2}$ *(2-6)*
32. $x = (1 \pm \sqrt{43})/3$ *(2-6)*   **33.** $x = -\frac{27}{8}, 64$ *(2-7)*   **34.** $m = \pm 2, \pm 3i$ *(2-7)*   **35.** $y = \frac{9}{4}, 3$ *(2-7)*   **36.** $x = 0.45$ *(2-1)*
37. $-2.24 \le x \le 1.12$ or $[-2.24, 1.12]$ *(2-3)*   **38.** $0.89 - 0.32i$ *(2-5)*   **39.** $x = -1.64, 0.89$ *(2-6)*
40. $x = 0.94, y = 1.02$ *(2-2)*   **41.** $M = P/(1 - dt)$ *(2-1)*   **42.** $I = (E \pm \sqrt{E^2 - 4PR})/(2R)$ *(2-6)*
43. $y = (5 - x)/(2x - 4)$ *(2-1)*   **44.** The correct answer is $x = -1$.   *(2-1)*
45. If $c < 9$, there are two distinct real roots; if $c = 9$, there is one real double root; and if $c > 9$, there are two distinct imaginary roots.   *(2-6)*

46. All real b and all negative a *(2-3)*   **47.** Less than 1   *(2-3)*   **48.** $x = 1/(1 - y)$ *(2-1)*

49. $6 - d < x < 6 + d, x \neq 6$ *(2-4)*   **50.** $x = \frac{1}{4}\sqrt{3} \pm \frac{1}{4}i$ *(2-6)*   **51.** $x = \pm\sqrt{(2 \pm \sqrt{3})/2}$ (four real roots)   *(2-7)*   **52.** 1   *(2-5)*

$(6 - d, 6) \cup (6, 6 + d)$

```
  ←————○———○————→ x
     6 − d  6  6 + d
```

53. No solution   *(2-8)*   **54.** Set of all real numbers   *(2-8)*

55. $x \leq -4$ or $-2 \leq x < 0$ or $0 < x \leq 2$ or $x \geq 4$; $(-\infty, -4] \cup [-2, 0) \cup (0, 2] \cup [4, \infty)$ *(2-8)*

56. $u = -31 + 5x - 7y, v = 13 - 2x + 3y$ *(2-2)*   **57.** (A) Infinite number of solutions   (B) No solution   *(2-2)*

58. $\frac{5}{3}$ or $-\frac{3}{5}$ *(2-6)*   **59.** (A) $H = 0.7(220 - A)$ (B) 140 beats/min   (C) 40 years old   *(2-1)*

60. 20 mL of 30% solution, 30 mL of 80% solution   *(2-1)*   **61.** 3 mph   *(2-6)*

62. (A) 3 km   (B) 16.2 km/h   (C) 20.6 min   *(2-1, 2-6)*   **63.** 85 bags of brand A and 45 bags of brand B *(2-2)*

64. (A) 2,000 and 8,000   (B) 5,000   *(2-6)*   **65.** $x = (13 \pm \sqrt{45})/2$ thousand, or approx. 3,146 and 9,854   *(2-6)*

66. $(13 - \sqrt{45})/2 < x < (13 + \sqrt{45})/2$ or approx. $3.146 < x < 9.854$, x in thousands   *(2-8)*   **67.** $|T - 110| \leq 5$ *(2-4)*

68. 20 cm by 24 cm   *(2-6)*   **69.** 6.58 ft or 14.58 ft   *(2-7)*

Cumulative Review Exercise: Chapters 1 and 2

1. $x = \frac{5}{2}$ *(2-1)*   **2.** (A) $ca + cb$ (B) $a + (b + c)$ (C) $c(a + b)$ *(1-1)*   **3.** $x = 1, y = -2$ *(2-2)*

4. $4x^2 - 6x + 10$ *(1-2)*   **5.** $3x^2 - 26x + 9$ *(1-2)*   **6.** $6x^2 - 7xy - 20y^2$ *(1-2)*   **7.** $4a^2 - 9b^2$ *(1-2)*

8. $25m^2 + 20mn + 4n^2$ *(1-2)*   **9.** $y \geq 5$ *(2-3)*   **10.** $-5 < x < 9$ *(2-4)*   **11.** $x \leq -5$ or $x \geq 2$ *(2-8)*

$[5, \infty)$ $(-5, 9)$ $(-\infty, -5] \cup [2, \infty)$

```
  [——→ y          ←——)    (——→ x       ←——]    [——→ x
  5                 −5    9              −5    2
```

12. Prime   *(1-3)*   **13.** $(3t + 5)(2t - 1)$ *(1-3)*   **14.** $\dfrac{2}{(x - 2)(x - 3)}$ *(1-4)*   **15.** $2y$ *(1-4)*

16. (A) $7 - 10i$ (B) $23 + 7i$ (C) $1 - i$ *(2-5)*   **17.** $3x^7y^{10}$ *(1-5)*   **18.** $\dfrac{2y^2}{x}$ *(1-5)*   **19.** $\dfrac{a^6}{b^9}$ *(1-6)*

20. $x = -4, 0$ *(2-6)*   **21.** $x = -\sqrt{5}, \sqrt{5}$ *(2-6)*   **22.** $x = 3 \pm \sqrt{7}$ *(2-6)*   **23.** $x = 3$ *(2-7)*   **24.** $5\sqrt[4]{a^3}$ *(1-7)*

25. $2x^{2/5}y^{3/5}$ *(1-7)*   **26.** $x^2y^4\sqrt[3]{xy^2}$ *(1-7)*   **27.** $2xy^2\sqrt{3x}$ *(1-7)*   **28.** $\frac{1}{2}\sqrt{2}$ *(1-7)*   **29.** $x \geq -\frac{2}{3}$ or $[-\frac{2}{3}, \infty)$ *(2-3)*

30. $\{2, 3, 5\}$ *(1-1, 1-2)*   **31.** (A) F   (B) T   (C) F   *(1-1)*   **32.** No solution   *(2-1)*   **33.** $x = \frac{1}{2}, 3$ *(2-6)*

34. $x = 1, \frac{5}{2}$ *(2-7)*   **35.** $x = 87/16, y = 5/8$ *(2-2)*

36. (a) is a second-degree polynomial.   (c) is a fourth-degree polynomial.   *(1-2)*

37. $10ab$ *(1-2)*   **38.** $6hx + 3h^2 - 4h$ *(1-2)*   **39.** $64m^3 + 144m^2n + 108mn^2 + 27n^3$ *(1-2)*   **40.** $(y + 4)(3y + 4)$ *(1-3)*

41. $(a + 3)(a + 2)(a - 2)$ *(1-3)*   **42.** $x^3(x + 1)^2(7x + 4)$ *(1-3)*   **43.** $\dfrac{x^3(x + 4)}{(x + 1)^4}$ *(1-4)*   **44.** $\dfrac{5}{a - 2b}$ *(1-4)*

45. $\dfrac{x + 2y}{x - 2y}$ *(1-4)*   **46.** Correct answer is $x + 1$ *(1-4)*   **47.** Correct answer is $x = 2$ *(2-1)*

48. $x < \frac{3}{2}$ or $x > 3$ *(2-4)*   **49.** $\frac{2}{3} \leq m \leq 2$ *(2-4)*   **50.** $-1 < x < 2$ or $x \geq 5$ *(2-8)*   **51.** $x \geq 2, x \neq 4$ *(2-3)*

$(-\infty, \frac{3}{2}) \cup (3, \infty)$ $[\frac{2}{3}, 2]$ $(-1, 2) \cup [5, \infty)$ $[2, 4) \cup (4, \infty)$

```
  ←——)    (——→ x     [———]——→ m     ←—)    [——→ x       [——○——→ x
     3/2   3          2/3  2        −1  2   5            2   4
```

52. (A) 0   (B) $\frac{6}{5}$ (C) $-i$ *(2-5)*   **53.** (A) $3 + 18i$ (B) $-2.9 + 10.7i$ (C) $-4 - 6i$ *(2-5)*

54. (A) 0.1767   (B) 1.434   (C) 1.435   (D) 5.724×10^{14} *(1-5, 1-7)*   **55.** $6x^2y^4\sqrt[3]{2x^2y}$ *(1-7)*   **56.** $\dfrac{1}{2b}\sqrt[5]{14a^3b}$ *(1-7)*

57. $y\sqrt{2y}$ *(1-7)*   **58.** $\sqrt{t + 9} + 3$ *(1-7)*   **59.** All three have the same value.   *(1-7)*

60. (A) $\frac{2}{5}x + 2x^{-2}$ (B) $\frac{1}{2} - \frac{5}{4}x^{-1/2}$ *(1-5, 1-7)*   **61.** $y = 3 \pm i\sqrt{5}$ *(2-6)*   **62.** $x = -\frac{1}{8}, \frac{27}{8}$ *(2-7)*   **63.** $u = \pm\sqrt{3}, \pm 2i$ *(2-7)*

64. $t = \frac{9}{4}$ *(2-7)*   **65.** $-18.36 \leq x < 16.09$; $[-18.36, 16.09)$ *(2-3)*   **66.** $x = -5.68, 1.23$ *(2-6)*   **67.** $x = 1.49, y = 0.55$ *(2-2)*

68. $y = \dfrac{3 - 3x}{x + 4}$ *(2-1)*   **69.** $s = 5 + x - 2y, t = -12 - 2x + 5y$ *(2-2)*   **70.** 0   *(1-7)*   **71.** 0   *(2-5)*   **72.** $6x^2 + 2$ *(1-2)*

73. All a and b such that $a < b$ *(2-3)*   **74.** $b^2(3b + 4a - 4)(3b - 4a + 4)$ *(1-3)*   **75.** $\dfrac{2 - m}{4m}$ *(1-4)*   **76.** $y = \dfrac{x + x^2}{x - 1}$ *(2-1)*

77. $x = (\sqrt{2} \pm i)/3$ *(2-6)*

78. If $b < -2$ or $b > 2$, there are two distinct real roots; if $b = -2$ or $b = 2$, there is one real double root; and if $-2 < b < 2$, there are two distinct imaginary roots.   *(2-6, 2-8)*

79. $x = -3$ *(2-7)*   **80.** (A) $a - b$ (B) $\dfrac{x^2 - y^2}{x^2 + y^2}$ *(1-5, 1-7)*   **81.** $\dfrac{1}{(8 + h)^{2/3} + 2(8 + h)^{1/3} + 4}$ *(1-7)*

82. $\dfrac{a^2 - b^2}{a^2 + b^2} + \dfrac{2ab}{a^2 + b^2}i$ *(2-5)*   **83.** $x < -1$ or $x > 2$; $(-\infty, -1) \cup (2, \infty)$ *(2-8)*   **84.** x^2 *(1-7)*   **85.** 2 or $-\frac{1}{2}$ *(2-6)*

86. 10.5 min *(2-1)*   **87.** 2.5 mph *(2-6)*   **88.** 12 gal *(2-1)*   **89.** 8,800 books *(2-2)*   **90.** $|p - 200| \leq 10$ *(2-4)*

91. \$16.50; 5,500 cheese heads *(2-2)*   **92.** $x^2 - (x - 1)^2 = 2x - 1$ ft³ *(1-2)*

93. (A) Profit: $5.5 < p < \$8$ or $(\$5.5, \$8)$ (B) Loss: $\$0 \leq p < 5.5$ or $p > \$8$ or $[\$0, \$5.5) \cup (\$8, \infty)$ *(2-8)*

94. 40 mi from A to B and 75 mi from B to C or 75 mi from A to B and 40 mi from B to C *(2-6)*

95. (A) $a = 2,500, b = -16$ (B) 2,500 ft   (C) 12.5 s *(2-1, 2-2, 2-6)*

CHAPTER 3   Exercise 3-1

1. The y axis   **3.** Quadrant III   **5.** Quadrant IV   **7.** Quadrant II and quadrant IV   **9.** Quadrant I and quadrant IV

11.

13. Symmetric with respect to the origin   **15.** Symmetric with respect to the x axis   **17.** Symmetric with respect to the x axis, y axis, and origin

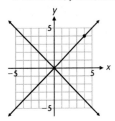

19. $\sqrt{106}$ **21.** $\sqrt{82}$ **23.** $x^2 + y^2 = 16$ **25.** $(x - 3)^2 + (y + 2)^2 = 1$ **27.** $(x - 2)^2 + (y - 6)^2 = 3$

29. (A) 3   (B) -2 (C) $-3, -1, 4$ (D) $-4, 1, 3$

31. (A)   (B)   (C)   (D)

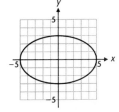

33. Symmetric with respect to the x axis   **35.** Symmetric with respect to the y axis   **37.** Symmetric with respect to the x axis, y axis, and origin   **39.** Symmetric with respect to the x axis, y axis, and origin

41. Symmetric with respect to the origin   **43.** Symmetric with respect to the y axis   **45.** Symmetric with respect to the y axis   **47.** Symmetric with respect to the y axis

49. Area $= 28$, perimeter $= 26.96$ **51.** $x = -12, 4$ **53.** $y = 4$

55. Center: $(-4, 2)$; radius: $\sqrt{7}$ **57.** Center: $(3, 2)$; radius: 7   **59.** Center: $(-4, 3)$; radius: $\sqrt{17}$ **61.**

63.   **65.** $y = \pm\sqrt{3 - x^2}$ **67.** $y = -1 \pm \sqrt{2 - (x + 3)^2}$ **69.** Center: $(1, 0)$; radius: 1;
$(x - 1)^2 + y^2 = 1$

            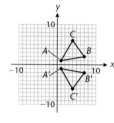

71. Center: $(2, 1)$; radius: 3;   **73.** Symmetric with respect   **75.** Symmetric with respect   **77.**
$(x - 2)^2 + (y - 1)^2 = 9$ to the y axis   to the origin

79. $5x + 3y = -2$ **81.** $(x - 4)^2 + (y - 2)^2 = 34$ **83.** $(x - 2)^2 + (y - 2)^2 = 50$ **85.** Yes
87. (A) 3,000 cases   (B) Demand decreases by 400 cases   (C) Demand increases by 600 cases
89. (A) 53°   (B) 68° at 3 P.M.   (C) 1 A.M., 7 A.M., 11 P.M.
91.   **93.** 2.5 ft   **95.** (A) $(x + 12)^2 + (y + 5)^2 = 26^2$; center: $(-12, -5)$; radius: 26   (B) 13.5 mi

Exercise 3-2

1. x intercept: -2; y intercept: 2; slope: 1; equation: $y = x + 2$ **3.** x intercept: -2; y intercept: -4; slope: -2; equation: $y = -2x - 4$
5. x intercept: 3; y intercept: -1; slope: $\frac{1}{3}$; equation: $y = \frac{1}{3}x - 1$
7. Slope $= -\frac{3}{5}$ **9.** Slope $= -\frac{3}{4}$ **11.** Slope $= \frac{2}{3}$ **13.** Slope $= \frac{4}{5}$

15. Slope = 2 **17.** Slope not defined **19.** Slope = 0 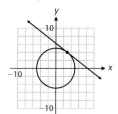 **21.** $x - y = 0$ **23.** $2x + 3y = -12$

25. $y = -2x + 3$ **27.** $y = \frac{3}{2}x + \frac{23}{2}$ **29.** $y = -4x + 13$ **31.** $x = -3$ **33.** $y = 2$ **35.** $3x + y = 5$
37. $2x + 3y = -12$ **39.** $y = 3$ **41.** $2x + 3y = 23$ **43.** $2x + 5y = 10$ **45.** $x = -2$ **47.** Trapezoid **49.** Rectangle
51. $y = -\frac{6}{7}x - \frac{5}{14}$ **53.** $3x + 4y = 25$ **55.** $x - y = 10$ **57.** $5x - 12y = 232$

59. (A) 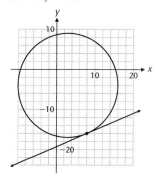 **61.** **63.** **65.**

71. $7x - 2y = -14$ **73.** (A)

x	0	5,000	10,000	15,000	20,000	25,000	30,000
B	212	203	194	185	176	167	158

(B)   The boiling point drops 9°F for each 5,000-ft increase in altitude.
75. The rental charges are \$25 per day plus \$0.25 per mile driven.
77. (A)

x	0	1	2	3	4
Sales	5.9	6.5	7.7	8.6	9.7
$f(x)$	5.7	6.7	7.7	8.6	9.6

(B)   (C)   \$10.6 billion, \$17.4 billion

79. (A) $F = \frac{9}{5}C + 32$ (B) 68°F; 30°C (C) $\frac{9}{5}$ **81.** (A) $V = -1,600t + 8,000, 0 \le t \le 5$ (B) \$3,200 (C) $-1,600$
83. (A) $T = -5A + 70, A \ge 0$ (B) 14,000 ft (C) -5; the temperature changes -5°F for each 1,000-ft rise in altitude
85. (A) $h = 1.13t + 12.8$ (B) 32.9 h
87. (A) $R = 0.001\ 52C - 0.159, C \ge 210$ (B) 0.236
(C) 0.001 52; coronary risk increases 0.001 52 per unit increase in cholesterol above the 210 cholesterol level

Exercise 3-3

1. Function **3.** Not a function **5.** Function **7.** Function; domain = {2, 3, 4, 5}; range = {4, 6, 8, 10} **9.** Not a function
11. Function; domain = {0, 1, 2, 3, 4, 5}; range = {1, 2} **13.** Function **15.** Not a function **17.** Not a function **19.** 4
21. 0 **23.** 10 **25.** 4 **27.** 5.25 **29.** 7 **31.** $-1, -5, 6$ **33.** A function with domain all real numbers
35. Not a function; for example, when $x = 0, y = \pm 2$ **37.** A function with domain all real numbers

39. Not a function; for example, when $x = 0$, y can be any real number   **41.** A function with domain all real numbers except 0

43. Domain: $x \geq 2$ **45.** Domain: all real numbers   **47.** Domain: all real numbers except 4

49. Domain: all real numbers except -2 and 1   **51.** Domain: $y \leq -1$ or $y \geq 3$ **53.** Domain: all real numbers

55. Domain: $y < -2$ or $y \geq 3$ **57.** $g(x) = 2x^3 - 5$ **59.** $G(x) = 2\sqrt{x} - x^2$

61. Function f multiplies the domain element by 2 and subtracts 3 from the result.

63. Function F multiplies the cube of the domain element by 3 and subtracts twice the square root of the domain element from the result.

65. 3   **67.** $-6 - h$ **69.** $-2h + 11$ **71.** $f(x) = 3x^2 - 5x + 9$ **73.** $m(t) = -2t^2 - 5\sqrt{t} - 2$ **75.** (A) 4   (B) 4

77. (A) $4x + 2h$ (B) $2x + 2a$ **79.** (A) $-8x + 3 - 4h$ (B) $-4x - 4a + 3$

81. (A) $3x^2 - 2 + 3xh + h^2$ (B) $x^2 + ax + a^2 - 2$ **83.** $P(w) = 2w + (128/w)$, $w > 0$ **85.** $h(b) = \sqrt{b^2 + 25}$, $b > 0$

87. $C(x) = 300 + 1.75x$

89. (A) $s(0) = 0$, $s(1) = 16$, $s(2) = 64$, $s(3) = 144$ (B) $64 + 16h$

    (C) Value of expression tends to 64; this number appears to be the speed of the object at the end of 2 s.

91. $V(x) = x(8 - 2x)(12 - 2x)$; domain: $0 < x < 4$

93. $F(x) = 8x + (250/x) - 12$;

x	4	5	6	7
$F(x)$	82.5	78	77.7	79.7

95. $C(x) = 10{,}000(20 - x) + 15{,}000\sqrt{x^2 + 64}$; domain: $0 \leq x \leq 20$ **97.** $C(v) = 100v + (200{,}000/v)$

Exercise 3-4

1. (A) $[-4, 4)$ (B) $[-3, 3)$ (C) 0   (D) 0   (E) $[-4, 4)$ (F) None   (G) None   (H) None

3. (A) $(-\infty, \infty)$ (B) $[-4, \infty)$ (C) $-3, 1$ (D) -3 (E) $[-1, \infty)$ (F) $(-\infty, -1]$ (G) None   (H) None

5. (A) $(-\infty, 2) \cup (2, \infty)$ (B) $(-\infty, -1) \cup [1, \infty)$ (C) None   (D) 1   (E) None   (F) $(-\infty, -2], (2, \infty)$

    (G) $[-2, 2)$ (H) $x = 2$

7. One possible answer: **9.** One possible answer: **11.** One possible answer:

13. Slope $= 2$, x intercept $= -2$, y intercept $= 4$ **15.** Slope $= -\frac{1}{2}$, x intercept $= -\frac{10}{3}$, y intercept $= -\frac{5}{3}$ **17.** $f(x) = -\frac{3}{2}x + 4$

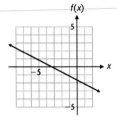

19. Min $f(x) = f(3) = 2$ **21.** Max $f(x) = f(-3) = -2$ **23.** x intercepts: $-1, 5$ **25.** x intercepts: 0, 6

Range $= [2, \infty)$ Range $= (-\infty, -2]$ y intercept $= -5$ y intercept $= 0$

27. Increasing on $[-3, \infty)$
Decreasing on $(-\infty, -3]$

29. Increasing on $(-\infty, 3]$
Decreasing on $[3, \infty)$

31. Domain: $[-1, 1]$
Range: $[0, 1]$

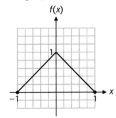

33. Domain: $[-3, -1) \cup (-1, 2]$
Range: $\{-2, 4\}$ (a set, not an interval)
Discontinuous at $x = -1$

35. Domain: all real numbers
Range: all real numbers
Discontinuous at $x = -1$

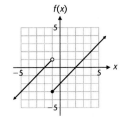

37. Domain: $x \neq 0$, or $(-\infty, 0) \cup (0, \infty)$
Range: $(-\infty, -1) \cup (1, \infty)$
Discontinuous at $x = 0$

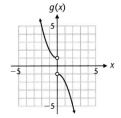

39. Min $f(x) = f(-2) = 1$
Range: $[1, \infty)$
No x intercepts
y intercept $= f(0) = 3$
Increasing on $[-2, \infty)$
Decreasing on $(-\infty, -2]$

41. Min $f(x) = f(\frac{3}{2}) = 0$
Range: $[0, \infty)$
x intercept $= \frac{3}{2}$
y intercept $= f(0) = 9$
Increasing on $[\frac{3}{2}, \infty)$
Decreasing on $(-\infty, \frac{3}{2}]$

43. Max $f(x) = f(-2) = 6$
Range: $(-\infty, 6]$
x intercepts: $-2 \pm \sqrt{3}$
y intercept $= f(0) = -2$
Increasing on $(-\infty, -2]$
Decreasing on $[-2, \infty)$

45. $f(x) = \begin{cases} -1 & \text{if } x < 0 \\ 1 & \text{if } x > 0 \end{cases}$
Domain: $x \neq 0$, or $(-\infty, 0) \cup (0, \infty)$
Range: $\{-1, 1\}$ (a set, not an interval)
Discontinuous at $x = 0$

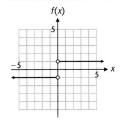

47. $f(x) = \begin{cases} x - 1 & \text{if } x < 1 \\ x + 1 & \text{if } x > 1 \end{cases}$
Domain: $x \neq 1$, or $(-\infty, 1) \cup (1, \infty)$
Range: $(-\infty, 0) \cup (2, \infty)$
Discontinuous at $x = 1$

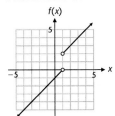

49. $f(x) = \begin{cases} 2 - 2x & \text{if } x < 0 \\ 2 & \text{if } 0 \le x < 2 \\ -2 + 2x & \text{if } x \ge 2 \end{cases}$
Domain: all real numbers
Range: $[2, \infty)$
No discontinuities

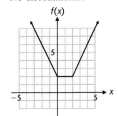

51. Domain: all real numbers
Range: all integers
Discontinuous at the even integers

$$f(x) = \begin{cases} \vdots \\ -2 & \text{if } -4 \le x < -2 \\ -1 & \text{if } -2 \le x < 0 \\ 0 & \text{if } 0 \le x < 2 \\ 1 & \text{if } 2 \le x < 4 \\ 2 & \text{if } 4 \le x < 6 \\ \vdots \end{cases}$$

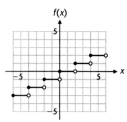

53. Domain: all real numbers
Range: all integers
Discontinuous at rational numbers of the form $k/3$, where k is an integer

$$f(x) = \begin{cases} \vdots \\ -2 & \text{if } -\frac{2}{3} \le x < -\frac{1}{3} \\ -1 & \text{if } -\frac{1}{3} \le x < 0 \\ 0 & \text{if } 0 \le x < \frac{1}{3} \\ 1 & \text{if } \frac{1}{3} \le x < \frac{2}{3} \\ 2 & \text{if } \frac{2}{3} \le x < 1 \\ \vdots \end{cases}$$

55. Domain: all real numbers
Range: [0, 1)
Discontinuous at all integers

$$f(x) = \begin{cases} \vdots \\ x+2 & \text{if } -2 \le x < -1 \\ x+1 & \text{if } -1 \le x < 0 \\ x & \text{if } 0 \le x < 1 \\ x-1 & \text{if } 1 \le x < 2 \\ x-2 & \text{if } 2 \le x < 3 \\ \vdots \end{cases}$$

57. Axis: $x = 2$; vertex: (2, 4); range: [4, ∞); no x intercepts

59. (A) One possible answer:

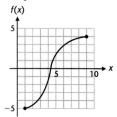

(B) The graph must cross the x axis exactly once.

61. (A) One possible answer:

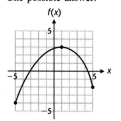

(B) The graph must cross the x axis at least twice. There is no upper limit on the number of times it can cross the x axis.

63. $y = 2x - 1$

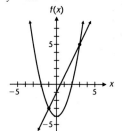

65. (A) $h + 1$ (B)

h	1	0.1	0.01	0.001
Slope	2	1.1	1.01	1.001

; approaching 1

67. Graphs of f and g

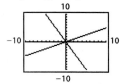

69. Graphs of f and g

71. Graphs of f and g

Graph of *m*

Graph of *m*

Graph of *m*

Graph of *n*

Graph of *n*

Graph of *n*

73. $m(x) = \max[f(x), g(x)]$

75. (A)

x	28	30	32	34	36
Mileage	45	52	55	51	47
f(*x*)	45.3	51.8	54.2	52.4	46.5

(B)

(C) $f(31) \approx 53.50$ thousand miles
$f(35) \approx 49.95$ thousand miles

77. (A) $s = f(w) = w/10$
(B) $f(15) = 1.5$ in, $f(30) = 3$ in
(C) Slope = $\frac{1}{10}$
(D)

79. $E(x) = \begin{cases} 200 & \text{if } 0 \le x \le 3{,}000 \\ 80 + 0.04x & \text{if } 3{,}000 < x < 8{,}000 \\ 180 + 0.04x & \text{if } x \ge 8{,}000 \end{cases}$

Discontinuous at $x = 8{,}000$
$E(5{,}750) = \$310, E(9{,}200) = \$548$

81. (A) $A(x) = 50x - x^2$
(B) Domain: $0 < x < 50$
(C)

(D) 25 by 25 ft

83.

x	4	−4	6	−6	24	25	247	−243	−245	−246
f(*x*)	0	0	10	−10	20	30	250	−240	−240	−250

; *f* rounds numbers to the tens place.

85. $f(x) = [\![0.5 + 100x]\!]/100$

87. (A) $C(x) = \begin{cases} 15 & 0 < x \le 1 \\ 18 & 1 < x \le 2 \\ 21 & 2 < x \le 3 \\ 24 & 3 < x \le 4 \\ 27 & 4 < x \le 5 \\ 30 & 5 < x \le 6 \end{cases}$

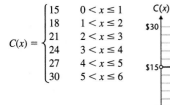

(B) No, since $f(x) \ne C(x)$ at $x = 1, 2, 3, 4, 5,$ or 6

89. \$50 per day; maximum income = \$12,500   **91.** (A) $32\sqrt{5} \approx 71.55$ ft/s   (B) 15 ft

Exercise 3-5

1. Domain: $[0, \infty)$; Range: $(-\infty, 0]$ **3.** Domain: *R*; Range: $(-\infty, 0]$ **5.** Domain: *R*; Range *R*

7. $(f + g)(x) = 5x + 1, (f - g)(x) = 3x - 1, (fg)(x) = 4x^2 + 4x, (f/g)(x) = 4x/(x + 1)$;
Domain of $f + g$ = Domain of $f - g$ = Domain of $fg = (-\infty, \infty)$, Domain of $f/g = (-\infty, -1) \cup (-1, \infty)$

9. $(f + g)(x) = 3x^2 + 1, (f - g)(x) = x^2 - 1, (fg)(x) = 2x^4 + 2x^2, (f/g)(x) = 2x^2/(x^2 + 1)$; Domain of each function $= (-\infty, \infty)$

11. $(f \circ g)(x) = x^4 - 8x^3 + 16x^2 + 3, (g \circ f)(x) = x^4 + 2x^2 - 3$; Domain of $f \circ g$ = Domain of $g \circ f = (-\infty, \infty)$

13. $(f \circ g)(x) = 2(x^3 - 1)^{2/3}$, $(g \circ f)(x) = 8x^2 - 1$; Domain of $f \circ g$ = Domain of $g \circ f = (-\infty, \infty)$

15.

17.

19.

21.

23. The graph of $y = |x|$ is shifted 2 units to the left and reflected in the x axis.

25. The graph of $y = x^2$ is shifted 2 units to the right and 4 units down.

27. The graph of $y = \sqrt{x}$ is vertically expanded by a factor of 2, reflected in the x axis, and shifted 4 units up.

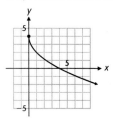

29. $(f + g)(x) = \sqrt{x + 2} + \sqrt{4 - x}$, $(f - g)(x) = \sqrt{x + 2} - \sqrt{4 - x}$, $(fg)(x) = \sqrt{8 + 2x - x^2}$, $(f/g)(x) = \sqrt{(x + 2)/(4 - x)}$; Domain of $f + g$ = Domain of $f - g$ = Domain of $fg = [-2, 4]$, Domain of $f/g = [-2, 4)$

31. $(f + g)(x) = 8 - 3\sqrt{x}$, $(f - g)(x) = 2 - \sqrt{x}$, $(fg)(x) = 15 - 11\sqrt{x} + 2x$, $(f/g)(x) = (5 - 2\sqrt{x})/(3 - \sqrt{x})$; Domain of $f + g$ = Domain of $f - g$ = Domain of $fg = [0, \infty)$, Domain of $f/g = [0, 9) \cup (9, \infty)$

33. $(f + g)(x) = \sqrt{x^2 + x - 2} + \sqrt{24 + 2x - x^2}$, $(f - g)(x) = \sqrt{x^2 + x - 2} - \sqrt{24 + 2x - x^2}$, $(fg)(x) = \sqrt{-x^4 + x^3 + 28x^2 + 20x - 48}$, $(f/g)(x) = \sqrt{(x^2 + x - 2)/(24 + 2x - x^2)}$; Domain of $f + g$ = Domain of $f - g$ = Domain of $fg = [-4, -2] \cup [1, 6]$, Domain of $f/g = (-4, -2] \cup [1, 6)$

35. $(f \circ g)(x) = \sqrt{4 - x} + 2$, $(g \circ f)(x) = \sqrt{2 - x}$; Domain of $f \circ g = (-\infty, 4]$, Domain of $g \circ f = (-\infty, 2]$

37. $(f \circ g)(x) = (3x - 5)/(x - 2)$, $(g \circ f)(x) = 1/(x + 1)$; Domain of $f \circ g = (-\infty, 2) \cup (2, \infty)$, Domain of $g \circ f = (-\infty, -1) \cup (-1, \infty)$

39. $(f \circ g)(x) = 3|(x - 2)/(x - 3)|$, $(g \circ f)(x) = |x + 2|/(|x + 2| - 3)$; Domain of $f \circ g = (-\infty, 3) \cup (3, \infty)$, Domain of $g \circ f = (-\infty, -5) \cup (-5, 1) \cup (1, \infty)$

41. $y = |x + 1| - 2$ **43.** $y = 3 - \sqrt[3]{x}$ **45.** $y = 1 - (x + 2)^3$

47. $y = \sqrt{x + 2} + 3$ **49.** $y = -|x - 3|$ **51.** $y = -(x + 2)^3 + 1$

53. Reversing the order does not change the result. **55.** Reversing the order can change the result.

57. Reversing the order does not change the result.

59. $h(x) = (f \circ g)(x)$; $f(x) = x^4$, $g(x) = 2x - 7$ **61.** $h(x) = (f \circ g)(x)$; $f(x) = x^{1/2}$, $g(x) = 4 + 2x$

63. $h(x) = (g \circ f)(x)$; $g(x) = 3x - 5$, $f(x) = x^7$ **65.** $h(x) = (g \circ f)(x)$; $g(x) = 4x + 3$, $f(x) = x^{-1/2}$

67. The graph of $y = |x|$ is reflected in the x axis and vertically expanded by a factor of 3. Equation: $y = -3|x|$.

69. The graph of $y = x^3$ is reflected in the x axis and vertically contracted by a factor of 0.5. Equation: $y = -0.5x^3$.

75. $(f + g)(x) = 2x$, $(f - g)(x) = 2/x$, $(fg)(x) = x^2 - (1/x^2)$, $(f/g)(x) = (x^2 + 1)/(x^2 - 1)$;
Domain of $f + g$ = Domain of $f - g$ = Domain of $fg = (-\infty, 0) \cup (0, \infty)$, Domain of $f/g = (-\infty, -1) \cup (-1, 0) \cup (0, 1) \cup (1, \infty)$

77. $(f + g)(x) = 2$, $(f - g)(x) = -2x/|x|$, $(fg)(x) = 0$, $(f/g)(x) = 0$;
Domain of $f + g$ = Domain of $f - g$ = Domain of $fg = (-\infty, 0) \cup (0, \infty)$, Domain of $f/g = (0, \infty)$

79. $(f \circ g)(x) = 9 - x$, $(g \circ f)(x) = \sqrt{9 - x^2}$; Domain of $f \circ g = (-\infty, 9]$, Domain of $g \circ f = [-3, 3]$

81. $(f \circ g)(x) = (3x + 1)/(8 - x)$, $(g \circ f)(x) = (4x - 3)/(7 - x)$; Domain of $f \circ g = (-\infty, 3) \cup (3, 8) \cup (8, \infty)$,
Domain of $g \circ f = (-\infty, 2) \cup (2, 7) \cup (7, \infty)$

83. $(f \circ g)(x) = \sqrt{x^2 + 1}$, $(g \circ f)(x) = \sqrt{x^2 + 1}$; Domain of $f \circ g = (-\infty, -2] \cup [2, \infty)$, Domain of $g \circ f = (-\infty, \infty)$

85. $P(p) = -70,000 + 6,000p - 200p^2$

87.

89.

91. (A) $r(h) = \frac{1}{2}h$ (B) $V(h) = \frac{1}{12}\pi h^3$ (C) $V(t) = \frac{0.125}{12}\pi t^{3/2}$

Exercise 3-6

1. One-to-one **3.** Not one-to-one **5.** One-to-one **7.** Not one-to-one **9.** One-to-one **11.** Not one-to-one

13. One-to-one **15.** One-to-one **17.** One-to-one **19.** Not one-to-one **21.** One-to-one

23. One-to-one **25.** Not one-to-one **27.** Not one-to-one **29.** One-to-one

31. Domain of $f^{-1} = [1, 5]$ **33.** Domain of $f^{-1} = [-3, 5]$ **35.**   **37.**
Range of $f^{-1} = [-4, 4]$ Range of $f^{-1} = [-5, 3]$

39.

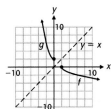

41. $f^{-1}(x) = 5x$ **43.** $f^{-1}(x) = \frac{1}{2}x - \frac{7}{2}$ **45.** $f^{-1}(x) = 5x - 2$ **47.** $f^{-1}(x) = 2/(3 - x)$

49. $f^{-1}(x) = x/(2 - x)$ **51.** $f^{-1}(x) = (5x + 4)/(2 - x)$ **53.** $f^{-1}(x) = 0.5\sqrt[3]{x + 5}$ **55.** $f^{-1}(x) = \frac{1}{3}(x - 2)^5 + \frac{7}{3}$
57. $f^{-1}(x) = 9 - \frac{1}{4}x^2, x \geq 0$ **59.** $f^{-1}(x) = -x^2 + 4x - 1, x \geq 2$
61. The x intercept of f is the y intercept of f^{-1}, and the y intercept of f is the x intercept of f^{-1}.
63. $f^{-1}(x) = 1 + \sqrt{x - 2}$ **65.** $f^{-1}(x) = -1 - \sqrt{x + 3}$
67. $f^{-1}(x) = \sqrt{9 - x^2}$ **69.** $f^{-1}(x) = -\sqrt{9 - x^2}$ **71.** $f^{-1}(x) = \sqrt{2x - x^2}$ **73.** $f^{-1}(x) = -\sqrt{2x - x^2}$
Domain of $f^{-1} = [-3, 0]$ Domain of $f^{-1} = [0, 3]$ Domain of $f^{-1} = [1, 2]$ Domain of $f^{-1} = [0, 1]$
Range of $f^{-1} = [0, 3]$ Range of $f^{-1} = [-3, 0]$ Range of $f^{-1} = [0, 1]$ Range of $f^{-1} = [-1, 0]$

      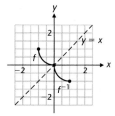

75. $f^{-1}(x) = (x - b)/a$ **77.** $a = 1$ and $b = 0$ or $a = -1$ and b arbitrary   **81.** (A) $f^{-1}(x) = 2 - \sqrt{x}$ (B) $f^{-1}(x) = 2 + \sqrt{x}$
83. (A) $f^{-1}(x) = 2 - \sqrt{4 - x^2}, 0 \leq x \leq 2$ (B) $f^{-1}(x) = 2 + \sqrt{4 - x^2}, 0 \leq x \leq 2$

Chapter 3 Review Exercise

1. (A) $\sqrt{45}$ (B) $-\frac{1}{2}$ (C) 2 *(3-1, 3-2)*   **2.** (A) $x^2 + y^2 = 7$ (B) $(x - 3)^2 + (y + 2)^2 = 7$ *(3-1)*
3. Center: $(-3, 2)$; radius $= \sqrt{5}$ *(3-1)*   **4.** Slope $= -\frac{3}{2}$ *(3-2)*   **5.** $2x + 3y = 12$ *(3-2)*   **6.** $y = -\frac{2}{3}x + 2$ *(3-2)*

7. Vertical: $x = -3$, slope not defined; horizontal: $y = 4$, slope $= 0$ *(3-2)*
8. (A) Function; domain $= \{1, 2, 3\}$, range $= \{1, 4, 9\}$ (B) Not a function
   (C) Function; domain $= \{-2, -1, 0, 1, 2\}$, range $= \{2\}$ *(3-3)*
9. (A) Not a function   (B) A function   (C) A function   (D) Not a function   *(3-3)*
10. Parts A and C specify functions.   *(3-3)*   **11.** 16 *(3-3)*   **12.** 1 *(3-3)*   **13.** 3 *(3-3)*   **14.** $-2a - h$ *(3-3)*
15. $9 + 3x - x^2$ *(3-5)*   **16.** $1 + 3x + x^2$ *(3-5)*   **17.** $20 + 12x - 5x^2 - 3x^3$ *(3-5)*   **18.** $(3x + 5)/(4 - x^2), x \neq \pm 2$ *(3-5)*
19. $17 - 3x^2$ *(3-5)*   **20.** $-21 - 30x - 9x^2$ *(3-5)*
21. (A)   (B)   (C)   (D)

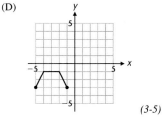

(3-5)

22. (A) g (B) m (C) n (D) f *(3-4, 3-5)*
23. (A) x intercepts: $-4, 0$; y intercept: 0 (B) Vertex: $(-2, -4)$ (C) Minimum: -4 (D) Range: $y \geq -4$ or $[-4, \infty)$
   (E) Increasing on $[-2, \infty)$ (F) Decreasing on $(-\infty, -2]$ *(3-4)*
24. Min $f(x) = f(3) = 2$; vertex: $(3, 2)$ *(3-4)*
25. (A) Reflected across x axis   (B) Shifted down 3 units   (C) Shifted left 3 units   *(3-5)*
26. (A) 0 (B) 1 (C) 2 (D) 0 *(3-4)*   **27.** (A) $-2, 0$ (B) 1 (C) No solution   (D) $x = 3$ and $x < -2$ *(3-4)*
28. Domain $= (-\infty, \infty)$, range $= (-3, \infty)$ *(3-4)*   **29.** $[-2, -1], [1, \infty)$ *(3-4)*   **30.** $[-1, 1)$ *(3-4)*   **31.** $(-\infty, -2)$ *(3-4)*
32. $x = -2, x = 1$ *(3-4)*   **33.** $f(x) = 4x^3 - \sqrt{x}$ *(3-3)*
34. The function f multiplies the square of the domain element by 3, adds 4 times the domain element, and then subtracts 6.   *(3-3)*
35. (A) $3x + 2y = -6$ (B) $\sqrt{52}$ *(3-1, 3-2)*   **36.** (A) $y = -2x - 3$ (B) $y = \frac{1}{2}x + 2$ *(3-2)*
37. It is symmetric with respect to all three.   *(3-1)*   **38.** $(-\infty, 3)$ *(3-3)*

39. Range = $[-4, \infty)$ *(3-4)*
Intercepts: $x = 1$ and $x = 5$, $y = 5$
Min $f(x) = f(3) = -4$

40. $[0, 16) \cup (16, \infty)$ *(3-3)*

Axis
$x = 3$

Vertex
$(3, -4)$

41. (A) $(f \circ g)(x) = \sqrt{|x|} - 8$, $(g \circ f)(x) = |\sqrt{x} - 8|$ (B)   Domain of $f \circ g = (-\infty, \infty)$, domain of $g \circ f = [0, \infty)$ *(3-5)*

42. Functions (A), (C), and (D) are one-to-one   *(3-6)*   **43.** (A) $(x + 7)/3$ (B) 4   (C) x (D) Increasing   *(3-6)*

44. Domain = $[-1, 1]$ *(3-4)*
Range = $[0, 1] \cup (2, 3]$
Discontinuous at $x = 0$

45. The graph of $y = x^2$ is vertically expanded by a factor of 2, reflected in the x axis, and shifted to the left 3 units.
Equation: $y = -2(x + 3)^2$.   *(3-5)*

46. $g(x) = 5 - 3|x - 2|$ *(3-5)*   **47.** $y = -(x - 4)^2 + 3$ *(3-4, 3-5)*

48. (A)

(B)

(C)   *(3-5)*

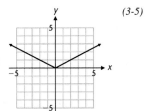

49. (A) $f^{-1}(x) = x^2 + 1$
(B) Domain of $f = [1, \infty) = $ Range of f^{-1}
Range of $f = [0, \infty) = $ Domain of f^{-1}
(C)   *(3-6)*

f^{-1}
f
$y = x$

50. $(x - 3)^2 + y^2 = 32$ *(3-1)*   **51.** Center: $(-2, 3)$, radius = 4   *(3-1)*

52. Symmetric with respect to the origin *(3-1)*   **53.** Decreasing *(3-2, 3-3)*

54. (A) Domain of $f = [0, \infty)$ = Range of f^{-1}; Range of $f = [-1, \infty)$ = Domain of f^{-1} (B) $\sqrt{x + 1}$ (C) 2   (D) 4
    (E) x *(3-6)*

55. The graph of $y = \sqrt[3]{x}$ is vertically expanded by a factor of 2, reflected in the x axis, shifted 1 unit left and 1 unit down.
    Equation $y = -2\sqrt[3]{x + 1} - 1$ *(3-4)*

56. It is the same as the graph of g shifted to the right 2 units, reflected in the x axis, and shifted down 1 unit. *(3-5)*

57.   *(3-5)*   **58.** $[-5, 5]$ *(2-8, 3-3)*

59. (A) $x^2\sqrt{1 - x}$, domain $= (-\infty, 1]$ (B) $x^2/\sqrt{1 - x}$, domain $= (-\infty, 1)$
    (C) $1 - x$, domain $= (-\infty, 1]$ (D) $\sqrt{1 - x^2}$, domain $= [-1, 1]$ *(3-5)*

60. (A) $(3x + 2)/(x - 1)$ (B) $\frac{11}{2}$ (C) x *(3-6)*

61. $f(x) = \begin{cases} -2 & \text{if } x < -1 \\ 2x & \text{if } -1 \le x < 1 \\ 2 & \text{if } x \ge 1 \end{cases}$; domain $= (-\infty, \infty)$, range $= [-2, 2]$ *(3-4)*

62. $x - y = 3$; a line *(3-1, 3-2)*   **65.** (A)   (B)   *(3-5)*

66. Domain: All real numbers except $x = 2$,   **67.** (A)   (B)   *(3-1, 3-5)*
    Range: $y > -3$ or $(-3, \infty)$,
    discontinuous at $x = 2$ *(3-4)*

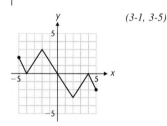

68. (A) The graph must cross the x axis exactly once.   (B) The graph can cross the x axis either once or not at all. *(3-4)*

69. (A) $V = -1,250t + 12,000$ (B) $5,750$ *(3-2)*   **70.** (A) $R = 1.6C$ (B) 168 *(3-2)*

71. $E(x) = \begin{cases} 200 & \text{if } 0 \le x \le 3,000 \\ 0.1x - 100 & \text{if } x > 3,000 \end{cases}$; $E(2,000) = 200$, $E(5,000) = 400$ *(3-4)*

72. (A)

x	0	5	10	15	20
Consumption	309	276	271	255	233
$f(x)$	303	286	269	252	234

(B)

(C) 217 in 1995, 200 in 2000   (D) Per capita egg consumption is dropping about 17 eggs every 5 years. *(3-4)*

73. (A) $C(x) = \begin{cases} 0.49x & \text{for } 0 \le x < 36 \\ 0.44x & \text{for } 36 \le x < 72 \\ 0.39x & \text{for } 72 \le x \end{cases}$ (B) Discontinuous at $x = 36$ and $x = 72$ *(3-4)*

74. (A) $C = 84{,}000 + 15x$; $R = 50x$ (B) $R = C$ at $x = 2{,}400$ units; $R < C$ for $0 \le x < 2{,}400$; $R > C$ for $x > 2{,}400$ *(3-2)*

75. $P(p) = -14{,}000 + 700p - 10p^2$ *(3-5)*   **76.** 5 ft   *(3-1)*
77. (A) $A(x) = 60x - \frac{3}{2}x^2$ (B) $0 < x < 40$ (C) $x = 20, y = 15$ *(3-4)*
78. (A) 0   (B) 1   (C) 2   (D) 0   (E) 1   (F) 0   *(3-4)*

CHAPTER 4   Exercise 4-1

1. c **3.** d **5.** h **7.** h, k **9.** $a - 2$; $R = 0$ **11.** $b - 3$; $R = -18$ **13.** $x^2 + 3$; $R = 5$ **15.** $4y^2 - 4y + 1$; $R = 0$

17. $x + 7 + \dfrac{6}{x - 3}$ **19.** $3x - 7 + \dfrac{7}{x + 2}$ **21.** $2x^2 - 3x + 1 - \dfrac{2}{x + 3}$ **23.** 4   **25.** -7 **27.** -2

29. $2x^4 + 2x^3 + 2x^2 - 3x - 3$; $R = 0$ **31.** $x^3 - 4x^2 + 16x - 64$; $R = 240$ **33.** $4x^3 + 3x^2 + x + 1$; $R = -4$

35. $x^5 + 2x^4 - x$; $R = 0$ **37.** $2x^3 + 6x^2 - 4x + 2$; $R = 0$ **39.** $3x^3 + 6x^2 - 3x + 9$; $R = 2$
41. $5x^3 - 3x^2 - 0.6x + 1.88$; $R = -4.624$ **43.** $5x^4 - x^3 + 4.6x^2 + 3.24x - 1.944$; $R = -4.8336$
45. The graph has three x intercepts and two turning points;   **47.** The graph has three x intercepts and two turning points;
$P(x) \to \infty$ as $x \to \infty$ and $P(x) \to -\infty$ as $x \to -\infty$ $P(x) \to \infty$ as $x \to \infty$ and $P(x) \to -\infty$ as $x \to -\infty$

49. The graph has one x intercept and two turning points;   **51.** The graph has one x intercept and no turning points;
$P(x) \to -\infty$ as $x \to \infty$ and $P(x) \to \infty$ as $x \to -\infty$ $P(x) \to -\infty$ as $x \to \infty$ and $P(x) \to \infty$ as $x \to -\infty$

53. $P(x) = x^3$ **55.** No such polynomial exists.   **57.** $x^2 + x + 2$; $R = 0$ **59.** $x^2 - 2x + 1$; $R = x + 3$
61. $x^3 + (2 + i)x^2 + (-3 + 2i)x - 3i$; $R = 0$ **63.** (A) -5 (B) $-40i$ (C) 0   (D) 0

65. The graph has two x intercepts and one turning point;
$P(x) \to \infty$ as $x \to \infty$ and as $x \to -\infty$

67. The graph has two x intercepts and three turning points;
$P(x) \to \infty$ as $x \to \infty$ and as $x \to -\infty$

69. The graph has four x intercepts and three turning points;
$P(x) \to -\infty$ as $x \to \infty$ and as $x \to -\infty$

71. The graph has five x intercepts and four turning points;
$P(x) \to \infty$ as $x \to \infty$ and $P(x) \to -\infty$ as $x \to -\infty$

73. (A) In both cases, the coefficient of x is a_2, the constant term is $a_2 r + a_1$, and the remainder is $(a_2 r + a_1)r + a_0$.
(B) The remainder expanded is $a_2 r^2 + a_1 r + a_0 = P(r)$.
75. $P(-2) = 81$; $P(1.7) = 6.2$

Exercise 4-2

1. -2 (multiplicity 5), 3 (multiplicity 4); degree of $P(x)$ is 9 **3.** -1, 1 (multiplicity 4), 7 (multiplicity 3); degree of $P(x)$ is 8
5. $P(x) = (x - 2)^3(x + 1)$; degree 4 **7.** $P(x) = x(x - 2)(x + 1 - \sqrt{3})(x + 1 + \sqrt{3})$; degree 4
9. $P(x) = (x + 2)^2(x - 2)^2(x - 3 + i)(x - 3 - i)$; degree 6 **11.** $(x + 2)(x - 1)(x - 3)$, degree 3 **13.** $(x + 2)^2(x - 1)^2$, degree 4
15. $(x + 3)(x + 2)x(x - 1)(x - 2)$, degree 5 **17.** Yes **19.** No **21.** $\pm 1, \pm 2, \pm 5, \pm 10$ **23.** $\pm 1, \pm 5, \pm \frac{1}{2}, \pm \frac{5}{2}$
25. $\pm 1, \pm 5, \pm 25, \pm \frac{1}{2}, \pm \frac{5}{2}, \pm \frac{25}{2}, \pm \frac{1}{3}, \pm \frac{5}{3}, \pm \frac{25}{3}, \pm \frac{1}{6}, \pm \frac{5}{6}, \pm \frac{25}{6}$ **27.** $P(x) = (x + 4)^2(x + 1)$ **29.** $P(x) = (x - 1)(x + 1)(x - i)(x + i)$
31. $P(x) = (2x - 1)[x - (4 + 5i)][(x - (4 - 5i)]$ **33.** $\frac{3}{2}, 1 \pm \sqrt{3}$ **35.** -1, (double root), $\pm\sqrt{3}$ **37.** $\pm 3, -1 \pm \sqrt{2}$
39. $-2, -\frac{1}{2}, 1, 1 \pm i$ **41.** $-4, -3, 2$ **43.** $-0.2, 0, 1.5, 2$ **45.** 3 (double zero), $-2 \pm \sqrt{3}$
47. -1 (double zero), $\frac{2}{3}, 1 \pm 2i$ **49.** $P(x) = (2x + 3)(3x - 1)(x + 2)$ **51.** $P(x) = (x + 3)[x - (2 + \sqrt{5})][x - (2 - \sqrt{5})]$
53. $P(x) = (x + 2)(2x + 1)(2x - 3)(x - 2)$ **55.** Inequality notation: $2 - \sqrt{3} \le x \le 2 + \sqrt{3}$; interval notation: $[2 - \sqrt{3}, 2 + \sqrt{3}]$
57. Inequality notation: $x \le -1$ or $1 \le x \le 3$; interval notation: $(-\infty, -1] \cup [1, 3]$
59. Inequality notation: $-3 \le x \le \frac{1}{2}$ or $x \ge 2$; interval notation: $[-3, \frac{1}{2}] \cup [2, \infty)$
61. $x^2 - 8x + 41$ **63.** $x^2 - 2ax + a^2 + b^2$ **65.** $1 - 2i, -2$ **67.** $3i, -4$ **69.** $3 + 2i, 1 \pm \sqrt{2}$
71. Inequality notation: $-\frac{5}{2} < x < -1$ or $x > 1$; interval notation: $(-\frac{5}{2}, -1) \cup (1, \infty)$
73. Inequality notation: $x \le -2$ or $-1 < x < 2$ or $3 < x \le 5$; interval notation: $(-\infty, -2] \cup (-1, 2) \cup (3, 5]$
75. $\frac{1}{3}, 6 \pm 2\sqrt{3}$ **77.** $-\frac{5}{2}, \frac{3}{2}, \pm 4i$ **79.** $\frac{3}{2}$ (double zero), $4 \pm \sqrt{6}$ **81.** (A) 3 (B) $-\frac{1}{2} - (\sqrt{3}/2)i, -\frac{1}{2} + (\sqrt{3}/2)i$
83. Max. $= n$, min. $= 1$ **85.** No, since $P(x)$ is not a polynomial with real coefficients (the coefficient of x is the imaginary number $2i$)
87. 2 ft **89.** 0.5 by 0.5 in or 1.59 by 1.59 in

Exercise 4-3

1. There is at least one x intercept in each of the intervals $(-5, -1)$, $(-1, 3)$, and $(5, 8)$
3. There is at least one x intercept in each of the intervals $(-6, -4)$, $(-4, 0)$, $(2, 4)$, and $(4, 7)$ **5.** Zeros in $(0, 1)$, $(3, 4)$, and $(4, 5)$
7. Zeros in $(-3, -2)$, $(-2, -1)$, and $(1, 2)$ **9.** Upper bound: 2; lower bound: -2
11. Upper bound: 3; lower bound: -2 **13.** Upper bound: 2; lower bound: -3
15. (A) Upper bound: 4; lower bound: -2; real zeros in $(-2, -1)$, $(0, 1)$, and $(3, 4)$ (B) 3.2
17. (A) Upper bound: 3; lower bound: -2; real zero in $(-2, -1)$ (B) -1.4
19. (A) Upper bound: 4; lower bound: -3; real zeros in $(-3, -2)$, $(-1, 0)$, $(1, 2)$, and $(3, 4)$ (B) 3.1
21. (A) Upper bound: 3; lower bound: -2; real zeros in $(-2, -1)$ and $(-1, 0)$ (B) -0.5
23. (A) Upper bound: 3; lower bound: -1 (B) 2.25 **25.** (A) Upper bound: 3; lower bound: -4 (B) $-3.51, 2.12$
27. (A) Upper bound: 2; lower bound: -3 (B) $-2.09, 0.75, 1.88$ **29.** (A) Upper bound: 1; lower bound: -1 (B) 0.83
31. (A) Upper bound: 5; lower bound: -2; real zeros in $(-2, -1)$, $(1, 2)$, and $(3, 4)$ (B) 3.22

Exercise 5-2

1.

x	y
−3	−0.05
−2	−0.14
−1	−0.37
0	−1
1	−2.72
2	−7.39
3	−20.09

3.

x	y
−5	3.68
−4	4.49
−3	5.49
−2	6.7
−1	8.19
0	10
1	12.21
2	14.92
3	18.22
4	22.26
5	27.18

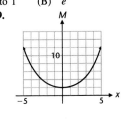

5.

t	$f(t)$
−5	164.87
−4	149.18
−3	134.99
−2	122.14
−1	110.52
0	100
1	90.48
2	81.87
3	74.08
4	67.03
5	60.65

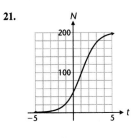

7. e^{-x} **9.** e^{3x} **11.** e^{3x-1} **13.** (A) $1 + 1/m$ is not equal to 1   (B) e

15.

17.

19.

21.

23. $\dfrac{e^{-2x}(-2x - 3)}{x^4}$ **25.** $2e^{2x} + 2e^{-2x}$ **27.** $2e^{2x}$ **29.** $x = 0$ **31.** $x = 0, 5$ **33.**

35. (A)

s	$f(s)$	s	$f(s)$
−0.5	4.0000	0.5	2.2500
−0.2	3.0518	0.2	2.4883
−0.1	2.8680	0.1	2.5937
−0.01	2.7320	0.01	2.7048
−0.001	2.7196	0.001	2.7169
−0.0001	2.7184	0.0001	2.7181

(B) $2.718 \cdots \approx e$ **37.**   **39.**

41. As $x \to \infty$, $f_n(x) \to 0$; as $x \to -\infty$, $f_n(x) \to \infty$ if n is even, and $f_n(x) \to -\infty$ if n is odd; $y = 0$ is a horizontal asymptote.

43. 7.1 billion   **45.** 2006

47.

49. (A) 62%   (B) 39%   **51.** (A) $10,691.81   (B) $36,336.69

53. Gill Savings: $1,230.60; Richardson S&L: $1,231.00; USA Savings: $1,229.03   **55.** $12,197.09

57. (A) 15 million   (B) 30 million

59. 40 boards **61.** 50°F **63.** 0.0009 coulomb **65.** 100 deer

67.

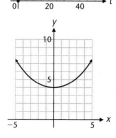

Exercise 5-3

1. $2^6 = 64$ **3.** $10^5 = 100{,}000$ **5.** $9^{1/2} = 3$ **7.** $\left(\frac{1}{4}\right)^{-3} = 64$ **9.** $\log_{10} 0.001 = -3$ **11.** $\log_8 4 = \frac{2}{3}$
13. $\log_{81}\left(\frac{1}{3}\right) = -\frac{1}{4}$ **15.** $\log_{125} 5 = \frac{1}{3}$ **17.** 1 **19.** 0 **21.** 3 **23.** $\frac{1}{2}$ **25.** -3 **27.** x^2 **29.** $(x-1)^3$
31. 8 **33.** $\frac{1}{25}$ **35.** 4 **37.** 64 **39.** -3 **41.** 4 **43.** 100 **45.** $6\log_b x + 9\log_b y$
47. $\frac{3}{2}\log_b u - \frac{5}{3}\log_b v$ **49.** $\log_b m + \log_b n - \log_b p - \log_b q$ **51.** $-4\log_b a$ **53.** $\frac{1}{2}\log_b(c^2 + d^2)$
55. $\frac{1}{2}\log_b u - \log_b v - 2\log_b w$ **57.** $\frac{2}{3}\log_b x + \frac{1}{6}\log_b y - \log_b z$ **59.** $\log_b(x^2/y)$ **61.** $\log_b(w/xy)$
63. $\log_b(x^3 y^2 / \sqrt[4]{z})$ **65.** $\log_b(\sqrt{u}/v^2)^5$ **67.** $\log_b \sqrt[5]{x^2 y^3}$ **69.** $5\log_b(x+3) + 2\log_b(2x-7)$
71. $7\log_b(x+10) - 2\log_b(1+10x)$ **73.** $2\log_b x - \frac{1}{2}\log_b(x+1)$ **75.** $2\log_b x + \log_b(x+5) + \log_b(x-4)$ **77.** $x = 4$
79. $x = \frac{1}{3}$ **81.** $x = \frac{8}{7}$ **83.** $x = 2$ **85.** $x = 2$ **87.** 3.40 **89.** -0.92 **91.** 3.30 **93.** 0.23 **95.** -0.05
97.

99.

101. (A)

(B) Domain $f = (-\infty, \infty) = $ Range f^{-1}
Range $f = (0, \infty) = $ Domain f^{-1}
(C) $f^{-1}(x) = \log_{1/2} x = -\log_2 x$

103. $f^{-1}(x) = \frac{1}{3}[1 + \log_5(x-4)]$ **105.** $g^{-1}(x) = \frac{1}{5}(e^{x/3} + 2)$ **107.** The reflection is not a function, since $y = 3^{x^2}$ is not one-to-one.
109. $x = 100e^{-0.08t}$

Exercise 5-4

1. 4.4408 **3.** -2.3644 **5.** -7.3324 **7.** 6.1242 **9.** 32.45 **11.** 0.1039 **13.** 0.055 68 **15.** 3,407 **17.** 1.238
19. 2.320 **21.** -51.083 **23.** 5.192 **25.** 35.779 **27.** -12.169 **29.** 4.6505×10^{21} **31.** 1.4925×10^{-5}

33.   **35.**   **37.**   **39.**

41. The inequality sign in the last step reverses because $\log \frac{1}{3}$ is negative.   **43.** (B) Domain $= (1, \infty)$; range $= (-\infty, \infty)$

45. $(0.90, -0.11), (38.51, 3.65)$ **47.** $(6.41, 1.86), (93.35, 4.54)$ **49.**   **51.**

53. (A) 0 decibel   (B) 120 decibels   **55.** 30 decibels more   **57.** 8.6   **59.** 1,000 times more powerful   **61.** 7.67 km/s

63. (A) 8.3, basic   (B) 3.0, acid   **65.** 6.3×10^{-6} mole/liter

Exercise 5-5

1. 1.44   **3.** 5.88   **5.** -1.68 **7.** 0.0967   **9.** 3.24   **11.** -3.43 **13.** $\frac{5}{2}$ **15.** $8/e^2$ **17.** $x = 10$ **19.** 86.7

21. -921 **23.** 5.18   **25.** ± 4.01 **27.** $x = 5$ **29.** $x = 2 + \sqrt{3}$ **31.** $x = \frac{1}{4}(1 + \sqrt{89})$ **33.** $x = 1, e^2, e^{-2}$

35. $x = e^e$ **37.** $x = 0.1, 100$ **39.** (B) 2   **41.** (B) $-1.252, 1.707$ **43.** 3.6776   **45.** -1.6094 **47.** -1.7372

49. $r = \dfrac{\ln (A/P)}{t}$ **51.** $I = I_0(10^{D/10})$ **53.** $I = I_0[10^{(6-M)/2.5}]$ **55.** $t = \dfrac{-L}{R} \ln \left(1 - \dfrac{RI}{E}\right)$ **57.** $x = \ln (y \pm \sqrt{y^2 - 1})$

59. $x = \dfrac{1}{2} \ln \dfrac{1 + y}{1 - y}$ **61.**   **63.**   **65.** 0.38   **67.** 0.55   **69.** 0.57

71. 0.85   **73.** 0.43   **75.** 0.27   **77.** Approx. 5 years   **79.** 9.16%   **81.** (A) 6   (B) 100 times brighter

83. Approx. 35 years   **85.** 18,600 years old   **87.** 7.52 s   **89.** $k = 0.40; t = 2.9$ h   **91.** 10 years

Chapter 5 Review Exercise

1. $\log m = n$ *(5-3)*   **2.** $\ln x = y$ *(5-3)*   **3.** $x = 10^y$ *(5-3)*   **4.** $y = e^x$ *(5-3)*   **5.** 7^{2x} *(5-1)*   **6.** e^{2x^2} *(5-1)*

7. $x = 8$ *(5-3)*   **8.** $x = 5$ *(5-3)*   **9.** $x = 3$ *(5-3)*   **10.** $x = 1.24$ *(5-3)*   **11.** $x = 11.9$ *(5-3)*   **12.** $x = 0.984$ *(5-3)*

13. $x = 103$ *(5-3)*   **14.** $x = 4$ *(5-3)*   **15.** $x = 2$ *(5-3)*   **16.** $x = -1, 3$ *(5-2)*   **17.** $x = 1$ *(5-1)*   **18.** $x = \pm 3$ *(5-2)*

19. $x = -2$ *(5-3)*   **20.** $x = \frac{1}{3}$ *(5-3)*   **21.** $x = 64$ *(5-3)*   **22.** $x = e$ *(5-3)*   **23.** $x = 33$ *(5-3)*   **24.** $x = 1$ *(5-3)*

25. 1.145 *(5-3)*   **26.** Not defined *(5-3)*   **27.** 2.211 *(5-3)*   **28.** 11.59 *(5-3)*   **29.** $x = 41.8$ *(5-1)*

30. $x = 1.95$ *(5-3)*   **31.** $x = 0.0400$ *(5-3)*   **32.** $x = -6.67$ *(5-3)*   **33.** $x = 1.66$ *(5-3)*   **34.** $x = 2.32$ *(5-5)*

35. $x = 3.92$ *(5-5)*   **36.** $x = 92.1$ *(5-5)*   **37.** $x = 2.11$ *(5-5)*   **38.** $x = 0.881$ *(5-5)*   **39.** $x = 300$ *(5-5)*

40. $x = 2$ *(5-5)*   **41.** $x = 1$ *(5-5)*   **42.** $x = \frac{1}{2}(3 + \sqrt{13})$ *(5-5)*   **43.** $x = 1, 10^3, 10^{-3}$ *(5-5)*   **44.** $x = 10^e$ *(5-5)*

45. $e^{-x} - 1$ *(5-2)*   **46.** $2 - 2e^{-2x}$ *(5-2)*

47. *(5-1)*   **48.** *(5-2)*   **49.** *(5-3)*

50.

(5-2) **51.** $y = -e^x$; $y = \left(\frac{1}{e}\right)^x = e^{-x}$ *(5-3)*

52. (A) $y = e^{-x/3}$ is decreasing, while $y = 4 \ln (x + 1)$ is increasing without bound. (B) 0.258 *(5-5)* **53.** 0.018, 2.187 *(5-3)*

54. (1.003, 0.010), (3.653, 4.502) *(5-4)* **55.** $I = I_0(10^{D/10})$ *(5-5)* **56.** $x = \pm\sqrt{-2 \ln (\sqrt{2\pi}y)}$ *(5-5)* **57.** $I = I_0(e^{-kx})$ *(5-5)*

58. $n = -\dfrac{\ln [1 - (Pi/r)]}{\ln (1 + i)}$ *(5-5)* **59.** $f^{-1}(x) = e^{x/2} + 1$ *(5-5, 3-7)* **60.** $f^{-1}(x) = \ln (x + \sqrt{x^2 + 1})$ *(5-5, 3-6)*

61. $y = ce^{-5t}$ *(5-3, 5-5)*

62. Domain $f = (0, \infty)$ = Range f^{-1} *(5-3)*
Range $f = (-\infty, \infty)$ = Domain f^{-1}

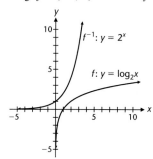

63. If $\log_1 x = y$, then we have $1^y = x$; that is, $1 = x$ for arbitrary positive x, which is impossible. *(5-3)* **65.** 23.4 years *(5-5)*

66. 23.1 years *(5-5)* **67.** 37,100 years *(5-5)* **68.** (A) $N = 2^{2t}$ or $N = 4^t$ (B) 15 days *(5-5)* **69.** $\$1.1 \times 10^{26}$ *(5-2)*

70. (A)

 (B) 0 *(5-2)* **71.** 6.6 *(5-4)* **72.** 7.08×10^{16} joules *(5-4)*

73. 50 decibels more *(5-4)* **74.** $k = 0.009\ 42$; 489 ft *(5-2)* **75.** 3 years *(5-5)*

Cumulative Review Exercise: Chapters 3–5

1. (A) $2\sqrt{5}$ (B) 2 (C) $-\frac{1}{2}$ *(3-1, 3-2)* **2.** (A) $x^2 + y^2 = 2$ (B) $(x + 3)^2 + (y - 1)^2 = 2$ *(3-1)*

3. Slope: $\frac{2}{3}$; y intercept: -2; x intercept: 3 *(3-2)*

4. (A) Function; domain: $\{1, 2, 3\}$; range: $\{1\}$ (B) Not a function
(C) Function; domain: $\{-2, -1, 0, 1, 2\}$; range: $\{-1, 0, 2\}$ *(3-3)*

5. (A) 20 (B) $x^2 + x + 3$ (C) $9x^2 - 18x + 13$ (D) $2a + h - 2$ *(3-2, 3-5)*

6. (A) $P(x) = (x + 1)^2(x - 1)(x - 2)$ (B) $P(x) \to \infty$ as $x \to \infty$ and as $x \to -\infty$ *(4-1)*

7. (A) Expanded by a factor of 2 (B) Shifted right 2 units (C) Shifted down 2 units *(3-5)*

8. (A)   (B)   *(3-5)*

9. $3x^3 + 5x^2 - 18x - 3 = (x + 3)(3x^2 - 4x - 6) + 15$ *(4-1)*   **10.** $-2, 3, 5$ *(4-2)*

11. $P(1) = -5$ and $P(2) = 5$ are of opposite sign *(4-3)*   **12.** $1, 2, -4$ *(4-2)*   **13.** $\dfrac{3}{x + 1} + \dfrac{2}{x - 2}$ *(4-5)*

14. (A) $x = \log y$ (B) $x = e^y$ *(5-4)*   **15.** (A) $8e^{3x}$ (B) e^{5x} *(5-2)*   **16.** (A) 9   (B) 4   (C) $\frac{1}{2}$ *(5-3)*

17. (A) 0.371   (B) 11.4   (C) 0.0562   (D) 15.6 *(5-4)*

18. (A) All real numbers   (B) $\{-2\} \cup [1, \infty)$ (C) 1   (D) $[-3, -2]$ and $[2, \infty)$ (E) $-2, 2$ *(3-3, 3-4)*

19. (A) $y = -\frac{3}{2}x - 8$ (B) $y = \frac{2}{3}x + 5$ *(3-2)*   **20.** $[-4, \infty)$ *(3-3)*

21. Range: $[-9, \infty)$ *(3-4)*
Intercepts: $x = -2$ and $x = 4$, $y = -8$
Min $f(x) = f(1) = -9$

22. $(f \circ g)(x) = \dfrac{x}{3 - x}$; Domain: $x \neq 0, 3$ *(3-5)*   **23.** $f^{-1}(x) = \frac{1}{2}x - \frac{5}{2}$ *(3-6)*   **24.** $f(x) = 3 \ln x - \sqrt{x}$ *(3-3, 5-3)*

25. The function f multiplies the base e raised to power of $\frac{1}{2}$ the domain element by 100 and then subtracts 50. *(3-3, 5-2)*

26. Domain: all real numbers *(3-4)*   **27.** (A)   (B)   *(3-5)*
Range: $(-\infty, -1) \cup [1, \infty)$
Discontinuous at $x = 0$

28. The graph of $y = |x|$ is contracted by $\frac{1}{2}$, reflected in the x axis, shifted 2 units to the right and 3 units up; $y = -\frac{1}{2}|x - 2| + 3$. *(3-5)*

29. (A) Domain: $x \neq -2$; x intercept: -4; y intercept: 4   **30.** (A) $f^{-1}(x) = x^2 - 4$, $x \geq 0$
   (B) Vertical asymptote: $x = -2$ (B) Domain $f = [-4, \infty) = $ Range f^{-1};
      Horizontal asymptote: $y = 2$ Range $f = [0, \infty) = $ Domain f^{-1}

   (C)   *(4-4)*   (C)   *(3-6)*

31. Center: $(3, -1)$; radius: $\sqrt{10}$ *(3-1)*   **32.** Symmetric with respect to the origin *(3-1)*   **33.** $y = (x + 2)^2 - 3$ *(3-4)*

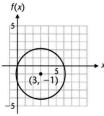

34. $P(\frac{1}{2}) = \frac{5}{2}$ *(4-2)*   **35.** (B) *(4-1)*

36. (A) The graph of $P(x)$ has four x intercepts and three turning points; $P(x) \to \infty$ as $x \to \infty$ and as $x \to -\infty$

(B) 2.8 *(4-1, 4-3)*

37. (A) Upper bound: 4; lower bound: -6 (B) 3.80   (C) $-5.68, 3.80$ *(4-3)*   **38.** $3, 1 \pm \frac{1}{2}i$ *(4-2)*

39. $-4, -1, \pm\sqrt{3}$; $P(x) = (x + 4)(x + 1)(x - \sqrt{3})(x + \sqrt{3})$ *(4-2)*   **40.** $x \le -2$ or $3 \le x \le 6$; $(-\infty, -2] \cup [3, 6]$ *(4-2)*

41. $\dfrac{1}{x} + \dfrac{2}{x + 1} - \dfrac{5}{(x + 1)^2}$ *(4-5)*   **42.** $-\dfrac{2}{x} + \dfrac{3x - 1}{x^2 - x + 1}$ *(4-5)*   **43.** $x = -2, 4$ *(5-1)*   **44.** $x = -1, \frac{1}{2}$ *(5-2)*

45. $x = 2.5$ *(5-3)*   **46.** $x = 10$ *(5-3)*   **47.** $x = \frac{1}{27}$ *(5-3)*   **48.** $x = 5$ *(5-5)*   **49.** $x = 7$ *(5-5)*   **50.** $x = 5$ *(5-5)*

51. $x = e^{0.1}$ *(5-4)*   **52.** $x = 1, e^{0.5}$ *(5-5)*   **53.** $x = 3.38$ *(5-5)*   **54.** $x = 4.26$ *(5-4)*   **55.** $x = 2.32$ *(5-4)*

56. $x = 3.67$ *(5-5)*   **57.** $x = 0.549$ *(5-5)*

58. *(5-1)*   **59.** *(5-4)*   **60.** *(5-2)*

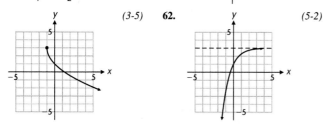

61. *(3-5)*   **62.** *(5-2)*

63. A reflection in the x axis transforms the graph of $y = \ln x$ into the graph of $y = -\ln x$. A reflection in the y axis transforms the graph of $y = \ln x$ into the graph of $y = \ln (-x)$. *(3-5, 5-3)*

64. (A) For $x > 0$, $y = e^{-x}$ decreases from 1 to 0 while $\ln x$ increases from $-\infty$ to ∞. Consequently, the graphs can intersect at exactly one point.

   (B) 1.31 *(5-3)*

65. $f(x) = \begin{cases} -2x & \text{if } x < -2 \\ 4 & \text{if } -2 \le x \le 2 \\ 2x & \text{if } x > 2 \end{cases}$

Domain: all real numbers; Range: $[4, \infty)$

(3-4)

66. (A) Domain g: $[-2, 2]$ (B) $\left(\dfrac{f}{g}\right)(x) = \dfrac{x^2}{\sqrt{4 - x^2}}$; Domain $\left(\dfrac{f}{g}\right)$: $(-2, 2)$ (C) $(f \circ g)(x) = 4 - x^2$; Domain $(f \circ g)$: $[-2, 2]$ *(3-5)*

67. (A) $f^{-1}(x) = 1 + \sqrt{x + 4}$ **68.** Vertical asymptote: $x = -2$
(B) Domain f^{-1}: $[-4, \infty)$;   Oblique asymptote: $y = x + 2$
Range f^{-1}: $[1, \infty)$

(C) *(3-6)*

(4-4)

69.

$f(x) = \begin{cases} \quad\vdots \\ 2x + 2 & \text{if } -1 \le x < -\frac{1}{2} \\ 2x + 1 & \text{if } -\frac{1}{2} \le x < 0 \\ 2x & \text{if } 0 \le x < \frac{1}{2} \\ 2x - 1 & \text{if } \frac{1}{2} \le x < 1 \\ 2x - 2 & \text{if } 1 \le x < \frac{3}{2} \\ 2x - 3 & \text{if } \frac{3}{2} \le x < 2 \\ \quad\vdots \end{cases}$

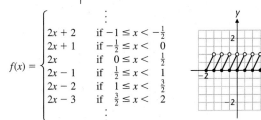

(3-4)

Domain: all real numbers; Range: $[0, 1)$; discontinuous at $x = k/2$, k an integer

70. $P(x) = (x + 1)^2 x^3 (x - 3 - 5i)(x - 3 + 5i)$; degree 7   *(4-2)*
71. Yes, for example: $P(x) = (x + i)(x - i)(x + \sqrt{2})(x - \sqrt{2}) = x^4 - x^2 - 2$ *(4-2)*
72. (A) Upper bound: 20; lower bound: -30 (B) $-26.68, -6.22, 7.23, 16.67$ *(4-3)*
73. -1 (double root), $2, 2 \pm i\sqrt{2}$; $P(x) = (x + 1)^2 (x - 2)(x - 2 - i\sqrt{2})(x - 2 + i\sqrt{2})$ *(4-2)*   **74.** -2 (double root)   *(4-3)*
75. $\dfrac{-2}{(x - 1)} + \dfrac{2}{(x - 1)^2} + \dfrac{2x + 3}{x^2 + x + 2}$ *(4-5)*   **76.** (A) $f^{-1}(x) = e^{x/3} + 2$ **77.** $n = \dfrac{\ln(1 + Ai/P)}{\ln(1 + i)}$ *(5-5)*
(B) Domain $f = (2, \infty) = $ Range f^{-1}
Range $f = $ Domain $f^{-1} = (-\infty, \infty)$

(C) *(3-6, 5-5)*

78. $y = Ae^{5x}$ *(5-5)*   **79.** $x = \ln(y + \sqrt{y^2 + 2})$ *(5-5)*   **80.** $x = -900p + 4{,}571$; 1,610 bottles   *(3-2)*

81. $C(x) = \begin{cases} 0.06x & \text{if } 0 \le x \le 60 \\ 0.05x + 0.6 & \text{if } 60 < x \le 150 \\ 0.04x + 2.1 & \text{if } 150 < x \le 300 \\ 0.03x + 5.1 & \text{if } 300 < x \end{cases}$

82. (A) $A(x) = 80x - 2x^2$ (B) $0 < x < 40$
(C) 20×40 ft

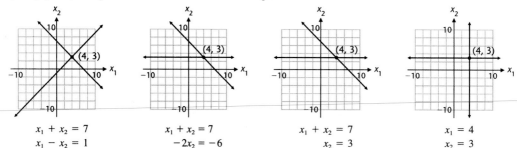

(3-4)

83. (A) $f(1) = f(3) = 1, f(2) = f(4) = 0$

(B) $f(n) = \begin{cases} 1 & \text{if } n \text{ is an odd integer} \\ 0 & \text{if } n \text{ is an even integer} \end{cases}$ *(3-4)*

84. $x = 2$ ft and $y = 2$ ft or $x = 1.3$ ft and $y = 4.8$ ft *(4-2)*   **85.** 1.8 by 3.3 ft *(4-3)*
86. (A) 46.8 million   (B) 103 million *(5-1)*   **87.** 10.2 years *(5-5)*   **88.** 9.90 years *(5-5)*
89. 63.1 times more powerful *(5-4)*   **90.** 6.31×10^{-4} W/m² *(5-4)*

CHAPTER 6   Exercise 6-1

1. B, no solution   **3.** D, $(1, -3)$ **5.** $(5, 2)$ **7.** $(2, -3)$ **9.** No solution (parallel lines)   **11.** $(2, -1)$ **13.** $(3, -1)$

15. $2 \times 3, 1 \times 3$ **17.** C **19.** B **21.** $-2, -6$ **23.** $-2, 6, 0$ **25.** $\begin{bmatrix} 4 & -6 & -8 \\ 1 & -3 & 2 \end{bmatrix}$ **27.** $\begin{bmatrix} -4 & 12 & -8 \\ 4 & -6 & -8 \end{bmatrix}$

29. $\begin{bmatrix} 1 & -3 & 2 \\ 8 & -12 & -16 \end{bmatrix}$ **31.** $\begin{bmatrix} 1 & -3 & 2 \\ 0 & 6 & -16 \end{bmatrix}$ **33.** $\begin{bmatrix} 1 & -3 & 2 \\ 2 & 0 & -12 \end{bmatrix}$ **35.** $\begin{bmatrix} 1 & -3 & 2 \\ 3 & -3 & -10 \end{bmatrix}$

37. $x_1 = 4, x_2 = 3$; each pair of lines has the same intersection point.

$x_1 + x_2 = 7$	$x_1 + x_2 = 7$	$x_1 + x_2 = 7$	$x_1 = 4$
$x_1 - x_2 = 1$	$-2x_2 = -6$	$x_2 = 3$	$x_2 = 3$

39. $x_1 = 2$ and $x_2 = 1$ **41.** $x_1 = 2$ and $x_2 = 4$ **43.** No solution   **45.** $x_1 = 1$ and $x_2 = 4$
47. Infinitely many solutions; for any real number s, $x_2 = s$, $x_1 = 2s - 3$
49. Infinitely many solutions; for any real number s, $x_2 = s$, $x_1 = \frac{1}{2}s + \frac{1}{2}$
51. $(1.12, 2.41)$ **53.** $(-2.24, -3.31)$ **55.** (A) $(-24, 20)$ (B) $(6, -4)$ (C) No solution   **57.** $(-23.125, 7.8125)$
59. $(3.225, -6.9375)$ **61.** 25 32¢ stamps, 50 23¢ stamps   **63.** \$107,500 in bond A and \$92,500 in bond B
65. 30 liters of 20% solution and 70 liters of 80% solution   **67.** 200 g of mix A and 80 g of mix B
69. Base price = \$17.95, surcharge = \$2.45/lb   **71.** 5,720 lb of the robust blend and 6,160 lb of the mild blend.

Exercise 6-2

1. No   **3.** Yes   **5.** No   **7.** Yes   **9.** $x_1 = -2, x_2 = 3, x_3 = 0$ **11.** $x_1 = 2t + 3, x_2 = -t - 5, x_3 = t$, t any real number

13. No solution   **15.** $x_1 = 2s + 3t - 5, x_2 = s, x_3 = -3t + 2, x_4 = t$, s and t any real numbers   **17.** $\begin{bmatrix} 1 & 0 & -7 \\ 0 & 1 & 3 \end{bmatrix}$

19. $\begin{bmatrix} 1 & 0 & 0 & -5 \\ 0 & 1 & 0 & 4 \\ 0 & 0 & 1 & -2 \end{bmatrix}$ **21.** $\begin{bmatrix} 1 & 0 & 2 & -\frac{5}{3} \\ 0 & 1 & -2 & \frac{1}{3} \\ 0 & 0 & 0 & 0 \end{bmatrix}$ **23.** $x_1 = -2, x_2 = 3, x_3 = 1$ **25.** $x_1 = 0, x_2 = -2, x_3 = 2$

27. $x_1 = 2t + 3, x_2 = t - 2, x_3 = t$, t any real number   **29.** $x_1 = 1, x_2 = 2$ **31.** No solution
33. $x_1 = 2t + 4, x_2 = t + 1, x_3 = t$, t any real number   **35.** $x_1 = s + 2t - 1, x_2 = s, x_3 = t$, s and t any real numbers
37. No solution   **39.** $x_1 = 2.5t - 4, x_2 = t, x_3 = -5$ for t any real number   **41.** $x_1 = 1, x_2 = -2, x_3 = 1$
43. (A) Dependent with two parameters   (B) Dependent with one parameter   (C) Independent   (D) Impossible
45. $x_1 = 2s - 3t + 3, x_2 = s + 2t + 2, x_3 = s, x_4 = t$ for s and t any real numbers   **47.** $x_1 = -0.5, x_2 = 0.2, x_3 = 0.3, x_4 = -0.4$

49. $x_1 = 2s - 1.5t + 1, x_2 = s, x_3 = -t + 1.5, x_4 = 0.5t - 0.5, x_5 = t$ for s and t any real numbers

51. 15¢ stamps: $3t - 100$, 20¢ stamps: $145 - 4t$, 35¢ stamps: t, where $t = 34, 35,$ or 36

53. 10% containers: $6t - 24$, 20% containers: $48 - 8t$, 50% containers: t, where $t = 4, 5,$ or 6

55. $a = 3, b = 2, c = 1$ **57.** $a = -2, b = -4, c = -20$ **59.** 20 one-person boats, 220 two-person boats, 100 four-person boats

61. One-person boats: $t - 80$, two-person boats: $-2t + 420$, four-person boats: $t, 80 \le t \le 210, t$ an integer

63. No solution; no production schedule will use all the labor-hours in all departments.   **65.** 8 oz food A, 2 oz food B, 4 oz food C

67. No solution   **69.** 8 oz food A, $-2t + 10$ oz food B, t oz food C, $0 \le t \le 5$ **71.** Company A: 10 h, company B: 15 h

Exercise 6-3

1. $(-12, 5), (-12, -5)$ **3.** $(2, 4), (-2, -4)$ **5.** $(5, -5), (-5, 5)$ **7.** $(4 + 2\sqrt{3}, 1 + \sqrt{3}), (4 - 2\sqrt{3}, 1 - \sqrt{3})$

9. $(2, 4), (2, -4), (-2, 4), (-2, -4)$ **11.** $(1, 3), (1, -3), (-1, 3), (-1, -3)$ **13.** $(1 + \sqrt{5}, -1 + \sqrt{5}), (1 - \sqrt{5}, -1 - \sqrt{5})$

15. $(\sqrt{2}, \sqrt{2}), (-\sqrt{2}, -\sqrt{2}), (2, 1), (-2, -1)$ **17.** $(2, 2i), (2, -2i), (-2, 2i), (-2, -2i)$ **19.** $(2, \sqrt{2}), (2, -\sqrt{2}), (-1, i), (-1, -i)$

21. $(3, 0), (-3, 0), (\sqrt{5}, 2), (-\sqrt{5}, 2)$ **23.** $(2, 1), (-2, -1), (i, -2i), (-i, 2i)$ **25.** $(-1, 4), (3, -4)$ **27.** $(0, 0), (3, 6)$

29. $(1, 4), (4, 1)$ **31.** $(-1, 3), (4, 8)$

33. (A)   The lines are tangent to the circle.   (B) $b = 5$, intersection point is $(2, -1)$; $b = -5$, intersection point is $(-2, 1)$

(C)   The line $x + 2y = 0$ is perpendicular to all the lines in the family and intersects the circle at the intersection points found in part B. Solving the system $x^2 + y^2 = 5$, $x + 2y = 0$ would determine the intersection points.

35. $(-5, -\frac{3}{5}), (-\frac{3}{2}, -2)$ **37.** $(0, -1), (-4, -3)$ **39.** $(2, 2), (-2, -2), (\sqrt{2}, -\sqrt{2}), (-\sqrt{2}, \sqrt{2})$

41. $(-3, 1), (3, -1), (-i, i), (i, -i)$ **43.** $(-1.41, -0.82), (-0.13, 1.15), (0.13, -1.15), (1.41, 0.82)$

45. $(-1.66, -0.84), (-0.91, 3.77), (0.91, -3.77), (1.66, 0.84)$ **47.** $(-2.96, -3.47), (-0.89, -3.76), (1.39, 4.05), (2.46, 4.18)$

49. $\frac{1}{2}(3 - \sqrt{5}), \frac{1}{2}(3 + \sqrt{5})$ **51.** 5 in. and 12 in.   **53.** 6 by 4.5 in.   **55.** 22 by 26 ft   **57.** Boat A: 30 mph; boat B: 25 mph

Exercise 6-4

1.

3.

5.

7.

9.

11. Region IV   **13.** Region I   **15.**

17.

19.

21. Region IV; corner points $(6, 4), (8, 0), (18, 0)$ **23.** Region I, corner points: $(0, 16), (6, 4), (18, 0)$

25. Corner points: $(0, 0)$, $(3, 0)$, $(0, 2)$
Bounded

27. Corner points: $(5, 0)$, $(0, 4)$
Unbounded

29. Corner points: $(0, 0)$, $(0, 4)$, $(\frac{12}{5}, \frac{16}{5})$, $(4, 0)$
Bounded

31. Corner points: $(0, 8)$, $(3, 4)$, $(9, 0)$
Unbounded

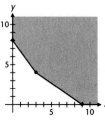

33. Corner points: $(0, 0)$, $(0, 5)$, $(4, 3)$, $(5, 2)$, $(6, 0)$
Bounded

35. Corner points: $(0, 14)$, $(2, 10)$, $(8, 4)$, $(16, 0)$
Unbounded

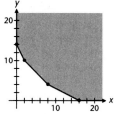

37. Corner points: $(2, 5)$, $(10, 1)$, $(1, 10)$
Bounded

39. The feasible region is empty.

41. Corner points: $(0, 3)$, $(5, 0)$, $(7, 3)$, $(2, 8)$
Bounded

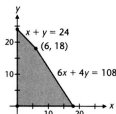

43. Corner points: $(1.27, 5.36)$, $(2.14, 6.52)$, $(5.91, 1.88)$ Bounded

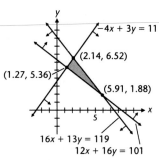

45. $6x + 4y \leq 108$
$x + y \leq 24$
$x \geq 0$
$y \geq 0$

47. (A) All production schedules in the feasible region that are on the graph of $50x + 60y = 1{,}100$ will result in a profit of $\$1{,}100$.

(B) There are many possible choices. For example, producing 5 trick and 15 slalom skis will produce a profit of $\$1{,}150$. The graph of the line $50x + 60y = 1{,}150$ includes all the production schedules in the feasible region that result in a profit of $\$1{,}150$.

49. $20x + 10y \geq 460$
$30x + 30y \geq 960$
$5x + 10y \geq 220$
$x \geq 0$
$y \geq 0$

(14, 18)
(20, 12)

51. $10x + 30y \geq 280$
$30x + 10y \geq 360$
$x \geq 0$
$y \geq 0$

(10, 6)
$10x + 30y = 280$

Exercise 6-5

1. Max $z = 16$ at $(7, 9)$ **3.** Max $z = 84$ at $(0, 12)$ and $(7, 9)$ (multiple optimal solutions)   **5.** Min $z = 32$ at $(0, 8)$

7. Min $z = 36$ at $(12, 0)$ and $(4, 3)$ (multiple optimal solutions)   **9.** Max $z = 18$ at $(4, 3)$ **11.** Min $z = 12$ at $(4, 0)$

13. Max $z = 52$ at $(4, 10)$ **15.** Min $z = 44$ at $(4, 4)$

17. Min $z = 1,500$ at $(60, 0)$; max $z = 3,000$ at $(60, 30)$ and $(120, 0)$ (multiple optimal solutions)

19. Min $z = 300$ at $(0, 20)$; max $z = 1,725$ at $(60, 15)$ **21.** Max $P = 5,507$ at $x = 6.62$ and $y = 4.25$

23. (A) $a > 2b$ (B) $\frac{1}{3}b < a < 2b$ (C) $b > 3a$ (D) $a = 2b$ (E) $b = 3a$

25. (A) 6 trick skis, 18 slalom skis; $780   (B) The maximum profit decreases to $720 when 18 trick and no slalom skis are produced.
(C) The maximum profit increases to $1,080 when no trick and 24 slalom skis are produced.

27. 9 model A trucks, 6 model B trucks, $279,000

29. (A) 40 tables, 40 chairs; $4,600   (B) The maximum profit decreases to $3,800 when 20 tables and 80 chairs are produced.

31. (A) Max $P = \$450$ when 750 gal is produced using the old process exclusively.
(B) The maximum profit decreases to $380 when 400 gal is produced using the old process and 700 gal using the new process.
(C) The maximum profit decreases to $288 when 1,440 gal is produced using the new process exclusively.

33. The nitrogen will range from a minimum of 940 lb when 40 bags of brand A and 100 bags of brand B are used to a maximum of 1,190 lb when 140 bags of brand A and 50 bags of brand B are used.

Chapter 6 Review Exercise

1. $(2, 3)$ *(6-1)*   **2.** No solution (inconsistent) *(6-1)*   **3.** Infinitely many solutions $(t, (4t + 8)/3)$, for any real number t *(6-1)*

4. $(-1, 3), (5, -3)$ *(6-3)*   **5.** $(1, -1), (\frac{7}{5}, -\frac{1}{5})$ *(6-3)*   **6.** $(1, 3), (1, -3), (-1, 3), (-1, -3)$ *(6-3)*

7.   *(6-1)*   **8.**   *(6-4)*   **9.**   *(6-4)*

(2, −1)

10. $\begin{bmatrix} 3 & -6 & | & 12 \\ 1 & -4 & | & 5 \end{bmatrix}$ *(6-1)*   **11.** $\begin{bmatrix} 1 & -4 & | & 5 \\ 1 & -2 & | & 4 \end{bmatrix}$ *(6-1)*   **12.** $\begin{bmatrix} 1 & -4 & | & 5 \\ 0 & 6 & | & -3 \end{bmatrix}$ *(6-1)*   **13.** $x_1 = 4$ *(6-2)*
$x_2 = -7$
$(4, -7)$

14. $x_1 - x_2 = 4$ *(6-2)*   **15.** $x_1 - x_2 = 4$ *(6-2)*   **16.** Min $z = 18$ at $(0, 6)$; max $z = 42$ at $(6, 4)$ *(6-5)*
$0 = 1$ $x_1 = t + 4, x_2 = t, t$ any real number
No solution

17. $x_1 = 2, x_2 = -2$; each pair of lines has the same intersection point. *(6-1)*

$$x_1 - x_2 = 4$$
$$2x_1 + x_2 = 2$$

$$x_1 - x_2 = 4$$
$$3x_2 = -6$$

$$x_1 - x_2 = 4$$
$$x_2 = -2$$

$$x_1 = 2$$
$$x_2 = -2$$

18. $(2.54, 2.15)$ *(6-1)*   **19.** $x_1 = -1, x_2 = 3$ *(6-2)*   **20.** $x_1 = -1, x_2 = 2, x_3 = 1$ *(6-2)*   **21.** $x_1 = 2, x_2 = 1, x_3 = -1$ *(6-2)*

22. Infinitely many solutions: $x_1 = -5t - 12, x_2 = 3t + 7, x_3 = t$, t any real number *(6-2)*   **23.** No solution *(6-2)*

24. Infinitely many solutions: $x_1 = -\frac{3}{7}t - \frac{4}{7}, x_2 = \frac{5}{7}t + \frac{9}{7}, x_3 = t$, t any real number *(6-2)*

25. $(2, \sqrt{2}), (2, -\sqrt{2}), (-1, i), (-1, -i)$ *(6-3)*   **26.** $(1, -2), (-1, 2), (2, -1), (-2, 1)$ *(6-3)*

27. $(2, -2), (-2, 2), (\sqrt{2}, \sqrt{2}), (-\sqrt{2}, -\sqrt{2})$ *(6-3)*

28. Corner points: $(0, 0), (0, 4),$
$(3, 2), (4, 0)$ *(6-4)*
Bounded

29. Corner points: $(0, 8), (\frac{12}{5}, \frac{16}{5}),$
$(12, 0)$ *(6-4)*
Unbounded

30. Corner points: $(4, 4), (10, 10),$
$(20, 0)$ *(6-4)*
Bounded

31. Max $z = 46$ at $(4, 2)$ *(6-5)*   **32.** Min $z = 75$ at $(3, 6)$ and $(15, 0)$ (multiple optimal solutions) *(6-5)*

33. Min $z = 44$ at $(4, 3)$; max $z = 82$ at $(2, 9)$ *(6-5)*   **34.** $x_1 = 1{,}000, x_2 = 4{,}000, x_3 = 2{,}000$ *(6-2)*

35. $(2, 2), (-2, -2), (\frac{4}{7}\sqrt{7}, -\frac{2}{7}\sqrt{7}), (-\frac{4}{7}\sqrt{7}, \frac{2}{7}\sqrt{7})$ *(6-3)*   **36.** Max $z = 26{,}000$ at $(600, 400)$ *(6-5)*

37. $(-2.16, -0.37), (-1.09, 5.59), (1.09, -5.59), (2.16, 0.37)$ *(6-3)*

38. (A) A unique solution   (B) No solution   (C) An infinite number of solutions *(6-2)*

39. $\frac{1}{2}$-lb packages: 48, $\frac{1}{3}$-lb packages: 72 *(6-1, 6-2)*   **40.** 6 by 8 m *(6-3)*   **41.** 40 g mix A, 60 g mix B, 30 g mix C *(6-2)*

42. (A) 22 nickels, 8 dimes   (B) $22 + 3t$ nickels, $8 - 4t$ dimes, and t quarters, $t = 0, 1, 2$ *(6-1, 6-2)*

43. (A) Maximum profit is $P = \$7{,}800$ when 80 regular and 30 competition sails are produced.
   (B) The maximum profit increases to $\$8{,}750$ when 70 competition and no regular sails are produced.
   (C) The maximum profit decreases to $\$7{,}200$ when no competition and 120 regular sails are produced. *(6-5)*

44. (A) The minimum cost is $C = \$13$ when 100 g of mix A and 150 g of mix B are used.
   (B) The minimum cost decreases to $\$9$ when 50 g of mix A and 275 g of mix B are used.
   (C) The minimum cost increases to $\$28.75$ when 250 g of mix A and 75 g of mix B are used. *(6-5)*

CHAPTER 7   Exercise 7-1

1. $\begin{bmatrix} 2 & 5 \\ 4 & -6 \end{bmatrix}$ **3.** $\begin{bmatrix} 3 & 2 \\ -2 & 8 \\ 12 & -5 \end{bmatrix}$ **5.** Not defined   **7.** $\begin{bmatrix} 3 & -5 & 6 \\ 1 & 1 & 8 \end{bmatrix}$ **9.** $\begin{bmatrix} 12 & -16 & 28 \\ -8 & 36 & 20 \end{bmatrix}$ **11.** $[41]$ **13.** $\begin{bmatrix} 3 \\ -13 \end{bmatrix}$

15. $\begin{bmatrix} 13 & 8 \\ 26 & 48 \end{bmatrix}$ **17.** $\begin{bmatrix} 16 & -18 \\ -10 & 18 \end{bmatrix}$ **19.** $[-36]$ **21.** $\begin{bmatrix} -6 & 12 \\ 15 & -30 \end{bmatrix}$ **23.** $[-13]$ **25.** $\begin{bmatrix} 5 & 0 & -3 \\ -10 & 0 & 6 \\ 30 & 0 & -18 \end{bmatrix}$ **27.** Not defined

29. $\begin{bmatrix} 3 & 2 & -6 \\ -31 & -10 & -8 \end{bmatrix}$ **31.** Not defined   **33.** $\begin{bmatrix} -19 & -2 & -14 \\ 15 & 4 & 6 \\ 25 & 10 & -4 \end{bmatrix}$ **35.** $\begin{bmatrix} 24 & -32 \\ -14 & 18 \\ -4 & 56 \end{bmatrix}$ **37.** Not defined

39. $\begin{bmatrix} 24 & 8 & 24 \\ -45 & 122 & -198 \\ -110 & 110 & -234 \end{bmatrix}$ **41.** $\begin{bmatrix} 6 & 4 & -12 \\ -189 & -62 & -42 \\ -226 & -76 & -38 \end{bmatrix}$ **43.** $\begin{bmatrix} 32 & 28 \\ 42 & -78 \end{bmatrix}$ **45.** $a = 4, b = 5, c = -9, d = -4$

47. $x = 5, y = 2$ **49.** $a = 1, b = -4, c = 0, d = 1$ **51.** $a = -3, b = 6, c = 2, d = -1$ **53.** True   **55.** True

57. True   **59.** True   **61.** True   **63.** False   **65.**

	Guitar	Banjo	
	$33	$26	Materials
	$57	$77	Labor

67.

	Basic car	Air	Markup AM/FM radio	Cruise control
Model A	$3,330	$77	$42	$27
Model B	$2,125	$93	$95	$50
Model C	$1,270	$113	$121	$52

69. (A) $11.80   (B) $30.30   (C) MN gives the labor costs per boat at each plant

(D)

$MN =$	Plant I	Plant II	
	$11.80	$13.80	One-person boat
	$18.50	$21.60	Two-person boat
	$26.00	$30.30	Four-person boat

71. (A) $A^2 = \begin{bmatrix} 0 & 0 & 2 & 0 & 0 \\ 1 & 0 & 0 & 0 & 1 \\ 0 & 1 & 0 & 2 & 0 \\ 1 & 0 & 0 & 0 & 1 \\ 0 & 0 & 1 & 0 & 0 \end{bmatrix}$; There is one way to travel from Baltimore to Atlanta with one intermediate connection; there are two ways to travel from Atlanta to Chicago with one intermediate connection. In general, the elements in A^2 indicate the number of different ways to travel from the ith city to the jth city with one intermediate connection.

(B) $A^3 = \begin{bmatrix} 2 & 0 & 0 & 0 & 2 \\ 0 & 1 & 0 & 2 & 0 \\ 0 & 0 & 3 & 0 & 0 \\ 0 & 1 & 0 & 2 & 0 \\ 1 & 0 & 0 & 0 & 1 \end{bmatrix}$; There is one way to travel from Denver to Baltimore with two intermediate connections; there are two ways to travel from Atlanta to El Paso with two intermediate connections. In general, the elements in A^3 indicate the number of different ways to travel from the ith city to the jth city with two intermediate connections.

(C) $A + A^2 + A^3 + A^4 = \begin{bmatrix} 2 & 3 & 2 & 5 & 2 \\ 1 & 1 & 4 & 2 & 1 \\ 4 & 1 & 3 & 2 & 4 \\ 1 & 1 & 4 & 2 & 1 \\ 1 & 1 & 1 & 3 & 1 \end{bmatrix}$; It is possible to travel from any origin to any destination with at most 3 intermediate connections.

73. (A) $3,550   (B) $6,000   (C) NM gives the total cost per town.

(D)

$NM =$	Cost per town	
	$3,550	Berkeley
	$6,000	Oakland

(E)

	Telephone call	House call	Letter
$[1 \quad 1]N =$	[3,000	1,300	13,000]

(F)

$N\begin{bmatrix}1\\1\\1\end{bmatrix} =$	Total contacts	
	6,500	Berkeley
	10,800	Oakland

Exercise 7-2

1. $\begin{bmatrix} 6 & -3 \\ 5 & 8 \end{bmatrix}$ **3.** $\begin{bmatrix} 2 & -5 \\ 8 & -4 \end{bmatrix}$ **5.** $\begin{bmatrix} 4 & -2 & 5 \\ 6 & 0 & -3 \\ -1 & 2 & 7 \end{bmatrix}$ **7.** $\begin{bmatrix} -3 & 2 & -5 \\ 4 & -1 & 6 \\ -3 & 4 & -2 \end{bmatrix}$ **15.** $\begin{bmatrix} 1 & -9 \\ 0 & 1 \end{bmatrix}$ **17.** $\begin{bmatrix} -5 & -2 \\ 2 & 1 \end{bmatrix}$

19. $\begin{bmatrix} -3 & -7 \\ -2 & -5 \end{bmatrix}$ **21.** $\begin{bmatrix} 0 & -1 & -1 \\ -1 & -1 & -1 \\ -1 & -1 & 0 \end{bmatrix}$ **23.** $\begin{bmatrix} -19 & 9 & -7 \\ 15 & -7 & 6 \\ -2 & 1 & -1 \end{bmatrix}$ **25.** Does not exist   **27.** $\begin{bmatrix} 5 & -3 \\ -3 & 2 \end{bmatrix}$ **29.** $\begin{bmatrix} 1 & 0 & 1 \\ \frac{1}{2} & \frac{1}{2} & 1 \\ 2 & 1 & 4 \end{bmatrix}$

31. Does not exist   **33.** $\begin{bmatrix} -9 & -15 & 10 \\ 4 & 5 & -4 \\ -1 & -1 & 1 \end{bmatrix}$ **37.** M^{-1} exists if and only if all the elements on the main diagonal are nonzero.

39. $\begin{bmatrix} -0.4 & 0.8 & 1.8 & -1.4 \\ 0.1 & -0.2 & -0.7 & 0.6 \\ -0.85 & 1.7 & 3.95 & -3.6 \\ -0.1 & -0.8 & -1.3 & 1.4 \end{bmatrix}$ **41.** $\begin{bmatrix} -0.75 & 3.75 & -4 & 0.75 & 3.5 \\ -0.5 & -0.5 & 1 & -0.5 & -1 \\ 2.5 & -10.5 & 11.5 & -1.5 & -9.5 \\ 0.125 & 1.375 & -1.75 & 0.375 & 1.5 \\ -0.75 & 0.75 & -0.5 & -0.25 & 0 \end{bmatrix}$

43. 14 5 195 74 97 37 181 67 49 18 121 43 103 41   **45.** GREEN EGGS AND HAM

47. 21 56 55 25 58 46 97 94 48 75 45 58 63 45 59 48 64 80 44 69 68 104 123 72 127

49. LYNDON BAINES JOHNSON

Exercise 7-3

1. $2x_1 - x_2 = 3$
$x_1 + 3x_2 = -2$

3. $-2x_1 + x_3 = 3$
$x_1 + 2x_2 + x_3 = -4$
$x_2 - x_3 = 2$

5. $\begin{bmatrix} 4 & -3 \\ 1 & 2 \end{bmatrix}\begin{bmatrix} x_1 \\ x_2 \end{bmatrix} = \begin{bmatrix} 2 \\ 1 \end{bmatrix}$

7. $\begin{bmatrix} 1 & -2 & 1 \\ -1 & 1 & 0 \\ 2 & 3 & 1 \end{bmatrix}\begin{bmatrix} x_1 \\ x_2 \\ x_3 \end{bmatrix} = \begin{bmatrix} -1 \\ 2 \\ -3 \end{bmatrix}$

9. $x_1 = -8, x_2 = 2$ **11.** $x_1 = 0, x_2 = 4$ **13.** (A) $x_1 = -20, x_2 = 9$ (B) $x_1 = 18, x_2 = -7$ (C) $x_1 = 19, x_2 = -8$

15. (A) $x_1 = 8, x_2 = 5$ (B) $x_1 = 4, x_2 = 4$ (C) $x_1 = -18, x_2 = -12$

17. (A) $x_1 = -3, x_2 = -4, x_3 = -2$ (B) $x_1 = 4, x_2 = 5, x_3 = 1$ (C) $x_1 = 2, x_2 = -1, x_3 = -1$

19. (A) $x_1 = -19, x_2 = 17, x_3 = -3$ (B) $x_1 = -104, x_2 = 82, x_3 = -11$ (C) $x_1 = 100, x_2 = -78, x_3 = 10$

21. $X = (A - B)^{-1}C$ **23.** $X = (I - A)^{-1}C$ **25.** $X = (3I - A)^{-1}C$

27. (A) $x_1 = 1, x_2 = 0$ (B) $x_1 = -5{,}000, x_2 = 1{,}000$ (C) $x_1 = 5{,}001, x_2 = -1{,}000$

29. $x_1 = 13, x_2 = 21, x_3 = -12$ **31.** $x_1 = 35, x_2 = -5, x_3 = 15, x_4 = 20$

33. Concert 1: 6,000 \$4 tickets and 4,000 \$8 tickets; Concert 2: 5,000 \$4 tickets and 5,000 \$8 tickets;
Concert 3: 3,000 \$4 tickets and 7,000 \$8 tickets

35. (A) $I_1 = 4, I_2 = 6, I_3 = 2$ (B) $I_1 = 3, I_2 = 7, I_3 = 4$ (C) $I_1 = 7, I_2 = 8, I_3 = 1$

37. (A) $a = 1, b = 0, c = -3$ (B) $a = -2, b = 5, c = 1$ (C) $a = 11, b = -46, c = 43$

39. Diet 1: 60 oz mix A and 80 oz mix B; Diet 2: 20 oz mix A and 60 oz mix B; Diet 3: 0 oz mix A and 100 oz mix B

Exercise 7-4

1. 7   **3.** -17 **5.** 9.79   **7.** $\begin{vmatrix} 4 & 6 \\ -2 & 8 \end{vmatrix}$ **9.** $\begin{vmatrix} 5 & -1 \\ 0 & -2 \end{vmatrix}$ **11.** $(-1)^{1+1}\begin{vmatrix} 4 & 6 \\ -2 & 8 \end{vmatrix} = 44$ **13.** $(-1)^{2+3}\begin{vmatrix} 5 & -1 \\ 0 & -2 \end{vmatrix} = 10$

15. 10   **17.** -21 **19.** -40 **21.** $(-1)^{1+1}\begin{vmatrix} a_{22} & a_{23} & a_{24} \\ a_{32} & a_{33} & a_{34} \\ a_{42} & a_{43} & a_{44} \end{vmatrix}$ **23.** $(-1)^{4+3}\begin{vmatrix} a_{11} & a_{12} & a_{14} \\ a_{21} & a_{22} & a_{24} \\ a_{31} & a_{32} & a_{34} \end{vmatrix}$ **25.** 22   **27.** -12 **29.** 0

31. 6   **33.** 60   **35.** 114   **37.** False   **39.** True

41. $\begin{vmatrix} a & b \\ c & d \end{vmatrix} = -\begin{vmatrix} c & d \\ a & b \end{vmatrix}$; interchanging the rows of this determinant changes the sign.

43. $\begin{vmatrix} ka & b \\ kc & d \end{vmatrix} = k\begin{vmatrix} a & b \\ c & d \end{vmatrix}$; multiplying a column of this determinant by a number k multiplies the value of the determinant by k.

45. $\begin{vmatrix} kc + a & kd + b \\ c & d \end{vmatrix} = \begin{vmatrix} a & b \\ c & d \end{vmatrix}$; adding a multiple of one row to the other row does not change the value of the determinant.

49. $49 = (-7)(-7)$ **51.** $f(x) = x^2 - 4x + 3$; 1, 3   **53.** $f(x) = x^3 + 2x^2 - 8x$; $-4, 0, 2$

Exercise 7-5

1. Theorem 1   **3.** Theorem 1   **5.** Theorem 2   **7.** Theorem 3   **9.** Theorem 5   **11.** $x = 0$ **13.** $x = 5$ **15.** -10

17. 10   **19.** -10 **21.** 25   **23.** -12 **25.** Theorem 1   **27.** Theorem 2   **29.** Theorem 5   **31.** $x = 5, y = 0$

33. $x = -3, y = 10$ **35.** -28 **37.** 106   **39.** 0   **41.** 6   **43.** 14

45. Expand the left side of the equation using minors.   **47.** Expand both sides of the equation and compare.

49. This follows from Theorem 4.

51. Expand the determinant about the first row to obtain $(y_1 - y_2)x - (x_1 - x_2)y + (x_1y_2 - x_2y_1) = 0$. Then show that the two points satisfy this linear equation.

53. If the determinant is 0, then the area of the triangle formed by the three points is 0. The only way this can happen is if the three points are on the same line—that is, the points are collinear.

Exercise 7-6

1. $x = 5, y = -2$ **3.** $x = 1, y = -1$ **5.** $x = -\frac{6}{5}, y = \frac{3}{5}$ **7.** $x = \frac{2}{17}, y = -\frac{20}{17}$ **9.** $x = 6{,}400, y = 6{,}600$

11. $x = 760, y = 760$ **13.** $x = 2, y = -2, z = -1$ **15.** $x = \frac{4}{3}, y = -\frac{1}{3}, z = \frac{2}{3}$ **17.** $x = -9, y = -\frac{7}{3}, z = 6$

19. $x = \frac{3}{2}, y = -\frac{7}{6}, z = \frac{2}{3}$ **21.** $x = 4$ **23.** $y = 2$ **25.** $z = \frac{5}{2}$

27. Since $D = 0$, the system has either no solution or infinitely many. Since $x = 0, y = 0, z = 0$ is a solution, the second case must hold.

31. (A) $R = 200p + 300q - 6p^2 + 6pq - 3q^2$
(B) $p = -0.3x - 0.4y + 180, q = -0.2x - 0.6y + 220, R = 180x + 220y - 0.3x^2 - 0.6xy - 0.6y^2$

Chapter 7 Review Exercise

1. $\begin{bmatrix} 4 & 8 \\ -12 & 18 \end{bmatrix}$ *(7-1)*   **2.** $[-11]$ *(7-1)*   **3.** $[-15 \;\; 19]$ *(7-1)*   **4.** $\begin{bmatrix} 16 \\ -6 \end{bmatrix}$ *(7-1)*   **5.** $\begin{bmatrix} 3 & 3 \\ -4 & 9 \end{bmatrix}$ *(7-1)*

6. Not defined *(7-1)*   **7.** Not defined *(7-1)*   **8.** $\begin{bmatrix} 13 & -29 \\ 20 & -24 \end{bmatrix}$ *(7-1)*   **9.** $[-5 \;\; 18]$ *(7-1)*   **10.** $\begin{bmatrix} 2 & 7 \\ -1 & -4 \end{bmatrix}$ *(7-2)*

11. (A) $x_1 = -1, x_2 = 3$ (B) $x_1 = 1, x_2 = 2$ (C) $x_1 = 8, x_2 = -10$ *(7-3)*   **12.** -17 *(7-4)*   **13.** 0 *(7-4, 7-5)*

14. $x = 2, y = -1$ *(7-6)*   **15.** (A) -2 (B) 6 (C) 2 *(7-5)*   **16.** $\begin{bmatrix} 7 & 16 & -9 \\ 28 & 40 & -30 \\ -21 & -8 & 17 \end{bmatrix}$ *(7-1)*

17. $\begin{bmatrix} 22 & 19 \\ 38 & 42 \end{bmatrix}$ *(7-1)*   **18.** $\begin{bmatrix} 12 & 24 & -6 \\ 0 & 0 & 0 \\ -8 & -16 & 4 \end{bmatrix}$ *(7-1)*   **19.** $[16]$ *(7-1)*   **20.** Not defined *(7-1)*

21. $\begin{bmatrix} 63 & -24 & -39 \\ -42 & 16 & 26 \end{bmatrix}$ *(7-1)*   **22.** $\begin{bmatrix} -1 & 1 & 1 \\ -2 & 3 & 2 \\ \frac{1}{2} & -\frac{1}{4} & -\frac{1}{4} \end{bmatrix}$ *(7-2)*

23. (A) $x_1 = 2, x_2 = 1, x_3 = -1$ (B) $x_1 = 1, x_2 = -2, x_3 = 1$ (C) $x_1 = -1, x_2 = 2, x_3 = -2$ *(7-3)*

24. $-\frac{11}{12}$ *(7-4)*   **25.** 35 *(7-4, 7-5)*   **26.** $y = \frac{10}{5} = 2$ *(7-6)*

27. (A) A unique solution   (B) Either no solution or an infinite number *(7-3)*

28. No. *(7-3)*   **29.** $X = (A - C)^{-1}B$ *(7-3)*   **30.** $\begin{bmatrix} -\frac{11}{12} & -\frac{1}{12} & 5 \\ \frac{10}{12} & \frac{2}{12} & -4 \\ \frac{1}{12} & -\frac{1}{12} & 0 \end{bmatrix}$ or $\dfrac{1}{12}\begin{bmatrix} -11 & -1 & 60 \\ 10 & 2 & -48 \\ 1 & -1 & 0 \end{bmatrix}$ *(7-2)*

31. $x_1 = 1{,}000, x_2 = 4{,}000, x_3 = 2{,}000$ *(7-3)*   **32.** 42 *(7-5)*

33. $\begin{vmatrix} u + kv & v \\ w + kx & x \end{vmatrix} = (u + kv)x - (w + kx)v = ux + kvx - wv - kvx = ux - wv = \begin{vmatrix} u & v \\ w & x \end{vmatrix}$ *(7-5)*

34. Theorem 4 in Section 7-5 implies that both points satisfy the equation. All other points on the line through the given points will also satisfy the equation. *(7-5)*

35. (A) 60 tons at Big Bend, 20 tons at Saw Pit   (B) 30 tons at Big Bend, 50 tons at Saw Pit
    (C) 40 tons at Big Bend, 40 tons at Saw Pit *(7-3)*

36. (A) $27   (B) Elements in LH give the total cost of manufacturing each product at each plant.
    (C)

$$LH = \begin{matrix} & \text{North} & \text{South} \\ & \text{Carolina} & \text{Carolina} \\ \begin{bmatrix} \$46.35 & \$41.00 \\ \$30.45 & \$27.00 \end{bmatrix} & \text{Desks} \\ & \text{Stands} \end{matrix}$$

(7-1)

37. (A) $\begin{bmatrix} 1{,}600 & 1{,}730 \\ 890 & 720 \end{bmatrix}$ (B) $\begin{bmatrix} 200 & 160 \\ 80 & 40 \end{bmatrix}$ (C) $\begin{bmatrix} 3{,}150 \\ 1{,}550 \end{bmatrix}$ Desks Stands
    Total production of each item in January *(7-1)*

38. GRAPHING UTILITY *(7-2)*

Cumulative Review Exercise: Chapters 6 and 7

1. $x = 2, y = -1$ *(6-1)*   **2.** $(-1, 2)$ *(6-1)*   **3.** $(-\frac{1}{5}, -\frac{7}{5}), (1, 1)$ *(6-3)*   **4.**

(6-4)

5. Maximum: 33; Minimum: 10 *(6-5)*

6. (A) $\begin{bmatrix} 0 & -3 \\ 3 & -9 \end{bmatrix}$ (B) Not defined   (C) $[3]$ (D) $\begin{bmatrix} 1 & 7 \\ 4 & -7 \end{bmatrix}$ (E) $[-1, 8]$ (F) Not defined *(7-1)*

7. -10 *(7-4)*   **8.** (A) $x_1 = 3, x_2 = -4$ (B) $x_1 = 2t + 3, x_2 = t, t$ any real number.   (C) No solution *(6-1)*

9. (A) $\left[\begin{array}{cc|c} 1 & 1 & 3 \\ -1 & 1 & 5 \end{array}\right]$ (B) $\left[\begin{array}{cc|c} 1 & 0 & -1 \\ 0 & 1 & 4 \end{array}\right]$ (C) $x_1 = -1, x_2 = 4$ *(6-1, 6-2)*

10. (A) $\begin{bmatrix} 1 & -3 \\ 2 & -5 \end{bmatrix}\begin{bmatrix} x_1 \\ x_2 \end{bmatrix} = \begin{bmatrix} k_1 \\ k_2 \end{bmatrix}$ (B) $A^{-1} = \begin{bmatrix} -5 & 3 \\ -2 & 1 \end{bmatrix}$ (C) $x_1 = 13, x_2 = 5$ (D) $x_1 = -11, x_2 = -4$ *(7-3)*

11. (A) 2   (B) $x = \frac{1}{2}, y = 0$ *(7-6)*

12. $x_1 = 1, x_2 = 3$; each pair of lines has the same intersection point.   *(6-1)*

   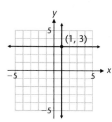

$$x_1 + 3x_2 = 10$$
$$2x_1 - x_2 = -1$$

$$x_1 + 3x_2 = 10$$
$$-7x_2 = -21$$

$$x_1 + 3x_2 = 10$$
$$x_2 = 3$$

$$x_1 = 1$$
$$x_2 = 3$$

13. (1.53, 3.35) *(6-1)*   **14.** (1, 0, −2) *(6-2)*   **15.** No solution *(6-2)*   **16.** $(t - 3, t - 2, t)$ t any real number *(6-2)*

17. $(1, 1), (-1, -1), (\sqrt{3}, \sqrt{3}/3), (-\sqrt{3}, -\sqrt{3}/3)$ *(6-3)*   **18.** $(0, i), (0, -i), (1, 1), (-1, -1)$ *(6-3)*

19. (A) $[-3]$ (B) $\begin{bmatrix} 1 & 2 & -1 \\ -1 & -2 & 1 \\ 2 & 4 & -2 \end{bmatrix}$ *(7-1)*   **20.** (A) $\begin{bmatrix} -1 & 2 \\ 2 & 3 \end{bmatrix}$ (B) Not defined *(7-1)*

21. *(6-4)*   **22.** 63 *(6-5)*

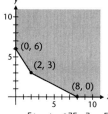

23. (A) $\begin{bmatrix} 1 & 4 & 2 \\ 2 & 6 & 3 \\ 2 & 5 & 2 \end{bmatrix}\begin{bmatrix} x_1 \\ x_2 \\ x_3 \end{bmatrix} = \begin{bmatrix} k_1 \\ k_2 \\ k_3 \end{bmatrix}$ (B) $A^{-1} = \begin{bmatrix} -3 & 2 & 0 \\ 2 & -2 & 1 \\ -2 & 3 & -2 \end{bmatrix}$ (C) $(7, -5, 6)$ (D) $(-6, 3, -2)$ *(7-3)*

24. (A) $D = 1$ (B) $z = 32$ *(7-5, 7-6)*   **25.** $(-1.35, 0.28), (-0.87, -1.60), (0.87, 1.60), (1.35, -0.28)$ *(6-3)*

26. (A) Infinite number of solutions   (B) No solution   (C) Unique solution *(6-2)*

27. $A = I$, the $n \times n$ identity *(7-3)*   **28.** $L, M,$ and P *(6-2)*   **32.** True *(7-1)*   **33.** True *(7-1)*   **34.** False *(7-1)*

35. False *(7-1)*   **36.** True *(7-2)*   **37.** True *(7-2)*   **38.** True *(7-4)*   **39.** True *(7-4)*

40. $8,000 at 8% and $4,000 at 14% *(6-1, 6-2)*   **41.** 60-g mix A, 50-g mix B, 40-g mix C *(6-2)*

42. 1 model A truck, 6 model B trucks, and 5 model C trucks; or 3 model A trucks, 3 model B trucks, and 6 model C trucks; or 5 model A trucks and 7 model C trucks. *(6-2)*

43. 8 by 4 m *(6-3)*

44. (A) Manufacturing 400 standard and 200 deluxe day packs produces a maximum weekly profit of $5,600.

     (B) The maximum weekly profit increases to $6,000 when 0 standard and 400 deluxe day packs are manufactured.

     (C) The maximum weekly profit increases to $6,600 when 600 standard and 0 deluxe day packs are manufactured. *(6-5)*

45. (A) $M\begin{bmatrix} 0.25 \\ 0.25 \\ 0.25 \\ 0.25 \end{bmatrix} = \begin{bmatrix} 82.25 \\ 83 \\ 92 \\ 83.75 \\ 82 \end{bmatrix} \begin{matrix} \text{Ann} \\ \text{Bob} \\ \text{Carol} \\ \text{Dan} \\ \text{Eric} \end{matrix}$ (B) $M\begin{bmatrix} 0.2 \\ 0.2 \\ 0.2 \\ 0.4 \end{bmatrix} = \begin{bmatrix} 83 \\ 84.8 \\ 91.8 \\ 85.2 \\ 80.8 \end{bmatrix} \begin{matrix} \text{Ann} \\ \text{Bob} \\ \text{Carol} \\ \text{Dan} \\ \text{Eric} \end{matrix}$

     Class averages

     (C)   Test 1   Test 2   Test 3   Test 4   *(7-1)*
$$[0.2 \quad 0.2 \quad 0.2 \quad 0.2 \quad 0.2]M = [84.4 \quad 81.8 \quad 85 \quad 87.2]$$

CHAPTER 8   Exercise 8-1

1. 7, 9, 11, 13, ...   **3.** $2, \frac{7}{3}, \frac{5}{2}, \frac{13}{5}, \ldots$ **5.** 1, 13, 19, 97, ...   **7.** 205   **9.** $\frac{601}{201}$ **11.** $1 + 2 + 3 + 4 + 5$

13. $\frac{1}{10} + \frac{1}{100} + \frac{1}{1,000}$ **15.** $-1 + 1 - 1 + 1$ **17.** $1, -4, 9, -16, 25$ **19.** 0.3, 0.33, 0.333, 0.3333, 0.333 33

21. $1, -\frac{1}{2}, \frac{1}{4}, -\frac{1}{8}, \frac{1}{16}$ **23.** $7, 3, -1, -5, -9$ **25.** $4, 1, \frac{1}{4}, \frac{1}{16}, \frac{1}{64}$ **27.** $a_n = 1 + 2n$ **29.** $a_n = (n + 1)^2$ **31.** $a_n = 2^{n+1}$

33. $a_n = (2n - 1)/n$ **35.** $a_n = 7(-1)^{n+1}$ **37.** $a_n = x^n/2^{n-1}$

39.   **41.**

43. $\frac{4}{1} - \frac{8}{2} + \frac{16}{3} - \frac{32}{4}$ **45.** $x^2 + \frac{x^3}{2} + \frac{x^4}{3}$ **47.** $x - \frac{x^2}{2} + \frac{x^3}{3} - \frac{x^4}{4} + \frac{x^5}{5}$ **49.** $\sum_{k=1}^{4} k^2$ **51.** $\sum_{k=1}^{5} \frac{1}{2^k}$ **53.** $\sum_{k=1}^{n} \frac{1}{k^2}$ **55.** $\sum_{k=1}^{n} (-1)^{k+1} k^2$

57. (A) 3, 1.83, 1.46, 1.415   (B) $\sqrt{2} \approx 1.4142$ (C) For $a_1 = 1$: 1, 1.5, 1.417, 1.414

59. The values of c_n are approximately 2.236 (i.e., $\sqrt{5}$) for large values of n.

61. Series approx. of $e^{0.2} = 1.221\ 400\ 0$; calculator value of $e^{0.2} = 1.221\ 402\ 8$

Exercise 8-2

1. 3   **3.** 5   **5.** P_1: $2 = 2 \cdot 1^2$; P_2: $2 + 6 = 2 \cdot 2^2$; P_3: $2 + 6 + 10 = 2 \cdot 3^2$

7. P_1: $a^5 a = a^{5+1}$; P_2: $a^5 a^2 = a^5(a^1 a) = (a^5 a)a = a^6 a = a^7 = a^{5+2}$; P_3: $a^5 a^3 = a^5(a^2 a) = a^5(a^1 a)a = [(a^5 a)a]a = a^8 = a^{5+3}$

9. P_1: $9^1 - 1 = 8$ is divisible by 4; P_2: $9^2 - 1 = 80$ is divisible by 4; P_3: $9^3 - 1 = 728$ is divisible by 4

11. P_k: $2 + 6 + 10 + \cdots + (4k - 2) = 2k^2$; P_{k+1}: $2 + 6 + 10 + \cdots + (4k - 2) + (4k + 2) = 2(k + 1)^2$

13. P_k: $a^5 a^k = a^{5+k}$; P_{k+1}: $a^5 a^{k+1} = a^{5+k+1}$ **15.** P_k: $9^k - 1 = 4r$; P_{k+1}: $9^{k+1} - 1 = 4s$; $r, s \in N$

23. $n = 4$, $p(x) = x^4 + 1$ **25.** $n = 23$ **43.** $2 + 4 + 6 + \cdots + 2n = n(n + 1)$

45. $1 + 2 + 3 + \cdots + (n - 1) = n(n - 1)/2$, $n \geq 2$ **51.** $3^4 + 4^4 + 5^4 + 6^4 \neq 7^4$

Exercise 8-3

1. (A) Arithmetic with $d = -5$; $-26, -31$ (B) Geometric with $r = -2$; $-16, 32$ (C) Neither

(D) Geometric with $r = \frac{1}{3}$; $\frac{1}{54}, \frac{1}{162}$

3. $a_2 = 11$, $a_3 = 16$, $a_4 = 21$ **5.** $a_{21} = -153$, $S_{21} = -1,743$ **7.** $S_{25} = 1,775$ **9.** $S_{15} = 6,375$

11. $a_2 = 3$, $a_3 = -\frac{3}{2}$, $a_4 = \frac{3}{4}$ **13.** $a_{10} = \frac{1}{243}$ **15.** $S_7 = 3,279$ **17.** $S_{50} = 3,925$ **19.** $a_{64} = 56.5$ **21.** $a_1 = 6$

23. $a_{10} = 3$, $d = -\frac{1}{9}$ **25.** $r = \pm\sqrt{10}$ **27.** $S_{10} = 42,625$ **29.** $a_2 = 60$, $a_3 = 90$ **31.** $S_{51} = 4,131$ **33.** $S_7 = 547$

35. $-1,071$ **37.** $\frac{1,023}{1,024}$ **39.** 4,446   **43.** $x = 2\sqrt{3}$ **45.** Yes   **47.** 66   **49.** 133   **51.** $S_\infty = \frac{9}{2}$ **53.** No sum

55. $S_\infty = \frac{8}{5}$ **57.** $\frac{7}{9}$ **59.** $\frac{6}{11}$ **61.** $3\frac{8}{37}$ or $\frac{119}{37}$ **65.** $a_n = (-2)(-3)^{n-1}$ **67.** *Hint:* $y = x + d$, $z = x + 2d$

71. $x = -1$, $y = 2$ **73.** Firm A: $501,000; firm B: $504,000   **75.** $4,000,000   **77.** $A = P(1 + r)^n$; approx. 12 yr

79. $700 per year; $115,500   **81.** 900   **83.** 1,250,000   **85.** (A) 336 ft   (B) 1,936 ft   (C) $16t^2$ **87.** $A = A_0 2^{2t}$

89. $r = 10^{-0.4} = 0.398$ **91.** 9.223×10^{16}; 1.845×10^{17} **93.** 0.0015 psi   **95.** 2   **97.** 3,420°

Exercise 8-4

1. 5,040   **3.** 210   **5.** 144   **7.** 56   **9.** 1   **11.** 720   **13.** 9!/8!   **15.** 8!/5!   **17.** 715   **19.** 3,432

21. 161,700   **25.** $8x^3 - 36x^2 y + 54xy^2 - 27y^3$ **27.** $x^4 - 8x^3 + 24x^2 - 32x + 16$

29. $32x^5 - 80x^4 y + 80x^3 y^2 - 40x^2 y^3 + 10xy^4 - y^5$ **31.** $5,005u^9 v^6$ **33.** $180m^2 n^8$ **35.** $392,445w^{16}$ **37.** $-48,384x^3 y^5$

39. 5   **41.** (A) $a_4 = 0.251$ (B) 1   **43.** 1.1046   **45.** True   **47.** True

Exercise 8-5

1. 2,730   **3.** 1   **5.** 84   **7.** 1,820   **9.** 6,720   **11.** 132,600   **13.** 1,287   **15.** 2,598,960   **17.** $5 \cdot 3 \cdot 4 \cdot 2 = 120$

19. $P_{10,3} = 10 \cdot 9 \cdot 8 = 720$ **21.** $C_{7,3} = 35$ subcommittees; $P_{7,3} = 210$ **23.** $C_{10,2} = 45$

25. No repeats: $6 \cdot 5 \cdot 4 \cdot 3 = 360$; with repeats: $6 \cdot 6 \cdot 6 \cdot 6 = 1,296$ **27.** $C_{13,5} = 1,287$

29. $26 \cdot 26 \cdot 26 \cdot 10 \cdot 10 \cdot 10 = 17,576,000$ possible license plates; no repeats: $26 \cdot 25 \cdot 24 \cdot 10 \cdot 9 \cdot 8 = 11,232,000$

31. $C_{13,5}C_{13,2} = 100,386$ **33.** $C_{8,3}C_{10,4}C_{7,2} = 246,960$

35. (B) $r = 0, 10$ (C) Each is the product of r consecutive integers, the largest of which is n for $P_{n,r}$, and r for $r!$

37. $12 \cdot 11 = 132$ **39.** (A) $C_{8,2} = 28$ (B) $C_{8,3} = 56$ (C) $C_{8,4} = 70$

41. 2 people: $P_{5,2} = 20$; 3 people: $P_{5,3} = 60$; 4 people: $P_{5,4} = 120$; 5 people: $P_{5,5} = 120$

43. (A) $P_{8,5} = 6,720$ (B) $C_{8,5} = 56$ (C) $2 \cdot C_{6,4} = 30$

45. There are $C_{4,1} \cdot C_{48,4} = 778,320$ hands that contain exactly one king, and $C_{39,5} = 575,757$ hands that contain no hearts, so the former is more likely.

47. 21

Chapter 8 Review Exercise

1. (A) Geometric   (B) Arithmetic   (C) Arithmetic   (D) Neither   (E) Geometric   *(8-1, 8-3)*
2. (A) 5, 7, 9, 11   (B) $a_{10} = 23$ (C) $S_{10} = 140$ *(8-1, 8-3)*
3. (A) 16, 8, 4, 2   (B) $a_{10} = \frac{1}{32}$ (C) $S_{10} = 31\frac{31}{32}$ *(8-1, 8-3)*
4. (A) $-8, -5, -2, 1$ (B) $a_{10} = 19$ (C) $S_{10} = 55$ *(8-1, 8-3)*
5. (A) $-1, 2, -4, 8$ (B) $a_{10} = 512$ (C) $S_{10} = 341$ *(8-1, 8-3)*   6. $S_\infty = 32$ *(8-3)*   7. 3,628,800   *(8-4)*
8. 17,100,720   *(8-4)*   9. 1,287   *(8-4)*   10. $P_{8,4} = 1,680$; $C_{8,4} = 70$ *(8-5)*
11. (A) 12 combined outcomes: (B) $6 \cdot 2 = 12$ *(8-5)*   12. $6 \cdot 5 \cdot 4 \cdot 3 \cdot 2 \cdot 1 = 720$ *(8-5)*   13. $P_{6,6} = 6! = 720$ *(8-5)*

14. P_1: $5 = 1^2 + 4 \cdot 1$; P_2: $5 + 7 = 2^2 + 4 \cdot 2$; P_3: $5 + 7 + 9 = 3^2 + 4 \cdot 3$ *(8-2)*
15. P_1: $2 = 2^{1+1} - 2$; P_2: $2 + 4 = 2^{2+1} - 2$; P_3: $2 + 4 + 8 = 2^{3+1} - 2$ *(8-2)*
16. P_1: $49^1 - 1 = 48$ is divisible by 6; P_2: $49^2 - 1 = 2,400$ is divisible by 6; P_3: $49^3 - 1 = 117,648$ is divisible by 6   *(8-2)*
17. P_k: $5 + 7 + 9 + \cdots + (2k + 3) = k^2 + 4k$; P_{k+1}: $5 + 7 + 9 + \cdots + (2k + 3) + (2k + 5) = (k + 1)^2 + 4(k + 1)$ *(8-2)*
18. P_k: $2 + 4 + 8 + \cdots + 2^k = 2^{k+1} - 2$; P_{k+1}: $2 + 4 + 8 + \cdots + 2^k + 2^{k+1} = 2^{k+2} - 2$ *(8-2)*
19. P_k: $49^k - 1 = 6r$ for some integer r; P_{k+1}: $49^{k+1} - 1 = 6s$ for some integer s *(8-2)*   20. $n = 31$ is a counterexample   *(8-2)*
21. $S_{10} = -6 - 4 - 2 + 0 + 2 + 4 + 6 + 8 + 10 + 12 = 30$ *(8-3)*   22. $S_7 = 8 + 4 + 2 + 1 + \frac{1}{2} + \frac{1}{4} + \frac{1}{8} = 15\frac{7}{8}$ *(8-3)*
23. $S_\infty = \frac{81}{5}$ *(8-3)*   24. $S_n = \sum_{k=1}^{n} \frac{(-1)^{k+1}}{3^k}$; $S_\infty = \frac{1}{4}$ *(8-3)*   25. $C_{6,3} = 20$ *(8-5)*   26. $d = 3$, $a_5 = 25$ *(8-3)*
27. $a_1 = -9$, $d = \frac{19}{5}$ *(8-3)*   28. 336; 512; 392   *(8-5)*   29. $\frac{8}{11}$ *(8-3)*   30. (A) $P_{6,3} = 120$ (B) $C_{5,2} = 10$ *(8-5)*
31. 190   *(8-4)*   32. 1,820   *(8-4)*   33. 1   *(8-4)*   34. $\frac{987!}{493!}$ *(8-4)*   35. $\binom{1000}{501}$ *(8-5)*

36. $x^5 - 5x^4y + 10x^3y^2 - 10x^2y^3 + 5xy^4 - y^5$ *(8-4)*   37. $-1,760x^3y^9$ *(8-4)*   41. 29   *(8-4)*   42. 26   *(8-1)*
43. $2^5 = 32$; 6   *(8-5)*   44. 49 g/2 ft; 625 g/2 ft   *(8-3)*   45. 12   *(8-5)*
46. $x^6 + 6ix^5 - 15x^4 - 20ix^3 + 15x^2 + 6ix - 1$ *(8-4)*   47. $P_{5,5} = 120$ *(8-5)*   53. $n = 5$ *(8-1)*
54. $\{a_n\}$ is neither arithmetic nor geometric; $\{b_n\}$ is geometric   *(8-3)*

CHAPTER 9   Exercise 9-1

1.

3.

5.

7.

9.

11.

13. (9.75, 0)   **15.** (0, −26.25)   **17.** (−19.25, 0)

19. $y^2 = -24x$ **21.** $x^2 = 20y$ **23.** $x^2 = -\frac{4}{3}y$ **25.** $y^2 = -100x$ **27.** $x^2 = -3y$ **29.** $y^2 = -\frac{1}{2}x$

31.   **33.**

35. (A) 2; $x = 0$ and $y = 0$ (B) (0, 0), $(4am, 4am^2)$

37. $(\pm 2a, a)$ **39.** False   **41.** True   **43.** $x^2 - 4x - 12y - 8 = 0$ **45.** $y^2 + 8x - 8x + 48 = 0$

47. (−0.78, 0.08), (40.78, 207.92)   **49.** (−6.84, −5.85), (0, 0)   **51.** $x^2 = -200y$

53. (A) $y = 0.0025x^2$, $-100 \le x \le 100$ (B) 25 ft

Exercise 9-2

1. Foci: $F'(-\sqrt{21}, 0)$, $F(\sqrt{21}, 0)$
Major axis length = 10
Minor axis length = 4

3. Foci: $F'(0, -\sqrt{21})$, $F(0, \sqrt{21})$
Major axis length = 10
Minor axis length = 4

5. Foci: $F'(-\sqrt{8}, 0)$, $F(\sqrt{8}, 0)$
Major axis length = 6
Minor axis length = 2

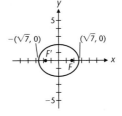

7. Foci: $F'(0, -4)$, $F(0, 4)$
Major axis length = 10
Minor axis length = 6

9. Foci: $F'(0, -\sqrt{6})$, $F(0, \sqrt{6})$
Major axis length = $2\sqrt{12} \approx 6.93$
Minor axis length = $2\sqrt{6} \approx 4.90$

11. Foci: $F'(-\sqrt{3}, 0)$, $F(\sqrt{3}, 0)$
Major axis length = $2\sqrt{7} \approx 5.29$
Minor axis length = 4

13. $x^2 + \frac{y^2}{9} = 1$ **15.** $\frac{x^2}{41} + \frac{y^2}{25} = 1$ **17.** $\frac{x^2}{143} + \frac{y^2}{144} = 1$ **19.** It does not pass the vertical line test.

21.   **23.**   **25.** (2.201, 4.403)

27. (3.565, 1.589)   **29.** False   **31.** True   **33.** $\frac{x^2}{16} + \frac{y^2}{12} = 1$; ellipse

35. $(-0.46, 2.57), (4.08, -1.06)$ **37.** $(\pm 3.64, \pm 9.50)$ **39.** $\dfrac{x^2}{400} + \dfrac{y^2}{144} = 1$; 7.94 ft   **41.** (A) $\dfrac{x^2}{576} + \dfrac{y^2}{15.9} = 1$ (B) 5.13 ft

Exercise 9-3

1. Foci: $F'(-\sqrt{13}, 0), F(\sqrt{13}, 0)$
Transverse axis length $= 6$
Conjugate axis length $= 4$

3. Foci: $F'(0, -\sqrt{13}), F(0, \sqrt{13})$
Transverse axis length $= 4$
Conjugate axis length $= 6$

5. Foci: $F'(-\sqrt{20}, 0), F(\sqrt{20}, 0)$
Transverse axis length $= 4$
Conjugate axis length $= 8$

7. Foci: $F'(0, -5), F(0, 5)$
Transverse axis length $= 8$
Conjugate axis length $= 6$

9. Foci: $F'(-\sqrt{10}, 0), F(\sqrt{10}, 0)$
Transverse axis length $= 4$
Conjugate axis length $= 2\sqrt{6} \approx 4.90$

11. Foci: $F'(0, -\sqrt{11}), F(0, \sqrt{11})$
Transverse axis length $= 4$
Conjugate axis length $= 2\sqrt{7} \approx 5.29$

13. $\dfrac{y^2}{25} - \dfrac{x^2}{81} = 1$ **15.** $\dfrac{x^2}{49} - \dfrac{y^2}{32} = 1$ **17.** $\dfrac{y^2}{36} - \dfrac{x^2}{36} = 1$

19. (A) Infinitely many; $\dfrac{x^2}{a^2} - \dfrac{y^2}{1 - a^2} = 1 \; (0 < a < 1)$ (B) Infinitely many; $\dfrac{x^2}{a^2} + \dfrac{y^2}{a^2 - 1} = 1 \; (a > 1)$ (C) One; $y^2 = 4x$

21.

23.

25. $(-1.389, 3.306), (6.722, 7.361)$

27. $(\pm 3.266, \pm 3.830)$ **29.** True   **31.** False   **33.** $\dfrac{x^2}{4} - \dfrac{y^2}{5} = 1$; hyperbola   **35.** $(-4.73, 2.88), (3.35, -0.90)$

37. $(\pm 1.39, \pm 2.96)$ **39.** $\dfrac{y^2}{16} - \dfrac{x^2}{8} = 1$; 5.38 ft above vertex   **41.** $y = \frac{4}{3}\sqrt{x^2 + 30^2}$

Exercise 9-4

1. (A) $x' = x - 3, y' = y - 5$ (B) $x'^2 + y'^2 = 81$ (C) Circle

3. (A) $x' = x + 7, y' = y - 4$ (B) $\dfrac{x'^2}{9} + \dfrac{y'^2}{16} = 1$ (C) Ellipse

5. (A) $x' = x - 4, y' = y + 9$ (B) $y'^2 = 16x'$ (C) Parabola

7. (A) $x' = x + 8, y' = y + 3$ (B) $\dfrac{x'^2}{12} + \dfrac{y'^2}{8} = 1$ (C) Ellipse   **9.** (A) $\dfrac{(x - 3)^2}{9} - \dfrac{(y + 2)^2}{16} = 1$ (B) Hyperbola

11. (A) $\dfrac{(x + 5)^2}{5} + \dfrac{(y + 7)^2}{6} = 1$ (B) Ellipse   **13.** (A) $(x + 6)^2 = -24(y - 4)$ (B) Parabola

15. $\dfrac{(x-2)^2}{9} + \dfrac{(y-2)^2}{4} = 1$; ellipse   **17.** $(x+4)^2 = -8(y-2)$; parabola

19. $(x+6)^2 + (y+5)^2 = 16$; circle   **21.** $\dfrac{(y-3)^2}{9} - \dfrac{(x+4)^2}{16} = 1$; hyperbola   **23.** $h = \dfrac{-D}{2A}$, $k = \dfrac{D^2 - 4AF}{4AE}$

    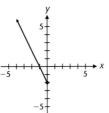

25. $x^2 - 10x - 32y + 121 = 0$ **27.** $36x^2 + 25y^2 + 216x - 200y - 176 = 0$ **29.** $3x^2 - y^2 - 24x + 2y + 44 = 0$
31. $4x^2 - 3y - 4 = 0$ **33.** $x^2 + 5y^2 + 4x + 10y = 0$ **35.** $F'(-\sqrt{5}+2, 2)$, $F(\sqrt{5}+2, 2)$ **37.** $F(-4, 0)$
39. $F'(-4, -2)$, $F(-4, 8)$ **41.** $(1.18, 1.98)$, $(6.85, -6.52)$ **43.** $(-1.72, -1.87)$, $(-0.99, 2.06)$

Exercise 9-5

1. $y = -2x - 2$;
straight line

3. $y = -2x - 2$, $x \le 0$;
a ray (part of a straight line)

5. $y = -\frac{2}{3}x$;
straight line

7. $y^2 = 4x$;
parabola

    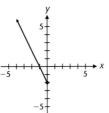

9. $y^2 = 4x$, $y \ge 0$;
parabola (upper half)

11. $\dfrac{x^2}{9} + \dfrac{y^2}{16} = 1$;
ellipse

13. $(x-2)^2 + (y-3)^2 = 4$;
circle

15. $y = -\dfrac{2}{x}$;
hyperbola

            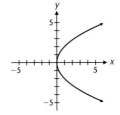

17. $y^2 = x + 1$, $y \ge 0$, $x \ge -1$;
parabola (upper half)

19. $x = t$, $y = \dfrac{At^2 + Dt + F}{-E}$, $-\infty < t < \infty$; parabola

21. $y = 1/x$, $x > 0$;
hyperbola (one branch)

23. $y^2 = -8(x - 1)$, $-1 \le x \le 1$;
part of a parabola

25. $x^2 + y^2 = 2x$, $x \ne 0$ or $(x - 1)^2 + y^2 = 1$, $x \ne 0$;
circle (note hole at origin)

27.

29. $\dfrac{(x - 3)^2}{36} + \dfrac{(y - 2)^2}{16} = 1$; ellipse with center $(3, 2)$

31. $\dfrac{(y + 1)^2}{25} - \dfrac{(x + 3)^2}{4} = 1$; hyperbola with center $(-3, -1)$

33. $5x - 2y = 0$

35. $\dfrac{(y + 1)^2}{25} - \dfrac{(x - 4)^2}{9} = 1$; hyperbola with center $(4, -1)$;

$x = 4 + 3 \tan t$, $y = -1 + 5 \sec t$, $-\dfrac{\pi}{2} < t < \dfrac{3\pi}{2}$, $t \ne \dfrac{\pi}{2}$

37. $\dfrac{(x - 3)^2}{49} + \dfrac{(y + 4)^2}{4} = 1$; ellipse with center $(3, -4)$;

$x = 3 + 7 \cos t$, $y = -4 + 2 \sin t$, $0 \le t \le 2\pi$

39. (A) $(4.8, -1.5)$
(B) $x^2 + y^2 = 25$; circle

41. (A) 43.292 s   (B) 9,183.619 m; 9.184 km   (C) 2,295.918 m

Chapter 9 Review Exercise

1. Foci: $F'(-4, 0)$, $F(4, 0)$ *(9-2)*
Major axis length = 10
Minor axis length = 6

2. *(9-1)*

3. Foci: $F'(0, -\sqrt{34})$, $F(0, \sqrt{34})$ *(9-3)*
Transverse axis length = 6
Conjugate axis length = 10

4. (A) $\dfrac{(y + 2)^2}{25} - \dfrac{(x - 4)^2}{4} = 1$ (B) Hyperbola *(9-4)*

5. (A) $(x + 5)^2 = -12(y + 4)$ (B) Parabola *(9-4)*

6. (A) $\dfrac{(x - 6)^2}{9} + \dfrac{(y - 4)^2}{16} = 1$ (B) Ellipse *(9-4)*

7. $y^2 = -7x$ *(9-1)*

8. $\dfrac{x^2}{4} + \dfrac{y^2}{965} = 1$ *(9-2)*

9. $\dfrac{x^2}{25} - \dfrac{y^2}{11} = 1$ *(9-3)*

10. *(9-2, 6-3)*

11. *(9-1, 9-2, 6-3)*

12. *(9-2, 6-3)*

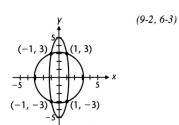

13. $\dfrac{(x+3)^2}{4} + \dfrac{(y-2)^2}{16} = 1$; ellipse *(9-4)*

14. $(x-2)^2 = 8(y+3)$; parabola *(9-4)*

15. $\dfrac{(x+3)^2}{9} - \dfrac{(y+2)^2}{4} = 1$; hyperbola *(9-4)*

16. $m = 0.2$; $x^2 = 50y$ is a magnification by a factor 50 of $x^2 = y$ *(9-1)*

17. $(y-4)^2 = -8(x-4)$, or $y^2 - 8y + 8x - 16 = 0$ *(9-1)*   **18.** $\dfrac{x^2}{4} - \dfrac{y^2}{12} = 1$; hyperbola *(9-3)*

19. $\dfrac{x^2}{36} + \dfrac{y^2}{20} = 1$; ellipse *(9-2)*   **20.** $F'(-3, -\sqrt{12}+2), F(-3, \sqrt{12}+2)$ *(9-4)*   **21.** $F(2, -1)$ *(9-4)*

22. $F'(-\sqrt{13}-3, -2), F(\sqrt{13}-3, -2)$ *(9-4)*

23. $(2.09, 2.50), (3.67, -1.92)$ *(9-4)*   **24.** 4 ft *(9-1)*   **25.** $\dfrac{x^2}{5^2} + \dfrac{y^2}{3^2} = 1$ *(9-2)*   **26.** 4.72 ft deep *(9-3)*

Cumulative Review Exercise: Chapters 8 and 9

1. (A) Neither (B) Arithmetic (C) Both (D) Geometric (E) Geometric (F) Neither *(8-3)*
2. (A) $-2, 4, -8, 16$ (B) $a_8 = 256$ (C) $S_8 = 170$ *(8-3)*
3. (A) $1, 7, 13, 19$ (B) $a_8 = 43$ (C) $S_8 = 176$ *(8-3)*
4. (A) $-20, -16, -12, -8$ (B) $a_8 = 8$ (C) $S_8 = -48$ *(8-3)*
5. (A) 5,040 (B) 13,800 (C) 210 *(8-4)* **6.** (A) 924 (B) 126 (C) 6,720 *(8-4, 8-5)*
7. Foci: $F'(-\sqrt{61}, 0), F(\sqrt{61}, 0)$ *(9-3)* **8.** Foci: $F'(-\sqrt{11}, 0), F(\sqrt{11}, 0)$ *(9-2)* **9.** *(9-1)*
Transverse axis length = 12   Major axis length = 12
Conjugate axis length = 10   Minor axis length = 10

10.   8 combined outcomes: **(B)** $2 \cdot 2 \cdot 2 = 8$ *(8-5)*   **11.**   **(A)** $4 \cdot 3 \cdot 2 \cdot 1 = 24$ **(B)** $P_{4,4} = 4! = 24$ *(8-5)*

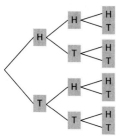

12. P_1: $1 = 1(1)$; P_2: $1 + 5 = 2(3)$ **13.** P_1: $1^2 + 1 + 2 = 4$ is divisible by 2
   P_3: $1 + 5 + 9 = 3(5)$ *(8-2)* P_2: $2^2 + 2 + 2 = 8$ is divisible by 2
                                                   P_3: $3^2 + 3 + 2 = 14$ is divisible by 2 *(8-2)*

14. P_k: $1 + 5 + 9 + \cdots + (4k - 3) = k(2k - 1)$
   P_{k+1}: $1 + 5 + 9 + \cdots + (4k - 3) + (4k + 1) = (k + 1)(2k + 1)$ *(8-2)*

15. P_k: $k^2 + k + 2 = 2r$ for some integer r
   P_{k+1}: $(k + 1)^2 + (k + 1) + 2 = 2s$ for some integer s *(8-2)*

16. $y = -2x^2$ *(9-1)*   **17.** $\dfrac{x^2}{25} + \dfrac{y^2}{16} = 1$ *(9-2)*   **18.** $\dfrac{x^2}{64} - \dfrac{y^2}{25} = 1$ *(9-3)*   **19.** $1 + 4 + 27 + 256 + 3{,}125 = 3{,}413$ *(8-1)*

20. $\displaystyle\sum_{k=1}^{6} \dfrac{(-1)^{k+1} 2^k}{(k + 1)!}$ *(8-1)*   **21.** 81 *(8-3)*   **22.** 360; 1,296; 750 *(8-5)*   **23.** $n = 22$ *(8-3)*

24. (A) 6,375,600   (B) 53,130   (C) 53,130 *(8-4, 8-5)*

25. $a^6 + 3a^5b + \frac{15}{4}a^4b^2 + \frac{5}{2}a^3b^3 + \frac{15}{16}a^2b^4 + \frac{3}{16}ab^5 + \frac{1}{64}b^6$ *(8-4)*   **26.** $153{,}090x^6y^4$; $-3{,}240x^3y^7$ *(8-4)*   **29.** 61,875 *(8-3)*

30. $\frac{27}{11}$ *(8-3)*   **31.** $a_{22} = 0.236$; 8 terms *(8-4)*

32. $(y - 2)^2 = 4(x + 3)$;   **33.** $\dfrac{(y + 2)^2}{4} - \dfrac{(x + 1)^2}{16} = 1$;   **34.** $\dfrac{(x - 2)^2}{9} + \dfrac{(y + 3)^2}{4} = 1$;
   parabola *(9-1)*   hyperbola *(9-3)*   ellipse *(9-2)*

35. $10^9 = 1{,}000{,}000{,}000$; 3,628,800 *(8-5)*   **36.** $(-2.26, -4.72)$, $(1.85, 3.09)$ *(9-4)*

38. $x^6 - 12ix^5 - 60x^4 + 160ix^3 + 240x^2 - 192ix - 64$ *(8-4)*   **39.** $(x - 6)^2 = -4(y - 2)$ or $x^2 - 12x + 4y + 28 = 0$ *(9-1)*

40. $\pm 2\sqrt{3}$ *(9-2)*   **41.** 8 *(9-3)*   **42.** Geometric *(8-3, 8-4)*   **43.** Neither *(8-3)*   **44.** Both *(8-3)*

45. Arithmetic *(8-3)*   **46.** False *(8-2, 8-4)*   **47.** True *(8-2, 8-4)*   **48.** False *(8-2, 8-4)*   **49.** True *(8-2, 8-4)*

50. True *(9-2)*   **51.** False *(9-1)*   **52.** False *(9-3)*   **53.** True *(9-2)*   **54.** False *(9-3)*

55. One parabola; $y^2 = 8x + 9$ *(9-4)*   **56.** $y + 4 = \pm\frac{1}{2}(x - 5)$ *(9-3, 9-4)*   **57.** $4\sqrt{10}$ *(9-2)*   **58.** $C_{7,3} = 35$ *(8-5)*

61. $x^2 - 8y^2 - 2x - 8y + 17 = 0$; hyperbola *(9-3)*   **62.** \$6,000,000 *(8-3)*   **63.** 4 in. *(9-1)*   **64.** 32 ft, 14.4 ft *(9-2)*

APPLICATIONS INDEX

SUBJECT INDEX

Hyperbolic paraboloid, 635
Hyperboloid, 635, 638

i, 120
Identity matrix, 499, 502
Imaginary numbers. *See* Complex numbers
Imaginary root, 131
Imaginary unit, 120
Imaginary zeros theorem, 302
Income analysis, 206
Income tax, 51
Increasing function, 225
Independent variable, 210
Index of a radial, 58
Inequalities. *See also* Inequality relation
    and absolute value, 111
    equivalent, 102
    and interval notation, 100
    linear, one variable, 102
    linear, two variables
        graphing of, 455
        half-planes, 455
    polynomial, 155
        standard form of, 153
    rational, 158
        standard form of, 158
    solution of, 102
    solution set of, 102
Inequality relation, 99
    double, 105
    graph of, 100
    greater than, 99
    less than, 99
    properties of, 103, 114
    trichotomy property of, 99
Infinite set, 2
Infinity, 100, 292, 323
Insecticides, 365
Installation charges, 205
Instantaneous rate of change, 162
Integer(s), 4
    exponent, 43
Intelligence quotient (IQ), 108
Intercept, 192, 223
Interest
    rate, 361
        per period, 361
Interpolating polynomial, 346
    and graphing utility, 347
Intersection of sets, 101
Interval, 99
    graph, 100
    left endpoint, 100
    notation, 100
    right endpoint, 100

Inventory value, 500
Inverse functions, 263
    domain of, 263
    graph of, 263
    hyperbolic, 403
    range of, 263
Inverse of a matrix, 504
Investment analysis, 431, 517
Irrational numbers, 4

Kirchhoff's laws, 521

Labor costs, 496, 500, 547
Lagrange's four square theorem, 566
Learning curve, 375
Learning theory, 336
Least common denominator (LCD), 37
Left endpoint, 100
Left half-plane, 455
Length of a line segment, 110
Less than, 99
Leukemic cells, 580
Light intensity, 374
Like terms, 15
Line(s)
    equation(s) of, 200
        intercept form of, 205
        point–slope form of, 197
        slope–intercept form of, 196
        standard form of, 191
    graph of, 191
    horizontal, 199
    parallel, 201
    perpendicular, 201
    rise, 193
    run, 193
    slope of, 193
    vertical, 199
    x intercept of, 192
    y intercept of, 192
Linear and quadratic factor theorem, 339
Linear equations
    graphing of, 191
    in one variable, 76
    solving, 76
    standard form of, 76, 191
    in two variables, 89, 191
Linear functions
    definition of, 225
    graph of, 225
    slope of, 225
Linear programming
    constant profit line, 471
    decision variables for, 469, 472

Circle (3-1)

$(x - h)^2 + (y - k)^2 = r^2$ Center at (h, k); radius r

$x^2 + y^2 = r^2$ Center at $(0, 0)$; radius r

Parabola (3-4, 12-1)

$y^2 = 4ax$, $a > 0$, opens right; $a < 0$, opens left
Focus: $(a, 0)$; Directrix: $x = -a$;
Axis: x axis

$x^2 = 4ay$, $a > 0$, opens up; $a < 0$, opens down
Focus: $(0, a)$; Directrix: $y = -a$;
Axis: y axis

Ellipse (12-2)

$\dfrac{x^2}{a^2} + \dfrac{y^2}{b^2} = 1$, $a > b > 0$
Foci: $F'(-c, 0)$, $F(c, 0)$; $c^2 = a^2 - b^2$

$\dfrac{x^2}{b^2} + \dfrac{y^2}{a^2} = 1$, $a > b > 0$
Foci: $F'(0, -c)$, $F(0, c)$; $c^2 = a^2 - b^2$

Hyperbola (12-3)

$\dfrac{x^2}{a^2} - \dfrac{y^2}{b^2} = 1$ Foci: $F'(-c, 0)$, $F(c, 0)$; $c^2 = a^2 + b^2$

$\dfrac{y^2}{a^2} - \dfrac{x^2}{b^2} = 1$ Foci: $F'(0, -c)$, $F(0, c)$; $c^2 = a^2 + b^2$

Translation Formulas (12-4)

$x = x' + h, y = y' + k;$ $x' = x - h, y' = y - k$
New origin (h, k)

Trigonometric Identities (6-5, 7-1–7-4)

Reciprocal Identities (6-5, 7-1)

$\csc x = \dfrac{1}{\sin x}$ $\sec x = \dfrac{1}{\cos x}$ $\cot x = \dfrac{1}{\tan x}$

Quotient Identities (6-5, 7-1)

$\tan x = \dfrac{\sin x}{\cos x}$ $\cot x = \dfrac{\cos x}{\sin x}$

Identities for Negatives (6-5, 7-1)

$\sin(-x) = -\sin x$ $\cos(-x) = \cos x$
$\tan(-x) = -\tan x$

Pythagorean Identities (6-5, 7-1)

$\sin^2 x + \cos^2 x = 1$ $\tan^2 x + 1 = \sec^2 x$
$1 + \cot^2 x = \csc^2 x$

Sum Identities (7-2)

$\sin(x + y) = \sin x \cos y + \cos x \sin y$
$\cos(x + y) = \cos x \cos y - \sin x \sin y$
$\tan(x + y) = \dfrac{\tan x + \tan y}{1 - \tan x \tan y}$

Difference Identities (7-2)

$\sin(x - y) = \sin x \cos y - \cos x \sin y$
$\cos(x - y) = \cos x \cos y + \sin x \sin y$
$\tan(x - y) = \dfrac{\tan x - \tan y}{1 + \tan x \tan y}$

Cofunction Identities (7-2)

$\sin\left(\dfrac{\pi}{2} - y\right) = \cos y$ $\tan\left(\dfrac{\pi}{2} - y\right) = \cot y$
$\sin(90° - \theta) = \cos \theta$ $\tan(90° - \theta) = \cot \theta$
$\cos\left(\dfrac{\pi}{2} - y\right) = \sin y$
$\cos(90° - \theta) = \sin \theta$

Double-Angle Identities (7-3)

$\sin 2x = 2 \sin x \cos x$

$\cos 2x = \begin{cases} \cos^2 x - \sin^2 x \\ 1 - 2\sin^2 x \\ 2\cos^2 x - 1 \end{cases}$

$\tan 2x = \dfrac{2 \tan x}{1 - \tan^2 x} = \dfrac{2 \cot x}{\cot^2 x - 1} = \dfrac{2}{\cot x - \tan x}$

Half-Angle Identities (7-3)

$$\sin \frac{x}{2} = \pm\sqrt{\frac{1 - \cos x}{2}}$$

Signs are determined by quadrant in which x/2 lies

$$\cos \frac{x}{2} = \pm\sqrt{\frac{1 + \cos x}{2}}$$

$$\tan \frac{x}{2} = \frac{1 - \cos x}{\sin x} = \frac{\sin x}{1 + \cos x} = \pm\sqrt{\frac{1 - \cos x}{1 + \cos x}}$$

Product-Sum Identities (7-4)

$$\sin x \cos y = \tfrac{1}{2}[\sin(x + y) + \sin(x - y)]$$
$$\cos x \sin y = \tfrac{1}{2}[\sin(x + y) - \sin(x - y)]$$
$$\sin x \sin y = \tfrac{1}{2}[\cos(x - y) - \cos(x + y)]$$
$$\cos x \cos y = \tfrac{1}{2}[\cos(x + y) + \cos(x - y)]$$

Sum-Product Identities (7-4)

$$\sin x + \sin y = 2 \sin \frac{x + y}{2} \cos \frac{x - y}{2}$$

$$\sin x - \sin y = 2 \cos \frac{x + y}{2} \sin \frac{x - y}{2}$$

$$\cos x + \cos y = 2 \cos \frac{x + y}{2} \cos \frac{x - y}{2}$$

$$\cos x - \cos y = -2 \sin \frac{x + y}{2} \sin \frac{x - y}{2}$$

Degrees and Radians (6-1)

$$\frac{\theta_D}{180°} = \frac{\theta_R}{\pi}$$

θ_D is degree measure;
θ_R is radian measure

$\frac{1}{360}$ circumference

1°

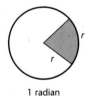

1 radian

Special Triangles (6-4)

30°–60° triangle

45° triangle

Trigonometric Functions (6-2, 6-3, 6-5)

$$\sin x = \frac{b}{r} \qquad \csc x = \frac{r}{b}$$

$$\cos x = \frac{a}{r} \qquad \sec x = \frac{r}{a}$$

$$\tan x = \frac{b}{a} \qquad \cot x = \frac{a}{b}$$

$r = \sqrt{a^2 + b^2} > 0$

(x in degrees or radians)

For x any real number and T any trigonometric function:

$$T(x) = T(x \text{ radians})$$

For a unit circle:

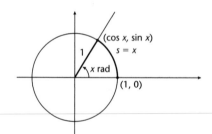

Inverse Trigonometric Functions (6-9)

$y = \sin^{-1} x$ means $x = \sin y$,   where $-\pi/2 \le y \le \pi/2$ and $-1 \le x \le 1$

$y = \cos^{-1} x$ means $x = \cos y$,   where $0 \le y \le \pi$ and $-1 \le x \le 1$

$y = \tan^{-1} x$ means $x = \tan y$,   where $-\pi/2 < y < \pi/2$ and x is any real number